Risk and Financial Management

Risk and Financial Management

Mathematical and Computational Methods

CHARLES TAPIERO
ESSEC Business School, Paris, France

John Wiley & Sons, Ltd

Telephone (+44) 1243 779777

Email (for orders and customer service enquiries): cs-books@wiley.co.uk
Visit our Home Page on www.wileyeurope.com or www.wiley.com

Other Wiley Editorial Offices

John Wiley & Sons Inc., 111 River Street, Hoboken, NJ 07030, USA

Jossey-Bass, 989 Market Street, San Francisco, CA 94103-1741, USA

Wiley-VCH Verlag GmbH, Boschstr. 12, D-69469 Weinheim, Germany

John Wiley & Sons Australia Ltd, 33 Park Road, Milton, Queensland 4064, Australia

John Wiley & Sons (Asia) Pte Ltd, 2 Clementi Loop #02-01, Jin Xing Distripark, Singapore 129809

John Wiley & Sons Canada Ltd, 22 Worcester Road, Etobicoke, Ontario, Canada M9W 1L1

Wiley also publishes its books in a variety of electronic formats. Some content that appears in print may not be available in electronic books.

Library of Congress Cataloging-in-Publication Data

Tapiero, Charles S.
Risk and financial management : mathematical and computational methods / Charles Tapiero.
p. cm.
Includes bibliographical references.
ISBN 0-470-84908-8
1. Finance–Mathematical models. 2. Risk management. I. Title.
HG106 .T365 2004
658.15′5′015192–dc22 2003025311

British Library Cataloguing in Publication Data

A catalogue record for this book is available from the British Library

ISBN 0-470-84908-8

Typeset in 10/12 pt Times by TechBooks, New Delhi, India
Printed and bound in Great Britain by Biddles Ltd, Guildford, Surrey
This book is printed on acid-free paper responsibly manufactured from sustainable forestry in which at least two trees are planted for each one used for paper production.

This book is dedicated to:

Daniel
Dafna
Oren
Oscar and
Bettina

Contents

Preface **xiii**

Part I: Finance and Risk Management

Chapter 1 Potpourri **03**
1.1 Introduction 03
1.2 Theoretical finance and decision making 05
1.3 Insurance and actuarial science 07
1.4 Uncertainty and risk in finance 10
1.4.1 Foreign exchange risk 10
1.4.2 Currency risk 12
1.4.3 Credit risk 12
1.4.4 Other risks 13
1.5 Financial physics 15
Selected introductory reading 16

Chapter 2 Making Economic Decisions under Uncertainty **19**
2.1 Decision makers and rationality 19
2.1.1 The principles of rationality and bounded rationality 20
2.2 Bayes decision making 22
2.2.1 Risk management 23
2.3 Decision criteria 26
2.3.1 The expected value (or Bayes) criterion 26
2.3.2 Principle of (Laplace) insufficient reason 27
2.3.3 The minimax (maximin) criterion 28
2.3.4 The maximax (minimin) criterion 28
2.3.5 The minimax regret or Savage's regret criterion 28
2.4 Decision tables and scenario analysis 31
2.4.1 The opportunity loss table 32
2.5 EMV, EOL, EPPI, EVPI 33
2.5.1 The deterministic analysis 34
2.5.2 The probabilistic analysis 34
Selected references and readings 38

Chapter 3 **Expected Utility** **39**
3.1 The concept of utility 39
3.1.1 Lotteries and utility functions 40
3.2 Utility and risk behaviour 42
3.2.1 Risk aversion 43
3.2.2 Expected utility bounds 45
3.2.3 Some utility functions 46
3.2.4 Risk sharing 47
3.3 Insurance, risk management and expected utility 48
3.3.1 Insurance and premium payments 48
3.4 Critiques of expected utility theory 51
3.4.1 Bernoulli, Buffon, Cramer and Feller 51
3.4.2 Allais Paradox 52
3.5 Expected utility and finance 53
3.5.1 Traditional valuation 54
3.5.2 Individual investment and consumption 57
3.5.3 Investment and the CAPM 59
3.5.4 Portfolio and utility maximization in practice 61
3.5.5 Capital markets and the CAPM again 63
3.5.6 Stochastic discount factor, assets pricing and the Euler equation 65
3.6 Information asymmetry 67
3.6.1 'The lemon phenomenon' or adverse selection 68
3.6.2 'The moral hazard problem' 69
3.6.3 Examples of moral hazard 70
3.6.4 Signalling and screening 72
3.6.5 The principal–agent problem 73
References and further reading 75

Chapter 4 **Probability and Finance** **79**
4.1 Introduction 79
4.2 Uncertainty, games of chance and martingales 81
4.3 Uncertainty, random walks and stochastic processes 84
4.3.1 The random walk 84
4.3.2 Properties of stochastic processes 91
4.4 Stochastic calculus 92
4.4.1 Ito's Lemma 93
4.5 Applications of Ito's Lemma 94
4.5.1 Applications 94
4.5.2 Time discretization of continuous-time finance models 96
4.5.3 The Girsanov Theorem and martingales* 104
References and further reading 108

Chapter 5 **Derivatives Finance** **111**
5.1 Equilibrium valuation and rational expectations 111

5.2 Financial instruments 113
5.2.1 Forward and futures contracts 114
5.2.2 Options 116
5.3 Hedging and institutions 119
5.3.1 Hedging and hedge funds 120
5.3.2 Other hedge funds and investment strategies 123
5.3.3 Investor protection rules 125
References and additional reading 127

Part II: Mathematical and Computational Finance

Chapter 6 Options and Derivatives Finance Mathematics 131
6.1 Introduction to call options valuation 131
6.1.1 Option valuation and rational expectations 135
6.1.2 Risk-neutral pricing 137
6.1.3 Multiple periods with binomial trees 140
6.2 Forward and futures contracts 141
6.3 Risk-neutral probabilities again 145
6.3.1 Rational expectations and optimal forecasts 146
6.4 The Black–Scholes options formula 147
6.4.1 Options, their sensitivity and hedging parameters 151
6.4.2 Option bounds and put–call parity 152
6.4.3 American put options 154
References and additional reading 157

Chapter 7 Options and Practice 161
7.1 Introduction 161
7.2 Packaged options 163
7.3 Compound options and stock options 165
7.3.1 Warrants 168
7.3.2 Other options 169
7.4 Options and practice 171
7.4.1 Plain vanilla strategies 172
7.4.2 Covered call strategies: selling a call and a share 176
7.4.3 Put and protective put strategies: buying a put and a stock 177
7.4.4 Spread strategies 178
7.4.5 Straddle and strangle strategies 179
7.4.6 Strip and strap strategies 180
7.4.7 Butterfly and condor spread strategies 181
7.4.8 Dynamic strategies and the Greeks 181
7.5 Stopping time strategies* 184
7.5.1 Stopping time sell and buy strategies 184
7.6 Specific application areas 195

7.7 Option misses 197
References and additional reading 204
Appendix: First passage time* 207

Chapter 8 Fixed Income, Bonds and Interest Rates 211
8.1 Bonds and yield curve mathematics 211
8.1.1 The zero-coupon, default-free bond 213
8.1.2 Coupon-bearing bonds 215
8.1.3 Net present values (NPV) 217
8.1.4 Duration and convexity 218
8.2 Bonds and forward rates 222
8.3 Default bonds and risky debt 224
8.4 Rated bonds and default 230
8.4.1 A Markov chain and rating 233
8.4.2 Bond sensitivity to rates – duration 235
8.4.3 Pricing rated bonds and the term structure risk-free rates* 239
8.4.4 Valuation of default-prone rated bonds* 244
8.5 Interest-rate processes, yields and bond valuation* 251
8.5.1 The Vasicek interest-rate model 254
8.5.2 Stochastic volatility interest-rate models 258
8.5.3 Term structure and interest rates 259
8.6 Options on bonds* 260
8.6.1 Convertible bonds 261
8.6.2 Caps, floors, collars and range notes 262
8.6.3 Swaps 262
References and additional reading 264
Mathematical appendix 267
A.1: Term structure and interest rates 267
A.2: Options on bonds 268

Chapter 9 Incomplete Markets and Stochastic Volatility 271
9.1 Volatility defined 271
9.2 Memory and volatility 273
9.3 Volatility, equilibrium and incomplete markets 275
9.3.1 Incomplete markets 276
9.4 Process variance and volatility 278
9.5 Implicit volatility and the volatility smile 281
9.6 Stochastic volatility models 282
9.6.1 Stochastic volatility binomial models* 282
9.6.2 Continuous-time volatility models 00
9.7 Equilibrium, SDF and the Euler equations* 293
9.8 Selected Topics* 295
9.8.1 The Hull and White model and stochastic volatility 296
9.8.2 Options and jump processes 297

9.9 The range process and volatility 299
References and additional reading 301
Appendix: Development for the Hull and White model (1987)* 305

Chapter 10 Value at Risk and Risk Management 309
10.1 Introduction 309
10.2 VaR definitions and applications 311
10.3 VaR statistics 315
10.3.1 The historical VaR approach 315
10.3.2 The analytic variance–covariance approach 315
10.3.3 VaR and extreme statistics 316
10.3.4 Copulae and portfolio *VaR* measurement 318
10.3.5 Multivariate risk functions and the principle of maximum entropy 320
10.3.6 Monte Carlo simulation and VaR 324
10.4 VaR efficiency 324
10.4.1 VaR and portfolio risk efficiency with normal returns 324
10.4.2 VaR and regret 326
References and additional reading 327

Author Index 329

Subject Index 333

Preface

Another finance book to teach what market gladiators/traders either know, have no time for or can't be bothered with. Yet another book to be seemingly drowned in the endless collections of books and papers that have swamped the economic literate and illiterate markets ever since options and futures markets grasped our popular consciousness. Economists, mathematically inclined and otherwise, have been largely compensated with Nobel prizes and seven-figures earnings, competing with market gladiators – trading globalization, real and not so real financial assets. Theory and practice have intermingled accumulating a wealth of ideas and procedures, tested and remaining yet to be tested. Martingale, chaos, rational versus adaptive expectations, complete and incomplete markets and whatnot have transformed the language of finance, maintaining their true meaning to the mathematically initiated and eluding the many others who use them nonetheless.

This book seeks to provide therefore, in a readable and perhaps useful manner, the basic elements or economic language of financial risk management, mathematical and computational finance, laying them bare to both students and traders. All great theories are based on simple philosophical concepts, that in some circumstances may not withstand the test of reality. Yet, we adopt them and behave accordingly for they provide a framework, a reference model, inspiring the required confidence that we can rely on even if there is not always something to stand on. An outstanding example might be complete markets and options valuation – which might not be always complete and with an adventuresome valuation of options. Market traders make seemingly risk-free arbitrage profits that are in fact model-dependent. They take positions whose risk and rewards we can only make educated guesses at, and make venturesome and adventuresome decisions in these markets based on facts, fancy and fanciful interpretations of historical patterns and theoretical–technical analyses that seek to decipher things to come.

The motivation to write this book arose from long discussions with a hedge fund manager, my son, on a large number of issues regarding markets behaviour, global patterns and their effects both at the national and individual levels, issues regarding psychological behaviour that are rendering markets less perfect than what we might actually believe. This book is the fruit of our theoretical and practical contrasts and language – the sharp end of theory battling the long and wily practice of the market gladiator, each with our own vocabulary and misunderstandings. Further, too many students in computational finance learn techniques, technical analysis and financial decision making without assessing the dependence of such analyses on the definition of uncertainty and the meaning of probability. Further, defining 'uncertainty' in specific ways, dictates the type of technical analysis and generally the theoretical finance practised. This book was written, both to clarify some of the issues confronting theory and practice and to explain some of the 'fundamentals, mathematical' issues that underpin fundamental theory in finance.

Fundamental notions are explained intuitively, calling upon many trading experiences and examples and simple equations-analysis to highlight some of the basic trends of financial decision making and computational finance. In some cases, when mathematics are used extensively, sections are starred or introduced in an appendix, although an intuitive interpretation is maintained within the main body of the text.

To make a trade and thereby reach a decision under uncertainty requires an understanding of the opportunities at hand and especially an appreciation of the underlying sources and causes of change in stocks, interest rates or assets values. The decision to speculate against or for the dollar, to invest in an Australian bond promising a return of five % over 20 years, are risky decisions which, inordinately amplified, may be equivalent to a gladiator's fight for survival. Each day, tens of thousands of traders, investors and fund managers embark on a gargantuan feast, buying and selling, with the world behind anxiously betting and waiting to see how prices will rise and fall. Each gladiator seeks a weakness, a breach, through which to penetrate and make as much money as possible, before the hordes of followers come and disturb the market's equilibrium, which an instant earlier seemed unmovable. Size, risk and money combine to make one richer than Croesus one minute and poorer than Job an instant later. Gladiators, too, their swords held high one minute, and history a minute later, have played to the arena. Only, it is today a much bigger arena, the prices much greater and the losses catastrophic for some, unfortunately often at the expense of their spectators.

Unlike in previous times, spectators are thrown into the arena, their money fated with these gladiators who often risk, not their own, but everyone else's money – the size and scale assuming a dimension that no economy has yet reached.

For some, the traditional theory of decision-making and risk taking has fared badly in practice, providing a substitute for reality rather than dealing with it. Further, the difficulty of problems has augmented with the involvement of many sources of information, of time and unfolding events, of information asymmetries and markets that do not always behave competitively, etc. These situations tend to distort the approaches and the techniques that have been applied successfully but to conventional problems. For this reason, there is today a great deal of interest in understanding how traders and financial decision makers reach decisions and not only what decisions they ought to reach. In other words, to make better decisions, it is essential to deal with problems in a manner that reflects reality and not only theory that in its essence, always deals with structured problems based on specific assumptions – often violated. These assumptions are sometimes realistic; but sometimes they are not. Using specific problems I shall try to explain approaches applied in complex financial decision processes – mixing practice and theory. The approach we follow is at times mildly quantitative, even though much of the new approach to finance is mathematical and computational and requires an extensive mathematical proficiency. For this reason, I shall assume familiarity with basic notions in calculus as well as in probability and statistics, making the book accessible to typical economics and business and maths students as well as to practitioners, traders and financial managers who are familiar with the basic financial terminology.

The substance of the book in various forms has been delivered in several institutions, including the MASTER of Finance at ESSEC in France, in Risk Management courses at ESSEC and at Bar Ilan University, as well as in Mathematical Finance courses at Bar Ilan University Department of Mathematics and Computer Science. In addition, the Montreal Institute of Financial Mathematics and the Department of Finance at Concordia University have provided a testing ground as have a large number of lectures delivered in a workshop for MSc students in Finance and in a PhD course for Finance students in the Montreal consortium for PhD studies in Mathematical Finance in the Montreal area. Throughout these courses, it became evident that there is a great deal of excitement in using the language of mathematical finance but there is often a misunderstanding of the concepts and the techniques they require for their proper application. This is particularly the case for MBA students who also thrive on the application of these tools. The book seeks to answer some of these questions and problems by providing as much as possible an interface between theory and practice and between mathematics and finance. Finally, the book was written with the support of a number of institutions with which I have been involved these last few years, including essentially ESSEC of France, the Montreal Institute of Financial Mathematics, the Department of Finance of Concordia University, the Department of Mathematics of Bar Ilan University and the Israel Port Authority (Economic Research Division). In addition, a number of faculty and students have greatly helped through their comments and suggestions. These have included, Elias Shiu at the University of Iowa, Lorne Switzer, Meir Amikam, Alain Bensoussan, Avi Lioui and Sebastien Galy, as well as my students Bernardo Dominguez, Pierre Bour, Cedric Lespiau, Hong Zhang, Philippe Pages and Yoav Adler. Their help is gratefully acknowledged.

PART I

Finance and Risk Management

CHAPTER 1

Potpourri

1.1 INTRODUCTION

Will a stock price increase or decrease? Would the Fed increase interest rates, leave them unchanged or decrease them? Can the budget to be presented in Transylvania's parliament affect the country's current inflation rate? These and so many other questions are reflections of our lack of knowledge and its effects on financial markets performance. In this environment, uncertainty regarding future events and their consequences must be assessed, predictions made and decisions taken. Our ability to improve forecasts and reach consistently good decisions can therefore be very profitable. To a large extent, this is one of the essential preoccupations of finance, financial data analysis and theory-building. Pricing financial assets, predicting the stock market, speculating to make money and hedging financial risks to avoid losses summarizes some of these activities. Predictions, for example, are reached in several ways such as:

- 'Theorizing', providing a structured approach to modelling, as is the case in financial theory and generally called fundamental theory. In this case, economic and financial theories are combined to generate a body of knowledge regarding trades and financial behaviour that make it possible to price financial assets.
- Financial data analysis using statistical methodologies has grown into a field called financial statistical data analysis for the purposes of modelling, testing theories and technical analysis.
- Modelling using metaphors (such as those borrowed from physics and other areas of related interest) or simply constructing model equations that are fitted one way or another to available data.
- Data analysis, for the purpose of looking into data to determine patterns or relationships that were hitherto unseen. Computer techniques, such as neural networks, data mining and the like, are used for such purposes and thereby make more money. In these, as well as in the other cases, the 'proof of the pudding is in the eating'. In other words, it is by making money, or at least making

Risk and Financial Management: Mathematical and Computational Methods. C. Tapiero
 ISBN: 0-470-84908-8

it possible for others to make money, that theories, models and techniques are validated.
- Prophecies we cannot explain but sometimes are true.

Throughout these 'forecasting approaches and issues' financial managers deal practically with uncertainty, defining it, structuring it and modelling its causes, explainable and unexplainable, for the purpose of assessing their effects on financial performance. This is far from trivial. First, many theories, both financial and statistical, depend largely on how we represent and model uncertainty. Dealing with uncertainty is also of the utmost importance, reflecting individual preferences and behaviours and attitudes towards risk. Decision Making Under Uncertainty (DMUU) is in fact an extensive body of approaches and knowledge that attempts to provide systematically and rationally an approach to reaching decisions in such an environment. Issues such as 'rationality', 'bounded rationality' etc., as we will present subsequently, have an effect on both the approach we use and the techniques we apply to resolve the fundamental and practical problems that finance is assumed to address. In a simplistic manner, uncertainty is characterized by probabilities. Adverse consequences denote the risk for which decisions must be taken to properly balance the potential payoffs and the risks implied by decisions – trades, investments, the exercise of options etc. Of course, the more ambiguous, the less structured and the more uncertain the situations, the harder it is to take such decisions. Further, the information needed to make decisions is often not readily available and consequences cannot be predicted. Risks are then hard to determine. For example, for a corporate finance manager, the decision may be to issue or not to issue a new bond. An insurance firm may or may not confer a certain insurance contract. A Central Bank economist may recommend reducing the borrowing interest rate, leaving it unchanged or increasing it, depending on multiple economic indicators he may have at his disposal. These, and many other issues, involve uncertainty. Whatever the action taken, its consequences may be uncertain. Further, not all traders who are equally equipped with the same tools, education and background will reach the same decision (of course, when they differ, the scope of decisions reached may be that much broader). Some are well informed, some are not, some believe they are well informed, but mostly, all traders may have various degrees of intuition, introspection and understanding, which is specific yet not quantifiable. A historical perspective of events may be useful to some and useless to others in predicting the future. Quantitative training may have the same effect, enriching some and confusing others. While in theory we seek to eliminate some of the uncertainty by better theorizing, in practice uncertainty wipes out those traders who reach the wrong conclusions and the wrong decisions. In this sense, no one method dominates another: all are important. A political and historical appreciation of events, an ability to compute, an understanding of economic laws and fundamental finance theory, use of statistics and computers to augment one's ability in predicting and making decisions under uncertainty are only part of the tool-kit needed to venture into trading speculation and into financial risk management.

1.2 THEORETICAL FINANCE AND DECISION-MAKING

Financial decision making seeks to make money by using a broad set of economic and theoretical concepts and techniques based on rational procedures, in a consistent manner and based on something more than intuition and personal subjective judgement (which are nonetheless important in any practical situation). Generally, it also seeks to devise approaches that may account for departures from such rationality. Behavioural and psychological reasons, the violation of traditional assumptions regarding competition and market forces and exchange combine to alter the basic assumptions of theoretical economics and finance.

Finance and financial instruments currently available through brokers, mutual funds, financial institutions, commodity and stock markets etc. are motivated by three essential problems:

- Pricing the multiplicity of claims, accounting for risks and dealing with the negative effects of uncertainty or risk (that can be completely unpredictable, or partly or wholly predictable)
- Explaining, and accounting for investors' behaviour. To counteract the effects of regulation and taxes by firms and individual investors (who use a wide variety of financial instruments to bypass regulations and increase the amount of money investors can make).
- Providing a rational framework for individuals' and firms' decision making and to suit investors' needs in terms of the risks they are willing to assume and pay for. For this purpose, extensive use is made of DMUU and the construction of computational tools that can provide 'answers' to well formulated, but difficult, problems.

These instruments deal with the uncertainty and the risks they imply in many different ways. Some instruments merely transfer risk from one period to another and in this sense they reckon with the *time phasing of events* to reckon with. One of the more important aspects of such instruments is to supply *'immediacy'*, i.e. the ability not to wait for a payment for example (whereby, some seller will assume the risk and the cost of time in waiting for that payment). Other instruments provide a *'spatial' diversification*, in other words, the distribution of risks across a number of independent (or almost independent) risks. For example, buying several types of investment that are less than perfectly correlated, maitaining *liquidity* etc. By liquidity, we mean the cost to instantly convert an asset into cash at its fair price. This liquidity is affected both by the existence of a market (in other words, buyers and sellers) and by the cost of transactions associated with the conversion of the asset into cash. As a result, risks pervading finance and financial risk management are varied; some of them are outlined in greater detail below.

Risk in finance results from the consequences of undesirable outcomes and their implications for individual investors or firms. A definition of risk involves their probability, individual and collective and consequences effects. These are relevant to a broad number of fields as well, each providing an approach to the measurement and the valuation of risk which *is motivated by their needs and by the set of questions they must respond to and deal with.* For these reasons, the problems of finance often transcend finance and are applicable to the broad areas of economics and decision-making. Financial economics seeks to provide approaches and answers to deal with these problems. The growth of theoretical finance in recent decades is a true testament to the important contribution that financial theory has made to our daily life. Concepts such as financial markets, arbitrage, risk-neutral probabilities, Black–Scholes option valuation, volatility, smile and many other terms and names are associated with a maturing profession that has transcended the basic traditional approaches of making decisions under uncertainty. By the same token, hedging which is an important part of the practice finance is the process of eliminating risks in a particular portfolio through a trade or a series of trades, or contractual agreements. Hedging relates also to the valuation-pricing of derivatives products. Here, a portfolio is constructed (the hedging portfolio) that eliminates all the risks introduced by the derivative security being analyzed in order to replicate a return pattern identical to that of the derivative security. At this point, from the investor's point of view, the two alternatives – the hedging portfolio and the derivative security – are indistinguishable and therefore have the same value. In practice too, speculating to make money can hardly be conceived without hedging to avoid losses.

The traditional theory of decision making under uncertainty, integrating statistics and the risk behaviour of decision makers has evolved in several phases starting in the early nineteenth century. At its beginning, it was concerned with collecting data to provide a foundation for *experimentation and sampling theory*. These were the times when surveys and counting populations of all sorts began. Subsequently, statisticians such as Karl Pearson and R. A. Fisher studied and set up the foundations of *statistical data analysis*, consisting of the assessment of the reliability and the accuracy of data which, to this day, seeks to represent large quantities of information (as given explicitly in data) in an aggregated and summarized fashion, such as probability distributions and moments (mean, variance etc.) and states how accurate they are. Insurance managers and firms, for example, spend much effort in collecting such data to estimate mean claims by insured clients and the propensity of certain insured categories to claim, and to predict future weather conditions in order to determine an appropriate insurance premium to charge. Today, financial data analysis is equally concerned with these problems, bringing sophisticated modelling and estimation techniques (such as linear regression, ARCH and GARCH techniques which we shall discuss subsequently) to bear on the application of financial analysis.

The next step, expounded and developed primarily by R. A. Fisher in the 1920s, went one step further with *planning experiments* that can provide effective information. The issue at hand was then to plan the experiments generating the information that can be analysed statistically and on the basis of which a decision could, justifiably, be reached. This important phase was used first in testing the agricultural yield under controlled conditions (to select the best way to grow plants, for example). It yielded a number of important lessons, namely that the procedure (statistical or not) used to collect data is intimately related to the kind of relationships we seek to evaluate. A third phase, expanded dramatically in the 1930s and the 1940s consisted in the construction of mathematical models that sought to bridge the gap between the process of data collection and the need of such data for specific purposes such as predicting and decision making. Linear regression techniques, used extensively in econometrics, are an important example. Classical models encountered in finance, such as models of stock market prices, currency fluctuations, interest rate forecasts and investment analysis models, cash management, reliability and other models, are outstanding examples.

In the 1950s and the 1960s the (Bayes) theory of *decision making under uncertainty* took hold. In important publications, Raiffa, Luce, Schlaiffer and many others provided a unified framework for integrating problems relating to data collection, experimentation, model building and decision making. The theory was intimately related to typical economic, finance and industrial, business and other problems. Issues such as the value of information, how to collect it, how much to pay for it, the weight of intuition and subjective judgement (as often used by behavioural economists, psychologists etc.) became relevant and integrated into the theory. Their practical importance cannot be understated for they provide a framework for reaching decisions under complex situations and uncertainty. Today, theories of decision making are an ever-expanding field with many articles, books, experiments and theories competing to provide another view and in some cases another vision of uncertainty, how to model it, how to represent certain facets of the economic and financial process and how to reach decisions under uncertainty. The DMUU approach, however, presumes that uncertainty is specified in terms of probabilities, albeit learned adaptively, as evidence accrues for one or the other event. It is only recently, in the last two decades, that theoretical and economic analyses have provided in some cases theories and techniques that provide an estimate of these probabilities. In other words, while in the traditional approach to DMUU uncertainty is exogenous, facets of modern and theoretical finance have helped 'endogenize' uncertainty, i.e. explain uncertain behaviours and events by the predictive market forces and preferences of traders. To a large extent, the contrasting finance fundamental theory and traditional techniques applied to reach decisions under uncertainty diverge in their attempts to represent and explain the 'making of uncertainty'. This is an important issue to appreciate and one to which we shall return subsequently when basic notions of fundamental theory including rational expectations and option pricing are addressed.

Today, DMUU is economics, finance, insurance and risk motivated. There are a number of areas of special interest we shall briefly discuss to better appreciate the transformations of finance, insurance and risk in general.

1.3 INSURANCE AND ACTUARIAL SCIENCE

Actuarial science is in effect one of the first applications of probability theory and statistics to risk analysis. Tetens and Barrois, already in 1786 and 1834 respectively, were attempting to characterize the 'risk' of life annuities and fire insurance and on that basis establish a foundation for present-day insurance. Earlier, the Gambling Act of 1774 in England (King George III) laid the foundation for life insurance. It is, however, to Lundberg in 1909, and to a group of Scandinavian actuaries (Borch, 1968; Cramer, 1955) that we owe much of the current mathematical theory of insurance. In particular, Lundberg provided the foundation for collective risk theory. Terms such as 'premium payments' required from the insured, 'wealth' or the 'firm liquidity' and 'claims' were then defined. In its simplest form, actuarial science establishes exchange terms between the insured, who pays the premium that allows him to claim a certain amount from the firm (in case of an accident), and the insurer, the provider of insurance who receives the premiums and invests and manages the moneys of many insured. The insurance terms are reflected in the 'insurance contract' which provides legally the 'conditional right to claim'. Much of the insurance literature has concentrated on the definition of the rules to be used in order to establish the terms of such a contract in a just and efficient manner. In this sense, 'premium principles' and a wide range of operational rules worked out by the actuarial and insurance profession have been devised. Currently, insurance is gradually being transformed to be much more in tune with market valuation of insurable contracts and financial instruments are being devised for this purpose. The problems of insurance are, of course, extremely complex, with philosophical and social undertones, seeking to reconcile individual with collective risk and individual and collective choices and interests through the use of the market mechanism and concepts of fairness and equity. In its proper time setting (recognizing that insurance contracts express the insured attitudes towards time and uncertainty, in which insurance is used to substitute certain for uncertain payments at different times), this problem is of course, conceptually and quantitatively much more complicated. For this reason, the quantitative approach to insurance, as is the case with most financial problems, is necessarily a simplification of the fundamental issues that insurance deals with.

Risk is managed in several ways including: '*pricing insurance, controls, risk sharing* and *bonus-malus*'. Bonus-malus provides an incentive not to claim when a risk materializes or at least seeks to influence insured behaviour to take greater care and thereby prevent risks from materializing. In some cases, it is used to discourage nuisance claims. There are numerous approaches to applying each of these tools in insurance. Of course, in practice, these tools are applied jointly, providing a capacity to customize insurance contracts and at the same time assuming a profit for the insurance firm.

In insurance and finance (among others) we will have to deal as well with special problems, often encountered in practical situations but difficult to analyse using statistical and analytical techniques. These essentially include *dependencies, rare events* and *man-made risks*. In insurance, correlated risks are costlier to assume while insuring rare and extremely costly events is difficult to assess. Earthquake and tornado insurance are such cases. Although, they occur, they do so with small probabilities. Their occurrence is extremely costly for the insurer, however. For this reason, insurers seek the participation of governments for such insurance, study the environment and the patterns in weather changes and turn to extensive risk sharing schemes (such as reinsurance with other insurance firms and on a global scale). Dependencies can also be induced internally (endogenously generated risks). For example, when trading agents follow each other's action they may lead to the rise and fall of an action on the stock market. In this sense, 'behavioural correlations' can induce cyclical economic trends and therefore greater market variability and market risk. Man-made induced risks, such as terrorists' acts of small and unthinkable dimensions, also provide a formidable challenge to insurance companies. John Kay (in an article in the *Financial Times*, 2001) for example states:

> The insurance industry is well equipped to deal with natural disasters in the developed world: the hurricanes that regularly hit the south-east United States; the earthquakes that are bound to rock Japan and California from time to time. Everyone understands the nature of these risks and their potential consequences. But we are ignorant of exactly when and where they will materialize. For risks such as these, you can write an insurance policy and assess a premium.
>
> But the three largest disasters for insurers in the past 20 years have been man-made, not natural. The human cost of asbestos was greater even than that of the destruction of the World Trade Center. The deluge of asbestos-related claims was the largest factor in bringing the Lloyd's insurance market to its knees.

By the same token, the debacle following the deregulation of Savings and Loans in the USA in the 1960s led to massive opportunistic behaviours resulting in huge losses for individuals and insurance firms. These disasters have almost uniformly involved government interventions and in some cases bail-outs (as was the case with airlines in the aftermath of the September 11th attack on the World Trade Center). Thus, risk in insurance and finance involves a broad range of situations, sources of uncertainty and a broad variety of tools that may be applied when disasters strike. There are special situations in insurance that may be difficult to assess from a strictly financial point of view, however, as in the case of man-made risks. For example, environmental risks have special characteristics that are affecting our approach to risk analysis:

- Rare events: Relating to very large disasters with very small probabilities that may be difficult to assess, predict and price.
- Spillover effects: Having behavioural effects on risk sharing and fairness since persons causing risks may not be the sole victims. Further, effects may be felt over long periods of time.
- International dimensions: having power and political overtones.

For these reasons, some of the questions raised in conjunction with environmental risk that are of acute interest today are numerous, including among others:

- Who pays for it?
- What prevention if at all?
- Who is responsible if at all?

By the same token, the future of genetic testing promises to reveal information about individuals that, hitherto has been unknown, and thereby to change the whole traditional approach to insurance. In particular, randomness, an essential facet of the insurance business, will be removed and insurance contracts could/would be tailored to individuals' profiles. The problems that may arise subsequent to genetic testing are tremendous. They involve problems arising over the power and information asymmetries between the parties to contracts. Explicitly, this may involve, on the one hand, moral hazard (we shall elaborate subsequently) and, on the other, adverse selection (which will see later as well) affecting the potential future/non-future of the insurance business and the cost of insurance to be borne by individuals.

1.4 UNCERTAINTY AND RISK IN FINANCE

Uncertainty and risk are everywhere in finance. As stated above, they result from consequences that may have adverse economic effects. Here are a few financial risks.

1.4.1 Foreign exchange risk

Foreign exchange risk measures the risk associated with unexpected variations in exchange rates. It consists of two elements: an internal element which depends on the flow of funds associated with foreign exchange, investments and so on, and an external element which is independent of a firm's operations (for example, a variation in the exchange rates of a country).

Foreign exchange risk management has focused essentially on short-term decisions involving accounting exposure components of a firm's working capital. For instance, consider the case of captive insurance companies that diversify their portfolio of underwriting activities by reinsuring a 'layer' of foreign risk. In this case, the magnitude of the transaction exposure is clearly uncertain, compounding the exchange and exposure risks. Bidding on foreign projects or acquisitions of foreign companies will similarly entail exposures whose magnitudes can be characterized at best subjectively. Explicitly, in big-ticket export transactions or large-scale construction projects, the exporter or contractor will first submit a bid $B(T)$ of say 100 million which is denominated in \$US (a foreign currency from the point of view of the decision maker) and which, if accepted, would give rise to a transaction exposure (asset or liability) maturing at a point in time T, say 2 years ahead. The bid will in turn be accepted or rejected at time t, say 6 months ahead $(0 < t < T)$, resulting in the transaction exposure which is uncertain until the resolution (time) standing at the full amount $B(T)$ if the bid is accepted, or being cancelled if the bid is rejected. Effective management of such uncertain exposures will require the existence of a futures market for foreign exchange allowing contracts to be entered into or cancelled at any time t over the bidding uncertainty resolution horizon $0 < t < T$. The case of foreign acquisition is a special case of the above more general problem with uncertainty resolution being arbitrarily set at $t = T$. Problems in long-term foreign exchange risk management – that is, long-term debt financing and debt refunding – in a multi-currency world, although very important, is not always understood and hedged. As global corporations expand operations abroad, foreign currency-denominated debt instruments become an integral part of the opportunities of financing options. One may argue that in a multi-currency world of efficient markets, the selection of the optimal borrowing source should be a matter of indifference, since nominal interest rates reflect inflation rate expectations, which, in turn, determine the pattern of the future spot exchange rate adjustment path. However, heterogeneous corporate tax rates among different national jurisdictions, asymmetrical capital tax treatment, exchange gains and losses, non-random central bank intervention in exchange markets and an ever-spreading web of exchange controls render the hypothesis of market efficiency of dubious operational value in the selection process of the least-cost financing option. How then, should foreign debt financing and refinancing decisions be made, since nominal interest rates can be misleading for decision-making purposes? Thus, a managerial framework is required, allowing the evaluation of the uncertain cost of foreign capital debt financing as a function of the 'volatility' (risk) of the currency denomination, the maturity of the debt instrument, the exposed exchange rate appreciation/depreciation and the level of risk aversion of the firm.

To do so, it will be useful to distinguish two sources of risk: internal and external. Internal risk depends on a firm's operations and thus that depends on the exchange rate while external risk is independent of a firm's operations (such as a devaluation or the usual variations in exchange rates). These risks are then expressed in terms of:

- Transaction risk, associated with the flow of funds in the firm
- Translation risk, associated with in-process, present and future transactions.
- Competition risk, associated with the firm's competitive posture following a change in exchange rates.

The actors in a foreign exchange (risk) market are numerous and must be considered as well. These include the firms that import and export, and the intermediaries (such as banks), or traders. Traders behave just as market makers do. At any instant, they propose to buy and sell for a price. Brokers are intermediaries that centralize buy and sell orders and act on behalf of their clients, taking the best offers they can get. Over all, foreign exchange markets are competitive and can reach equilibrium. If this were not the case, then some traders could engage in arbitrage, as we shall discuss later on. This means that some traders will be able to make money without risk and without investing any money.

1.4.2 Currency risk

Currency risk is associated with variations in currency markets and exchange rates. A currency is not risky because its depreciation is likely. If it were to depreciate for sure and there were to be no uncertainty as to its magnitude and timing-there would not be any risk at all. As a result, a weak currency can be less risky than a strong currency. Thus, the risk associated with a currency is related to its randomness. The problems thus faced by financial analysts consist of defining a reasonable measure of exposure to currency risk and managing it. There may be several criteria in defining such an exposure. First, it ought to be denominated in terms of the relevant amount of currency being considered. Second, it should be a characteristic of any asset or liability, physical or financial, that a given investor might own or owe, defined from the investor's viewpoint. And finally, it ought to be practical. Currency risks are usually associated with macroeconomic variables (such as the trade gap, political stability, fiscal and monetary policy, interest rate differentials, inflation, leadership, etc.) and are therefore topics of considerable political and economic analysis as well as speculation. Further, because of the size of currency markets, speculative positions may be taken by traders leading to substantial profits associated with very small movements in currency values. On a more mundane level, corporate finance managers operating in one country may hedge the value of their contracts and profits in another foreign denominated currency by assuming financial contracts that help to relieve some of the risks associated with currency (relative or absolute) movements and shifts.

1.4.3 Credit risk

Credit risk covers risks due to upgrading or downgrading a borrower's creditworthiness. There are many definitions of credit risk, however, which depend on the potential sources of the risk, who the client may be and who uses it. Banks in particular are devoting a considerable amount of time and thoughts to defining and managing credit risk. There are basically two sources of uncertainty in credit risk: default by a party to a financial contract and a change in the present value (PV) of future cash flows (which results from changes in financial market conditions, changes in the economic environment, interest rates etc.). For example, this can take the form of money lent that is not returned. Credit risk considerations underlie capital adequacy requirements (CAR) regulations that are required by financial institutions. Similarly, credit terms defining financial borrowing and lending transactions are sensitive to credit risk. To protect themselves, firms and individuals turn to rating agencies such as Standard & Poors, Moody's or others (such as Fitch Investor Service, Nippon Investor Service, Duff & Phelps, Thomson Bank Watch etc.) to obtain an assessment of the risks of bonds, stocks and financial papers they may acquire. Furthermore, even after a careful reading of these ratings, investors, banks and financial institutions proceed to reduce these risks by risk management tools. The number of such tools is of course very large. For example, limiting the level of obligation, seeking collateral, netting, recouponing, insurance, syndication, securitization, diversification, swaps and so on are some of the tools a financial service firm or bank might use.

An exposure to credit risk can occur from several sources. These include an exposure to derivatives products (such as options, as we shall soon define) in exposures to the replacement cost (or potential increases in future replacement costs) due to default arising from market adverse conditions and changes. Problems of credit risk have impacted financial markets and global deflationary forces. 'Wild money' borrowed by hedge funds faster than it can be reimbursed to banks has created a credit crunch. Regulatory distortions are also a persistent theme over time. Over-regulation may hamper economic activity. The creation of wealth, while 'under-regulation' (in particular in emerging markets with cartels and few economic firms managing the economy) can lead to speculative markets and financial distortions. The economic profession has been marred with such problems. For example:

> One of today's follies, says a leading banker, is that the Basle capital adequacy regime provides greater incentives for banks to lend to unregulated hedge funds than to such companies as IBM. The lack of transparency among hedge funds may then disguise the bank's ultimate exposure to riskier markets. Another problem with the Basle regime is that it forces banks to reinforce the economic cycle – on the way down as well as up. During a recovery, the expansion of bank profits and capital inevitably spurs higher lending, while capital shrinkage in the downturn causes credit to contract when it is most needed to business. (*Financial Times*, 20 October 1998, p. 17)

Some banks cannot meet international standard CARs. For example, Daiwa Bank, one of Japan's largest commercial banks, is withdrawing from all overseas business partly to avoid having to meet international capital adequacy standards. For Daiwa, as well as other Japanese banks, capital bases have been eroded by growing pressure on them to write off their bad loans and by the falling value of shares they hold in other companies, however, undermining their ability to meet these capital adequacy standards.

To address these difficulties the Chicago Mercantile Exchange, one of the two US futures exchanges, launched a new bankruptcy index contract (for credit default) working on the principle that there is a strong correlation between credit charge-off rates and the level of bankruptcy filings. Such a contract is targeted at players in the consumer credit markets – from credit card companies to holders of car loans and big department store groups. The data for such an index will be based on bankruptcy court data.

1.4.4 Other risks

There are other risks of course, some of which are defined below while others will be defined, explained and managed as we move along to define and use the tools of risk and computational finance management.

Market risk is associated with movements in market indices. It can be due to a stock price change, to unpredictable interest rate variations or to market liquidity, for example.

Shape risk is applicable to fixed income markets and is caused by non-parallel shifts of interest rates on straight, default-free securities (i.e. shifts in the term structure of interest rates). In general, rates risks are associated with the set of relevant flows of a firm that depend on the fluctuations of interest rates. The debt of a firm, the credit it has, indexed obligations and so on, are a few examples.

Volatility risk is associated with variations in second-order moments (such as process variance). It reduces our ability to predict the future and can induce preventive actions by investors to reduce this risk, while at the same time leading others to speculate wildly. Volatility risk is therefore an important factor in the decisions of speculators and investors. Volatility risk is an increasingly important risk to assess and value, owing to the growth of volatility in stocks, currency and other markets.

Sector risk stems from events affecting the performance of a group of securities as a whole. Whether sectors are defined by geographical area, technological specialization or market activity type, they are topics of specialized research. Analysts seek to gain a better understanding of the sector's sources of uncertainty and their relationship to other sectors.

Liquidity risk is associated with possibilities that the bid–ask spreads on security transactions may change. As a result, it may be impossible to buy or sell an asset in normal market conditions in one period or over a number of periods of time. For example, a demand for an asset at one time (a house, a stock) may at one time be oversubscribed such that no supply may meet the demand. While a liquidity risk may eventually be overtaken, the lags in price adjustments, the process at hand to meet demands, may create a state of temporary (or not so temporary) shortage.

Inflation risk: inflation arises when prices increase. It occurs for a large number of reasons. For example, agents, traders, consumers, sellers etc. may disagree on the value of products and services they seek to buy (or sell) thereby leading to increasing prices. Further, the separation of real assets and financial markets can induce adjustment problems that can also contribute to and motivate inflation. In this sense, a clear distinction ought to be made between financial inflation (reflected in a nominal price growth) and real inflation, based on the real terms value of price growth. If there were no inflation, discounting could be constant (i.e. expressed by fixed interest rates rather than time-varying and potentially random) since it could presume that future prices would be sustained at their current level. In this case, discounting would only reflect the time value of money and not the predictable (and uncertain) variations of prices. In inflationary states, discounting can become nonstationary (and stochastic), leading to important and substantial problems in modelling, understanding how prices change and evolve over time.

Importantly inflation affects economic, financial and insurance related issues and problems. In the insurance industry, for example, premiums and benefits calculations induced by real as well as nominal price variations, i.e. inflation, are difficult to determine. These variations in prices alter over time the valuation of premiums in insurance contracts introducing a risk due to a lack of precise knowledge about economic activity and price level changes. At the same time, changes in the nominal value of claims distributions (by insurance contract holders), increased costs of living and lags between claims and payment render insurance even more risky. For example, should a negotiated insurance contract include inflation-sensitive clauses? If not, what would the implications be in terms of consumer protection, the time spans of negotiated contracts and, of course, the policy premium? In this simple case, a policyholder will gradually face declining payments but also a declining protection. In case of high inflation, it is expected that the policyholder will seek a renegotiation of his contract (and thereby increased costs for the insurer and the insured). The insurance firm, however, will obtain an unstable stream of payments (in real terms) and a very high cost of operation due to the required contract renegotiation. Unless policyholders are extremely myopic, they would seek some added form of protection to compensate on the one hand for price levels changes and for the uncertainty in these prices on the other. In other words, policyholders will demand, and firms will supply, inflation-sensitive policies. Thus, inflation clearly raises issues and problems that are important for both the insurer and the insured. For this reason, protection from inflation risk, which is the loss at a given time, given an uncertain variation of prices, may be needed. Since this is not a 'loss' per se, but an uncertainty regarding the price, inflation-adjusted loss valuation has to be measured correctly. Furthermore, given an inflation risk definition, the apportioning of this risk between the policyholder and the firm is also required, demanding an understanding of risk attitudes and behaviours of insured and insurer alike. Then, questions such as: who will pay for the inflation risk? how? (i.e. what will be the insurance policy which accounts expressly for inflation) and how much? These issues require that insurance be viewed in its inter-temporal setting rather than its static actuarial approach.

To clarify these issues, consider whether an insurance firm should *a priori* absorb the inflation risk pass it on to policyholders by an increased load factor (premium) or follow a posterior procedure where policyholders increase payments as a function of the published inflation rate, cost of living indices or even the value of a given currency. These are questions that require careful evaluation.

1.5 FINANCIAL PHYSICS

Recently, domains such Artificial Intelligence, Data Mining and Computational Tools, as well as the application of constructs and themes reminiscent of financial problems, have become fashionable. In particular, a physics-like approach has been devised to deal with selected financial problems (in particular with option valuation, volatility smile and so on). The intent of physical models is to explain (and thereby forecast) phenomena that are not explained by the fundamental theory. For example, trading activity bursts, bubbles and long and short cycles, as well as long-run memory, that are poorly explained or predicted by fundamental theory and traditional models are typical applications. The physics approach is essentially a modelling approach, using metaphors and processes/equations used in physics and finding their parallel in economics and finance. For example, an individual consumer might be thought to be an atom moving in a medium/environment which might correspond in economics to a market. The medium results from an infinite number of atoms acting/interacting, while the market results from an infinite number of consumers consuming and trading among themselves. Of course, these metaphors are quite problematic, modelling simplifications, needed to render intractable situations tractable and to allow aggregation of the many atoms (consumers) into a whole medium (market). There are of course many techniques to reach such aggregation. For example, the use of Brownian motion (to represent the uncertainty resulting from many individual effects, individually intractable), originating in Bachelier's early studies in 1905, conveniently uses the Central Limit Theorem in statistics to aggregate events presumed independent. However, this 'seeming normality', resulting from the aggregation of many independent events, is violated in many cases, as has been shown in many financial data analyses. For example, data correlation (which cannot be modelled or explained easily), distributed (stochastic) volatility and the effects of long-run memory not accounted for by traditional modelling techniques, etc. are such cases. In this sense, if there is any room for financial physics it can come only after the failure of economic and financial theory to explain financial data. The contribution of physics to finance can be meaningful only by better understanding of finance – however complex physical notions may be. The true test is, as always, the 'proof of the pudding'; in other words, whether models are supported by the evidence of financial data or making money where no one else thought money could be made.

SELECTED INTRODUCTORY READING

Bachelier, L. (1900) *Théorie de la spéculation*, Thèse de Mathématique, Paris.

Barrois, T. (1834) Essai sur l'application du calcul des probabilités aux assurances contre l'incendie, *Mem. Soc. Sci. De Lille*, 85–282.

Beard, R.E., T. Pentikainen and E. Pesonen (1979) *Risk Theory* (2nd edn), Methuen, London.

Black, F., and M. Scholes (1973) The pricing of options and corporate liabilities, *Journal of Political Economy*, **81**, 637–659.

Borch, K.H. (1968) *The Economics of Uncertainty*, Princeton University Press, Princeton, N. J.

Bouchaud, J.P., and M. Potters (1997) *Théorie des Risques Financiers*, Aléa-Saclay/Eyrolles, Paris.

Cootner, P.H. (1964) *The Random Character of Stock Prices*. MIT Press, Cambridge, MA.

Cramer, H. (1955) *Collective Risk Theory* (Jubilee Volume), Skandia Insurance Company.

Hull, J. (1993) *Options, Futures and Other Derivatives Securities* (2nd edn), Prentice Hall, Englewood Cliffs, NJ.

Ingersoll, J.E., Jr (1987) *Theory of Financial Decision Making*, Rowman & Littlefield, New Jersey.

Jarrow, R.A. (1988) *Finance Theory*, Prentice Hall, Englewood Cliffs, NJ.

Kalman, R.E. (1994) Randomness reexamined, *Modeling, Identification and Control*, **15**(3), 141–151.

Lundberg, F. (1932) Some supplementary researches on the collective risk theory, *Skandinavisk Aktuarietidskrift*, **15**, 137–158.
Merton, R.C. (1990) *Continuous Time Finance*, Cambridge, M.A, Blackwell.
Modigliani, F., and M. Miller (1958) The cost of capital and the theory of investment, *American Economic Review*, **48**(3), 261–297.
Tetens, J.N. (1786) *Einleitung zur Berchnung der Leibrenten und Antwartschaften*, Leipzig.

CHAPTER 2

Making Economic Decisions under Uncertainty

2.1 DECISION MAKERS AND RATIONALITY

Should we invest in a given stock whose returns are hardly predictable? Should we buy an insurance contract in order to protect ourselves from theft? How much should we be willing to pay for such protection? Should we be rational and reach a decision on the basis of what we know, or combine our prior and subjective assessment with the unfolding evidence? Further, do we have the ability to use a new stream of statistical news and trade intelligently? Or 'bound' our procedures? This occurs in many instances, for example, when problems are very complex, outpacing our capacity to analyse them, or when information is so overbearing or so limited that one must take an educated or at best an intuitive guess. In most cases, steps are to be taken to limit and 'bound' our decision processes for otherwise no decision can be reached in its proper time. These 'bounds' are varied and underlie theories of 'bounded rationality' based on the premise that we can only do the best we can and no better! However, when problems are well defined, when they are formulated properly – meaning that the alternatives are well-stated, the potential events well-established, and their conditional consequences (such as payoffs, costs, etc.) are determined, we can presume that a rational procedure to decision making can be followed. If, in addition, the uncertainties inherent in the problem are explicitly stated, a rational decision can be reached.

What are the types of objectives we may consider? Although there are several possibilities (as we shall see below) it is important to understand that no criterion is the objectively correct one to use. The choice is a matter of economic, individual and collective judgement – all of which may be imbued with psychological and behavioural traits. Utility theory, for example (to be seen in Chapter 3), provides an approach to the selection of a 'criterion of choice' which is both consistent and rational, making it possible to reconcile (albeit not always) a decision and its

Risk and Financial Management: Mathematical and Computational Methods. C. Tapiero
 ISBN: 0-470-84908-8

economic and risk justifications. It is often difficult to use, however, as we shall see later on for it requires parameters and an understanding of human decision making processes that might not be available.

To proceed rationally it is necessary for an individual decision-maker (an investor for example) to reach a judgement about: the alternatives available, the sources of uncertainties, the conditional outcomes and preferences needed to order alternatives. Then, combine them without contradicting oneself (i.e. by being rational) in selecting the best course of action to follow. Further, to be rational it is necessary to be self-consistent in stating what we believe or are prepared to accept and then accept the consequences of our actions. Of course, it is possible to be 'too rational'. For example, a decision maker who refuses to accept any dubious measurements or assumptions will simply never make a decision! He then incurs the same consequences as being irrational. To be a practical investor, one must accept that there is a 'bounded rationality' and that an investment will in the end bear some risk one did not plan on assuming. This understanding is an essential motivation for financial risk management. That is, we can only be satisfied that we did the best possible analysis we could, given the time, the information and the techniques available at the time the decision to invest (or not) was made. Appropriate rational decision-making approaches, whether these are based on theoretical and/or practical considerations, would thus recognize both our capacities and their limit.

2.1.1 The principles of rationality and bounded rationality

Underlying rationality is a number of assumptions that assume (Ariel Rubinstein, 1998):

- knowledge of the problem,
- clear preferences,
- an ability to optimize,
- indifference to equivalent logical descriptions of alternative and choice sets.

Psychologists and economists have doubted these. The fact that decisions are not always rational does not mean that there are no underlying procedures to the decision-making process. A systematic approach to departures from rationality has been a topic of intense economic and psychological interest of particular importance in finance, emphasizing 'what is' rather than 'what ought to be'.

For example, decision-makers often have a tendency to 'throw good money after bad', also known as sunk costs. Although it is irrational, it is often practised. Here are a few instances: Having paid for the movie, I will stay with it, even though it is a dreadful and time-consuming movie. An investment in a stock, even if it has failed repeatedly, may for some irrational reason generate a loyalty factor. The reason we are so biased in favour of bringing existing projects to fruition irrespective of their cost is that such behaviour is imbedded in our brains. We resist the conceptual change that the project is a failure and refuse to change our decision process to admit such failure. The problem is psychological: once we have made an irreversible investment, we imbue it with extra value, the price of our emotional 'ownership'. There are many variations of this phenomenon. One is the 'endowment effect' in which a person who is offered $10 000 for a painting he paid only $1000 for refuses the generous offer. The premium he refuses is accounted for by his pride in an exceptionally good judgement—truly, perhaps the owner's wild fantasy that make such a painting wildly expensive. Similarly, once committed to a bad project one becomes bound to its outcome. This is equivalent to an investor to being OTM (on the money) in a large futures position and not exercising it. Equivalently, it is an alignment, not bounded by limited responsibility, as would be the case for stock options traders; and therefore it leads to maintaining an irrational risky position.

Currently, psychology and behavioural studies focus on understanding and predicting traders' decisions, raising questions regarding markets' efficiency (meaning: being both rational and making the best use of available information) and thereby raising doubts regarding the predictive power of economic theory. For example, aggregate individual behaviour leading to herding, black sheep syndrome, crowd psychology and the tragedy of the commons, is used to infuse a certain reality in theoretical analyses of financial markets and investors' decisions. It is with such intentions that funds such as ABN AMRO Asset Management (a fund house out of Hong Kong) are proposing mutual investment funds based on 'behavioural finance principles' (IHT Money Report, 24–25 February 2001, p. 14). These funds are based on the assumption that investors make decisions based on multiple factors, including a broad range of identifiable emotional and psychological biases. This leads to market mechanisms that do not conform to or are not compatible with fundamental theory (as we shall see later on) and therefore, provide opportunities for profits when they can be properly apprehended. The emotional/psychological factors pointed out by the IHT article are numerous. '*Investors' mistakes are not due to a lack of information but because of mental shortcuts inherent in human decision-making that blinds investors. For example, investors overestimate their ability to forecast change and they inefficiently process new information. They also tend to hold on to bad positions rather than admit mistakes.*' In addition, image bias can keep investors in a stock even when this loyalty flies in the face of balance sheet fundamentals. Over-reaction to news can lead investors to dump stocks when there is no rational reason for doing so. Under-reaction is the effect of people's general inability to admit mistakes. This is a trait that is also encountered by analysts and fund managers as much as individual investors. These factors are extremely important for they underlie financial practice and financial decision-making, drawing both on theoretical constructs and an appreciation of individual and collective (market) psychology. Thus, to construct a rational approach to making decisions, we can only claim to do the best we can and recognize that, however thorough our search, it is necessarily bounded.

Rationality is also a 'bounded' qualitative concept that is based on essentially three dimensions: analysis of information, perception of risk and decision-making. It may be defined and used in different ways. 'Classical rationality', underlying important economic and financial concepts such as 'rational expectations' and 'risk-neutral pricing' (we shall attend to this later on in great detail), suppose that the investor/decision maker uses all available information, perceives risk without bias and makes the best possible investment decision he can (given his ability to compute) with the information he possesses at the time the decision is made. By contrast, a 'Bayesian rationality', which underlies this chapter, has a philosophically different approach. Whereas 'rational expectations' supposes that an investor extrapolates from the available information the true distribution of payoffs, Bayesian rationality supposes that we have a prior subjective distribution of payoffs that is updated through a learning mechanism with unfolding new information. Further, 'rational expectations' supposes that this prior or subjective distribution is the true one, imbedding future realizations while the Bayes approach supposes that the investor's belief or prior distribution is indeed subjective but evolving through learning about the true distribution. These 'differences of opinion' have substantive impact on how we develop our approach to financial decision making and risk management. For 'rational expectations', the present is 'the present of the future' while Bayesian rationality incorporates learning from one's bias (prejudice or misconception) into risk measurement and hence decision making, the bias being gradually removes uncertainty as learning sets in. In this chapter we shall focus our attention on Bayes decision making under uncertainty.

2.2 BAYES DECISION MAKING

The basic elements of Bayes rational decision making involve behaviours including:

(1) A decision to be taken from a set of known alternatives.
(2) Uncertainty defined in terms of events with associated known (subjective) probabilities.
(3) Conditional consequences resulting from the selection of a decision and the occurrence of a specific event (once uncertainty, ex-post, is resolved).
(4) A preference over consequences, i.e. there is a well-specified preference function or procedure for selecting a specific alternative among a set of given alternatives.

An indifferent decision maker does not really have a problem. A problem arises when certain outcomes are preferred over others (such as making more money over less) and when preferences are sensitive to the risks associated with such outcomes. What are these preferences? There are several possibilities, each based on the information available – what is known and not known and how we balance the two and our attitude toward risk (or put simply, how we relate to the probabilities of uncertain outcomes, their magnitude and their adverse consequences). For these reasons, *risk management* in practice is very important, impacting events' desirability and their probabilities. There are many ways to do so, as we shall see below.

2.2.1 Risk management

Risk results from the direct and indirect adverse consequences of outcomes and events that were not accounted for, for which we are ill-prepared, and which effects individuals, firms, financial markets and society at large. It can result from many reasons, both internally induced and occurring externally. In the former case, consequences are the result of failures or misjudgements, while, in the latter, these are the results of uncontrollable events or events we cannot prevent. As a result, a definition of risk involves (i) consequences, (ii) their probabilities and their distribution, (iii) individual preferences and (iv) collective, market and sharing effects. These are relevant to a broad number of fields as well, each providing an approach to measurement, valuation and minimization of risk which *is motivated by psychological needs and the need to deal with problems that result from uncertainty and the adverse consequences they may induce.*

Risk management is broadly applied in finance. Financial economics, for example, deals intensively with hedging problems in to order eliminate risks in a particular portfolio through a trade or a series of trades, or through contractual agreements reached to share and induce a reduction of risk by the parties involved. Risk management consists then in using financial instruments to negate the effects of risk. It might mean a judicious use of options, contracts, swaps, insurance contracts, investment portfolio design etc. so that risks are brought to bearable economic costs. These tools cost money and, therefore, risk management requires a careful balancing of the numerous factors that affect risk, the costs of applying these tools and a specification of (or constraints on) tolerable risks an economic optimization will be required to fulfil. For example, options require that a premium be paid to limit the size of losses just as the insured are required to pay a premium to buy an insurance contract to protect them in case an adverse event occurs (accidents, thefts, diseases, unemployment, fire, etc.). By the same token, 'value at risk' (see Chapter 10) is based on a quantile risk constraint, which provides an estimate of risk exposure. Each profession devises the tools it can apply to manage the more important risks to which it is subjected.

The definition of risk, risk measurement and risk management are closely related, one feeding the other to determine the proper/optimal levels of risk. In this process a number of tools are used based on:

- ex-ante risk management,
- ex-post risk management and
- robustness.

Ex-ante risk minimization involves the application of *preventive controls;* preventive actions of various forms; information seeking, statistical analysis and forecasting; design for reliability; insurance and financial risk management etc. Ex-post risk minimization involves by contrast control audits, the design of optional, flexible-reactive schemes that can deal with problems once they have occurred and limit their consequences. Robust design, unlike ex-ante and ex-post risk minimization, seeks to reduce risk by rendering a process insensitive to its adverse consequences. Thus, *risk management consists of altering the states a system many reach in a desirable manner (financial, portfolio, cash flow etc.), and their probabilities or reducing their consequences to planned or economically tolerable levels*. There are many ways to do so, however, each profession devises the tools it can apply or create a market for. For example, insurance firms use reinsurance to share the risks insured while financial managers use derivative products to contain unsustainable risks.

Risk management tools are applied in insurance and finance in many ways. *Control* seeks to ascertain that 'what is intended occurs'. It is exercised in a number of ways rectifying decisions taken after a nonconforming event or problem has been detected. For example, auditing a trader, controlling a portfolio performance over time etc. are such instances. The disappearance of $750 million at AIB (Allied Irish Bank) in 2002 for example, accelerated implementation of control procedures within the bank and its overseas traders.

Insurance is a medium or a market for risk, substituting payments now for potential damages (reimbursed) later. The size of such payments and the potential damages that may occur with various probabilities, can lead to widely distributed market preferences and thereby to a possible exchange between decision-makers of various preferences. Insurance firms have recognized the opportunities of such differences and have, therefore, provided mechanisms for pooling, redistributing and capitalizing on the 'willingness to pay to avoid losses'. It is because of such attitudes, combined with goals of personal gain, social welfare and economic efficiency, that markets for fire and theft insurance, as well as sickness, unemployment, accident insurance, etc., have come to be as important as they are today. It is because of persons' or institutions' desires to avoid too great a loss (even with small probabilities), which would have to be borne alone, that markets for reinsurance (i.e., sub-selling portions of insurance contracts) and mutual protection insurance (based on the pooling of risks) have also come into being. Today, risk management in insurance has evolved and is much more in tune with the valuation of insurance risks by financial markets. Understanding the treatment of risk by financial markets; the 'law of the single price' (which we shall consider below); risk diversification (when is is possible) and risk transfer techniques using a broad set of financial instruments currently used and traded in financial markets; the valuation of risk premiums and the estimation of yield curves (see also Chapter 8); mastering financial statistical and simulation techniques; and finally devising applicable risk metrics and measurement approaches for insurance firms – all have become essential for insurance risk management.

While insurance is a *passive form of risk management*, based on exchange mechanisms only (or, equivalently, 'passing the buck' to some willing agent), loss prevention and technological innovations are *active means of managing risks*. Loss prevention is a means of altering the probabilities and the states of undesirable, damaging states. For example, maintaining one's own car properly is a form of loss prevention seeking to alter the chances of having an accident. Similarly, driving carefully, locking one's own home effectively, installing fire alarms, etc. are all forms of loss prevention. Of course, insurance and loss prevention are, in fact, two means to the similar end of risk protection. Car insurance rates tend, for example, to be linked to a person's past driving record. Certain clients (or areas) might be classified as 'high risk clients', required to pay higher insurance fees. Inequities in insurance rates will occur, however, because of an imperfect knowledge of the probabilities of damages and because of the imperfect distribution of information between the insured and insurers. Thus, situations may occur where persons might be 'over-insured' and have no motivation to engage in loss prevention. Such outcomes, known as 'moral hazard' (to be seen in greater detail in Chapter 3), counter the basic purposes of insurance. It is a phenomenon that can recur in a society in widely different forms, however. Over-insuring unemployment may stimulate persons not to work, while under-insuring may create uncalled-for social inequities. Low car insurance rates (for some) can lead to reckless driving, leading to unnecessary damages inflicted on others, on public properties, etc. Risk management, therefore, seeks to ensure that risk protection does not become necessarily a reason for not working. More generally, risk management in finance considers both risks to the investor and their implications for returns, 'pricing one at the expense of the other'. In this sense, finance, has gone one step further in using the market to price the cost an investor is willing to sustain to prevent the losses he may incur. Financial instruments such as options provide a typical example. For this reason, given the importance of financial markets, many insurance contracts have to be reassessed and valued using basic financial instruments.

Technological innovation means that the structural process through which a given set of inputs is transformed into an output is altered. For example, building a new six-lane highway can be viewed as a way for the public to change the 'production-efficiency function' of transport servicing. Environmental protection regulation and legal procedures have, in fact, had a technological impact by requiring firms to *change* the way in which they convert inputs into outputs, by considering as well the treatment of refuse. Further, pollution permits have induced companies to reduce their pollution emissions in a given by-product and sell excess pollution to less efficient firms.

Forecasting, learning, information and its distribution is also an essential ingredient of risk management. Banks learn every day how to price and manage risk better, yet they are still acutely aware of their limits when dealing with complex portfolios of structured products. Further, most non-linear risk measurement and assessment are still 'terra incognita' asymmetries. Information between insured and insurers, between buyers and sellers, etc., are creating a wide range of opportunities and problems that provide great challenges to risk managers and, for some, 'computational headaches' because they may be difficult to value. These problems are assuming added importance in the age of internet access for all and in the age of 'total information accessibility'. Do insurance and credit card companies have access to your confidential files? Is information distribution now swiftly moving in their favour? These are issues creating 'market inefficiencies' as we shall see in far greater detail in Chapter 9.

Robustness expresses the insensitivity of a process to the randomness of parameters (or mis-specification of the model) on which it is based. The search for robust solutions and models has led to many approaches and techniques of optimization. Techniques such as VaR (Value at Risk), scenario optimization, regret and ex-post optimization, min-max objectives and the like (see Chapter 10) seek to construct robust systems. These are important tools for risk management; we shall study them here at length. They may augment the useful life of a portfolio strategy as well as provide a better guarantee that 'what is intended will likely occur', even though, as reality unfolds over time, working assumptions made when the model was initially constructed turn out to be quite different.

Traditional decision problems presume that there are *homogeneous* decision makers, deciding as well what information is relevant. In reality, decision makers may be heterogeneous, exhibiting broadly varying preferences, varied access to information and a varied ability to analyse (forecast) and compute it. In this environment, decision-making becomes an extremely difficult process to understand and decisions become difficult to make. For example, when there are few major traders, the apprehension of each other's trades induces an endogenous uncertainty, resulting from a mutual assessment of intentions, knowledge, knowhow etc. A game may set in based on an appreciation of strategic motivations and intentions. This may result in the temptation to collude and resort to opportunistic behaviour.

2.3 DECISION CRITERIA

The selection of a decision criterion is an essential part of DMUU, expressing decision-makers' impatience and attitudes towards uncertain outcomes and valuing them. Below we shall discuss a few commonly used approaches.

2.3.1 The expected value (or Bayes) criterion

Preferences for decision alternatives are expressed by sorting their expected outcomes in an increasing order. For monetary values, the Expected Monetary Value (or EMV) is calculated and a choice is made by selecting the greatest EMV. For example, given an investment of 3 million dollars yielding an uncertain return one period hence (with a discount rate of 7%), and given in the returns in the table below, what is the largest present expected value of the investment? For the first, alternative we calculate the EMV of the investment one period hence and obtain: EMV = 4.15. The current value of the investment is thus equal to the present value of the expected return (EMV less the cost of the investment) or:

$$V(I) = (1+r)^{-1}EMV - I = 4.15 * (1+0.07)^{-1} - 3 = 0.878$$

Probability	0.10	0.20	0.30	0.15	0.15	0.10
Return	−4	−1	5	7	8	10

When there is more than one alternative (measured by the initial outlay and forecasting of future cash flows), a decision is then reached by comparing the economic properties of each investment alternative. For example, consider another investment proposal consisting of an initial outlay of 1 million dollars only (rather than 3) with a prospective cash flow given by the following:

Probability	0.10	0.20	0.30	0.15	0.15	0.10
Return	−8	−3	5	3	4	8

If we maintain the same EMV criterion, we note that:

$$V(I) = (1+r)^{-1}EMV - I = 1.95 * (1+0.07)^{-1} - 1 = 0.822$$

which clearly ranks the first investment alternative over the second (in terms of the EMV criterion). In both cases the EMV is positive and therefore both projects seem to be economically worthwhile. There may be other considerations, for example, an initial outlay of 3 (rather than 1) million dollars for sure compared to an uncertain cash flow in the future (with prospective potential losses, albeit probabilistic, in the future). The attidude towards these losses are often important considerations to consider as well. Such considerations require the application of other criteria for decision making, as we shall briefly outline below. Note that it is noteworthy that such an individual approach does not deal with the market valuation of such cash flow streams and expresses only an individual's judgement (and not market valuation of the cash flow, that is the consensus of judgements of participants on a market price). Financial analysis, as we shall see subsequently, provides a market-sensitive discounting to these uncertain streams of cash.

2.3.2 Principle of (Laplace) insufficient reason

The Laplace principle states that, when the probabilities of the states of nature in a given problem are not known, we assume they are equally likely. In other words, a state of utmost ignorance will be replaced by assigning to each potential state the same probability! In this case, when we return to our first investment project, we are faced with the following prospect:

Probability	0.166	0.166	0.166	0.166	0.166	0.166
Return	−4	−1	5	7	8	10

and its present EMV is,

$$V(I) = (1+r)^{-1}EMV - I = 5 * (1+0.07)^{-1} - 3 = 1.672$$

which implies that 'not knowing' can be worth money! This is clearly not the case, since reaching a decision on this basis can lead to losses since the probability we have assumed are not necessarily the true ones. Gathering information in these cases may be useful, since it may be used to reduce the potential (miscalculated) expected losses.

2.3.3 The minimax (maximin) criterion

The criterion consists in selecting the decision that will have the least maximal loss regardless of what future (state) may occur. It is used when we seek protection from the worst possible events and expresses generally an attitude of abject pessimism. Consider again the two investment projects with cash flows I and II specified below and for simplicity, assume that they require initially the same investment outlay. The flows to compare are:

Probability	0.10	0.20	0.30	0.15	0.15	0.10
Return I	−4	−1	−5	−7	8	10
Return II	−8	−1	5	7	8	10

The worst prospect in the first project is −7 million dollars while it is –8 million dollars in the second project. The minimum of the maximum loss is therefore −7 million dollars, which provides a criterion (albeit very pessimistic) justifying the selection of the first investment project.

The minimax criterion takes the smallest of the available maximums. In this case, the projects have an equal maximum value and the investor is indifferent between the two. It is a second-best objective. Who cares about getting the gold medal as long as we get the silver! Honour is safe and the player satisfied. This criterion can be extended using this sporting analogy. A bronze is third best, good enough; while fourth best may be just participating, providing a reward in itself. Maximin is a loss-averse mindset. As long as we do get the best of all worst possible outcomes the investor is satisfied.

2.3.4 The maximax (minimin) criterion

This is an optimist's criterion, banking on the best possible future, yielding the hoped for largest possible profits. It is based on the belief or the urge to profit as much as possible, regardless of the probability of desirable or other events. Again, returning to our previous example, we note that both projects have a maximal gain of 10 million dollars and therefore the maximum–maximum gain (maximax criterion) will indicate indifference in selecting one or the other project, as was the case for the minimax criterion. As Voltaire's Candide would put it: 'We live in the best of all possible worlds' as he travelled in a world ravaged by man, as a prelude story to the French Revolution.

The minimin criterion is a pessimist's point of view. Regardless of what happens, only the worst case can happen. On the upside, such a point of view, leads only to upbeat news. My house has not burned today! Amazing!

2.3.5 The minimax regret or Savage's regret criterion

The previous criteria involving maximums and minimums were evaluated ex-ante. In practice, payoffs and probabilities are not easily measured. Thus, these criteria express a philosophical outlook rather than an objective to base a decision on. Ex-post, unlike ex-ante, decision-making is reached once information is revealed and uncertainty is resolved. Each decision has then a regret defined by the difference between the gain made and the gain that could have been realized had we selected the best decision (associated with the event that actually occurs). An expected 'regret' decision-maker would then seek to minimize the expectation of such a regret, while a minimax regret decision-maker would seek to select the decision providing the least maximal regret.

The cost of a decision's regret represents the difference between the ex-ante payoffs that would be received with a given outcome compared to the maximum possible ex-post payoff received. Savage, Bell and Loomes and Sugden (see references) have pointed out the relevance of this criterion to decision-making under uncertainty by suggesting that decision makers may select an act by minimizing the regrets associated with potential decisions. Behaviourally, such a criterion would be characteristic of people attached to their past. Their past mistakes haunt their present day, hence, they do the best they can to avoid them in the future. Specifically, assume that we select an action (decision) and some event occurs. The decision/event combination generates a *payoff table*, expressing the conditional consequences of that decision when, ex-post, the event occurs. For example, the following table gives the payoff on a portfolio dependent on two different decisions on the portfolio allocation.

	Event A	Event B	Event C	Event D	Event E	Event F
Probability	0.10	0.20	0.30	0.15	0.15	0.10
Return I	−3	−1	−5	−7	8	10
Return II	−8	1	6	7	8	12

The decision/event combination may then generate a 'regret' for the decision – for it is possible that we could have done better! Was decision 1 the better one? This is an opportunity loss, since a profit could have been made – had we known what events were to occur. If event B is the one that happens then clearly, based on an ex-post basis, decision 2 is the better one. If decisions were reversible then it might be possible to compensate (at least partially) for the fact that we took, a posteriori, a 'wrong' decision. Such a characteristic is called 'flexibility' and is worth money that decision makers are willing to pay for. What would I be willing to pay to have taken effectively decision 1 instead of 2 when event B happens? Options for example, provide such an opportunity, as we shall see in Chapter 6. An option would give us the right but not the obligation to make a decision in the future, once uncertainty is resolved. In most cases, these are decisions to sell or buy. But applications to real world problems have led to options to switch from one technology to another for example.

For example, say that we expect the demand for a product to grow significantly, and as a result we decide to expand the capacity of our plants. Assume that in fact, this expectation for demand growth does not materialize and we are left with a large excess capacity, unable to reduce it except at a substantial loss. What can we do then, except regret our decision! Similarly, assume that we expect peace to come on earth and decide to spend less on weapons development. Optimism, however much it may be wanted, may not, unfortunately, be justified and instead we find ourselves facing a war for which we may be ill-prepared. What can we do? Not much, except regret our decision. The *regret* (also called the *Savage regret*) criterion, then, seeks to minimize the regret we may have in adopting a decision. This explains why some actions are taken to reduce the possibility of such extreme regrets (as with the buying of insurance, steps taken to reduce the risks of bankruptcy, buying options to limit downside risk, in times of peace prepare for war – Sun-Tze and so on). Examples to this effect will be considered below using the opportunity loss table in the next section (Table 2.3).

Example: Regret and the valuation of firms

Analysts' valuation of stocks are growing in importance. Analyst recommendations have a great impact on investors, but their effects are felt particularly when analysts are 'disappointed' by a stock performance and revise their recommendations downwards. In these cases, the effects can be disastrous for the stock price in consideration. In practice, analysts use a number of techniques that are based on firms' reports. Foremost is the net return multiple factor. It is based on the ratio of the stock value of the firm to its net return. The multiple factor is then selected by comparing firms that have the same characteristics. It is then believed that *the larger the risk, the smaller the multiple factor.* In practice, analysts price stocks quite differently. A second technique is based on the firm's future discounted (at the firm's internal rate of return) cash flow. In practice, the future cash flow is based on forecasts that may not be precise. Finally, the third technique is based on assets value (which is the most conservative one). In other words, there is not a uniform agreement regarding which objective to use in valuing a firm's stock. Financial fundamental theory has made an important contribution by providing a set of proper circumstances to resolve this issue. This will be considered in Chapter 6 in particular.

Example: The firm and risk management

Consider a firm operating in a given industry. Evidently, competition with other firms, as well as explicit (or implicit) government intervention through regulation, tax rebates for special environmental protection investments, grants or subsidized capital budgets in distress areas, etc., are instances where firms are required to be sensitive to uncertainty and risk. Managers, of course, will seek to reduce and manage the risk implied by such uncertainty and seek ways to augment the market control (by vertical integration, acquisition of competition, etc.), or they may diversify risks by seeking activities in unrelated markets.

In the example Table 2.1 we have constructed a list of uncertainties and risks faced by firms and how these may be met. The list provided is by no means exhaustive and provides only an indication of the kind of problems that we can address. For example, competition can be an important source of risk which may be met by many means such as strategic M & A, collusion practices, diversification an so on.

Table 2.1 Sources of uncertainty and risks.

Uncertainty and risks	Protective actions taken
Long-range changes in market growth	Research and development on new products, diversification to other markets
Inflation	Indexation of assets, and accounts receivable
Price uncertainty of input materials	Building up inventory, contract with suppliers (essentially futures), buying options and hedging techniques
Competition	Mergers and acquisition, cartels, price-fixing, advertising and marketing effort, diversification

2.4 DECISION TABLES AND SCENARIO ANALYSIS

Decision tables and trees are simple mechanisms for structuring some decision problems involving uncertainty and solving them. It requires that an objective, the problem's states and probabilities be given. To construct a payoff table we proceed as follows:

- Identify the alternative courses of action, mutually exclusive, and collectively exhaustive, which are variables (at least two) we can control directly.
- Consider all possible and relevant states of a problem. Each state represents one and only one potential event; each state may itself be defined in terms of multiple other states, however; states represent events which are mutually exclusive; they are collectively exhaustive; one and only one state will actually result.
- Assign to each state a probability of occurrence. This probability should be based on the information we have regarding the problem and, since states are mutually exclusive and collectively exhaustive, these probabilities (summed over all states) should be equal to one.

All conditional (payoff or cost) consequences are then assembled in a table format – see Table 2.2 where $[c_{ij}]$ are the conditional costs of alternative i if event j occurs.

Table 2.2 The payoff table.

States	1	2	3	...	...	...	n
Probabilities	p_1	p_2	p_3	...	...	...	p_n
A_1	c_{11}	c_{12}	c_{13}	...	...	...	c_{1n}
A_2	c_{21}	c_{22}	c_{23}	...	...	...	c_{2n}
A_3	c_{31}	c_{32}	c_{33}	...	...	...	c_{3n}
...	...	...	...	...	...	...	...
A_m	c_{m1}	c_{m2}	c_{m3}				c_{mn}

For example, for a credit manager, what are the relevant states to consider when a customer comes in and demands a loan? Simply grant the loan (state 1) or not (state 2). If the loan is not reimbursed on time (and reimbursement delays are introduced) there may be other ways to express these states. For example, a first state would stand for no delay, a second, would stand for a one-period delay, a third state for a two-period delay, and so on. The entries in the table tell us what will be the conditional payoffs (or costs) associated with each action. The sample Table 2.2 specifies n states $1, 2, 3, 4, \ldots n$ and m alternatives. When alternative $A_i, i = 1, 2, \ldots, m$ is taken and say state j occurs with probability $p_j, j = 1, 2, \ldots, n$, then the cost (or payoff) is c_{ij} (or π_{ij}). Thus, in such a decision problem, there are:

(1) n potential, mutually exclusive and exhaustive states,
(2) m alternative actions, one of which only can be selected,
(3) nm conditional consequences we should be able to define.

If we use an expected cost (or expected payoff) criterion, then the decision selected would be the one yielding the least expected cost (or, equivalently, the largest expected payoff). The expected monetary cost of alternative i is then:

$$EMC_i = \sum_{j=1}^{n} p_{ij} c_{ij}$$

while the least cost alternative k selected is:

$$k \in Min_{i\in[1,..n]} \{EMC_i\}$$

Problem
Cash management consists of managing the short-term flow of funds in order to meet a potential need or demand for cash. Cash is kept primarily because of its need in the future. Assume, for example, that an investor has the following needs for money:

Quantities	100	300	500	700	900
Probabilities	0.05	0.25	0.50	0.15	0.05

(1) What are the potential courses of action?
(2) What are the problem states? And their probabilities?
(3) What are the conditional costs if the bank rate is 20 % yearly?

2.4.1 The opportunity loss table

Say that action i has been selected and event j occurs and thus payoff π_{ij} is gained. If we were equipped with this knowledge prior to making a decision, it is possible that another decision would bring greater profits. Assume such knowledge and let the maximum payoff, based on the best decision be

$$\underset{j}{\text{Max}}\,[\pi_{ij}]$$

Table 2.3 Opportunity loss table.

States	1	2	3	...	...	...	n
Probabilities	p_1	p_2	p_3	...	...	...	p_n
A_1	l_{11}	l_{12}	l_{13}	...	...	...	l_{1n}
A_2	l_{21}	l_{22}	l_{23}	...	...	...	l_{2n}
A_3	l_{31}	l_{32}	l_{33}	...	...	...	l_{3n}
...	...	...	...	...	...	...	...
A_m	l_{m1}	l_{m2}	l_{m3}				l_{mn}

The difference between this maximum payoff and the payoff obtained by taking any other decision is called the opportunity loss, denoted by:

$$l_{ij} = \operatorname*{Max}_{j} [\pi_{ij}] - \pi_{ij}$$

The opportunity loss table is therefore a matrix as given in Table 2.3.

Thus, the opportunity loss is the difference between the costs or profits actually realized and the costs or profits which would have been realized if the decision had been the best one possible. A project might seem like a good investment, but it means that we have lost the opportunity to do something else that might be more profitable. This loss may be likened to the additional income a trader would have realized had he been an inside trader, benefiting from information regarding stock prices before they reach the market! As a result, we can verify that the difference between the expected profits of any two acts is equal in magnitude but opposite in sign to the difference between their expected losses. By the same token, the difference between the expected costs of any two acts is equal in magnitude and identical in sign to the difference between their expected opportunity losses. With these definitions on hand we can also state that: *the cost of uncertainty is the expected opportunity loss of the best possible decision under a given probability distribution.*

2.5 EMV, EOL, EPPI, EVPI

EMV, EOL, EPPI and EVPI are terms associated with a decision; they will be elucidated through an application. Assume that data supplied by a Port Authority points to a number of development alternatives for the port. Uncertainty regarding the economic state of the country, geopolitical developments and so on, lead to a number of scenarios to be considered and against which each of these alternatives must be assessed. Each alternative can generate, ex-post, a sense of satisfaction at having followed the proper course of action as well as a sense that a suboptimal alternative was taken. Four scenarios are assumed each to lead to the following results, summarized in the table below where entries are payoffs (losses):

Scenario	1	2	3	4	5	6
Probability	0.1	0.15	0.25	0.05	0.3	0.15
Alternative 1	–200	–100	150	400	–300	700
Alternative 2	300	–150	300	600	100	500
Alternative 3	–500	300	400	–100	400	100
Alternative 4	400	600	–100	–250	–300	100

2.5.1 The deterministic analysis

An alternative is selected irrespective of the probabilities of forthcoming events. Given a number of alternatives and specified events, a decision can be taken. A number of criteria are used, such as maximax, maximin, minimax regret and the 'equally likely' (Laplace) criteria as stated earlier. Under these criteria, we see that alternatives 1 and 2 are always better than alternatives 3 and 4. Explicitly, the following results are obtained:

Criterion	Decision	Payoff
Maximax	Alternative 1	700
Maximin	Alternative 2	−150
Minimax regret	Alternative 1	700
Equally likely	Alternative 2	275

2.5.2 The probabilistic analysis

Probabilistic analysis characterizes the likelihood of forthcoming events by associating a probability with each event. It uses a number of potential criteria but we shall be concerned essentially with the EMV – expected monetary value index of performance. The results for our example are given by the following:

Probabilistic analysis: The Port Authority
Expected value – Summary report

Decision	Expected payoff
Alternative 1	37.50
Alternative 2	217.50
Alternative 3	225.00 *
Alternative 4	17.50

Calculations were made as follows:

Alternative 1: $0.1(-200) + 0.15(-100) + 0.25(150) + 0.05(400) + 0.3(-300) + 0.15(700) = 37.50$

Alternative 2: $0.1(300) + 0.15(-150) + 0.25(300) + 0.05(600) + 0.3(100) + 0\ .15(500) = 217.50$

Alternative 3: $0.1(-500) + 0.15(300) + 0.25(400) + 0.05(-100) + 0.3(400) + 0.15(100) = 225.00$

Alternative 4: $0.1(400) + 0.15(600) + 0.25(-100) + 0.05(-2500) + 0.3(-300) + 0.15(100) = 17.50$

The EMV (expected monetary value) consists of valuing each alternative by its EMV. The 'best' choice (in an EMV context) is 225. In other words, ex-ante, the best decision we can take is alternative 3. By contrast, if a decision could be taken ex-post, once uncertainty is revealed and removed, the cost of each decision is given by its opportunity loss, whose expectation is the EOL (expected opportunity loss). This value is calculated explicitly through the opportunity loss table below:

Table of opportunity losses, calculations

Scenario	1	2	3	4	5	6
Probability	0.1	0.15	0.25	0.05	0.3	0.15
Alternative 1	$400 - (-200)$	$600 - (-100)$	$400 - 150$	$600 - 400$	$400 - (-300)$	$700 - 700$
Alternative 2	$400 - 300$	$600 - (-150)$	$400 - 300$	$600 - 600$	$400 - 100$	$700 - 500$
Alternative 3	$400 - (-500)$	$600 - 300$	$400 - 400$	$600 - (-100)$	$400 - 400$	$700 - 100$
Alternative 4	$400 - 400 = 0$	$600 - 600$	$400 - (-100)$	$600 - (-250)$	$400 - (-300)$	$700 - 100$

Table of opportunity losses

Scenario	1	2	3	4	5	6
Probability	0.1	0.15	0.25	0.05	0.3	0.15
Alternative 1	60	105	62.5	10	210	0
Alternative 2	10	110.5	25	0	90	30
Alternative 3	90	45	0	35	0	90
Alternative 4	0	0	125	42.5	210	90

Entries are calculated as follows. Say that scenario 1 realizes itself. The best alternative would then be alternative 4 yielding a payoff of 400. We replace in the table the entry 400 by 0 and then calculate in the first column corresponding to Scenario 1 the relative losses had we selected a suboptimal alternative. Now compute for each alternative the expected opportunity loss, which is the sum of columns for each row. Verify that the sums EMV + EOL are equal for each alternative, called the EPPI, or the Expected Profit under Perfect Information. Further, note that the recommended alternative under an EOL criterion is also alternative 3 as in the expected payoff (EMV) case. This is always the case and should not come as any surprise, since selecting the largest EMV is equivalent to the smallest EOL. Since,

$$\text{EMV} + \text{EOL} = \text{EPPI}$$

Note that the EOL for the third alternative equals 260 and, therefore, note that the EPPI is 485, which is the same for all alternatives. The EPPI means that if, ex-post, we always have the best alternative, then in expectation our payoff would be 485. Since, ex-ante, it is only 225 (=EMV), the potential for improving the ex-ante payoff EMV by better forecasts of the scenarios, by a better management of uncertainty (through contracts of various sorts that manage risk) cannot be larger than the EOL or 260. Such an approach would be slightly more complex if we were to introduce sample surveys, information guesses etc. used to improve our assessment of the states, the probabilities and the economic value of such an assessment.

Optimal Decision: Alternative 3; Expected payoff : 225.00
Probabilistic analysis
Expected value of perfect information

State	Prob.	Decision	Payoff	Prob.*Payoff
Scenario 1	0.1000	Alternative 4	400.00	40.00
Scenario 2	0.1500	Alternative 4	600.00	90.00
Scenario 3	0.2500	Alternative 3	400.00	100.00
Scenario 4	0.0500	Alternative 2	600.00	30.00
Scenario 5	0.3000	Alternative 3	400.00	120.00
Scenario 6	0.1500	Alternative 1	700.00	105.00

Expected payoff with perfect information (EPPI)	485.00
Expected payoff without perfect information (EMV)	225.00
Expected value of perfect information (EVPI)	260.00

If we integrate other sources of information, it is possible to improve the probability estimates and, therefore, improve the optimal decision. The value of information, of a sample on the basis of which such information is available, is called EVSI (or the expected value of sample information). It is a gain obtained by improving our assessment of the events/states probabilities. Finally, if gains and losses are weighted in a different manner, then we are led to approaches based on disappointment (giving greater weight to losses, relative to gains) and elation (when the prospects of 'doing better than expected' is more valued because of the self-gratification it produces). Avoidance of losses, motivated by disappointment, can also lead to selecting alternatives that have smaller gain expectation but reduce the probability of having made the 'wrong choice', in the sense of ending the development project with losses. We shall return to this approach in Chapter 3.

Problem

The Corporate Financial Officer Vice of HardKoor Co. has the problem of raising some additional capital. To do so, it is possible to sell 10 000 convertible bonds. A preliminary survey of the capital market indicates that they could be sold at the present time for $100 per bond. However, the company is currently engaged in a union contract dispute and there is a possibility of a strike. If the strike were to take place, the selling price of the bonds would be decreased by 20 %. There is also a possibility of winning a large, exclusive contract which, if obtained, would mean the bonds could be sold for 30 % more. The VP Finance would like to raise the maximum amount of capital, and so must decide whether to offer the bonds now or wait for the situation to become clearer.

(a) What are the alternatives?
(b) What are the sources and the types of uncertainty?
(c) What action should be taken if an EMV criterion is used? – if a minimax criterion is taken? – if a maximin criterion is taken? – if a regret criterion is taken?

(d) If the probability of a strike is felt to be 0.4, while the probability of the contract being awarded is 0.8, what action is best if the EMV criterion is applied (note that it is necessary to calculate the proceeds for the various outcomes), and if the expected opportunity loss (EOL) criterion is used?

(e) Give one example of how the principle of bounded rationality was apparently used in formulating the problem?

SELECTED REFERENCES AND READINGS

Bell, D.E. (1982) Regret in decision making under uncertainty, *Operations Research*, **30**, 961–981.

Bell, D.E. (1983) Risk premiums for decision regrets, *Management Science*, **29**, 1156–1166.

Loomes, G., and R. Sugden (1982) Regret theory: An alternative to rational choice under uncertainty, *Economic Journal*, **92**, 805–824.

Loomes, G., and R. Sugden (1987) Some implications of a more general form of regret theory, *Journal of Economic Theory*, **41**, 270–287.

Luce, R.D., and H. Raiffa (1958) *Games and Decisions*, John Wiley & Sons, Inc., New York.

Raiffa, H., and R. Schlaiffer (1961) *Applied Statistical Decision Theory*, Division of Research, Graduate School of Business, Harvard University, Boston, MA.

Rubinstein, A. (1998) *Modeling Bounded Rationality*, MIT Press, Boston, MA.

Savage, L.J. (1954) *The Foundations of Statistics*, John Wiley & Sons, Inc., New York.

Winkler, R.L. (1972) *Introduction to Bayesian Inference and Decision*, Holt, Rinehart & Winston, New York.

CHAPTER 3

Expected Utility

3.1 THE CONCEPT OF UTILITY

When the expected monetary value (EMV) is used as the sole criterion to reach a decision under uncertainty, it can lead to results we might not have intended. Outstanding examples to this effect are noted by observing people gambling in a casino or acquiring insurance. For example, in Monte Carlo, Atlantic City or Las Vegas, we might see people gambling (investing!) their wealth on ventures (such as putting \$100 on number 8 in roulette), knowing that these ventures have a negative expected return. To explain such an 'irrational behaviour', we may argue that not all people value money evenly. Alternatively we may rationalize that the prospect of winning $36 * 100 = \$3600$ in a second at the whim of the roulette is worth taking the risk. After all, someone will win, so it might as well be me! Both an attitude towards money and the willingness to take risks, originating in a person's initial wealth, emotional state and the pleasure to be evoked in some way by such risk, are reasons that may justify a departure from the Bayes EMV criterion. If all people were 'straight' expected payoff decision-makers, then there would be no national lotteries and no football or basketball betting. Even the mafia might be much smaller! People do not always use straight expected payoffs to reach decisions, however. The subjective valuation of money and people's attitudes towards risk and gambling provide the basic elements that characterize gambling and the utility of money associated with such gambling. Utility theory seeks to represent how such subjective valuation of wealth and attitude towards risk can be quantified so that it may provide a rational foundation for decision-making under uncertainty.

Just as in Las Vegas we might derive 'pleasure from gambling', we may be also concerned by the loss of our wealth, even if it can happen with an extremely small probability. To protect ourselves from large losses, we often turn to insurance. Do we insure our house against fire? Do we insure our belongings against theft? Should we insure our exports against currency fluctuations or against default payment by foreign buyers? Do we invest in foreign lands without seeking insurance against national takeovers? And so on. In these situations and in order to avoid large losses, we willingly pay money to an insurance firm – the premium needed to buy such insurance. In other words, we transfer our risk to the insurer who in

Risk and Financial Management: Mathematical and Computational Methods. C. Tapiero
 ISBN: 0-470-84908-8

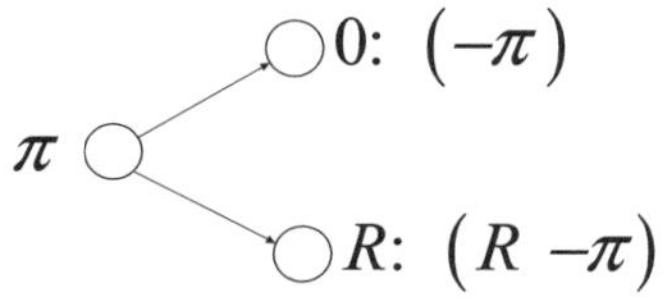

Figure 3.1 A lottery.

turn makes money by collecting the premium. Of course, how much premium to pay for how much risk insured underscores our ability to sustain a great loss and our attitude towards risk. Thus, just as our gambler was willing to pay a small amount of money to earn a very large one (albeit with a very small probability), we may be willing to pay a small amount (the premium) to prevent and protect ourselves from having to face a large loss, even if it occurs with a very small probability. In both cases, the Bayes expected payoff (EMV) criterion breaks down, for otherwise there would be no casinos and no insurance firms. Yet, they are here and provide an important service to society. Due to the importance of utility theory to economics and finance, providing a normative framework for decision-making under uncertainty and risk management, we shall outline its basic principles. Subsequently, we shall see how the concepts of expected utility have been used importantly in financial analysis and financial decision-making.

3.1.1 Lotteries and utility functions

Lotteries consist of the following: we are asked to pay a price π (say it is \$5) for the right to participate in a lottery and earn, potentially, another amount, R, called the reward (which is say \$1 000 000), with some probability, p. If we do not win the lottery, the loss is π. If we win, the payoff is R. This lottery is represented graphically in Figure 3.1 where all cash expenditures are noted. Lotteries of this sort appear in many instances. A speculator buys a stock expecting to make a profit (in probability) or losing his investment. Speculators are varied, however, owning various lotteries and possessing varied preferences for these lotteries. It is the exchange between speculators and investors that create a 'financial market' which, once understood, can provide an understanding and a valuation of lotteries pricing.

If we use an EMV criterion for valuing the lottery, as seen in the previous chapter, then the value of the lottery would be:

$$\text{Expected value of lottery} = p(R - \pi) - (1 - p)\pi = pR - \pi < 0$$

By participating in the lottery, we will be losing money in an expected sense. In other words, if we had 'an infinite amount of money' and were to play the lottery forever, then in the long run we would lose \$$(\pi - pR)$! Such odds for lotteries are not uncommon, and yet, however irrational they may seem at first, many people play such lotteries. For example, people who value the prospect of 'winning big' even with a small probability much more than the prospect of 'losing small' even with a large probability, buy lottery tickets. This uneven valuation of money means that we may not be able to compare two sums of money easily. People are different in many ways, not least in their preferences for outcomes that are uncertain. An understanding of human motivations and decision making is thus needed to reconcile observed behaviour in a predictable and theoretical framework. This is in essence what expected utility theory is attempting to do. Explicitly, it seeks to define a scale that values money by some function, called the utility function $U(.)$, whose simple expectation provides the scale for comparing alternative financial and uncertain prospects. The larger the expected utility, the 'better it is'.

More precisely, the function $U(.)$ is a transformation of the value of money that makes lotteries of various sums comparable. Namely, the two sums $(R-\pi)$ and $(-\pi)$, can be transformed into $U(R-\pi)$ and $U(-\pi)$, and then the lottery would be,

- Make $U(R-\pi)$ with a probability p.
- Make (lose) $U(-\pi)$ with a probability $1-p$.

while its expected value, which tells how valuable it is compared to other lotteries, is:

$$\text{Expected utility} = EU = pU(R-\pi) + (1-p)U(-\pi)$$

This means that:

- If $EU = 0$, we are indifferent whether we participate in the lottery or not.
- If $EU > 0$, we are better off participating in the lottery.
- If $EU < 0$, we are worse off participating in the lottery.

Thus, participation in a lottery is measured by its expected utility. Further, the price $\$\pi$ we will be willing to pay – the premium, for the prospect of winning $\$R$ with probability p – is the price that renders the expected utility null, or $EU = 0$, found by the solution to

$$EU = 0 = pU(R-\pi) + (1-p)U(-\pi)$$

which can be solved for π when the utility function is specified.

By the same token, expected utility can be used by an investor to compare various lotteries, various cash flows and payments, noting that the value of each has an expected utility, known for certain and used to scale the uncertain prospects. The 'expected utility' approach to decision-making under uncertainty is thus extremely useful, providing a rational approach 'eliminating the uncertainty from decision-making' and bringing it back to a problem under certainty, which we can solve explicitly and numerically. But there remains the nagging question: how can we obtain such utility functions? And how justified are we in using them? Von Neumann and Morgenstern, two outstanding mathematicians and economists, concluded in the late 1940s, that for expected utility to be justified as a scaling function for uncertain prospects the following holds:

(1) The higher the utility the more desirable the outcome. This makes it possible to look for the best decision by seeking the decision that makes the expected utility largest.
(2) If we have three possibilities (such as potential investment alternatives), then if possibility '1' is 'better' than '2' and '2' is better than '3', then necessarily '1' is better than '3'. This is also called the transitivity axiom.
(3) If we are indifferent between two outcomes or potential acts, then necessarily the expected utilities will be the same.

These three assumptions, underlie the rational framework for decision making under uncertainty that expected utility theory provides.

3.2 UTILITY AND RISK BEHAVIOUR

An expected utility provides a quantitative expression of a decision makers' desires for higher rewards as well as his attitude towards the 'risks' of such rewards. Say that $\{R, P(.)\}$ is a set of rewards R assumed to occur with probability $P(.)$ and let $u(.)$ define a utility function. The basic utility theorem states that the expected utility provides an objective index to evaluate the desirability of rewards, or:

$$E\,(u(R)) = \int u(R)P(R)\,\mathrm{d}R;\ R \in \Re$$

Given uncertain prospects, a rational decision-maker will then select that prospect whose expected utility is largest. For example, the EU of an alternative prospect i with probability outcomes (π_{ij}, p_{ij}) is:

$$EU_i = \sum_{j=1}^{n} p_{ij}u(\pi_{ij})$$

and the optimal alternative k is found by:

$$k \in \mathrm{Max}_{i\in[1,n]}\,\{EU_i\}$$

In this decision approach, the function $u(.)$, stands for the investor's psychology. For example, we might construe that $u'(.) > 0$ implies *greed*, $u''(.) < 0$ implies *fear*, while *risk tolerance and prudence* are implied by the signs of the third derivative $u'''(.) > 0$ and $u'''(.) < 0$ respectively. Given a probability distribution for rewards, $P(R)$, the basic assumptions regarding continuous utility functions are that alternative rewards:

(1) can be compared (comparability).
(2) can be ranked such that preferred alternatives have greater utility.
(3) have strong independence.
(4) have transitive preferences (transitivity).
(5) are indifferent if their utilities are equal.

3.2.1 Risk aversion

Expected utility provides an investor preference for uncertain payoffs, expressing thereby his attitude toward the risk associated with such payoffs. Three attitudes are defined: (1) risk aversion (2) risk loving and (3) risk neutrality. Risk aversion expresses a risk-avoidance preference and thus a preference for more conservative gambles. For example, a risk-averse investor may be willing to pay a premium to reduce risk. A risk lover would rather enjoy the gamble that an investment risk provides. Finally, risk neutrality implies that rewards are valued at their objective value by the expectation criterion (EMV). In other words, the investor would be oblivious to risk. For risk-averse investors, the desire for greater rewards with smaller probabilities will decrease (due to the increased risk associated with such rewards); such an attitude will correspond to a negative second derivative of the utility function or equivalently to an assumption of concavity, as we shall see below. And, vice versa, for a risk loving decision-maker the second derivative of the utility function will be positive. To characterize quantitatively a risk attitude, two approaches are used:

- Risk aversion directly relates to the risk premium, expressed by the difference between the expected value of a decision and its certainty (riskless) equivalent reward.
- Risk aversion is expressed by a decreasing preference for an increased risk, while maintaining a mean preserving spread.

These two definitions are equivalent for concave utility functions, as we shall see below.

Certainty equivalence and risk premium

Assume an uncertain reward $\tilde{R}$ whose expected utility is $E(u(\tilde{R}))$. Its equivalent sure amount of money, given by the expected utility of that amount, is called the *certainty equivalent* which we shall denote here by $\bar{R}$ and is given by

$$u(\bar{R}) = E(u(\tilde{R})) \qquad \text{and} \qquad \bar{R} = u^{-1}\{E[u(\tilde{R})]\}$$

Note that the certainty equivalent is not equal to the expected value $\hat{R} = E(\tilde{R})$ for it embodies as well the cost of risk associated with the uncertain prospect valued by its expected utility. The difference $\rho = \hat{R} - \bar{R}$, expresses the *risk premium* a decision maker would be willing to pay for an outcome that provides for sure the expected return compared to the certainty equivalent. It can be null, positive or negative. In other words, the risk premium is:

$$\text{Risk premium}\,(\rho) = \text{Expected return}\ (\hat{R}) - \text{The certainty equivalent}\ (\bar{R})$$

An alternative representation of the risk premium can be reached by valuing the expected utility of the random payoff: $\tilde{R} = \hat{R} + \tilde{\varepsilon}$ where $E(\tilde{\varepsilon}) = 0$, $\text{var}(\tilde{\varepsilon}) = \sigma^2$ and σ^2 denotes the payoff spread. In this case, note that a Taylor series expansion around the mean return yields:

$$Eu(\tilde{R}) = Eu(\hat{R} + \tilde{\varepsilon}) = u(\hat{R}) + \sigma^2 \frac{u''(\hat{R})}{2}$$

Similarly, a first-order Taylor series expansion of the certainty equivalent utility around the mean return (since there are no uncertain elements associated with it) yields:

$$u(\bar{R}) = u(\hat{R} - \rho) = u(\hat{R}) - \rho u'(\hat{R})$$

Equating these two equations, we obtain the risk premium calculated earlier but expressed in terms of the derivatives of the utility function and the return variance, or:

$$\rho = -\frac{1}{2}\sigma^2 \frac{u''(\hat{R})}{u'(\hat{R})}$$

This risk premium can be used as well to define the index of risk behaviour suggested by Arrow and Pratt. In particular, Pratt defines an *index of absolute risk aversion* expressing the quantity by which a fair bet must be altered by a risk-averse decision maker in order to be indifferent between accepting and rejecting the bet. It is given by:

$$\rho_a(\tilde{R}) = \frac{\rho}{\sigma^2/2} = -\frac{u''(\hat{R})}{u'(\hat{R})}$$

Prudence and robustness

When a decision-maker's expected utility is not (or is mildly) sensitive to other sources of risk, we may state that the expected utility is 'robust' or expresses a prudent attitude by the decision-maker. A prudent investor, for example, who adopts a given utility function to reach an investment decision, expresses both his desire for returns and the prudence he hopes to assume in obtaining these returns, based on the functional form of the utility function he chooses. Thus, an investor with a precautionary (prudence) motive will tend to save more to hedge against the uncertainty that arises from additional sources of risk not accounted for by the expected utility of uncertain returns. This notion of prudence was first defined by Kimball (1990) and Eeckoudt and Kimball (1991) and is associated with the optimal utility level (measured by the relative marginal utilities invariance), which is, or could be, perturbed by other sources of risk. Explicitly, say that $(w, \tilde{R})$ is the wealth of a person and the random payoff which results from some investment. If we use the expected marginal utility, then at the optimum investment decision:

$$Eu'(w + \tilde{R}) > u'(w) \quad \text{if } u' \text{ is convex}$$
$$Eu'(w + \tilde{R}) < u'(w) \quad \text{if } u' \text{ is concave}$$

The risk premium ψ that the investor pays for 'prudence' is thus the amount of money required to maintain the marginal utility for sure at its optimal wealth level. Or:

$$u'(w - \psi) = Eu'(w + \tilde{R}) \quad \text{and} \quad \psi = w - u'^{-1}[Eu'(w + \tilde{R})]$$

Proceeding as before (by using a first term Taylor series approximation on the marginal utility), we find that:

$$\psi = \frac{1}{2}\,\text{var}(\tilde{R})\left[-\frac{u'''(w)}{u''(w)}\right]$$

The square bracket term is called the *degree of absolute prudence*. For a risk-averse decision maker, the utility second-order derivative is negative ($u' \leq 0$) and therefore prudence will be positive (negative) if the third derivative u''' is positive (negative). Further, Kimball also shows that if the risk premium is positive and decreases with wealth w, then $\psi > \pi$. As a result, $\psi - \pi$ is a premium an investor would pay to render the expected utility of an investment invariant under other sources of risks.

The terms expected utility, certainty equivalent, risk premium, Arrow–Pratt index of risk aversion and prudence are used profusely in insurance, economics and financial applications, as we shall see later on.

3.2.2 Expected utility bounds

In many instances, calculating the expected utility can be difficult and therefore bounds on the expected utility can be useful, providing a first approximation to the expected utility. For risk-averse investors with utility function $u(.)$ and $u''(.) \leq 0$, the expected utility has a bound from above, known as *Jensen's inequality.* It is given by:

$$Eu(\tilde{R}) \leq u(\hat{R}) \quad \text{when} \quad u''(.) \leq 0$$
$$Eu(\tilde{R}) \geq u(\hat{R}) \quad \text{when} \quad u''(.) \geq 0$$

and vice versa when it is the utility function of a risk-loving investor (i.e. $u''(.) \geq 0$). When rewards have known mean and known variance however, Willasen (1981, 1990) has shown that for risk-averse decision-makers, the expected utility can be bounded from below as well. In this case, we can bound the expected utility above and below by:

$$u(\hat{R}) \geq Eu(\tilde{R}) \geq \hat{R}^2 u(\alpha_2/\hat{R})/\alpha_2;\ \alpha_2 = E(\tilde{R}^2)$$

The first bound is, of course, Jensen's inequality, while the second inequality provides a best lower bound. It is possible to improve on this estimate by using the best upper and lower Tchebycheff bounds on expected utility (Willasen, 1990). This inequality is particularly useful when we interpret and compare the effects of uncertainty on the choice of financial decisions, as we shall see in the example below. Further, it is also possible to replace these bounds by polynomials such that:

$$Eu(\tilde{R}) \leq EA(\tilde{R});\ Eu(\tilde{R}) \geq EB(\tilde{R})$$

where $A(.)$ and $B(.)$ are polynomials of the third degree. To do so, second- and third-order Taylor series approximations are taken for the utility functions (using thereby the decision-makers' prudence). For example, consider the following portfolio prospect with a mean return of $\hat{R}$ and a variance σ^2. Say that mean returns are also a function of the variance, expressing the return-risk substitution, with:

$$\hat{R} = \hat{R}(\sigma), \partial \hat{R}(\sigma)/\partial\sigma >, \hat{R}(0) = R_f$$

where R_f denotes the riskless rate of return. It means that the larger the returns uncertainty, the larger the required expected payoff. Using the Jensen and Willasen inequalities, we have for any portfolio, the following bounds on the expected utility:

$$\frac{u(\hat{R}(\sigma)(1+\nu))}{1+\nu} \leq Eu(\tilde{R}) \leq u(\hat{R}(\sigma)); \; \nu = \frac{\sigma^2}{\hat{R}^2}; \quad E(\tilde{R}^2) = \hat{R}^2 + \sigma^2$$

Thus, lower and upper bounds of the portfolio expected utility can be constructed by maximizing (minimizing) the lower (upper) bounds over feasible $(\hat{R}, \sigma)$ portfolios. Further, if we set $\hat{R} = R_f + \lambda\sigma$ where λ is used as a measure for the price of risk (measured by the return standard deviation and as we shall see subsequently), we have equivalently the following bounds:

$$\frac{u((R_f + \lambda\sigma)(1+\nu))}{1+\nu} \leq Eu(\tilde{R}) \leq u(R_f + \lambda\sigma)$$

The definition of an appropriate utility function is in general difficult. For this reason, other means are often used to express the desirability of certain outcomes. For example, some use targets, expressing the desire to maintain a given level of cash, deviations from which induce a dis-utility. Similarly, constraints (as they are defined by specific regulation) as well as probability constraints can also be used to express a behavioural attitude towards outcomes and risks. Such an approach has recently been found popular in financial circles that use 'value at risk' (VaR) as an efficiency criterion (see Chapter 10 in particular). Such assumptions regarding decision-makers' preferences are often used when we deal with practical problems.

3.2.3 Some utility functions

A utility function is selected because it represents the objective of an investor faced with uncertain payoffs and his attitude towards risk. It can also be selected for its analytical convenience. In general, such a selection is difficult and has therefore been one of the essential reasons in practice for seeking alternative approaches to decision making under uncertainty. Below we consider a number of analytical utility functions often used in theoretical and practical applications.

(1) The exponential utility function: $u(w) = 1 - e^{-aw}, a > 0$ is a concave function. For this function, $u'(w) = a\,e^{-aw} > 0$, $u''(w) = -a^2 e^{-aw} < 0$ while the index of absolute risk aversion R_A is constant and given by: $R_A(w)$

$= -u''/u' = a > 0$. Further $u'''(w) = a^3\, e^{-aw} > 0$ and therefore the degree of prudence is a while the prudence premium is, $\psi = \frac{1}{2} a \operatorname{var}(\tilde{R})$.

(2) The logarithmic utility function: $u(w) = \log(\beta + \gamma w)$, with $\beta > 0, \gamma > 0$ is strictly increasing and strictly concave and has a strictly decreasing absolute risk aversion. Note that, $u'(w) = \gamma/(\beta + \gamma w) > 0, u''(w) = -\gamma^2/(\beta + \gamma w)^2 < 0$ while, $R_A(w) = \gamma/(\beta + \gamma w) = u'(w)$ which is decreasing in wealth.

(3) The quadratic utility function: $u(w) = w - \rho w^2$ is a concave function for all $\rho \geq 0$ since $u' = 1 - 2\rho w, u' \geq 0 \to w \leq 1/2\rho$ and $u'' = -2\rho \leq 0$. As a result, the Arrow–Pratt index of absolute risk aversion is

$$R_A(w) = -\frac{u''[E(w)]}{u'[E(w)]} = \frac{2\rho}{1 - 2\rho w}$$

and the prudence is null (since the third derivative is null).

(4) The cubic utility function: $u(w) = w^3 - 2kw^2 + (k^2 + g^2)w,\ k^2 > 3g^2$ is strictly increasing and strictly concave and has a decreasing absolute risk aversion if $0 \leq w \leq \frac{2}{3}k - \frac{1}{2}\sqrt{k^2 - 3g^2}$.

(5) The power utility function: $u(w) = (w - \delta)^\beta$, $0 < \beta < 1$ is strictly increasing and has a strictly absolute risk aversion on $[\delta, \infty)$ since: $u' = \beta\,(w - \delta)^{\beta-1}$, and $u'' = -(1 - \beta)\beta\,(w - \delta)^{\beta-2}$. The risk aversion index is thus, $R_A(w) = -(1 - \beta)\,(w - \delta)^{-1}$.

(6) The HARA (hyperbolic absolute risk aversion) has a utility function given by:

$$u(w) = \frac{1 - \gamma}{\gamma}\left[\frac{aw}{1 - \gamma} + b\right]^\gamma$$

while its first and second derivatives as well as its index of absolute risk aversion are given by:

$$u' = a\left[\frac{aw}{1 - \gamma} + b\right]^{\gamma-1} > 0,\ u'' = -a^2\left[\frac{aw}{1 - \gamma} + b\right]^{\gamma-2} < 0;$$

$$R_a(w) = -\frac{u''}{u'} = \frac{a}{b + aw/(1 - \gamma)} > 0$$

This utility function includes a number of special cases. In particular, when γ tends to one, we obtain the logarithmic utility.

3.2.4 Risk sharing

Two firms sign an agreement for a joint venture. A group of small firms organize a cooperative for marketing their products. The major aerospace companies in the US west coast set up a major research facility for deep space travel. A group of 70 leading firms form a captive insurance firm in the Bahamas to insure their managers against kidnappings, and so on. These are all instances of risk sharing. Technically, when we combine together a number of (independent) participants and split among them a potential loss or gain, the resulting variance of the loss or gain for each of the participants will be smaller. Assuming that this variance is an indicator of the 'risk', and if decision makers are assumed to be risk averse, then the more partners in the venture the smaller the individual risk sustained by each partner. Such arguments underly the foundations of insurance firms (that create the means for risk sharing), of major corporations based on numerous shareholders etc. Assuming that our preference is well defined by a utility function $U(.)$, how would we know if it is worthwhile to share risk? Say that the net benefits (profits less costs) of a venture is $\$\tilde{X}$ whose probability distribution is $p(\tilde{X})$. If we do nothing, nothing is gained and nothing is lost and therefore the 'value' of doing nothing is $U(0)$. The venture with its n participants, however, will have an expected utility $EU(\tilde{X}/n)$. Thus, if sharing is worthwhile the expected utility of the venture ought to be greater than the utility of doing nothing! Or, $EU(\tilde{X}/n) > U(0)$.

Problems

(1) Formulate the problem of selecting the optimal size of a risk-sharing pool.
(2) How much does a member of the pool benefit from participating in sharing.

3.3 INSURANCE, RISK MANAGEMENT AND EXPECTED UTILITY

How much would it be worth paying for car insurance (assuming that there is such a choice)? This simple question highlights an essential insurance problem. If we are fully insured and the premium is $\$\pi$, then the expected utility is, for sure, $U(w - \pi)$ where w is our initial wealth. If we-self-insure for a risk whose probability distribution is $p(\tilde{X})$, then using the expected utility theory paradigm, we should be willing to pay a premium π as long as $U(w - \pi) > EU(w - \tilde{X})$. In fact the largest premium we would be willing to pay solves the equation above, or

$$\pi^* = w - U^{-1}(EU(w - \tilde{X}))$$

Thus, if the utility function is known, we can find out the premium π^* above which we would choose to self-insure.

Problem

For an exponential, HARA and logarithmic utility function, what is the maximal premium an individual will be willing to pay for insurance?

3.3.1 Insurance and premium payments

Insurance risk is not reduced but is transferred from an individual to an insurance firm that extracts a payment in return called the premium and profits from it by investing the premium and by risk reducing aggregation. In other words, it is the difference in risk attitudes of the insurer and the insured, as well as the price insured, and insurers are willing to pay for that to create an opportunity for the insurance business.

Say that $\tilde{X}$ is a risk to insure (a random variable) whose density function is $F(\tilde{X})$. Insurance firms, typically, seek some rule to calculate the premium they ought to charge policyholders. In other words, they seek a 'rule' Υ such that a premium can be calculated by:

$$P = \Upsilon(F(\tilde{X}))$$

Although there are alternative ways to construct this rule, the more prominent ones are based on the application of the expected utility paradigm and traditionally based on a factor loading the mean risk insured. The expected utility approach seeks a 'fair' premium P which increases the firm expected utility, or:

$$U(W) \leq EU(W + P - \tilde{X})$$

where W is the insurance firm's capital. The loading factor approach seeks, however, to determine a loading parameter λ providing the premium to apply to the insured and calculated by $P/n = (1 + \delta)E(\tilde{x})$, where $\tilde{x}$ denotes the individual risk in a pool of n insured, i.e. $\tilde{x} = \tilde{X}/n$ and P/n is an individual premium share. For the insured, whose utility function is $u(.)$ and whose initial wealth is w, the expected utility of insurance ought to be greater than the expected utility of self-insurance. As a result, a premium P is feasible if the expected utilities of both the insurer and the insured are larger with insurance, or:

$$u(w - P/n) \geq Eu(w - \tilde{x}_i),\ \tilde{X} = \sum_{i=1}^{n} \tilde{x}_i;\ U(W) \leq EU(W + P - \tilde{X})$$

Note that in this notation, the individual risk is written as $\tilde{x}_i$ which is assumed to be identically and independently distributed for all members of the insurance pool. Of course, since an insurance firm issues many policies, assumed independent, it will profit from risk aggregation. However, if risks are correlated, the variance of $\tilde{X}$ will be much greater, prohibiting in some cases the insurance firm's ability or willingness to insure (as is the case in natural disaster, agricultural and weather related insurance).

Insurance 'problems' arise when it is necessary to resolve the existing disparities between the insured and the insurer, which involves preferences and insurance terms that are specific to both the individual and the firm. These lead to extremely rich topics for study, including the important effects of moral hazard, adverse selection resulting from information asymmetry which will be studied subsequently, risk correlation, rare events with substantive damages, insurance against human-inspired terrorists acts etc.

Risk sharing, risk transfer, reinsurance and other techniques of risk management are often used to spread risk and reduce its economic cost. For example, let $\tilde{x}$ be the insured risk; the general form of reinsurance schemes associated with an insurer (I), an insured (i) and a reinsurer (r) and consisting in sharing risk can be written as follows:

$$\text{Insured:} \quad R_i(\tilde{x}\,|a\ ,q) = \begin{cases} \tilde{x} & \tilde{x} \le a \\ (1-q)\tilde{x} & \tilde{x} \ge a \end{cases}$$

$$\text{Insurer:} \quad R_I(\tilde{x}\,|a,b,c,q) = \begin{cases} 0 & \tilde{x} \le a \\ q(\tilde{x}-a) & a < \tilde{x} \le b \\ c & b > \tilde{x} \end{cases}$$

$$\text{Reinsurer:} \quad R_r(\tilde{x}\,|b,c,q) = \begin{cases} 0 & \tilde{x} \le b \\ q(b-\tilde{x}) - c & \tilde{x} \ge b \end{cases}$$

Here, if a risk materializes and it is smaller than 'a', then no payment is made by the insurance firm while the insured will be self-insured up to this amount. When the risk is between the lower level 'a' and the upper one 'b', then only a proportion q is paid where $1-q$ is a co-participation rate assumed by the insured. Finally, when the risk is larger than 'b', then only c is paid by the insurer while the remaining part $\tilde{x} - c$ is paid by a reinsurer. In particular for a proportional risk scheme we have $R(\tilde{x}) = q\tilde{x}$ while for an excess-loss reinsurance scheme we have:

$$R_I(\tilde{x}\,|a) = \begin{cases} 0 & \tilde{x} \le a \\ \tilde{x} - a & \tilde{x} > a \end{cases}$$

where a is a deductible specified by the insurance contract. A reinsurance scheme is thus economically viable if the increase in utility is larger than the premium P_r to be paid to the reinsurer by the insurance firm. In other words, for utility functions $u_I(.)$, $u_i(.)$, $u_r(.)$ for the individual, the insurance and the reinsurance firms with premium payments: P_i, P_I, P_r, the following conditions must be held:

$$u_i(w - R_i\,(\tilde{x}\,|a,q) - P_i) \ge Eu_i(w - \tilde{x}) \text{ (the individual condition)}$$

$$u_I(W) \le Eu_I(W + P_i - P_I - R_I\,(\tilde{x}\,|a,b,c,q)) \text{ (the insurance firm condition)}$$

$$u_r(W_r) \le Eu_r(W_r + P_I - R_r\,(\tilde{x}\,|b,c,q)) \text{ (the reinsurer condition)}$$

Other rules for premium calculation have also been suggested in the insurance literature. For example, some say that in insurance 'you get what you give'. In this sense, the premium payments collected from an insured should equal what he has claimed plus some small amounts to cover administrative expenses. These issues are in general much more complex because the insurer benefits from risk aggregation over the many policies he insures, a concept that is equivalent to portfolio risk diversification. In other words, if the insurance firm is large enough it might be justified in using a small (risk-free) discount rate in valuing its cash flows, compared to an individual insured, sensitive with the uncertain losses associated with the risk insured. For this reason, the determination of the loading rate is often a questionable parameter in premium determination. Recent research has greatly improved the determination of insurance premiums by indexing insurance risk to market risk and using derivative markets (such as options) to value insurance contracts (and thereby the cost of insurance or premium).

3.4 CRITIQUES OF EXPECTED UTILITY THEORY

Theory and practice do not always concur when we use expected utility theory. There are many reasons for such a statement. Are decision-makers irrational? Are they careless? Are they uninformed or clueless? Do they lack the proper incentives to reach a rational decision. Of course, the axioms of rationality that underlie expected utility theory may be violated. Empirical and psychological research has sought to test the real premises of decision-makers under uncertainty. To assess potential violations, we consider a number of cases. Consider first the example below called the St Petersburg Paradox that has motivated the development of the utility approach.

3.4.1 Bernoulli, Buffon, Cramer and Feller

Daniel Bernoulli in the early 1700s suggested a problem whose solution was not considered acceptable in practice, albeit it seemed to be appropriate from a theoretical viewpoint. This is called the St Petersburg Paradox. The paradox is framed in a tossing game stating how much one would be willing to pay for a game where a fair coin is thrown until it falls 'heads'. If it occurs at the rth throw, the player receives 2^r dollars from the bank. Thus, the gain doubles at each throw. In an expected sense, the probability of obtaining 'heads' at the kth throw is $1/2^k$, since the pay-out is also equal to 2^k, the expected value of the game is:

$$\sum_{k=1}^{\infty}(1/2^k)2^k = 1 + 1 + \cdots = \infty$$

Thus, the fair amount to pay to play this game is infinite, which clearly does not reflect the decision makers' behaviour. Bernoulli thus suggested a logarithmic utility function whose expected utility:

$$Eu(x) = \sum_{i=1}^{\infty} p(i)u(2^{i-1}) = \sum_{i=1}^{\infty}\left(\frac{1}{2^i}\right)\log(2^{i-1}) = \sum_{i=1}^{\infty}\left(\frac{i-1}{2^i}\right)\log(2)$$

Since

$$\sum_{i=1}^{\infty}\left(\frac{1}{2^i}\right) = 1 \quad \text{and} \quad \sum_{i=1}^{\infty}\left(\frac{i-1}{2^i}\right) = 1$$

The expected utility of the game equals $Eu(x) = \log(2)$.

Mathematicians such as Buffon, Cramer, Feller and others have attempted to provide a solution that would seem to be appropriate. Buffon and Cramer suggest that the game be limited (in the sense that the bank has a limited amount of money and, therefore, it can only pay a limited amount). Say that the bank has only a million dollars. In this case, we will have the following amounts,

$$\sum_{k=1}^{19}(1/2^k)2^k + 10^6\sum_{k=20}^{\infty}(1/2^k) = 1 + 1 + \cdots 1 + 1.19 \approx 21$$

Therefore, the fair amount to play this game is 21 dollars only. Any larger amount would be favourable to the bank.

Gabriel Cramer, on the other hand, suggested a square root utility function. Then for the St Petersburg game, we have:

$$Eu(x) = \sum_{i=1}^{\infty} p(i)u(i) = \sum_{i=1}^{\infty} \left(\frac{1}{2^i}\right) \sqrt{2^i - 1}$$
$$= \frac{1}{2} + \frac{\sqrt{2}}{2^2} + \frac{(\sqrt{2})^2}{2^2} + \cdots + \frac{(\sqrt{2})^j}{2^{j+1}} + \cdots = \frac{1}{2}\frac{1}{1-\sqrt{2}/2} = \frac{1}{2-\sqrt{2}}$$

And therefore, for Cramer, the value of the game is $1/(2-\sqrt{2})$. Feller suggests another approach however, seeking a mechanism for the gains and payments to be equivalent in the long run. In other words, a lottery will be fair if:

$$\frac{\text{Accumulated gains}}{\text{Accumulated fees}} = \frac{N_n}{R_n} \to 1 \text{ as } n \to \infty$$

$$\text{or} \quad P\left\{\left|\frac{N_n}{R_n} - 1\right| < \varepsilon\right\} \to \infty \text{ as } n \to \infty$$

Feller noted that the game is fair if $R_n = n \log_2(n)$. Thus if the accumulated entrance fee to the game is proportional to the number of games, it will not be a fair game.

3.4.2 Allais Paradox

The strongest attack on expected utility theory can be found in Allais' Paradox, which doubts the strong independence assumption needed for consistent choice in expected utility. Allais proved that the assumption of linearity in probabilities applied in calculating the expected utility is often doubtful in practice. Explicitly, the independence axiom also called the 'sure-thing' principle asserts that two alternatives that have a common outcome under a particular state of nature should imply that ordering should be independent of the value of their common outcome. This is not always the case and counter examples abound, in particular due to Allais.

For example, let us confront people with two lotteries. First, we have to pick one of the two gambles given by (p_1, p_2) below. The first gamble consists of \$100 000 for sure (probability 1) while the other is \$5 000 000 with probability 0.1, 1 million with probability 0.89 and nothing with probability 0.01 as stated below.

$$p_1 \Rightarrow \{1 \quad 100\,000 \quad \text{and} \quad p_2 \Rightarrow \begin{cases} 0.1 & 5\,000\,000 \\ 0.89 & 1\,000\,000 \\ 0.01 & 0 \end{cases}$$

A second set of gambles (p_3, p_4) consists of:

$$p_3 \Rightarrow \begin{cases} 0.1 & 5\,000\,000 \\ 0.9 & 0 \end{cases} \quad \text{and} \quad p_4 \Rightarrow \begin{cases} 0.11 & 1\,000\,000 \\ 0.89 & 0 \end{cases}$$

Confronted with choosing between p_1 and p_2, people chose p_1 while confronting p_1, p_3 and p_4, people preferred p_3 which is in contradiction with the strong independence axiom of utility theory. In other words, if gamble 1 is selected over gamble 2, while in presenting people with gambles 1, 3 and 4 results in their selecting 3 over 1, there is necessarily a contradiction, since if we were to compare gambles 3 and 2, clearly, 3 is not as good as 2. This contradiction means, therefore, that application of expected utility theory does not always represent investors' and decision-maker's psychology. Since then, a large number of studies have been done, seeking to bridge a gap between investors' psychology and the concepts of utility theory. Some essential references include Kahnemann and Tverski (1979), Machina on anticipated utility (1982, 1987) and Quiggin (1985) as well as many other others. In these approaches, the expected utility framework is 'extended' by stating that an uncertain prospect can be measured by an 'expected utility' $u(.)$ interpreted either as the choice of a utility function (as was the case in traditional expected utility) or by a preferred probability distribution (P) (or a function $g(P)$ assumed over the probability distribution). Then, the probabilities used to calculate the expected utility would be 'subjective' estimates, or beliefs, about the probabilities of returns, imbedding 'something else' above and beyond the objective assessment of uncertain prospects. Thus, the objective index used to value the relative desirability of uncertain prospects is also a function of the model used for probabilities $P(.)$. This is in contrast to the utility function, expressing a behaviour imbedded in the choice of the function $u(.)$ only, which stands solely for the investor's psychology, as we saw above. For example, what if we were to determine probabilities $P^*(.)$ such that the price of random prospects $\tilde{R}$ could be uniquely defined by the following expected value?

$$\pi = \int_{\Re} \tilde{R}\, \mathrm{d}P^*(\tilde{R})$$

In this case, once such a well-defined transformation of these probabilities is reached, all uncertain prospects may be valued uniquely, thereby simplifying greatly the problem of financial valuation of risk assets such as stocks, default bonds and the like. This approach, defined in terms of economic exchange market mechanisms (albeit subject to specific assumptions regarding markets and individual behaviours), underlies the modern theories of finance and 'risk-neutral pricing'. This is also an essential topic of our study in subsequent chapters.

3.5 EXPECTED UTILITY AND FINANCE

Finance provides many opportunities for applications of expected utility. For example, portfolio management consists essentially of selecting an allocation strategy among n competing alternatives (stocks, bonds, etc.), each yielding an uncertain payoff. Each stock purchase is an alternative which can lead to a (speculative) profit or loss with various (known or unknown) probabilities. When selecting several stocks and bonds to invest in, balancing the potential gains with the risks of losing part or all of the investment, the investor in effect constructs a portfolio with a risk/reward profile which is preferred in an expected utility sense. Similarly, to evaluate projects, contracts, investments in real estate, futures and forward contracts, etc. financial approaches have been devised based directly on (or inspired by) expected utility. Below, we shall review some traditional techniques for valuing cash flows and thereby introduce essential notions of financial decision making using expected utility. Typical models include the CAPM (Capital Asset Pricing Model) as well the SDF (Stochastic Discount Factor) approach.

3.5.1 Traditional valuation

Finance values money and cash flow, the quantity of it, the timing of it and the risk associated with it. A number of techniques and approaches that are subjective – defined usually by corporate financial officers or imposed by managerial requirements – have traditionally been used. For example, let $C_0, C_1, C_2, \ldots, C_n$ be a prospective cash flow in periods $i = 1, 2, \ldots, n$. Such a cash flow may be known for sure, may be random, payments may be delayed unexpectedly, defaulted etc. To value these flows, various techniques can be used, each assuming a body of presumptions regarding the cash flow and its characteristics. Below, we consider first a number of 'traditional' approaches including 'the payback period', 'the accounting rate of return' and the traditional 'NPV' (Net Present Value).

Payback period

The payback period is the number of years required for an investment to be recovered by a prospective cash flow. CFOs usually specify the number of years needed for recovery. For example, if 4 years is the specified time to recover an investment, then any project with a prospective cash flow of recovery less or equal to 4 years is considered acceptable. While, any investment project that does not meet this requirement is rejected. This is a simple and an arbitrary approach, although in many instances it is effective in providing a first cut approach to multiple investment opportunities. For example, say that we have an investment of \$100 000 with a return cash flow (yearly and cumulatively) given by the following table:

Year	1	2	3	4	5	6	7	8	9
Return	−5	5	10	20	40	30	20	10	10
Cumulative	−5	0	10	30	70	100	120	130	140

Thus, only after the sixth year is the investment is recovered. If management specifies a period of 4 years payback, then of course the investment will be rejected.

The accounting rate of return (ARR)

The ARR is a ratio of average profit after depreciation and average investment book value. There are, of course, numerous accounting procedures in calculating these terms, making this approach as arbitrary as the payback period.

A decision may be made to specify a required ARR. Any such ratio 'better' than the ARR selected would imply that the investment project is accepted.

The internal rate of return (IRR)

The net present value (NPV) approach uses a discount rate R for the time value of money, usually called the rate of return. This discount rate need not be the risk-free rate (even if future cash flows are known for sure). Instead, they are specified by CFOs and used to provide a firm's valuation of the prospective cash flow. Typically, it consists of three components: the real interest rate, the inflation rate and a component adjusting for investment risk.

Discount (Nominal) Rate = (Real + Inflation + Risk Compensating) Rates

Each of these rates is difficult to assess and, therefore, much of finance theory and practice seeks to calculate these rates. The NPV of a cash flow over n periods is given by:

$$NPV = C_0 + \frac{C_1}{1+R} + \frac{C_2}{(1+R)^2} + \frac{C_3}{(1+R)^3} + \cdots + \frac{C_n}{(1+R)^n}$$

where R is the discount rate applied to value the cash flow. The IRR, however, is found by finding the rate that renders the NPV null (NPV = 0), or by solving for R^* is:

$$0 = C_0 + \frac{C_1}{1+R^*} + \frac{C_2}{(1+R^*)^2} + \frac{C_3}{(1+R^*)^3} + \cdots + \frac{C_n}{(1+R^*)^n}$$

Each project may therefore have its own IRR which in turn can be used to rank alternative investment projects. In fact, one of the essential problems CFOs must deal with is selecting an IRR to enable them to select/accept investment alternatives. If the IRR is larger than a strategic discount factor, specified by the CFO, then investment is deemed economical and therefore can be made. There are, of course, many variants of the IRR, such as the FIE (fixed equivalent rate of return) which assumes a fixed IRR with funds, generated by the investment, reinvested at the IRR. Such an assumption is not always realistic, however, tending to overvalue investment projects. Additional approaches based on 'risk analysis' and the market valuation of risk have therefore been devised, seeking to evaluate the probabilities of uncertain costs and uncertain payoffs (and thereby uncertain cash flows) of the investment at hand. In fact, the most significant attempt of fundamental finance has been to devise a mechanism that takes the 'arbitrariness' out of investment valuation by letting the market be the mechanism to value risk (i.e. by balancing supply and demand for risky assets at an equilibrium price for risky assets). We shall turn to this important approach subsequently.

Net present value (NPV) and random cash flows

When cash flows are random we can use the expected utility of the random quantities to calculate the NPV. We consider first a simple two-period example. Let $\tilde{C}$ be an uncertain cash flow whose expected utility is $Eu(\tilde{C})$. Its certainty equivalent, is CE where $Eu(\tilde{C}) = u(CE)$ or $CE = u^{-1}(Eu(\tilde{C}))$. Since CE is a sure quantity, the discount rate applied to value the reception of such a quantity for one period hence is the risk free rate R_f. In other words, for a one-period model the *PV* is:

$$PV = \frac{CE}{1+R_f} = \frac{u^{-1}(Eu(\tilde{C}))}{1+R_f} = \frac{E(\tilde{C})-P}{1+R_f}$$

where P is the risk premium. Equivalently we can calculate the *PV* by using the expected cash flow but discounted at a rate k (incorporating the risk inherent in the cash flow). Namely,

$$PV = \frac{E(\tilde{C})-P}{1+R_f} = \frac{E(\tilde{C})}{1+k}$$

As a result, we see that the risk-free rate and the risk premium combine to determine the risk adjusted rate as follows:

$$1-\frac{P}{E(\tilde{C})} = \frac{1+R_f}{1+k} \to k - R_f = (1+k)\frac{P}{E(\tilde{C})} \quad \text{or} \quad k = \frac{1+R_f}{1-P/E(\tilde{C})} - 1$$

In particular note that $k - R_f$ defines the 'excess discount rate'. It is the rate of return needed to compensate for the uncertainty in the cash flow $\tilde{C}$. To calculate the appropriate discount rate to apply, a concept of equilibrium reflecting investors' homogeneity is introduced. This is also called 'the capital market equilibrium' which underlies the CAPM as we shall see below. Over multiple periods of time, calculations of the *PV* for an uncertain future cash stream, yields similarly:

$$PV = C_0 + \frac{E(\tilde{C}_1)}{(1+k)^1} + \frac{E(\tilde{C}_2)}{(1+k)^2} + \frac{E(\tilde{C}_3)}{(1+k)^3} + \cdots$$

However, interest rates (and similarly, risk-free rates, risk premiums, etc.) may vary over time – reflecting the effects of time (also called the term structure) either in a known or unknown manner. Thus, discounting must reflect the discount adjustments to be applied to both uncertainties in the cash flow and the discount rate to apply because of the timing of payments associated with these cash flows. If the discount (interest) rates vary over time, and we recognize that each instant of time discounting accounts for both time and risk of future cash flows, the present value is then given by:

$$PV = C_0 + \frac{E(\tilde{C}_1)}{(1+k_1)^1} + \frac{E(\tilde{C}_2)}{(1+k_2)^2} + \frac{E(\tilde{C}_3)}{(1+k_3)^3} + \cdots = \sum_{i=0} \frac{E(\tilde{C}_i)}{(1+k_i)^i}$$

where $k_1, k_2, k_3, \ldots, k_n \ldots$ express the term structure of the risk-adjusted rates. Finally, note that if we use the certainty cash equivalents C_i, associated with a utility $u(C_i) = Eu(\tilde{C}_i)$, then:

$$PV = \sum_{i=0} \frac{C_i}{(1+R_{f,i})^i}$$

where $R_{f,i}$ is the risk-free term structure of interest rate for a discount over i periods. In Chapter 7, we shall discuss these issues in greater detail.

3.5.2 Individual investment and consumption

A number of issues in finance are stated in terms of optimal consumption problems and portfolio holdings. Say that an individual maximizes the expected utility of consumption, separable in time and state and is constrained by his wealth accumulation equation (the returns on savings and current wage income). Technically, assume that an individual investor has currently a certain amount of money invested in a portfolio consisting of N_0 shares of a stock whose current price is p_0 and a riskless investment in a bond whose current price is B_0. In addition, the investor has a wage income of s_0. Thus, current wealth is:

$$W_0 = N_0 p_0 + B_0 + s_0$$

Let c_0 be a planned current consumption while the remaining part $W_0 - c_0$ is reinvested in a portofolio consisting of N_1 shares of a stock whose price is p_0 and a bond whose current price is B_1. At the next time period, time '1', the investor consumes all available income. Disposable savings $W_0 - c_0$ are thus invested in a portfolio whose current wealth is $N_1 p_0 + B_1$, or initially:

$$W_0 - c_0 = N_1 p_0 + B_1$$

At the end of the period, the investor's wealth is random due to a change in the stock price $\Delta\tilde{p}$ and is wholly consumed, or:

$$\tilde{W}_1 = N_1(p_0 + \Delta\tilde{p}) + B_1(1 + R_f) \quad \text{and} \quad c_1 = \tilde{W}_1$$

Given the investor's utility function, there are three decisions to reach: how much to consume now, how many shares of stock to buy and how much to invest in bonds. The problem can be stated as the maximization of:

$$U = u_0(c_0) + \frac{1}{1+R} E u_1(\tilde{W}_1)$$

with, $u_0(.)$ and $u_1(.)$ the utilities of the current and next (final) period consumption. An individual's preference is expressed here twice. First we use an individual discount rate R for the expected utility of consumption at retirement $Eu_1(\tilde{W}_1)$. And, second, we have used the expected utility as a mechanism to express the effects of uncertainty on the value of such uncertain payments. Define a cash (certainty) equivalent to such expected utility by C_1 or $C_1 = u_1^{-1}(Eu_1(\tilde{W}_1))$. Since this is a 'certain cash equivalent', we can also write in terms of cash worth:

$$U = C_0 + \frac{C_1}{1+R_f}, \quad C_0 = u_0^{-1}(u_0(c_0)) = c_0$$

Note that once we have used a certain cash amount we can use the risk-free rate R_f to discount that amount. As a result,

$$\frac{1+R_f}{1+R} = \frac{u_1{}^{-1}(Eu_1(\tilde{W}_1))}{Eu_1(\tilde{W}_1)}$$

which provides a relationship between the discounted expected utility and the risk-free discount rate for cash. Using the expected utility discount rate, we have:

$$U = u_0(N_0 p_0 + B_0 + s_0 - N_1 p_0 - B_1) + \frac{1}{1+R} E u_1(N_1(p_0 + \Delta\tilde{p}) + B_1(1 + R_f))$$

A maximization of the current utility provides the investment strategy (N_1, B_1), found by the solution of:

$$p_0 \frac{\partial u_0(c_0)}{\partial N_1} = \frac{1}{1+R} E\left(\frac{\partial u_1(c_1)}{\partial N_1}(p_0 + \Delta\tilde{p})\right), \frac{\partial u_0}{\partial B_1} = E\left(\frac{\partial u_1}{\partial B_1}\right)$$

This portfolio allocation problem will be dealt with subsequently in a general manner but has already interesting implications. For example, if the utility of consumption is given by a logarithmic function, $u_0(c) = \ln(c)$ and $u_1(c) = \ln(c)$, we have:

$$\frac{p_0}{c_0} = \frac{1}{1+R} E\left(\frac{p_0 + \Delta\tilde{p}}{c_1}\right) \quad \text{or} \quad \eta_0 = \frac{1}{1+R} E(\eta_1), \eta_i = \frac{p_i}{c_i}, i = 0, 1$$

meaning that the price per unit consumption η_i is an equilibrium whose value is the rate R. Further, the condition for the bond yields

$$\frac{\partial u_0(c_0)}{\partial c_0} = E\left(\frac{\partial u_1(c_1)}{\partial c_1}\right) \quad \text{or} \quad \frac{1}{c_0} = E\left(\frac{1}{c_1}\right)$$

It implies that the current marginal utility of consumption equals the next period's expected marginal consumption. The investment policy is thus a solution of the following two equations for (N_1, B_1) (where $\tilde{W}_1 = N_1(p_0 + \Delta\tilde{p}) + B_1(1 + R_f)$):

$$\frac{p_0}{(N_0 p_0 + B_0 + s_0 - N_1 p_0 - B_1)} = \frac{1}{1+R} E\left(\frac{p_0 + \Delta\tilde{p}}{\tilde{W}_1}\right)$$

$$\frac{1}{N_0 p_0 + B_0 + s_0 - N_1 p_0 - B_1} = E\left(\frac{1}{\tilde{W}_1}\right)$$

or $p_0 R E(1/\tilde{W}_1) = E(\Delta\tilde{p}/\tilde{W}_1)$. To solve this equation numerically, we still need to specify the probability of the stock price.

Problem

Assume that

$$\Delta\tilde{p} = \begin{cases} H p_0 & \text{w.p. } \pi \\ L p_0 & \text{w.p. } 1 - \pi \end{cases}$$

where w.p. means with probability, and find the optimal portfolio. In particular, set $\pi = 0.6, H = 0.3, L = -0.2$ and $R_f = 0.1$, then show that the optimal investment policy is an all-bond investment. However, if $H = 0.5$, we have:

$$B_1^* = 0.085(N_0 p_0 + B_0 + s_0),\ N_1^* = -0.6444\left(N_0 + \frac{B_0 + s_0}{p_0}\right)$$

3.5.3 Investment and the CAPM

Say that an investor has an initial wealth level W_0. Let $\tilde{k}$ be a random rate of return of a portfolio with known mean and known variance given respectively by $\hat{k}, \sigma_k^2$ respectively. Say that part of the individual wealth, S_1, is invested in the risky asset while the remaining part $B_1 = W_0 - S_1$ is invested in a non-risky asset whose rate of return is the risk-free rate R_f. The wealth one period hence is thus:

$$\tilde{W}_1 = B_1(1 + R_f) + S_1(1 + \tilde{k})$$

The demand for the risky asset is thus given by optimizing the expected utility function:

$$\underset{S_1 \geq 0}{\text{Max}}\, Eu[B_1(1 + R_f) + S_1(1 + \tilde{k})] \quad \text{or} \quad \underset{S_1 \geq 0}{\text{Max}}\, Eu[W_0(1 + R_f) + S_1(\tilde{k} - R_f)]$$

The first two derivatives conditions are:

$$\frac{\mathrm{d}(Eu(\tilde{W}_1))}{\mathrm{d}S_1} = E(u'(\tilde{W}_1)(\tilde{k} - R_f)) = 0 \frac{\mathrm{d}^2(Eu(\tilde{W}_1))}{\mathrm{d}(S_1)^2} = E(u''(\tilde{W}_1)(\tilde{k} - R_f)^2) < 0$$

This is always satisfied when the investor is risk-averse. Consider the first-order condition, which we rewrite for convenience as follows:

$$E[u'(\tilde{W}_1)\tilde{k}] = E[u'(\tilde{W}_1)R_f]$$

By definition of the covariance, we have:

$$E(u'(\tilde{W}_1)\tilde{k}) = \hat{k}E(u'(\tilde{W}_1)) + \operatorname{cov}(u'(\tilde{W}_1), \tilde{k})$$

Thus,

$$\hat{k}E(u'(\tilde{W}_1)) + \operatorname{cov}(u'(\tilde{W}_1), \tilde{k}) = R_f E(u'(\tilde{W}_1))$$

Since the derivative of the expected utility is not null, we divide this expression by it and obtain:

$$\hat{k} + \frac{\operatorname{cov}(u'(\tilde{W}_1), \tilde{k})}{E(u'(\tilde{W}_1))} = R_f$$

which clearly outlines the relationship between the expected returns of the risky and the non-risky asset and provides a classical result called the Capital Asset Pricing Model (CAPM). If we write the CAPM regression equation (to be seen below) by:

$$\hat{k} = R_f + \beta(R_m - R_f)$$

then, we can recuperate the beta factor often calculated for stocks and risky investments:

$$\beta = -\frac{\operatorname{cov}(u'(\tilde{W}_1), \tilde{k})}{(R_m - R_f)E(u'(\tilde{W}_1))}$$

where R_m is the expected rate of return of a market portfolio or the stock market index. However, the beta found in the previous section implies:

$$\beta = \frac{\operatorname{cov}(\tilde{R}_m, \tilde{k})}{\sigma_m^2} = -\frac{\operatorname{cov}(u'(\tilde{W}_1), \tilde{k})}{(R_m - R_f)E(u'(\tilde{W}_1))}$$

and therefore:

$$\hat{k} - R_f = (R_m - R_f)\frac{\operatorname{cov}(\tilde{R}_m, \tilde{k})}{\sigma_m^2} = -\frac{\operatorname{cov}(u'(\tilde{W}_1), \tilde{k})}{E(u'(\tilde{W}_1))}$$

This sets a relationship between individuals' utility of wealth, the market mechanism and a statistical estimate of market parameters.

An equivalent approach consists in constructing a portfolio consisting of a proportional investment $y_i = S_i/W_0$ in a risky asset i with a rate of return $\tilde{k}_i$ while the remaining part is invested in a market index whose rate of return is $\tilde{R}_m$ (rather than investing in a riskless bond). The rate of return of the portfolio is then $\tilde{k}_p = y_i\tilde{k}_i + (1 - y_i)\tilde{R}_m$ with mean and variance:

$$\hat{k}_p = y_i\hat{k}_i + (1 - y_i)R_m,\ R_m = E\tilde{R}_m$$
$$\sigma_p^2 = y_i^2\sigma_i^2 + (1 - y_i)^2\sigma_m^2 + 2y_i(1 - y_i)\sigma_{im};\ \sigma_{im} = \operatorname{cov}(\tilde{k}_i, \tilde{k}_m)$$

where variances σ^2 are appropriately indexed according to the return variable they represent. The returns–risk substitution is found by calculating by chain differentiation:

$$\frac{d\hat{k}_p}{d\sigma_p} = \frac{d\hat{k}_p/dy_i}{d\sigma_p/dy_i}$$

where:

$$\frac{d\hat{k}_p}{dy_i} = \hat{k}_i - R_m;\quad \frac{d\sigma_p}{dy_i} = \frac{y_i\left(\sigma_i^2 + \sigma_m^2 - 2\sigma_{im}\right) + \sigma_{im} - \sigma_m^2}{\sigma_p}$$

A portfolio invested only on the market index (i.e. $y_i = 0$) will lead to:

$$\left(\frac{d\sigma_p}{dx_i}\right)_m = \left(\frac{d\sigma_p}{dx_i}\right)_{x_i=0} = \frac{\sigma_{im} - \sigma_m^2}{\sigma_p} \quad \text{and} \quad \left(\frac{d\hat{k}_p}{d\sigma_p}\right)_m = \frac{(\hat{k}_i - R_m)\sigma_m}{\sigma_{im} - \sigma_m^2}$$

However since for all investors, the preference for returns is a linear function of the returns' standard deviation, (assuming they all maximize a quadratic utility function!), we have:

$$\hat{k} = R_f + \lambda\sigma$$

where σ is the volatility of the portfolio. For all assets we have equivalently: $\hat{k}_i = R_f + \lambda\sigma_i$ and for the portfolio as well, or:

$$\hat{k}_p = R_f + \lambda\sigma_p \rightarrow \frac{d\hat{k}_i}{d\sigma_i} = \lambda \quad \text{and} \quad \left(\frac{d\hat{k}_p}{d\sigma_p}\right)_m = \frac{(\hat{k}_i - R_m)\sigma_m}{\sigma_{im} - \sigma_m^2}$$

and therefore,

$$\left(\frac{d\hat{k}_p}{d\sigma_p}\right)_m = \frac{(\hat{k}_i - R_m)\sigma_m}{\sigma_{im} - \sigma_m^2} = \lambda \text{ which leads to: } \hat{k}_i = R_m - \lambda\sigma_m + \frac{\lambda\sigma_{im}}{\sigma_m}$$

But also $R_m = R_f + \lambda\sigma_m$ which we insert in the previous equation, leading thereby to a linear expression for risk discounting which assumes the form of the (CAPM) or:

$$\hat{k}_i = R_f + \frac{\lambda\sigma_{im}}{\sigma_m} \quad \text{and explicitly,} \quad \hat{k}_i = R_f + \frac{\lambda \operatorname{cov}(\tilde{k}_i, \tilde{R}_m)}{\sigma_m}$$

as seen earlier. Note that this expression can be written in a form easily amenable to a linear regression in returns and providing an estimate for the β_i factor:

$$\hat{k}_i = R_f + \beta_i(R_m - R_f);\ \beta_i = \frac{\lambda \operatorname{cov}(\tilde{k}_i, \tilde{R}_m)}{(R_m - R_f)\sigma_m}$$

Since $\lambda = (R_m - R_f)/\sigma_m$ is the market price of risk, we obtain also the following expression for the beta factor,

$$\beta_i = \frac{\operatorname{cov}(\tilde{k}_i, \tilde{R}_m)}{\sigma_m^2}$$

With this 'fundamental' identity on hand, we can calculate the risk premium of an investment as well as the betas for traded securities.

3.5.4 Portfolio and utility maximization in practice

Market valuation of a portfolio and individual valuation of a portfolio are not the same. The latter is based on an individual preference for the assets composition of the portfolio and responding to specific needs. For example, denote by $\$W$ a budget to be invested and let $y_i, i = 1, 2, 3, \ldots, n$ be the dollars allocated to each of the available alternatives with a resulting uncertain payoff

$$\tilde{R} = \sum_{i=1}^{n} \tilde{r}_i(y_i)$$

The portfolio investment problem is then formulated by solving the following expected utility maximization problem:

$$\underset{y_1, y_2, y_3, \ldots, y_n}{\text{Maximize}}\ Eu(\tilde{R}) = Eu\left(\sum_{i=1}^{n} \tilde{r}_i(y_i)\right) \text{ subject to : } \sum_{i=1}^{n} y_i \leq W, y_i \geq 0, i = 1, 2, \cdots, n$$

where $u(.)$ is the individual utility function, providing a return risk ordering over all possible allocations. This problem has been solved in many ways. It clearly sets up a transformation of an uncertain payoffs problem into a problem which is deterministic and to which we can apply well-known optimization and numerical techniques. If $u(.)$ is a quadratic utility function and the rates of return are linear in the assets allocation (i.e. $\tilde{r}_i(y_i) = \tilde{r}_i y_i$), then we have:

$$\underset{y_1, y_2, y_3, \ldots, y_n}{\text{Maximize}}\ Eu(\tilde{R}) = Eu\left(\sum_{i=1}^{n} \tilde{r}_i(y_i)\right) = \sum_{i=1}^{n} \hat{r}_i y_i - \mu \sum_{j=1}^{n}\sum_{i=1}^{n} \rho_{ij} y_i y_j;$$

$$\rho_{ij} = \text{cov}(\tilde{r}_i, \tilde{r}_j)$$

where $\hat{r}_i$ is the mean rate of return on asset i, ρ_{ij} is the covariance between the returns on two assets (i, j) and μ is a parameter expressing the investor's risk aversion ($\mu > 0$) or risk loving ($\mu < 0$). This defines a well known quadratic optimization problem that can be solved using standard computational software when the index of risk aversion is available. There are many other formulations of this portfolio problem due to Harry Markowitz, as well as many other techniques for solving it, such as scenario optimization, multi-criterion optimization and others.

The Markowitz approach had a huge impact on financial theory and practice. Its importance is due to three essential reasons. First, it justifies the well-known belief that it is not optimal to put all one's eggs in one basket (or the 'principle of diversification'). Second, a portfolio value is expressed in terms of its mean return and its variance, which can be measured by using statistical techniques. Further, the lower the correlation, the lower the risk. In fact, two highly and negatively correlated assets can be used to create an almost risk-free portfolio. Third and finally, for each asset there are two risks, one diversifiable through a combination of assets and the other non-diversifiable to be borne by the investor and for which there may be a return compensating this risk.

Markowitz suggested a creative approach to solving the quadratic utility portfolio problem by assuming a specific index of risk aversion. The procedure consists in solving two problems. The first problem consists in maximizing the expected returns subject to a risk (variance) constraint (Problem 1 below) and the second problem consists in minimizing the risk (measured by the variance) subject to a required expected return constraint (Problem 2 below). In other words,

Problem 1:

$$\underset{y_1, y_2, y_3, \ldots, y_n}{\text{Maximize}} \sum_{i=1}^{n} \hat{r}_i y_i$$

Subject to:

$$\sum_{j=1}^{n}\sum_{i=1}^{n} \rho_{ij} y_i y_j \leq \lambda$$

Problem 2:

$$\underset{y_1, y_2, y_3, \ldots, y_n}{\text{Minimize}} \sum_{j=1}^{n}\sum_{i=1}^{n} \rho_{ij} y_i y_j$$

Subject to:

$$\sum_{i=1}^{n} \hat{r}_i y_i \geq \mu$$

An optimization of these problems provides the efficient set of portfolios defined in the (λ, μ) plane.

The importance of Markowitz's (1959) seminal work cannot be overstated. It laid the foundation for portfolio theory whereby rational investors determine the optimal composition of their portfolio on the basis of the expected returns, the standard deviations of returns and the correlation coefficients of rates of return. Sharpe (1964) and Lintner (1965) gave it an important extension leading to the CAPM to measure the excess premium paid to hold a risky financial asset, as we saw earlier.

3.5.5 Capital markets and the CAPM again

The CAPM for the valuation of assets is essentially due to Markowitz, James Tobin and William Sharpe. Markowitz first set out to show that 'diversification pay', in other words an investment in more than one security, provides an opportunity to 'make money with less risk', compared to the prior belief that the optimal investment strategy consists of putting all of one's money in the 'best basket'. For risk-averse investors this was certainly a strategy to avoid. Subsequently, Tobin (1956) showed that when there is a riskless security, the set of efficient portfolios can be characterized by a 'two-fund separation theorem' which showed that an efficient portfolio can be represented by an investor putting some of his money in the riskless security and the remaining moneys invested in a representative fund constructed from the available securities. This led to the CAPM, stated explicilty by Sharpe in 1964 and Lintner in 1965. Both assumed that investors are homogeneous and mean-variance utility maximizers (or, alternatively, investors are quadratic utility maximizers). These assumptions led to an equilibrium of financial markets where securities' risks are measured by a linear function (due to the quadratic utility function) given by the risk-free rate and a beta multiplied by the relative returns of the 'mutual fund' (usually taken to be the market average rate of return). In this sense, the CAPM approach depends essentially on quadratic utility maximizing agents and a known risk-free rate. Thus, the returns of an asset i can be estimated by a linear regression given explicitly by:

$$k_i = \alpha_i + \beta_i R_m + \varepsilon_i$$

where R_m is the market rate of return calculated by the rate of return of the stock market as a whole, β_i is an asset specific parameter while ε_i is the statistical error. The CAPM has, of course, been subject to criticism. Its assumptions may be too strong: for example, it implies that all investors hold the same portfolio – which is the market portfolio, by definition fully diversified. In addition, in order to invest, it is sufficient to know the beta associated with a stock, since it is the parameter that fully describes the asset/stock return. Considerable effort has been devoted to estimating this parameter through a statistical analysis of stocks' risk–return history. The statistical results obtained in this manner should then clarify whether such a theory is applicable or not. For example, for *one-factor models (market premium in the CAPM)* the following regression is run:

$$(k_j - R_f) = \alpha_j + \beta_j[R_m - R_f] + \varepsilon_j$$

where k_j is the stock (asset) j rate of return at time t, when the risk-free rate at that time was R_f, (α_j, β_j) are regression parameters, while R_m is the market rate of return. Finally, ε_j is the residual value, an error term, assumed to be normally distributed with mean zero and known variance. Of course, if $(\alpha_j \neq 0)$ then this will violate a basic assumption of no excess returns of the CAPM. In addition, R_m must also, according to the CAPM theory, capture the market portfolio. If, again, this is not the case, then it will also violate a basic assumption of the CAPM. Using the regression equation above, the risk consists now of (assuming perfect diversification, or equivalently, no correlation):

$$\text{var}(k_j - R_f) = \beta_j^2\sigma_m^2 + \sigma_\varepsilon^2, \sigma_m^2 = \text{var}(R_m);\ \sigma_\varepsilon^2 = \text{var}(\varepsilon_j)$$

expressing risk as a summation of beta-squared times the index-market plus the residual risk.

The problems with a one-factor model, although theory-independent (since it is measured by simple financial statistics), are its assumptions. Namely, it assumes that the regression is stable and that nonstationarities and residual risks are known as well. Further, to estimate the regression parameters, long time series are needed, which renders their estimate untimely (and, if used carefully, it is of limited value). The generalization of the one-factor model into a *multiple-factor model* is also known as *APT*, or *arbitrage pricing theory*. Dropping the time index, it leads to:

$$(k_j - R_f) = \alpha_j + \beta_{j1}[R_m - R_f] + \cdots + \beta_{jK}[R_K - R_f] + \varepsilon_j$$

where $R_K - R_f$ is the expected risk premium associated with factor K. The number of factors that can be used is large, including, among many others, the yield, interest-rate sensitivity, market capitalization, liquidity, leverage, labour intensity, recent performance (momentum), historical volatility, inflation, etc. This model leads also to risks defined by the matrices calculated (the orthogonal factors of APT) using the multivariate regression above, or

$$\mathbf{\Sigma} = \mathbf{B}'\mathbf{\Gamma}\mathbf{B} + \mathbf{\Phi}$$

where $\mathbf{\Sigma}$ is the variance–covariance matrix of assets returns, $\mathbf{B}$ is the matrix of assets' exposures to the different risk factors, $\mathbf{\Gamma}$ is the vector of factor risk premiums (i.e. in excess of the risk-free rate for period t) and finally, $\mathbf{\Phi}$ is the (diagonal) matrix of asset residual risks. These matrices, unfortunately involve many parameters and are therefore very difficult to estimate.

A number of approaches based on the APT are available, however. One approach, the *fundamental factor model* assumes that the matrix $\mathbf{B}$ is given and proceeds to estimate the vector $\mathbf{\Gamma}$. A second approach, called the *macroeconomic approach*, takes $\mathbf{\Gamma}$ as given and estimates the matrix $\mathbf{B}$. Finally, the *statistical approach* seeks to estimate $(\mathbf{B}, \mathbf{\Gamma})$ simultaneously. These techniques each have their problems and are therefore used in varying circumstances, validated by the validity of the data and the statistical results obtained.

3.5.6 Stochastic discount factor, assets pricing and the Euler equation

The financial valuation of assets is essentially based on defining an approach accounting for the time and risk of future payoffs. To do so, financial practice and theories have sought to determine a discounting mechanism that would, appropriately, reflect the current value of uncertain payoffs to be realized at some future time. Thus, techniques such as classical discounting based on a pre-specified discount rate (usually a borrowing rate of return provided by banks or some other interest rate) were used. Subsequently, a concept of risk-adjusted discount rate applied to discounting the mean value of a stream of payoffs was used. A particularly important advance in determining an appropriate discount mechanism was ushered in, first by the CAPM approach, as we saw above, and subsequently by the use of risk-neutral pricing, to be considered in forthcoming chapters. Both approaches use 'the market mechanism' to determine the appropriate discounting process to value an asset's future payoffs. We shall consider these issues at length when we seek a risk-neutral approach to value options and derivatives in general. These approaches are not always applicable, however, in particular when markets are incomplete and the value of a portfolio may not be determined in a unique manner. In these circumstances, attempts have been made to maintain the framework inspired by rational expectations and at the same time be consistent theoretically and empirically verifiable. The SDF approach, or the generalized method of moments, seeks to value an asset generally in terms of its future values using a stochastic discount factor. This development follows risk-neutral pricing, which justified a risk-free rate discounting process with respect to some probability measure, as we shall see later on (in Chapter 6).

Explicitly, the SDF approach for a single asset states that the price of an asset equals the expected value of the asset payoff, times a stochastic discount factor. This approach has the advantage that it leads to some of the classical results of financial economics and at the same time it can be used by applying financial statistics in asset pricing by postulating such a relationship. Define:

p_t = asset price at time t that an investor may wish to buy

$\tilde{x}_t$ = asset returns, a random variable

$\tilde{m}_t$ = a stochastic discount factor to be defined below

The SDF postulates:

$$p_t = E(\tilde{m}_{t+1}\tilde{x}_{t+1}),\ \tilde{m}_{t+1} = \frac{1}{1+\tilde{R}_{t+1}}$$

where $\tilde{R}_{t+1}$ is a random discounting. The rationality for such a postulate is based on the expected utility of a consumption-based model. Say that an investor has a utility function for consumption $u(.)$, which remains the same at times t and $t+1$. We let the discount factor be ρ, expressing the subjective discount rate of the consumer. Current consumption is certain, while next period's consumption is uncertain and discounted. Thus, in terms of expected utility we have:

$$U(c_t, c_{t+1}) = u(c_t) + \rho E_t u(c_{t+1})$$

where E_t is an expectation operator based on the information up to time t. Now assume that s_t is a consumer's salary at time t, part of which may be invested for future consumption in an asset whose price is p_t (for example, buying stocks). Let y be the quantity of an asset bought (say a stock). Current consumption left over after such an investment equals: $c_t = s_t - yp_t$. If the asst price one period hence is $\tilde{x}_{t+1}$, then the next period consumption is simply equal to the sum of the period's current income and the return from the investment, namely $c_{t+1} = s_{t+1} + y\tilde{x}_{t+1}$. As a result, the consumer problem over two periods is reduced to:

$$U(c_t, c_{t+1}) = u(s_t - yp_t) + \beta E_t u(s_{t+1} + y\tilde{x}_{t+1})$$

The optimal quantity to invest (i.e. the number of shares to buy), found by maximizing the expected utility with respect to y, leads to:

$$\begin{aligned} \frac{\partial U}{\partial y} &= -p_t u'(s_t - p_t y) + \beta E_t[\tilde{x}_{t+1} u'(s_{t+1} + \tilde{x}_{t+1} y)] \\ &= -p_t u'(c_t) + \beta E_t[\tilde{x}_{t+1} u'(c_{t+1})] = 0 \end{aligned}$$

which yields for an optimum portfolio:

$$p_t = E_t\left[\beta \frac{u'(c_{t+1})}{u'(c_t)} \tilde{x}_{t+1}\right]$$

Thus, if we set the stochastic discount factor, $\tilde{m}_{t+1}$, expressing the inter-temporal substitution of current and future marginal utilities of consumption, then:

$$\tilde{m}_{t+1} = \beta \frac{u'(c_{t+1})}{u'(c_t)} \quad \text{and therefore } \tilde{R}_{t+1} = \frac{1 - \beta \frac{u'(c_{t+1})}{u'(c_t)}}{\beta \frac{u'(c_{t+1})}{u'(c_t)}}$$

the pricing equation becomes $p_t = E_t[\tilde{m}_{t+1}\tilde{x}_{t+1}]$, which is the desired SDF asset-pricing equation. This equation is particularly robust and therefore it is also very appealing. For example, if the utility function is of the logarithmic type, $u(c) = \ln(c)$, then, $u'(c) = 1/c$ and $\tilde{m}_{t+1} = \beta c_t / c_{t+1}$, or $\tilde{R}_{t+1} = \beta[(1/\beta)c_{t+1} - c_t]/c_t$ and further, $p_t/c_t = \beta E_t[\tilde{x}_{t+1}/c_{t+1}]$. In other words, if we write $\pi_t = p_t/c_t; \tilde{\pi}_{t+1} = \tilde{x}_{t+1}/c_{t+1}$, then we have: $\pi_t = \beta E_t(\tilde{\pi}_{t+1})$.

The results of such an equation can be applied to a broad a number of situations, which were summarized by Cochrane (2001) and are given in Table 3.1. This approach is extremely powerful and will be considered subsequently in Chapter 8, since it is applicable to a broad number of situations and financial products. For example, the valuation of a call option would be (following the information in Table 3.1) given as follows (where the time index is ignored):

$$C = E\,[m\,\text{Max}(S_T - K, 0)]$$

This approach can be generalized in many ways, notably by considering multiple periods and various agents (heterogeneous or not) interacting in financial markets

Table 3.1 Selected examples.

	Price p_t	Payoff x_{t+1}
Dividend paying stock	p_t	$p_{t+1} + d_{t+1}$
Investment return	1	R_{t+1}
Price/dividend ratio	$\frac{p_t}{d_t}$	$\left(\frac{p_{t+1}}{d_{t+1}} + 1\right)\frac{d_{t+1}}{d_t}$
Managed portfolio	z_t	$z_t R_{t+1}$
Moment condition	$E(p_t z_t)$	$x_{t+1} z_t$
One-period bond	p_t	1
Risk-free rate	1	R_f
Option	C	$\text{Max}[S_T - K, 0]$

to buy, sell and transact financial assets. Cochrane (2001) in particular suggests many such situations. An inter-temporal framework uses the Euler conditions for optimality to generate an equilibrium discount factor and will be considered in Chapter 9.

Example
Consider a one-dollar investment in a risk-free asset whose payoff is R_f. Thus, $1/\left(1 + R_f\right) = E(\tilde{m} * 1)$ and therefore we have $1/(1 + R_f) = E(\tilde{m})$ as expected.

3.6 INFORMATION ASYMMETRY

Uncertainty and information asymmetry have special importance because of their effects on decision-makers. These result also in markets being 'incomplete' since the basic assumptions regarding 'fair competition' are violated. In general, the presumption that information is commonly shared is also, often, violated. Some information may be truthful, some may not be. Truth-in-lending for example, is an important legislation passed to protect consumers, which is, in most cases, difficult to enforce. Courts are filled with litigation on claims and counter-claims, leading to a battle of experts on what the truth is and where it may lie. Environmental litigation has often led to a 'battle of PhDs' expounding alternative and partial pearls of knowledge. In addition, positive, negative, informative, partial, asymmetric, etc. information has different effects on both decision-makers and markets. For example, firms and funds are extremely sensitive to negative information regarding their stock, their products as well as their services. Pharmaceutical firms may be bankrupted upon adverse publicity, whether true or not, regarding one of their products. For example, the Food and Drug Administration warning on the content of benzene in Perrier's sparkling water has more than tainted the company's image, its bottom-line profits and, at a certain time, its future prospects. Of course, the tremendous gamble Perrier has taken to meet these claims (that were not entirely verified) is a sign of the importance Perrier attached to its reputation and to the effects of negative information.

Information asymmetry and uncertainty can open up the possibility of cheating, however. For example, some consumer journals may receive money in various forms (mostly advertising dollars) not to publish certain articles and thereby manage information in a way that does not benefit the public. For this reason, regulatory authorities are needed in certain areas. A used-car salesman may be tempted to sell a car with defects unknown to the prospective buyer. In some countries, importers are not required to inform clients of the origin and the quality (state) of the product and parts used in the product sold. As a result, a product claimed to be new by the seller may not in fact be new, opening up many possibilities for cheating legally. These questions arise on Wall Street in many ways. In an article in the *Wall Street Journal* the question was raised by Hugo Dixon with respect to analysts' claims and the conflict of interest when they act according to their own edicts:

> Shouldn't analysists put their money where their mouth is? That is the contrarian response to Merrill Lynch's decision to ban its analysts from buying shares of companies they cover. It might be said that researchers will have an incentive to give better opinions if they stand to make money if they are right – and lose money if they are wrong. Clients might also be comforted to know that the analysts who are peppering them with 'buy' recommendations are following their own advice. Under this contrarian position, the fact that some analysts buy shares in companies they follow isn't a conflict of interest all. Quite the reverse! It is an alignment of interests . . . this contrarian view cannot be dismissed as a piece of errant nonsense. But it is nevertheless misconceived. There is a better way of aligning analyst's financial interests with those of their clients. And there are potential conflicts caused by an analyst trading stocks he or she covers!

These 'information problems' are the subject of extensive study, both for practical and theoretical reasons. A number of references are included at the end of this chapter. Below, we only consider some of the outstanding implications information asymmetry may create.

3.6.1 'The lemon phenomenon' or adverse selection

In a seminal paper, Akerlof (1970) pointed out that goods of different qualities may be uniformly priced when buyers cannot realize that there are quality differences. This is also called 'adverse selection' because some of the information associated with the choice problem may be hidden. For example, one may buy a used car, not knowing its true state, and therefore be willing to pay a price that would not truly reflect the value of the car. In fact, we may pay an agreed-upon market price even though this may be a lemon. The used-car salesman may have such information but, for some obvious reason, he may not be amenable to revealing the true state of the car. In such situations, price is not an indicator of quality and informed sellers can resort to opportunistic behaviour (the used-car salesman phenomenon stated above). While Akerlof demonstrated that average quality might still be a function of price, individual units may not be priced at that level. By contrast, people who discovered in the 1980s that they had AIDS were very quick in taking out very large life insurance (before insurance firms knew what it really entailed and therefore were at first less informed than the insured). Bonds or stocks of various qualities (but of equal ratings) are sometimes difficult to discern for an individual investor intending to buy. For this reason, rating agencies, tracking and following firms have an important role to play, compensating for the problems of information asymmetry and making markets more efficient. This role is not always properly played, as evidenced by the Enron debacle, where changing accounting practices had in fact hidden information from the public.

Information asymmetry and uncertainty can largely explain the desires of consumers to buy service or product warranties to protect themselves against failures or to favour firms who possess service organizations (in particular when the products are complex or involve some up-to-date technologies). Generally in transactions between producers and suppliers, uncertainty leads to constructing long-term trustworthy relationships and contractual engagements to assure that 'what is contracted is also delivered'. The potential for adverse selection may also be used to protect national markets. Anti-dumping laws, non-tariff trade barriers, national standards and approval of various sorts are some of the means used to manage problems of adverse selection on the one hand and to manage market entries to maintain a competitive advantage on the other. Finance and insurance are abound with applications and examples where asymmetry induces an uncertainty which has nothing to do with 'what nature does' but with 'what people do'.

Problems of adverse selection can sometimes be overcome by compulsory insurance regulation requiring all homeowners to insure their homes or requiring everyone to take out medical insurance, for example. Some employers insure all their employees as one package to avoid adverse selection problems. If everyone is insured, high-risk individuals will be better off (since they will be insured at a lower premium than justified by the risks they have in fact). Whether low-risk individuals are better off under this scheme depends on how risk-averse they are as the insurance they are offered is not actuarially fair.

3.6.2 'The moral hazard problem'

For many situations, the cost of providing a product or a service depends on the behaviour of the purchaser. For example, the cost of insurance depends on the amount of travel done by the purchaser and by the care he takes in driving. Similarly, the cost of warranties depends on the care of the purchaser in using the commodity. Such behaviour cannot always be observed directly by the supplier/seller. As a result, the price cannot depend on the behaviour of the purchaser that affects costs. In this case, equilibrium cannot always be the first-best-optimum and some intervention is required to reach the best solution. Questions are of course, how and how much. For example, should car insurance be obligatory? How do purchasers react after buying insurance? How do markets behave when there is moral hazard and how can it be compensated?

Imperfect monitoring of fund managers, for example, can lead to moral hazard. What does it mean? It implies that when the fund manager cannot be observed, there is a possibility that the provider, the fund manager, will use that fact to his advantage and not deliver the right level of performance. Of course, if we contract the delivery of a given level of returns and if the fund manager knowingly does not maintain the terms of the contract, he would be cheating. We can deal with such problems with various sorts of controls combined with incentive contracts to create an incentive not to cheat or lie and to perform in the interest of the investor. If a fund manager were to cheat or lie, and if he were detected, he would then be penalized accordingly (following the terms agreed on by the contract).

For some, transparency (i.e. sharing information) is essential to provide a 'signal' that they operate with the best of intentions. For example, some restaurants might open their kitchen to their patrons to convey a message of truthfulness in so far as cleanliness is concerned. A supplier would let the buyer visit the manufacturing facilities as well as reveal procedures relating to quality, machining controls and the production process in general. A fund manager will provide regulators with truthful reports regarding the fund's state and strategy.

Moral hazard pervades some of the most excruciating problems of finance. The problem of deposit insurance and the 'too big to fail' syndrome encourages excessive risk taking. As a result of implicit governmental guarantee, banks enjoy a lower cost of capital, which leads to the consistent under-pricing of credit. Swings in economic cycles are thus accentuated. The Asian financial crisis of 1998 is a case in point. The extent of its moral hazard is difficult to measure but with each bail out by governments and the IMF, the trend for excessive risk taking is reinforced.

3.6.3 Examples of moral hazard

(1) An over-insured driver may drive recklessly. Thus, while the insured motorist is protected against any accident, this may induce him to behave in a nonrational manner and cause accidents that are costly to society.

(2) In 1998, the NYSE at last, belatedly (since the practice was acknowledged to have been going on since 1992), investigated charges against floor brokers for 'front-running' or 'flipping'. This is a practice in which the brokers used information obtained on the floor to trade and earn profits on their own behalf. One group of brokers was charged with making \$11 millions. (Financial Times, 20 February 2001).

(3) The de-responsibilization of workers in factories also induces a moral hazard. It is for this reason that incentives, performance indexation and responsibilization are so important and needed to minimize the risks from moral hazard (whether these are tangibles or intangibles). For example, decentralization of the workplace and getting people involved in their jobs may be a means to make them care a little more about their job and deliver the required performance in everything they do.

Throughout these examples, there are negative inducements to good performance. To control or reduce these risks, it is necessary to proceed in a number of ways. Today's concern for firms' organizational design, the management of traders and their compensation packages, is a reflection of the need to construct relationships that do not induce counterproductive acts. Some of the steps that can be followed include:

(1) Detecting signals of various forms and origins to reveal agents' behaviours, rationality and performance. A greater understanding of agents' behaviour can lead to a better design of the workplace and to appropriate inducements for all parties involved in the firm's business.
(2) Managing and controlling the relationship between business partners, employees and workers. This means that no relationship can be taken for granted. Earlier, we saw that information asymmetry can lead to opportunistic behaviour such as cheating, lying and being counterproductive, just because there may be an advantage in doing so without having to sustain the consequences of such behaviour.
(3) Developing an environment which is cooperative, honest and open, and which leads to a frank exchange of information and optimal performance.

All these actions are important. It is therefore not surprising that many of the concerns of managers deal with people, communication, simplification and the transparency of everything firms do.

Example: Genetic testing and insurance
In a *Financial Times* article of 7–8 November 1998, it was pointed out that genetic testing can give early warning of disease and that those results could have serious consequences for those seeking insurance. The problem at hand, therefore, indicates an important effect of information on insurance. How can such problems be resolved? Should genetic testing be a requirement imposed by insurers or should they not? If accurate genetic tests do become widely available, they could encourage two trends that would undermine the present economic basis of the insurance industry:

(1) Adverse selection: people who know they are at high risk take out insurance. This drives up the price of premiums, so low-risk people are deterred from taking out policies and withdrawing from the insurance 'pool'.
(2) Cherry-picking: insurers identify people at lower risk than average and offer them reduced premiums. If they join the preferred pool, this increases the average risk in the standard pool and premiums have to rise. For example, the insurance industry suffered from adverse selection in the 1980s when individuals who knew they had HIV/AIDS took out extra insurance cover without disclosing their HIV status. A more respectable name for cherry-picking is market segmentation, as applied in general insurance for house contents and motor vehicles, where policies favouring the (lesser) better risks are common. These are becoming increasingly important issues due to the improved databases available about insured and insurance firms ability to tap these databases.

3.6.4 Signalling and screening

In conditions of information asymmetry, one of the parties may have an incentive to reveal some of the information it has. The seller of an outstanding concept for a start-up to invest in will certainly have an interest in making his concept transparent to the potential VC (venture capitalist) investor. He may do so in a number of ways, such as pricing it high and therefore conveying the message to the potential VC that it is necessarily (at that price) a dream concept that will realize an extraordinary profit in an IPO (but then, the concept seller may also be cheating!). The seller may also spend heavily on advertising the concept, claiming that it is an outstanding one with special technology that it is hard to verify (but then, the seller may again be lying!). Claiming that a start-up concept is just 'great' may be insufficient. Not all VCs are gullible. They require and look for signals that reveal the true potential of a concept and its potential for making large profits. Pricing, warranties, advertising, are some of the means used by well-established firms selling a product to send signals. For example, the seller of a lemon with a warranty will eventually lose money. Similarly, a firm that wants to limit the entry of new competitors may signal that its costs are very low (and so if they decide to enter, they are likely to lose money in a price battle). Advertising heavily may be recuperated only through repeat purchase and, therefore, over-advertising may be used as a signal that the over-advertised products are of good quality. For start-ups, the game is quite different, VCs look for the signals leading to potential success, such as good management, proven results, patentable ideas and a huge potential combined with hefty growth rate in sales. Still, these are only signals, and more sophisticated investors actually get involved in the start-ups they invest in to reduce further the risks of surprise.

Uninformed parties, however, have an incentive to look for and obtain information. For example, shop and compare, search for a job etc. are instances of information-seeking by uninformed parties. Such activities are called *screening*. A life insurer requires a medical record history; a driver who has a poor accident record history is likely to pay a greater premium (if he can obtain insurance at all). If characteristics of customers are unobservable, firms can use *self-selection* constraints as an aid in screening to reveal private information. For example, consider the phenomenon of rising wage profiles where workers get paid an increasing wage over their careers. An explanation may be that firms are interested in hiring workers who will stay for a long time. Especially if workers get training or experience, which is valuable elsewhere, this is a valid concern. Then they will pay workers below the market level initially so that only 'loyal' workers will self-select to work for the firm. The classic example of 'signalling' was first analysed by Spence (1974) who pointed out that high-productivity individuals try to differentiate themselves from low-productivity ones by the amount of education they acquire. In other words, only the most productive workers invest in education. This is the case because the signalling cost to the productive workers is lower than to low-productivity workers and therefore firms can differentiate between these two types of workers because they make different choices.

Uninformed parties can screen by offering a menu of choices or possible contracts to prospective (informed) trading partners who 'self-select' one of these offerings. Such screening was pointed out in insurance economics for example, showing that, if the insurer offers a menu of insurance policies with different premiums and amounts of cover, the high-risk clients self-select into a policy with high cover. This can lead to insurance firms portfolios where bad risks crowd out the good ones. Insurance companies that are aware of these problems create risk groups and demand higher premium from members in the 'bad' risk portfolio, as well as introducing a number of clauses that will share responsibility for payments in case claims are made (Reyniers, 1999).

3.6.5 The principal–agent problem

Consider a business or economic situation involving two parties: a principal and an agent. For example the manager of a company may be the 'agent' for a stockholder who acts as a 'principal', trusting the manager to perform his job in the interest of stockholders. Similar situations arise between a fund and its traders. The fund 'principal' seeks to provide incentives motivating the traders – 'agents' – to perform in the interest of the fund. In these situations, the actions taken by the agent may be observed only imperfectly. That is, the performance observed by the principal is the outcome of the agent actions (known only by the agent) and some random variable, which may be known, or unknown by the agent at the time an action is assumed and taken. The principal – agent problem consists then in determining the rules for sharing the outcomes obtained through such an organization. This asymmetry of information leads of course to a situation of potential moral hazard. There are several approaches to this problem, which we consider below. For example, designing appropriate incentive systems is of great practical importance. CAR (capital adequacy requirements), health-care compensation etc. are only some of the tools that are used and widely practised to mitigate the effects of moral hazard through agency.

The principal–agent problem is well researched, and there are many research papers using assumptions leading to what we may call normative behaviours and normative compensations. Here we consider a simple example based on the *first-order approach*. Let $\tilde{x}$ be a random variable, which represents the gross return, obtained by a hedge fund manager – the principal. The distribution of this return is influenced by the variable a, which is under the control of the trader and not observed by the principal. Now assume that the sharing rule is given by $F(\tilde{x}, a)$ while the probability distribution of the outcome is $f(\tilde{x}, a)$ which is independent statistically of a. The principal–agent problem consists in determining the amount transferred to the agent by the principal in order to compensate him for the efforts he performs on behalf of the principal. To do so, we assume that the agent utility is separable and given by:

$$V(y, a) = v(y) - w(a); v' > 0, v'' \leq 0, w' > 0, w'' > 0$$

In order to assure the agent's participation, it is necessary to provide at least an expected utility:

$$EV(y, a) \geq 0 \text{ or } \mathrm{E}\,v(y) \geq w(a)$$

In this case, the utility of the principal is:

$$u(\tilde{x} - y), u' > 0, u'' \leq 0$$

The problems we formulate depends then on the information distribution between the principal and the agent.

Assume that the agent's effort 'a' is observable by the principal. In this case, the problem of the principal is formulated by optimizing both 'a' and of course the transfer. That is,

$$\underset{a,y(.)}{\text{Max}}\, Eu(\tilde{x} - y(\tilde{x})) \text{ subject to: } Ev(y(\tilde{x}) - w(a)) \geq 0$$

By applying the conditions for optimality, the optimal solution is found to be:

$$\frac{u'(\tilde{x} - y(\tilde{x}))}{v'(y(\tilde{x}))} = \lambda$$

This yields a sharing rule based on the agent and the principal marginal utility functions, a necessary condition for Pareto optimal risk sharing. A differentiation of the sharing rule, indicates that:

$$\frac{\mathrm{d}y}{\mathrm{d}x} = \frac{u''/u'}{u''/u' + v''/v'}$$

For example, if we assume exponential utility functions given by:

$$u(w) = \frac{-\mathrm{e}^{-aw_u}}{a}; v(w) = \frac{-\mathrm{e}^{-bw_v}}{b}$$

where initial wealth is given by (w_u, w_v) then, we have *a linear risk-sharing rule*, given by:

$$\frac{\mathrm{e}^{-a(w_u-(x-y(x))}}{\mathrm{e}^{-b(w_v-y(x))}} = \lambda \Rightarrow y(x) = \left(\frac{a}{a+b}\right)x + \left(\frac{bw_v - aw_a}{a+b}\right)$$

In other words, the share of the first party is proportional to the risk tolerance, which is given by λ/a and λ/b respectively.

Similarly, assume a fund manager can observe the trader's effort. Or, alternatively, consider a manager–trader relationship where there is a direct relationship between performance and effort. For example, for salesmen of financial products, there is a direct relationship between the performance of the salesmen (quantity of contracts sold) and his effective effort. Let e be the effort and $P(e)$ the profit function. The employee's cost is $C(e)$ and the employee's reservation utility is u. There are a number of simple payment schemes that can motivate a trader/worker to work and provide the efficient amount of effort. These are payments based on effort, forcing contracts and franchises considered below (Reyniers (1999))

(1) *Payments based on effort:* The worker is paid based on his effort, e, according to the wage payment $(we + K)$, the manager's problem is then to solve:

$$\text{Max } \pi = P(e) - (we + K) - C(e)$$
$$\text{Subject to: } we + K - C(e) \geq u$$

The inequality constraint is called the 'participation constraint' or the 'individual rationality constraint'. The employer has no motivation to give more money to the trader than his reservation utility. In this case, the effort selected by the manager will be at the level where marginal cost equals the marginal profit of effort, or $P'(e^*) = C'(e^*)$. The trader has to be encouraged to provide the optimal effort level that leads to the *incentive compatibility constraint.* In other words, the worker's net payoff should be maximized at the optimal effort or $w = C'(e^*)$. Thus, the trader is paid a wage per unit time equal to his marginal disutility of effort and a lump sum K that leaves him with his reservation utility.

(2) *Forcing contracts*: The manager could propose to pay the trader a lump sum L which gives him his reservation utility if he makes effort e^*, i.e. $L = u + C(e^*)$ and zero otherwise. Clearly, the participation and incentive compatibility constraints are satisfied under this simple payment scheme. This arrangement is called a forcing contract because the trader is forced to make effort e^* (while above, the trader is left to select his effort level).

(3) *Franchises*: Now assume that the trader can keep the profits of his effort in return for a certain payment to the principal/manager. This can be interpreted as a franchise structure (similar in some ways to a fund of funds). To set the franchise fee, the trader proceeds as follows. First the trader maximizes $P(e) - C(e) - F$ and therefore chooses the same optimal effort as before such that: $P'(e^*) = C'(e^*)$. The principal/manager can charge a franchise fee which leaves the trader with his reservation utility: $F = P(e^*) - C(e^*) - u$. When the effort cannot be observed, the problem is more difficult. In this case, payment based on effort is not possible. If we choose to pay based on output, then the employer would choose a franchise structure. However, if the employee is risk-averse, he will seek some payment to compensate the risk he is assuming. If the manager is risk-neutral, he may be willing to assume the trader's risk and therefore the franchise solution will not be possible in its current form!

REFERENCES AND FURTHER READING

Akerlof, G. (1970) The market for lemons: Quality uncertainty and the market mechanism, *Quarterly Journal of Economics*, **84**, 488–500.

Allais, M. (1953) Le Comportement de l'homme rationnel devant le risque: Critique des postulats et axiomes de l'ecole americaine, *Econometrica*, **21**, 503–546.

Allais, M. (1979) The foundations of a positive theory of choice involving risk and a criticism of the postulates and axioms of the American School, in M. Allais, and O. Hagen (Eds), *Expected Utility Hypothesis and the Allais Paradox*, D. Reidel, Dordrecht.
Arrow, K.J. (1951) Alternative approaches to the theory of choice in risk-taking situations, *Econometrica*, October.
Arrow, K.J. (1965) *Aspects of the Theory of Risk-Bearing*, Yrjo Jahnssonin Säätiö, Helsinki.
Arrow, K.J. (1982) Risk perception in psychology and in economics, *Economics Inquiry*, January, 1–9.
Bawa, V. (1978) Safety first, stochastic dominance and optimal portfolio choice, *Journal of Financial and Quantitative Analysis*, **13**(2), 255–271.
Beard, R.E., T. Pentikainen and E. Pesonen (1979) *Risk Theory* (2nd edn), Methuen, London.
Bell, D. (1982) Regret in decision making under uncertainty, *Operations Research*, **30**, 961–981.
Bell, D. (1985) Disappointment in decision making under uncertainty, *Operations Research*, **33**, 1–27.
Bell, D. (1995) Risk, return and utility, *Management Science*, **41**, 23–30.
Bernoulli, D. (1954) Exposition of a new theory on the measurement of risk, *Econometrica*, January.
Bierman, H., Jr (1989) The Allais paradox: A framing perspective, *Behavioral Science*, **34**, 46–52.
Borch, K. (1968) *The Economics of Uncertainty*, Princeton University Press, Princeton, NJ.
Borch, K. (1974) *The Mathematical Theory of Insurance*, Lexington Books, Lexington, MASS.
Borch, K., and J. Mossin (1968) *Risk and Uncertainty*, Proceedings of the Conference on Risk and Uncertainty of the International Economic Association, Macmillan, London.
Buhlmann, H. (1970) *Mathematical Methods in Risk Theory*, Springer-Verlag, Bonn.
Chew, Soo H., and Larry G. Epstein (1989) The structure of preferences and attitudes towards the timing of the resolution of uncertainty, *International Economic Review*, **30**, 103–117.
Christ, Marshall (2001) *Operational Risks*, John Wiley & Sons, Inc., New York.
Cochrane, John H., (2001) *Asset Pricing*, Princeton University Press, Princeton, New Jersey.
Dionne, G. (1981) Moral hazard and search activity, *Journal of Risk and Insurance*, **48**, 422–434.
Dionne, G. (1983) Adverse selection and repeated insurance contracts, *Geneva Papers on Risk and Insurance*, **29**, 316–332.
Dreze, Jacques, and Franco Modigliani (1966) Epargne et consommation en avenir aleatoire, *Cahiers du Seminaire d'Econometrie*.
Dyer, J.S., and J. Jia (1997) Relative risk–value model, *European Journal of Operations Research*, **103**, 170–185.
Eeckoudt, L., and M. Kimball (1991) Background risk prudence and the demand for insurance, in *Contributions to Insurance Economics*, G. Dionne (Ed.), Kluwer Academic Press, Boston, MA.
Ellsberg, D. (1961) Risk, ambuguity and the Savage axioms, *Quarterly Journal of Economics*, November, 643–669.
Epstein, Larry G., and Stanley E. Zin (1989) Substitution, risk aversion and the temporal behavior of consumption and asset returns: A theoretical framework, *Econometrica*, **57**, 937–969.
Epstein, Larry G., and Stanley E. Zin (1991) Substitution, risk aversion and the temporal behavior of consumption and asset returns: An empirical analysis, *Journal of Political Economy*, **99**, 263–286.
Fama, Eugene F. (1992) The cross-section of expected stock returns, *The Journal of Finance*, **47**, 427–465.
Fama, Eugene F. (1996) The CAPM is wanted, dead or alive, *The Journal of Finance*, **51**, 1947.
Fishburn, P.C. (1970) *Utility Theory for Decision Making*, John Wiley & Sons, Inc. New York.
Fishburn, P.C. (1988) *Nonlinear Preference and Utility Theory*, The Johns Hopkins University Press, Baltimore, MD.

Friedman, M., and L.J. Savage (1948) The utility analysis of choices involving risk, *Journal of Political Economy*, August.

Friedman, M., and L.J. Savage (1952) The expected utility hypothesis and the measurability of utility, *Journal of Political Economy*, December.

Grossman, S., and O. Hart (1983) An analysis of the principal agent model, *Econometrica*, **51**, 7–46.

Gul, Faruk (1991) A theory of disappointment aversion, *Econometrica*, **59**, 667–686.

Hadar, Josef, and William R. Russell (1969) Rules for ordering uncertain prospects, *American Economic Review*, **59**, 25–34.

Holmstrom, B. (1979) Moral hazard and observability, *Bell Journal of Economics*, **10**, 74–91.

Holmstrom, B. (1982) Moral hazard in teams, *Bell Journal of Economics*, **13**, 324–340.

Hirschleifer, J. (1970) Where are we in the theory of information, *American Economic Review*, **63**, 31–39.

Hirschleifer, J., and J.G. Riley (1979) The analysis of uncertainty and information: An expository survey, *Journal of Economic Literature*, **17**, 1375–1421.

Holmstrom, B. (1979) Moral hazard and observability, *Bell Journal of Economics*, **10**, 74–91.

Jacque, L., and C.S. Tapiero (1987) Premium valuation in international insurance, *Scandinavian Actuarial Journal*, 50–61.

Jacque, L., and C.S. Tapiero (1988) Insurance premium allocation and loss prevention in a large firm: A principal–agency analysis, in M. Sarnat and G. Szego (Eds), *Studies in Banking and Finance*.

Jia, J., J.S. Dyer and J.C. Butler (2001) Generalized disappointment models, *Journal of Risk and Uncertainty*, **22**, 159–178.

Kahnemann, D., and A. Tversky (1979) Prospect theory: An analysis of decision under risk, *Econometrica*, March, 263–291.

Kreps, D. (1979) A representation theorem for preference for flexibility, *Econometrica*, **47**, 565–577.

Kimball, M. (1990) Precautionary saving in the small and in the large, *Econometrica*, **58**, 53–78.

Knight, F.H. (1921) *Risk, Uncertainty and Profit*, Houghton Mifflin, New York.

Kreps, David M., and Evan L. Porteus (1978) Temporal resolution of uncertainty and dynamic choice theory, *Econometrica*, **46**, 185–200.

Kreps, David M., and Evan L. Porteus (1979) Dynamic choice theory and dynamic programming, *Econometrica*, **47**, 91–100.

Laibson, David (1997) 'Golden eggs and hyperbolic discounting', *Quarterly Journal of Economics*, **112**, 443–477.

Lintner, J. (1965) The valuation of risky assets and the selection of risky investments in stock portfolios and capital budgets, *Review of Economic and Statistics*, **47**, 13–37.

Lintner, J. (1965) Security prices, risk and maximum gain from diversification, *Journal of Finance*, **20**, 587–615.

Loomes, Graham, and Robert Sugden (1986) Disappointment and dynamic consistency in choice under uncertainty, *Review of Economic Studies*, **53**, 271–282.

Lucas, R.E. (1978) Asset prices in an exchange economy, *Econometrica*, **46**, 1429–1446.

Machina, M.J. (1982) Expected utility analysis without the independence axiom, *Econometrica*, March, 277–323.

Machina, M.J. (1987) Choice under uncertainty, problems solved and unsolved, *Economic Perspectives*, Summer, 121–154.

Markowitz, Harry M. (1959) *Portfolio Selection*; *Efficient Diversification of Investments*, John Wiley & Sons, Inc., New York.

Mossin, Jan (1969) A note on uncertainty and preferences in a temporal context, *American Economic Review*, **59**, 172–174.

Pauly, M.V. (1974) Overinsurance and the public provision of insurance: The roles of moral hazard and adverse selection, *Quarterly Journal of Economics*, **88**, 44–74.

Pratt, J.W. (1964) Risk aversion in the small and in the large, *Econometrica*, **32**, 122–136.

Pratt, J.W. (1990) The logic of partial-risk aversion: Paradox lost, *Journal of Risk and Uncertainty*, **3**, 105–113.
Quiggin, J. (1985) Subjective utility, anticipated utility and the Allais paradox, *Organizational Behavior and Human Decision Processes*, February, 94–101.
Rabin, Matthew (1998) Psychology and economics, *Journal of Economic Literature*, **36** 11–46.
Reyniers, D. (1999) Lecture Notes in Microeconomics, London School of Ecomomics, London.
Riley, J. (1975) Competitive signalling, *Journal of Economic Theory*, **10**, 174–186.
Rogerson, W.P. (1985) The first order approach to principal agent problems, *Econometrica*, **53**, 1357–1367.
Ross, S. (1973) On the economic theory of agency: The principal's problem, *American Economic Review*, **63**(2), 134–139.
Ross, Stephen A. (1976) The arbitrage theory of capital asset pricing, *Journal of Monetary Economics*, **13**(3), 341–360.
Samuelson, Paul A. (1963) Risk and uncertainty: A fallacy of large numbers, *Scientia*, **98**, 108–163.
Sharpe, W.F. (1964) Capital asset prices: A theory of market equilibrium under risk, *The Journal of Finance*, **19**, 425–442.
Siegel, Jeremy J., and Richard H. Thaler (1997) The Equity Premium Puzzle, *Journal of Economic Perspectives*, **11**, 191–200.
Spence, M. (1974) *Market Signaling*, Harvard University Press, Cambridge, MA.
Spence, Michael, and Richard Zeckhauser (1972) The effect of the timing of consumption decisions and the resolution of lotteries on the choice of lotteries, *Econometrica*, **40**, 401–403.
Sugden, R. (1993) An axiomatic foundation of regret theory, *Journal of Economic Theory*, **60**,150–180.
Tapiero, C.S. (1983) The optimal control of a jump mutual insurance process, *Astin Bulletin*, **13**, 13–21.
Tapiero, C.S. (1984) A mutual insurance diffusion stochastic control problem, *Journal of Economic Dynamics and Control*, **7**, 241–260.
Tapiero, C.S. (1986) The systems approach to insurance company management, in *Developments of Control Theory for Economic Analysis*, C. Carraro and D. Sartore (Eds), Martinus Nijhoff, Dordrecht.
Tapiero, C.S. (1988) *Applied Stochastic Models and Control in Management*, North Holland, Amsterdam.
Tapiero, C.S., and L. Jacque (1987) The expected cost of ruin and insurance premiums in mutual insurance, *Journal of Risk and Insurance*, **54** (3), 594–602.
Tapiero, C.S., and D. Zuckerman (1982) Optimum excess-loss reinsurance: A dynamic framework, *Stochastic Processes and Applications*, **12**, 85–96.
Tapiero, C.S., and D. Zuckerman (1983) Optimal investment policy of an insurance firm, *Insurance Mathematics and Economics*, **2**, 103–112.
Tobin, J. (1956) The interest elasticity of the transaction demand for cash, *Review of Economics and Statistics*, **38**, 241–247.
Willasen, Y. (1981) Expected utility, Chebychev bounds, mean variance analysis, *Scandinavian Economic Journal*, **83**, 419–428.
Willasen, Y. (1990) Best upper and lower Tchebycheff bounds on expected utility, *Review of Economic Studies*, **57**, 513–520.

CHAPTER 4

Probability and Finance

4.1 INTRODUCTION

Probability modelling in finance and economics provides a means to rationalize the unknown by imbedding it into a coherent framework, clearly distinguishing what we know and what we do not know. Yet, the assumption that we can formalize our lack of knowledge is both presumptuous and essential at the same time. To appreciate the problems of probability modelling it is essential to distinguish between randomness, uncertainty and chaos. These terms are central to an important polemic regarding 'modelling cultures' in probability, finance and economics. Kalman (1994) states that 'the majority of observed phenomena of randomness in nature (always excluding games of chance) cannot and should not be explained by conventional probability theory; there is little or no experimental evidence in favour of (conventional) probability but there is massive, accumulating evidence that explanations and even descriptions should be sought outside the conventional framework'. This means that randomness might be defined without the use of probabilities. Kolmogorov, defined randomness in terms of non-uniqueness and non-regularity. For example, a die has six faces and therefore it has non-uniqueness. Further, the expansion of $\sqrt{2}$ or of π provides an infinite string of numbers that appear irregularly, and can therefore be thought of as 'random'. The Nobel Laureate, Born, in his 1954 inaugural address also stated that randomness occurs when 'determinacy lapses into indeterminacy' without any logical, mathematical, empirical or physical argumentation, preceding thereby an important research effort on chaos. Kalman, seeking to explain these approaches to modelling defined chaos as 'randomness without probability'. Statements such as 'we might have trouble forecasting the temperature of coffee one minute in advance, but we should have little difficulty in forecasting it an hour ahead' by Edward Lorenz, a weather forecaster and one of the co-founders of chaos theory, reinforces the many dilemmas we must deal with in modelling uncertain phenomena. In weather modelling and forecasting for example, involving in many cases as many as 50 000 equations and more, it is presumed that if small models can predict well, it is only natural to expect that bigger and more sophisticated ones can do better. This turned out not to be the case, however. Bigger does not always turn out to be better; more sophisticated does not always mean improved accuracy.

Risk and Financial Management: Mathematical and Computational Methods. C. Tapiero
 ISBN: 0-470-84908-8

In weather forecasting it soon became evident that no matter what the size and sophistication of the models used, forecasting accuracy decreased considerably beyond two to three days and provided no better predictions than using the average weather conditions of similar days of previous years to predict temperature, rainfall, or snow. What came to be known as the 'butterfly effect' (meaning in fact an insensitivity to initial conditions) or the effects of a flying butterfly exerting an unlikely and unpredictable critical influence on future weather patterns (Lorenz, 1966). In the short term too, the accuracy of weather forecasting could not improve much beyond the use of the naive approach which predicts that tomorrow's or the next day's weather will be exactly the same as today's. Subsequent studies have amplified the importance of chaos in biology. Similar issues are raised in financial forecasting: the time scale of data, whether it is tickertape data or daily, weekly or monthly stock quotations, alters significantly the meaningfulness of forecasts.

In economic and business forecasting, the accuracy of predictions did not turn out to be any better than those of weather forecasting seen above. Further, accrued evidence points out that assumptions made by probability models are in practice violated. Long-run memory undermines the existence of martingales in finance. Further, can stock prices uncertainty or 'noise' be modelled by Brownian motion? This is one of the issues we must confront and deal with in financial modelling. The index of Hurst, entropy and chaos, which we shall discuss subsequently, are important concepts because they stimulate and highlight that there may be other approaches to be reckoned with and thereby stimulate economic and financial theoretical and empirical thinking.

The study of nonlinear financial time series and in particular chaos has assumed recently an added importance. Traditionally, it has been assumed for mathematical convenience that time series have a number of characteristics including:

- Existence of an equilibrium (or fixed point or a stationary state), or equivalently an insensitivity to initial conditions in the long run.
- Periodicity.
- Structural stability which allows the transformation of equations which are hard to study to some other forms which are stable and amenable to analysis.

There are a number of physical and economic phenomena that do not share all these properties. When this is the case, we call these series chaotic, implying that both indeterminacy and our inability to predict what the state of a system may be. Chaos can thus occur in both deterministic time series as well as in stochastic ones. For this reason, it has re-ignited the age-old confrontation of a deterministic versus a probabilistic view of nature and the world as well as mathematical modelling between externally and internally induced disturbances (which are the source of nonlinearities).

Commensurate analysis of nonlinear time series has also followed its course in finance. ARCH and GARCH type models used to estimate volatility are also nonlinear models expressed as a function (linear or not) of past variations in stocks.

Their analysis and estimation is the more difficult, the greater the nonlinearities assumed in representing the process. Current research is diverted towards the study of various nonlinear (non-Gaussian) and leptokurtic distributions, seeking to bridge a gap between traditional probability approaches in finance (based on the normal probability distribution) and systems exhibiting a chaotic behaviour and 'fat tails' in their distributions. For example, a great deal of research effort is devoted to explain why probability distributions are leptokurtic. Some approaches span herding behaviour in financial markets, reflecting the interaction of traders, imitation of investor groups and the following of gurus or opinion leaders. In such circumstances, collective behaviour can be 'irrational' leading to markets crashing. A paper 'Turbulent cascades in foreign exchange markets' by Ghashghaie *et al.*, published in a letter in *Nature* in 1996, has also pointed to the statistical observation that a similar behaviour is seen in financial exchange markets and hydrodynamic turbulence. Such behaviour clearly points to a 'non-Gaussian-Normal noise' and thereby to invalidating the assumption of 'normal noise' implicit in the underlying random walk models used in finance.

In general, in order to model uncertainty we seek to distinguish the known from the unknown and find some mechanisms (such as theories, common sense, metaphors and more often intuition) to reconcile our knowledge with our lack of it. For this reason, modelling uncertainty is not merely a collection of techniques but an art in blending the relevant aspects of a situation and its unforeseen consequences with a descriptive, yet theoretically justifiable and tractable, economic and mathematical methodology. Of course, we conveniently use probabilities to describe quantitatively the set of possible events that may unfold over time. Specification of these probabilities and their associated distributions are important and based on an understanding of the process at hand and the accrued evidence we can apply to estimate these probabilities. Any model is rationally bounded and also has its own sources of imperfections that we may (or may not) be aware of. However, 'at the end of the day', probabilities and their quantitative assessment, remain essential and necessary to provide a systematic approach to constructing a model of uncertainty. For this reason, it is important to know some of the assumptions we use in building probability models, as we shall briefly outline below. The approach we shall use is informal, however, emphasizing a study of models' implications at the expense of formality.

4.2 UNCERTAINTY, GAMES OF CHANCE AND MARTINGALES

Games of chance, such as betting in Monte Carlo or any casino, are popular metaphors to represent the ongoing exchanges of stock markets, where money is thrown to chance. Its historical origins can be traced to Girolamo Cardano who proposed an elementary theory of gambling in 1565 (*Liber de Ludo Aleae* – The Book of Games of Chance). The notion of 'fair game' was clearly stated: 'The most fundamental principle of all in gambling is simply equal conditions, e.g. of opponents, of bystanders, of money, of situation, of the dice box, and of the die itself. The extent to which you depart from that equality, if it is in your opponent's favour, you are a fool, and if in your own, you are unjust'. *This is the essence of the Martingale* (although Cardano did not use the word 'martingale'). It was in Bachelier's thesis in 1900 however that a mathematical model of a fair game, the martingale, was proposed. Subsequently J. Ville, P. Levy, J.L. Doob and others have constructed stochastic processes. The 'concept of a fair game' or martingale, in money terms, states that the expected profit at a given time given the total past capital is null with probability one. Gabor Szekely points out that a martingale is also a paradox. Explicitly,

> If a share is expected to be profitable, it seems natural that the share is worth buying, and if it is not profitable, it is worth selling. It also seems natural to spend all one's money on shares which are expected to be the most profitable ones. Though this is true, in practice other strategies are followed, because while the expected value of our money may increase (our expected capital tends to infinity), our fortune itself tends to zero with probability one. So in Stock Exchange business, we have to be careful: shares that are expected to be profitable are sometimes worth selling.

Games of dice, blackjack, roulette and many other games, when they are fair, corrected for the bias each has, are thus martingales. 'Fundamental finance theory' subsumes as well that under certain probability measures, asset prices turn out to have the martingale property. Intuitively, what does a martingale assume?

- Tomorrow's price is today's best forecast.
- Non-overlapping price changes are uncorrelated at all leads and lags.

The martingale is considered to be a necessary condition for an efficient asset market, one in which the information contained in past prices is instantly, fully and perpetually reflected in the asset's current price. A technical definition of a martingale can be summarized as the presumption that each process event (such as a new price) is independent and can be summed (i.e. it is integrable) and has the property that its conditional expectation remains the same (i.e. it is time-invariant). That is, if $\mathbf{\Phi}_t = \{p_0, p_1, \ldots, p_t\}$ are an asset price history at time $t = 0, 1, 2, \ldots$ expressing the relevant information we have at this time regarding the time series, also called the filtration. Then the expected next period price at time $t + 1$ is equal to the current price

$$E\left(p_{t+1}\left|p_0, p_1, p_2, \ldots, p_t\right.\right) = p_t$$

which we also write as follows:

$$E\left(p_{t+1}\left|\mathbf{\Phi}_t\right.\right) = p_t \quad \text{for any time } t$$

If instead asset prices decrease (or increase) in expectation over time, we have a super-martingale (sub-martingale):

$$E\left(p_{t+1}\left|\mathbf{\Phi}_t\right.\right) \leq (\geq) p_t$$

Martingales may also be defined with respect to other processes. In particular, if $\{p_t, t \geq 0\}$ and $\{y_t, t \geq 0\}$ are two processes denoting, say, price and interest rate processes, we can then say that $\{p_t, t \geq 0\}$ is a martingale with respect to $\{y_t, t \geq 0\}$ if:

$$E\{|p_t|\} < \infty \quad \text{and} \quad E(p_{t+1}\,|y_0, y_1, \ldots, y_t) = p_t,\ \forall t$$

Of course, by induction, it can be easily shown that a martingale implies an invariant mean:

$$E(p_{t+1}) = E(p_t) = \cdots = E(p_0)$$

For example, given a stock and a bond process, the stock process may turn out to be a martingale with respect to the bond (a deflator) process, in which case the bond will serve as a numeraire facilitating our ability to compute the value of the stock.

Martingale techniques are routinely applied in financial mathematics and are used to prove many essential and theoretical results. For example, the first 'fundamental theorem of asset pricing', states that if there are no arbitrage opportunities, then properly normalized security prices are martingales under some probability measure. Furthermore, efficient markets are defined when the relevant information is reflected in market prices. This means that at any one time, the current price fully represents all the information, i.e. the expected future price $p(t+T)$ conditioned by the current information and using a price process normalized to a martingale equals the current price. 'The second fundamental theorem of asset pricing' states in contrast that if markets are complete, then for each numeraire used there exists one and only one pricing function (which is the martingale measure). Martingales and our ability to construct price processes that have the martingale properties are thus extremely useful to price assets in theoretical finance as we shall see in Chapter 6.

Martingales provide the possibility of using a risk-neutral pricing framework for financial assets. Explicitly, when and if it can be used, it provides a mechanism for valuing assets 'as if investors were risk neutral'. It is indeed extremely convenient, allowing the pricing of securities by using their expected returns valued at the risk-free rate. To do so, one must of course, find the probability measure, or equivalently find a discounting mechanism that renders the asset values a martingale. Equivalently, it requires that we determine the means to replicate the payoff of an uncertain stream by an equivalent 'sure' stream to which a risk-free discounting can be applied. Such a risk-neutral probability exists if there are no arbitrage opportunities. The martingale measures are therefore associated with a pricing of an asset which is unique only if markets are complete. This turns out to be the case when the assumptions made regarding market behaviours include:

- rational expectations,
- law of the single price,
- no long-term memory,
- no arbitrage.

The problem in applying rational expectations to financial valuation is that it may not be always right, however. The interaction of markets can lead to instabilities due to very rapid and positive feedback or to expectations that are becoming trader- and market-dependent. Such situations lead to a growth of volatility, instabilities and perhaps, in some special cases, to bubbles and chaos. George Soros, the hedge fund financier has also brought attention to the concept of 'reflexivity' summarizing an environment where conventional traditional finance theory no longer holds and therefore theoretical finance does not apply. In these circumstances, 'there is no hazard in uncertainty'. A trader's ability to 'identify a rational behaviour' in what may seem irrational to others can provide great opportunities for profit making.

The 'law of the single price', claiming that two cash flows of identical characteristics must have, necessarily, the same price (otherwise there would be an opportunity for arbitrage) is not always satisfied as well. Information asymmetry, for example, may violate such an assumption. Any violation of these assumptions perturbs the basic assumptions of theoretical finance, leading to incomplete markets. In particular, we apply this 'law' in constructing portfolios that can replicate risky assets. By hedging, i.e. equating these portfolios to a riskless asset, it becomes possible to value the assets 'as if they were riskless'. This approach will be developed here in greater detail and for a number of situations.

We shall attend to these issues at some length in subsequent chapters. At this point, we shall turn to defining terms often used in finance: random walks and stochastic processes.

4.3 UNCERTAINTY, RANDOM WALKS AND STOCHASTIC PROCESSES

A stochastic process is an indexed pair {events, time} expressed in terms of a function – a random variable indexed to time. This defines a sample path, i.e. a set of values that the process can assume over time. For example, it might be a stock price denoting events, indexed to a time scale. The study of stochastic processes has its origin in the study of the kinetic behaviour of molecules in gas by physicists in the nineteenth century. It was only in the twentieth century, following work by Einstein, Kolmogorov, Levy, Wiener and others, that stochastic processes were studied in some depth. In finance, however, Bachelier, in his dissertation in 1900, had already provided a study of stock exchange speculation using a fundamental stochastic process we call the 'random walk', establishing a connection between price fluctuations in the stock exchange and Brownian motion – a continuous-time expression of the random walk assumptions.

4.3.1 The random walk

The random walk model of price change is based on two essential behavioural hypotheses.

(1) In any given time interval, prices may increase with a known probability $0 < p < 1$, or decrease with probability $1 - p$.
(2) Price changes from period to period are statistically independent.

Denote by $\Delta\xi(t)$ the random event denoting the price change (of size Δx) in a small time interval Δt:

$$\Delta\xi(t) = \begin{cases} +\Delta x & \text{w.p.} \quad p \\ -\Delta x & \text{w.p.} \quad 1-p \end{cases}$$

Thus, if $x(t)$ is the price at the discrete time t, and if it is *only* a function of the last price $x(t - \Delta t)$ and price changes $\Delta\xi(t)$ in $(t - \Delta t, t)$, then an evolution of prices is given by:

$$x(t) = x(t - \Delta t) + \Delta\xi(t)$$

Prices are thus assuming values $x(t)$ at times $\ldots,\ t - \Delta t, t,\ t + \Delta t, \ldots$ These values denote a stochastic process $x(t)$ which is also written as $\{x(t), t \geq 0\}$. The price at time t, $x(t)$, assumes in this case a binomial distribution since events are independent and of fixed probability, as we shall see next. Say that we start at a given price x_0 at time $t_0 = 0$. At time $t_1 = t_0 + \Delta t$, either the price increases by Δx with probability $0 < p < 1$ or it decreases with probability $1 - p$. Namely,

$$x(t_1) = x(t_1 - \Delta t) + \Delta\xi_1, \quad \text{or} \quad x(t_1) = x(t_0) + \Delta\xi_1.$$

We can also write this equation in terms of the number of times i_1 the price increases. In our case, prices either increase or decrease in Δt, or

$$x(t_1) = x_0 + i_1\Delta x - (1 - i_1)\Delta x,\ i_1 \sim B(1, p)$$

where i_1 assumes two values $i_1 = 0, 1$ given by the binomial probability distribution

$$B(1, p) = \binom{1}{i_1} p^{i_1}(1 - p)^{1-i_1}$$

$i_1 = 0, 1$ and parameter $(1, p), 0 < p < 1$. An instant of time later $t_2 = t_1 + \Delta t = t_0 + 2\Delta t$, we have:

$$x(t_2) = x(t_2 - \Delta t) + \Delta\xi_2 \text{ or } x(t_2) = x(t_1) + \Delta\xi_2 \text{ or } x(t_2) = x(t_0) + \Delta\xi_1 + \Delta\xi_2$$

which we can write as follows (see also Figure 4.1):

$$x(t_2) = x_0 + i_2\Delta x - (2 - i_2)\Delta x,\ i_2 \sim B(2, p)$$

and generally, for n successive intervals of time ($t_n = n\Delta t$), the price is defined by:

$$x(t_n) = x_0 + i_n\Delta x - (n - i_n)\Delta x,\ i_n \sim B(n, p)$$

where:

$$B(n, p) = \binom{n}{i_n} p^{i_n}(1 - p)^{n-i_n};\ i_n = 0, 1, 2, \ldots, n$$

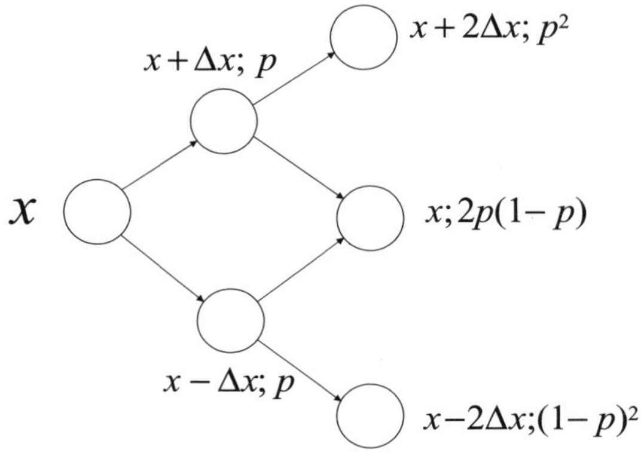

Figure 4.1 A two-period tree.

The price process can thus be written by:

$$x(t_n) = x(t_{n-1}) + \Delta\xi_n \quad \text{or} \quad x(t_n) = x(t_0) + \sum_{i=0}^{n} \Delta\xi_j$$

where $x(t_n) - x(t_0)$ has the probability distribution of the sum $\Delta\xi_j (j = 1, \ldots, n)$. Since price changes are of equal size, we can state that the number of times prices have increased is given by the binomial distribution $B(n, p)$. The expected price and its variance can now be calculated easily. The expected price at time t_n is:

$$E(x(t_n)) = x_0 + \Delta x E(i_n) - \Delta x E(n - i_n);\ E(i_n) = np;\ E(n - i_n) = n(1 - p)$$

Set $d = [i_n - (n - i_n)]\,\Delta x$. The mean distance and its variance, given by $E(d)$ and var(d), with $q = 1 - p$ are then,

$$E(d) = n(p - q)\Delta x \quad \text{and} \quad \text{var}(d) = 4npq(\Delta x)^2$$

This is easily proved. Note that $E(i) = np$ and var(i) $= npq$, with i replacing i_n for simplicity. Thus,

$$E(d) = E\,[i - (n - i)]\,\Delta x = E\,[2i - n]\,\Delta x = (2np - n)\Delta x = n(p - q)\Delta x$$

Also

$$\text{var}(d) = (\Delta x)^2\ \text{var}\,[i - (n - i)]\ = (\Delta x)^2\ \text{var}\,[2i - n]\ = 4npq(\Delta x)^2$$

The results above are expressed in terms of small distance (which we shall henceforth call states) increments Δx and small increments of time Δt. Letting these increments be very small, we can obtain continuous time and continuous state limits for the equation of motion. Explicitly, in a time interval $[0, t]$, let the number of jumps be n and be given by $n = [t/\Delta t]$. When Δt is a small time increment, then (with $t/\Delta t$ integer):

$$E(d) = \frac{t(p-q)\Delta x}{\Delta t}, \quad \text{var}(d) = \frac{4tpq(\Delta x)^2}{\Delta t}$$

For the problem to make sense, the limits of $\Delta x/\Delta t$ and $(\Delta x)^2/\Delta t$ as $\Delta x \to 0$ must exist, however. In other words, we are specifying a priori that the stochastic process, has at the limit, finite mean and finite variance growth rates. If we let these limits be:

$$\lim_{\substack{\Delta t \to 0 \\ \Delta x \to 0}} \frac{\Delta x}{\Delta t} = 2C, \quad \lim_{\substack{\Delta t \to 0 \\ \Delta x \to 0}} \frac{(\Delta x)^2}{\Delta t} = 2D$$

It is also possible to express the probability of a price increase in terms of these parameters which we choose for convenience to be:

$$p = \frac{1}{2} + \frac{C}{2D}\Delta x,$$

Inserting this probability in the mean and variance equations, and moving to the limit, we obtain the mean and variance functions $m(t)$ and $\sigma^2(t)$ which are linear in time:

$$m(t) = 2Ct; \; \sigma^2(t) = 2Dt$$

where C is called the 'drift' of the process expressing its tendency over time while D is its diffusion, expressing the process variability. The proof of these is simple to check. First note that:

$$E(d) = n(p-q)\Delta x = n\frac{C}{D}(\Delta x)^2$$

However, $(\Delta x)^2 = 2D\Delta t$ and therefore, $E(d) = 2Cn\Delta t = 2Ct$. By the same token,

$$\text{var}(d) = 4npq(\Delta x)^2 = n\left[1 + \frac{C}{D}\Delta x\right]\left[1 - \frac{C}{D}\Delta x\right](\Delta x)^2$$

However, we also have $(\Delta x)^2 = 4C^2(\Delta t)^2$, $n = t/\Delta t$ which is inserted in the equation above to lead to:

$$\text{var}(d) = t\frac{(\Delta x)^2}{\Delta t}\left[1 - \frac{C^2}{D^2}(\Delta x)^2\right] = 2Dt\left[1 - \frac{2(\Delta x)^2(\Delta t)}{4(\Delta t)^2(\Delta x)^2}(\Delta x)^2\right]$$

since,

$$\left[1 - \frac{2(\Delta x)^2(\Delta t)}{4(\Delta t)^2(\Delta x)^2}(\Delta x)^2\right] = [1 - D\Delta t]$$

At the time limit, we obtain the variance $\text{var}(d) = \sigma^2(t) = 2Dt$ stated above.

Since this limit results from limiting arguments to the underlying binomial process describing the random walk, we can conclude that the parameters $(m(t), \sigma^2(t))$ are normally distributed, or:

$$f(x,t) = \frac{1}{\sqrt{2\pi}\sigma(t)} \exp\left\{-\frac{1}{2}\frac{[x-m(t)]^2}{\sigma^2(t)}\right\}$$

This equation turns out to be also a particular solution of a partial differential equation expressing the continuous time–state evolution of the process probabilities and called the Fokker–Planck equation. Using the elementary observation that a linear transformation of normal random variables are also normal, we can write the price equation in terms of its drift $2C$ and diffusion $\sqrt{2D}$ (also called volatility), by:

$$\Delta x(t) = 2C\Delta t + \sqrt{2D}\Delta w(t), \quad \text{with } E(\Delta w(t)) = 0, \ \text{var}(\Delta w(t)) = \Delta t$$

where $\Delta w(t)$ is a normal probability distribution with zero mean and variance Δt. Such processes, in continuous time are called stochastic differential equations (SDEs) while the process $\Delta x(t) = \Delta w(t)$ is called a Wiener (Levy) process. Finally, the integral of $\Delta w(t)$, or $W(t)$, is also known as Brownian motion which is essentially a zero mean normally distributed random variable with independent increments and a linear variance in time t. It is named after Robert Brown (1773–1858), a botanist who discovered the random motion of colloid-sized particles found in experiments performed in June–August 1825 with pollen. If we were to take a stock price, it would be interesting to estimate both the drift and the diffusion of the process. Would it fit? Would the residual error be indeed a normal probability distribution with mean zero, and a linear time variance with no correlation? Such a study would compare stock data taken every minute (tickertape), daily, weekly and monthly. Probably, results will differ according to the time scale taken for the estimation and thereby violate the assumptions of the model. Such studies are important in financial statistics when they seek to justify the assumption of 'error normality' in financial time series.

The Wiener process is of fundamental importance in mathematical finance because it is used to model the uncertainty associated with many economic processes. However, it is well known in finance that such a process underestimates the probability of the price not changing, and overestimates the mid-range value price fluctuations. Further, extreme price jumps are grossly underestimated by the Wiener (normal) process. The search for distributions that can truly reflect stock market behaviour has thus became an important preoccupation. Mandelbrot and Fama for example have suggested that we use Pareto–Levy distributions as well as leptokurtic distributions to describe the statistics of price fluctuations. Explicitly, say that a distribution has mean m and variance σ^2 and define the following coefficients $\zeta_1 = m_3/\sigma^3$ and $\zeta_2 = m_4/\sigma^4 - 3$ where m_3 and m_4 are the third and the fourth moment respectively. The first index is an index of asymmetry pointing to leptokurtic distributions while the second is 'an excess coefficient' point to platokurtic distributions. For the Normal distribution we have $\zeta_1 = 0$ and $\zeta_2 = 0$, thus any departure from these reference values will also indicate a departure from normality. Pareto–Levy stable distributions exhibit, however, an infinite variance, practically referred to as 'fat tail distributions' that also violate the underlying assumptions of 'Normal–Wiener' processes. When weekly or monthly data is used (rather than daily and intraday data), a smoothing of the data allows the use of the Normal distribution. This observation thus implies that the time scale we choose to characterize uncertainty is an important factor to deal with. When the time scale increases, the use of Normal distributions is justified because in such cases, we gradually move from leptokurtic to Normal distributions. What statistical distribution can one assume over different periods of consideration? The random walk is by far the most used and the easiest to work with and agrees well for larger periods of time. Other distributions are mathematically more challenging, especially since different results are seen for various assets. Part of the problem can be explained by the deviations from the *efficient markets hypothesis* and external influences on the market, as we shall see in subsequent chapters.

Formally, it is a Markov stochastic process $x = \{x(t); t \geq 0\}$ whose non-overlapping increments Δx_t and Δx_s

$$\Delta x(\tau) = x(\tau + \Delta t) - x(\tau);\ \tau = t, s$$

are *stationary*, *independently* and *normally distributed* with mean zero and variance Δt, i.e. with zero drift and volatility 1. In continuous time, this equation is often written as:

$$\mathrm{d}x = 2C\,\mathrm{d}t = \sqrt{2D}\,\mathrm{d}w(t)$$

Such equations are known as *stochastic differential equations*. Generalization to far more complex movements can also be constructed by changing the modelling hypotheses regarding the drift and the diffusion processes. When the diffusion–volatility is also subject to uncertainty, this leads to processes we call stochastic volatility models, leading to incomplete markets (as will be seen in Chapter 5). In many cases, volatility can be a function of the process itself. For example, say that $\sigma = \sigma(x)$, then evidently,

$$\Delta x = \sigma(x)\Delta w$$

which need not lead, necessarily, to a Normal probability distribution for x. For example, in some cases, it is convenient to presume that rates of returns are Normal, meaning that $\Delta x/x$ can be represented by a process with known drift (the expected rate of return) and known diffusion (the rates of returns volatility). Thus, the following hypothesis is stated:

$$\frac{\Delta x}{x} = \alpha \Delta t + \sigma \Delta w.$$

This is equivalent to stating that the log of return $y = \ln(x)$ has a Normal probability distribution:

$$\Delta y = \alpha\Delta t + \sigma\Delta w, \;\; y = \ln(x)$$

with mean αt and variance $\sigma^2 t$ and therefore, x has a lognormal probability distribution.

In many economic and financial applications stochastic processes are driven by a Wiener process leading to models of the form:

$$x(t + \Delta t) = x(t) + f(x, t)\Delta t + \sigma(x, t)\Delta w(t)$$

Of course, if the time interval is $\Delta t = 1$, this is reduced to a difference equation,

$$X_{t+1} = X_t + f_t(X) + \sigma_t(X)\varepsilon_t, \;\; \varepsilon_t \sim N(0, 1), \;\; t = 0, 1, 2, \ldots$$

where ε_t is a zero mean, unit variance and normally distributed random variable. When the time interval is infinitely small, in continuous time, we have a stochastic differential equation:

$$\mathrm{d}x(t) = f(x, t)\,\mathrm{d}t + \sigma(x, t)\,\mathrm{d}w(t), \;\; x(0) = x_0, \;\; 0 \le t \le T$$

The variable $x(t)$ is defined, however, only if the above equation is meaningful in a statistical sense. In general, existence of a solution for the stochastic differential equation cannot be taken for granted and conditions have to be imposed to guarantee that such a solution exists. Such conditions are provided by the Lipschitz conditions assuming that: f, σ and the initial condition $x(0)$ are real and continuous and satisfy the following hypotheses:

- f and σ satisfy uniform Lipschitz conditions in x. That is, there is a $K > 0$ such that for x_2 and x_1,

$$|f(x_2, t) - f(x_1, t)| \le K\,|x_2 - x_1|$$
$$|\sigma(x_2, t) - \sigma(x_1, t)| \le K\,|x_2 - x_1|$$

- f and σ are continuous in t on $[0, T]$, $x(0)$ is any random variable with $E\,(x(0))^2 < \infty$, independent of the increment stochastic process. Then:
  (1) The stochastic differential equation has, in the mean square limit sense, a solution on

$$t \in [0, T]\,, x(t) - x(0) = \int_0^t f(x, \tau)\,\mathrm{d}\tau + \int_0^t \sigma(x, \tau)\,\mathrm{d}w(\tau)$$

  (2) $x(t)$ is mean square continuous on $[0, T]$
  (3) $E\,(x(0))^2 < M$, for all $t \in [0, T]$ and arbitrary M,

$$\int_0^T E((x(t))^2)\,\mathrm{d}t < \infty$$

(4) $x(t) - x(0)$ is independent of the stochastic process $\{dw(\tau);\ \tau > t\}$ for $t \in [0, T]$.

The stochastic process $x(t),\ t \in [0, T]$, is then a Markov process and, in a mean square sense, is uniquely determined by the initial condition $x(0)$. The Lipschitz and the growth conditions, meaning $(f(x,t))^2 + (\sigma(x,t))^2 \leq K^2(1 + |x|^2)$, provide both a uniqueness and existence non-anticipating solution $x(t)$ of the stochastic differential equation in the appropriate range $[0, T]$. In other words, if these conditions are not guaranteed, as is the case when the variance of processes increases infinitely, a solution to the stochastic differential equation cannot be assured.

Clearly, there is more than one way to conceive and formalize stochastic models of prices. In this approach, however, the evolution of prices was entirely independent of their past history. And further, a position at an instant of time depends only on the position at the previous instant of time. Such assumptions, compared to the real economic, financial and social processes we usually face, are extremely simplistic. They are, however, required for analytical tractability and we must therefore be aware of their limitations. The stringency of the assumptions required to construct stochastic processes, thus, point out that these can be useful to study systems which exhibit only small variations in time. Models with large and unpredictable variations must be based therefore on an intuitive understanding of the problem at hand or some other modelling techniques.

4.3.2 Properties of stochastic processes

The characteristics of time series are mostly expressed in terms of, 'stationarity, ergodicity, correlation and independent increments'. These terms are often encountered in the study of financial time series and we ought therefore to understand them.

Stationarity

A time series is stationary when the evolution of its mean (drift) and variance (volatility–diffusion) are not a function of time. If $f(x,t)$ is the probability distribution of x at time t, then: $f(x,t) = f(x, t+\tau) = f(x)$ for all t and τ. This property is called *strict stationarity*. In this case, for a two random variables process, we have:

$$f(x_1, x_2, t_1, t_2) = f(x_1, x_2, t_1, t_1 + \tau) = f(x_1, x_2, t_2 - t_1) = f(x_1, x_2, \tau)$$

That is, for the joint distribution of a strict stationary process, the distribution is a function of the time difference τ of the two (prices) random variables. As a result, the correlation function $B(t_1, t_2) = E(x(t_1)x(t_2))$, describing the correlation between (x_1, x_2) at instants of time (t_1, t_2), is a function of the time difference $t_2 - t_1 = \tau$ only. The autocovariance function (the correlation function about the mean) is then given by $K(t_1, t_2)$, with $K(t_1, t_2) = B(t_1, t_2) - Ex_1(t_1)Ex_2(t_2)$. By the same token, the correlation coefficient $R_1(\tau)$ of the random variable x_1 is a function of the time difference τ only, or

$$R_1(\tau) = \frac{\text{cov}[x_1(t), x_1(t+\tau)]}{\text{var}[x_1(t)]\ \text{var}[x_1(t+\tau)]}$$

For stationary processes we have necessarily $\text{var}[x(t)] = \text{var}\,[x(t+\tau)]$ and therefore the correlation coefficient is a function of the time difference only, or

$$R(\tau) = \frac{[B(\tau) - m]^2}{\text{var}[x(t)]} = \frac{K(\tau)}{K(0)}$$

Independent increments

Increments $\Delta x(t) = x(t+1) - x(t)$ are stationary and independent if non-overlapping $\Delta x(t)$ and $\Delta x(s)$ are statistically, identically and independently distributed. This property leads to well-known processes such as the Poisson Jump and the Wiener process we saw earlier and can, sometimes, be necessary for the mathematical tractability of stochastic processes. The first two moments of non-overlapping independent and stationary increments point to a linear function of time (hence the term of linear finance, associated with using Brownian motion in financial model building). This is shown by the simple equalities:

$$E[X(t)] = tE[X(1)] + (1-t)E[X(0)];$$
$$\text{var}[X(t)] = t\ \text{var}[X(1)] + (1-t)\ \text{var}[X(0)]$$

The proof is straightforward and found by noting that if we set $f(t) = E[X(t)] - E[X(0)]$, then, non-overlapping stationary increments imply that:

$$\begin{aligned} f(t+s) &= E[X(t+s)] - E[X(0)] = E[X(t+s) - X(t)] + E[X(t) - X(0)] \\ &= E[X(s) - X(0)] + E[X(t) - X(0)] \\ &= f(t) + f(s) \end{aligned}$$

And the only solution is $f(t) = tf(1)$, which is used to prove the result for the expectation. The same technique applies to the variance.

4.4 STOCHASTIC CALCULUS

Financial and computational mathematics use stochastic processes extensively and thus we are called to manipulate equations of this sort. To do so, we mostly use Ito's stochastic calculus. The ideas of this calculus are simple and are based on the recognition that the magnitudes of second-order terms of asset prices are not negligible. Many texts deal with the rules of stochastic calculus, including Arnold (1974), Bensoussan (1982, 1985), Bismut (1976), Cox and Miller (1965), Elliot (1982), Ito (1961), Ito and McKean (1967), Malliaris and Brock (1982) and my own (Tapiero, 1988, 1998). For this reason, we shall consider here these rules in an intuitive manner and emphasize their application. Further, for simplicity, functions of time such as $x(t)$ and $y(t)$ are written by x and y except when the time specification differs.

The essential feature of Ito's calculus is Ito's Lemma. It is equivalent to the total differential rule in deterministic calculus. Explicitly, state that a functional relationship $y = F(x, t)$, continuous in x and time t, expresses the value of some economic variable y measured in terms of another x (for example, an option price measured in terms of the underlying stock price on which the option is written, the value of a bond measured as a function of the underlying stochastic interest rate process etc.) whose underlying process is known. We seek $\Delta y = y(t + \Delta t) - y(t)$. If x is deterministic, then application of the total differential rule in calculus, resulting from an application of Taylor series expansion of $F(x, t)$, provides the following relationship:

$$\Delta y = \frac{\partial F}{\partial t}\Delta t + \frac{\partial F}{\partial x}\Delta x$$

Of course, having higher-order terms in the Taylor series development yields:

$$\Delta y = \frac{\partial F}{\partial t}\Delta t + \frac{1}{2}\frac{\partial^2 F}{\partial t^2}[\Delta t]^2 + \frac{\partial F}{\partial x}\Delta x + \frac{1}{2}\frac{\partial^2 F}{\partial x^2}[\Delta x]^2 + \frac{\partial^2 F}{\partial t\,\partial x}[\Delta t\ \Delta x]$$

If the process Δx is deterministic, then obviously, terms of the order $[\Delta t]^2$, $[\Delta x]^2$ and $[\Delta t\ \Delta x]$ are negligible relative to Δt and Δx, which leads us to the previous first-order development. However, when Δx is stochastic, with variance of orderΔt, terms of the order $[\Delta x]^2$ are non-negligible (since they are also of orderΔt). As a result, the appropriate development of $F(x, t)$ leads to:

$$\Delta y = \frac{\partial F}{\partial t}\Delta t + \frac{\partial F}{\partial x}\Delta x + \frac{1}{2}\frac{\partial^2 F}{\partial x^2}[\Delta x]^2$$

This is essentially Ito's differential rule (also known as Ito's Lemma), as we shall see below for continuous time and continuous state stochastic processes.

4.4.1 Ito's Lemma

Let $y = F(x, t)$ be a continuous, twice differentiable function in x and t, or $\partial F/\partial t, \partial F/\partial x, \partial^2 F/\partial x^2$ and let $\{x(t), t \geq 0\}$ be defined in terms of a stochastic differential equation with drift $f(x, t)$ and volatility (diffusion) $\sigma(x, t)$,

$$\mathrm{d}x = f(x, t)\,\mathrm{d}t + \sigma(x, t)\,\mathrm{d}w,\ x(0) = x_0,\ 0 \leq t \leq T$$

then:

$$\mathrm{d}F = \frac{\partial F}{\partial t}\,\mathrm{d}t + \frac{\partial F}{\partial x}\,\mathrm{d}x + \frac{1}{2}\frac{\partial^2 F}{\partial x^2}(\mathrm{d}x)^2.$$

Or

$$\mathrm{d}F = \frac{\partial F}{\partial t}\,\mathrm{d}t + \frac{\partial F}{\partial x}[f(x, t)\,\mathrm{d}t + \sigma(x, t)\,\mathrm{d}w] + \frac{1}{2}\frac{\partial^2 F}{\partial x^2}[f(x, t)\,\mathrm{d}t + \sigma(x, t)\,\mathrm{d}w]^2$$

Neglecting terms of higher order than $\mathrm{d}t$, we obtain Ito's Lemma:

$$\mathrm{d}F = \left\{\frac{\partial F}{\partial t} + \frac{\partial F}{\partial x}f(x, t) + \frac{1}{2}\sigma^2(x, t)\frac{\partial^2 F}{\partial x^2}\right\}\mathrm{d}t + \frac{\partial F}{\partial x}\sigma(x, t)\,\mathrm{d}w$$

This rule is a 'work horse' of mathematical finance in continuous time. Note in particular, that when the function $F(.)$ is not linear, the volatility affects the process drift. Applications to this effect will be considered subsequently. Generalizing to multivariate processes is straightforward. For example, for a two-variable process, $y = F(x_1, x_2, t)$ where $\{x_1(t), x_2(t); t \geq 0\}$ are two stochastic processes while F admits first- and second-order partial derivatives, then the stochastic total differential yields:

$$\begin{aligned} dF = {} & \frac{\partial F}{\partial t}\,dt + \frac{\partial F}{\partial x_1}\,dx_1 + \frac{1}{2}\frac{\partial^2 F}{\partial x_1^2}(dx_1)^2 + \frac{\partial F}{\partial x_2}\,dx_2 + \frac{1}{2}\frac{\partial^2 F}{\partial x_2^2}(dx_2)^2 \\ & + \frac{\partial^2 F}{\partial x_1 x_2}(dx_1\,dx_2) \end{aligned}$$

in which case we introduce the appropriate processes $\{x_1(t), x_2(t); t \geq 0\}$ and maintain all terms of order dt. For example, define $y = x_1 x_2$, then for this case:

$$\frac{\partial F}{\partial t} = 0;\ \frac{\partial F}{\partial x_1} = x_2;\ \frac{\partial^2 F}{\partial x_1^2} = 0;\ \frac{\partial F}{\partial x_2} = x_1;\ \frac{\partial^2 F}{\partial x_2^2} = 0;\ \frac{\partial^2 F}{\partial x_1 x_2} = 1$$

which means that:

$$dF = x_2\,dx_1 + x_1\,dx_2 + dx_1\,dx_2$$

Other examples will be highlighted through application in this and subsequent chapters. Below, a number of applications in economics and finance are considered.

4.5 APPLICATIONS OF ITO'S LEMMA

4.5.1 Applications

The examples below can be read after Chapter 6, in particular the applications of the Girsanov Theorem and Girsanov and the binomial process.

(a) The Ornstein–Uhlenbeck process

The Ornstein-Uhlenbeck process is a process used in many circumstances to model mean returning processes. It is given by the following stochastic differential equation:

$$dx = -ax\,dt + \sigma\,dw(t),\ a > 0$$

We shall show first that the process has a Normal probability distribution and solve the equation by an application of Ito's Lemma. Let $y(t) = e^{at}x(t)$ and apply Ito's differential rule to lead to:

$$dy = \sigma\,e^{at}\,dw(t)$$

An integration of the above equation with substitution of y yields the solution:

$$x(t) = x(0)\,\mathrm{e}^{-at} + \sigma \int_0^t \mathrm{e}^{-a(t-\tau)}\,\mathrm{d}w(\tau)$$

The meaning of this equation is that the process $x(t)$ is an exponentially weighted function of past noise. Note that the transformed process $y(t)$ has a constant mean since its mean growth rate is null, or $E(\mathrm{d}y) = E[\sigma\,\mathrm{e}^{at}\mathrm{d}w(t)] = \sigma\,\mathrm{e}^{at}E[\mathrm{d}w(t)] = 0$. In other words, the exponential growth process $y(t) = \mathrm{e}^{at}\,x(t)$ is a constant mean process.

(b) The wealth process of a portfolio of stocks

Let x be the invested wealth of an investor at a given time t and suppose that in the time interval $(t, t + \mathrm{d}t)$, $c\,\mathrm{d}t$ is consumed while $y\,\mathrm{d}t$ is the investor's income from both investments and other sources. In this case, the rate of change in wealth is equal $x(t + \mathrm{d}t) = x + [y - c]\,\mathrm{d}t$. In order to represent this function in terms of investment assets, say that all our wealth is invested in stocks. The price of a stock, denoted by S_i, $i = 1, 2, \ldots, n$ and the number of stocks N_i held of each type i, determines wealth as well as income. If income is measured only in terms of price changes (i.e. we do not include at this time borrowing costs, dividend payments for holding shares etc.), then income $y\,\mathrm{d}t$ in $\mathrm{d}t$ is necessarily:

$$y\,\mathrm{d}t = \sum_{i=1}^{n} N_i\,\mathrm{d}S_i$$

and therefore the investor's worth is:

$$\mathrm{d}x = \sum_{i=1}^{n} N_i\,\mathrm{d}S_i - c\,\mathrm{d}t$$

For example, assume that prices are lognormal, given by:

$$\frac{\mathrm{d}\,S_i}{S_i} = \alpha_i\,\mathrm{d}t + \sigma_i\,\mathrm{d}w_i,\ S_i(0) = S_{0,i} \quad \text{given}, i = 1, 2, \ldots, n$$

where $w_i(t)$ are standard Wiener processes (that may be independent or not, in which case they are assumed to be correlated). Inserting into the wealth process equation, we obtain:

$$\mathrm{d}x = \sum_{i=1}^{n} N_i\,[\alpha_i S_i\,\mathrm{d}t + \sigma_i S_i\,\mathrm{d}w_i] - c\,\mathrm{d}t = \sum_{i=1}^{n} [\alpha_i N_i S_i - c]\,\mathrm{d}t + \sum_{i=1}^{n} [\sigma_i N_i S_i\,\mathrm{d}w_i]$$

This is of course, a linear stochastic differential equation. Simplifications to this equation can be reached, allowing a much simpler treatment. For example, say that a proportion θ_i of the investor wealth is invested in stock i. In other words,

$$N_i S_i = \theta_i x \text{ or } N_i = \theta_i x / S_i \quad \text{with} \sum_{i=1}^{n} \theta_i = 1$$

Further, say that the Wiener processes are uncorrelated, in which case:

$$\sum_{i=1}^{n} [\sigma_i N_i S_i \, \mathrm{d}w_i] = \sum_{i=1}^{n} [\sigma_i \theta_i x \, \mathrm{d}w_i] = x \sqrt{\sum_{i=1}^{n} [\sigma_i^2 \theta_i^2]} \, \mathrm{d}w$$

which leads to:

$$\mathrm{d}x = \left[\sum_{i=1}^{n} \alpha_i \theta_i x - c \right] \mathrm{d}t + x \sqrt{\sum_{i=1}^{n} [\sigma_i^2 \theta_i^2]} \, \mathrm{d}w, \sum_{i=1}^{n} \theta_i = 1$$

Thus, by selecting trading and consumption strategies represented by $\theta_i, i = 1, 2, \ldots, n$ we will, in fact, also determine the evolution of the (portfolio) wealth process.

4.5.2 Time discretization of continuous-time finance models

When the underlying model is given in a continuous time and in a continuous state framework, it is often useful to use discrete models as an approximation. There are a number of approaches to doing so. Discretization can be reached by discretizing the state space, the time or both. Assume that we are given a stochastic differential equation (SDE). A time (process) discretization might lead to a simple stochastic difference equation or to a stochastic difference equation subject to multiple sources of risk, as we shall see below. A state discretization means that the underlying process is represented by discrete state probability models (such as a binomial random walk, a trinomial walk, or Markov chains and their like). These approaches will be considered below.

(a) Probability approximation (discretizing the states)

In computational finance, numerical techniques are sought that make it possible also to apply risk-neutral pricing, in other words approximate the process by using binomial models or other models with desirable mathematical characteristics that allow the application of fundamental finance theories. Approximations by binomial trees are particularly important since many results in fundamental finance are proved and explained using the binomial model. It would therefore be useful to define a sequence of binomial processes that converge weakly (at least) to diffusion–stochastic differential equation models. Nelson and Ramaswamy (1990) have suggested such approximations, which are in fact similar to a drift–volatility approximation. Let x be the current price of a stock and let its next price (in the discretized model with time intervals h) be either $X^+(x, t)$ or $X^-(x, t)$.

We denote by $P(x, t)$, the probability of transition to state $X^+(x, t)$. Or

$$X_t = x;\ X_{t+1} = \begin{cases} [X^+(x,t)] & \text{with Pr } P(x,t) \\ [X^-(x,t)] & \text{with Pr } 1 - P(x,t) \end{cases}$$

Thus,

$$X_{t+1} - X_t = \Delta x = \begin{cases} [X^+(x,t) - x] & \text{with Pr } P(x,t) \\ [X^-(x,t) - x] & \text{with Pr } 1 - P(x,t) \end{cases}$$

For a financial process defined by a stochastic differential equation with drift $\mu(x, t)$ and volatility $\sigma(x, t)$ we then have:

$$\begin{aligned} \mu(x,t)h &= P(x,t)[X^+(x,t) - x] + [1 - P(x,t)][X^-(x,t) - x] \\ \sigma^2(x,t)h &= P(x,t)[X^+(x,t) - x]^2 + [1 - P(x,t)][X^-(x,t) - x]^2 \end{aligned}$$

This is a system of two equations in the three values $P(x, t)$, $[X^-(x, t)]$ and $[X^+(x, t)]$ that the discretized scheme requires. These values can be defined in several ways. Explicitly, we can write:

$$\begin{aligned} &X^+(x,t) \equiv x + \sigma\sqrt{h};\ X^-(x,t) \equiv x - \sigma\sqrt{h}; \\ &P(x,t) = 1/2 + \sqrt{h}\mu(x,t)/2\sigma(x,t) \end{aligned}$$

And, therefore, an approximate binomial tree can be written as follows:

$$\Delta x = \begin{cases} \sigma\sqrt{h} & \text{with Pr } [1/2 + \sqrt{h}\mu(x,t)/2\sigma(x,t)] \\ -\sigma\sqrt{h} & \text{with Pr } [1/2 - \sqrt{h}\mu(x,t)/2\sigma(x,t)] \end{cases}$$

For example, say that $Hx = X^+(x, t)$ and $Dx = X^-(x, t)$ as well as $p = P(x, t)$. Then:

$$X_{t+1} = \begin{cases} Hx & \text{with Pr } p \\ Lx & \text{with Pr } 1 - p \end{cases};\ X_t = x$$

Let i be the number of times the price (process) increases over a period of time T, then, the price distribution at time T is:

$$\frac{X_T}{x} = (H)^i (L)^{T-i} \text{ w.p. } P_i = \binom{T}{i} p^i (1-p)^{T-i}\ i = 0, 1, 2, 3, \ldots, T$$

Consider now the mean reverting (Ornstein–Uhlenbeck) process often used to model interest rate and volatility processes:

$$\mathrm{d}x = \beta(\alpha - x)\,\mathrm{d}t + \sigma\,\mathrm{d}w,\ x(0) = x_0 > 0,\ \beta > 0, x \in [0, 1]$$

Then, we define the following binomial process as an approximation:

$$\begin{aligned} &X^+(x,t) \equiv x + \sigma\sqrt{h};\ X^-(x,t) \equiv x - \sigma\sqrt{h}; \\ &P(x,t) = 1/2 + \sqrt{h}\mu(x,t)/2\sigma(x,t) \end{aligned}$$

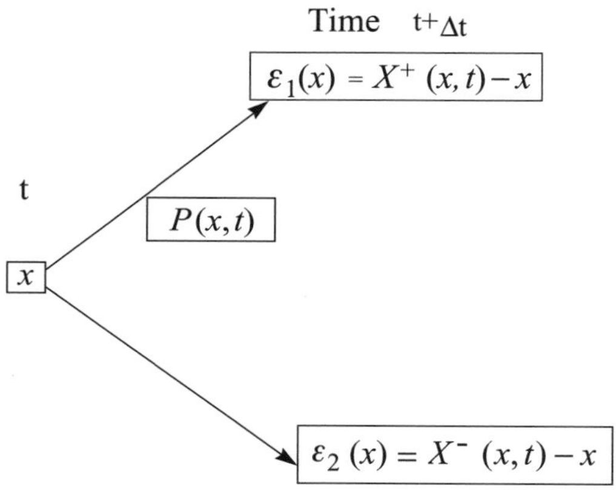

Figure 4.2 A binomial tree.

with the explicit transition probability given by:

$$P(x,t) = \begin{cases} 1/2 + \sqrt{h}\beta(\alpha - x)/2\sigma & \text{if} \quad 0 \le 1/2 + \sqrt{h}\beta(\alpha - x)/2\sigma \le 1 \\ 0 & \text{if} \quad 1/2 + \sqrt{h}\beta(\alpha - x)/2\sigma < 0 \\ 1 & \text{otherwise} \end{cases}$$

The probability $P(x, t)$ is chosen to match the drift, it is censored if it falls outside the boundaries [0, 1]. As a result, the basic building block of the binomial process will be given as shown in Figure 4.2. In order to construct a simple (plain vanilla) binomial process it is essential that the volatility be constant, however. Otherwise the process will exhibit conditional heteroscedasticity. In the example above, this was the case and therefore we were able to define the states and the transition probability simply. When this is not the case, we can apply Ito's differential rule and find the proper transformation that will 'purge' this heteroscedasticity. Namely, we consider the transformation $y(x, t)$ to which we apply Ito's Lemma:

$$\mathrm{d}y = \left\{ \frac{\partial y}{\partial t} + \frac{\partial y}{\partial x} f(x,t) + \frac{1}{2}\sigma^2 \frac{\partial^2 y}{\partial x^2} \right\} \mathrm{d}t + \frac{\partial y}{\partial x}\sigma(x,t)\,\mathrm{d}w$$

and choose:

$$y(x,t) = \int^{x} \frac{\mathrm{d}z}{\sigma(z,t)}$$

in which case, the term

$$\left(\sigma(x,t) \frac{\partial y(x,t)}{\partial x} \right) \mathrm{d}w$$

is replaced by $\mathrm{d}w$ and the instantaneous volatility of the transformed process is constant. This allows us to obtain a computationally simple binomial tree. To see how to apply this technique, we consider another example. Consider the CEV stock price:

$$dx = \mu x \, dt + \sigma x^{\gamma} \, dw; \; 0 < \gamma < 1 \quad \text{and} \quad y(x,t) \equiv \sigma^{-1} \int^{x} z^{-\gamma} \, dz = \frac{x^{1-\gamma}}{\sigma(1-\gamma)}$$

which has the effect of reducing the stochastic differential equation to a constant volatility process, thus making it possible to transform it into a simple binomial process with an inverse transform given by:

$$x(y,t) = \begin{cases} [\sigma(1-\gamma)x]^{1/(1-\gamma)} & \text{if } x > 0 \\ 0 & \text{otherwise} \end{cases}$$

(b) The Donsker Theorem

A justification of this approximation based on binomial trees can be made using the Donsker Theorem which is presented intuitively below. Given that the process we wish to represent is a trinomial random walk (or a simple random walk), we represent for convenience the transition probabilities as follows:

$$p_n = \frac{1-r}{2} + \frac{\alpha}{2n}; \; q_n = \frac{1-r}{2} - \frac{\alpha}{2n}$$

with $\alpha \in R$, $0 \leq r < 1$ real numbers and n is a parameter representing the number of segments n^2 used in dividing a given time interval, assumed to be large. Note that these (Markov) transition probabilities satisfy: $p_n \geq 0$, $q_n \geq 0$, $r + p_n + q_n = 1$. For each partitioning of the process, we associate the random walk $(X_t^{(n)}; t \in T)$ which is a piecewise linear approximation of the stochastic process where the time interval $[0, t]$ is divided into equal segments each of width $1/n^2$. At the kth segment, we have the following states:

$$X_{k/n^2}^{(n)} = \frac{1}{\sqrt{n^2}} X_k = X^{(k)}([n^2 t])$$

$$X_t^{(n)} = X_{k/n^2}^{(n)} + (n^2 t - [n^2 t])\big(X_{(k+1)/n^2}^{(n)} - X_{k/n^2}^{(n)}\big) \forall t \in \left[\frac{k+1}{n^2}, \frac{k}{n^2}\right]$$

Note that the larger n, the greater the number of states and thereby, the more refined the approximation. The corresponding continuous (Brownian motion) process (a function of n) is then assumed given by:

$$B^n(t) = \frac{1}{n}\big(X^{(n)}([n^2 t]) + (n^2 t - [n^2 t])\varepsilon_{([n^2 t+1])}\big); \; t \geq 0$$

[..] denotes the integer value of its argument and $\varepsilon_{(.)}$ is the random walk with transition probabilities (p_n, q_n). Note that by writing the continuous process in this form, we essentially divide the time scale in widths of $1/n^2$. Of course, when n is large, then time will become approximately continuous. The Donsker Theorem essential statement is that when n is large, the approximating random walk converge to a Brownian motion. Note that we can easily deduce that for n

large, the following moments:

$$E\left[\frac{1}{n}X^{(n)}([n^2t])\right] = \frac{\alpha}{n^2}[n^2t];$$

$$\text{var}\left\{\frac{1}{n}X^{(n)}([n^2t])\right\} = \frac{[n^2t]}{n^2}\left(r(1-r)+(1-r)^2-\frac{\alpha^2}{n^2}\right)$$

Consequently, for a fixed time, at the limit ($[n^2t] \underset{t\to\infty}{\approx} n^2t,\ t$ fixed):

$$\operatorname*{Lim}_{n\to+\infty} E\left[\frac{1}{n}X^{(n)}([n^2t])\right] = \alpha t$$

$$\operatorname*{Lim}_{n\to\infty} \text{Var}\left\{\frac{1}{n}X^{(n)}([n^2t])\right\} = t(r(1-r)+(1-r)^2) = (1-r)t$$

Furthermore, the last term in $B^{(n)}(t)$ disappears since $|(n^2t-[n^2t])\varepsilon_{([n^2t])+1}| \le 1$.

These allow the application of the Donsker Theorem stating that the process $B^{(n)}$ converges in distribution to $(\sqrt{1-r}B_t+\alpha t;\ t\ge 0)$ where $(B_t;\ t\ge 0)$ designates a real and standardized Brownian motion starting at 0. This result, which is derived formally, justifies the previous continuous approximation for the random walk. Note that when r (the probability of remaining in the same state) is large, then we obtain a process with drift α and zero variance. However, when n is small (close to zero or equal to zero) we obtain a Brownian motion with drift. Finally, we can proceed in a similar manner and calculate the empirical variance process of a trinomial random walk, providing therefore a volatility approximation as well. This is left as an exercise.

(c) Discrete time approximations

In some cases, we approximate an underlying price stochastic differential equation by a stochastic difference equation. There are a number of ways to do so, however. As a result, a unique (price) valuation process based on a continuous-time model might no longer be unique in its 'difference' form. This will be seen below by using an approximation due to Milshtein (1974) which uses principles of stochastic integration in the relevant approximate discretized time interval. Thus, unlike application of the Donsker Theorem which justified the partitioning of the price (state) process, we construct below a discrete time process and apply Ito's differential rule within each time interval in order to estimate the evolution of prices (states) within each interval. The approximation within a time interval is achieved by a Taylor series approximation which can be linear, or of higher order. Integration using Ito's calculus provides then an estimate of the states in the discretized scheme. As we shall see below, this has the effect of introducing a process uncertainty which is not normal, thereby leading to stochastic volatility. The higher the order approximation the larger the number of uncertainty sources. To see how this is defined, we consider first the following Ito differential equation:

$$\mathrm{d}x = f(x,t)\,\mathrm{d}t + \sigma(x,t)\,\mathrm{d}w(s)$$

Consider next two subsequent instants of time t and $r, r > t$, then a discretized process in the interval $r - t, r > t$ can be written as follows:

$$x_r = x_t + \alpha(r, x_t, r - t) + \beta(r, x_t, r - t, w_r - w_t)$$

where $\alpha(r, x_t, r - t)$, $\beta(r, x_t, r - t, w_r - w_t)$ are a drift, a function of the beginning state and the end state as well as as a function of time, the time interval and the volatility which is, in addition, a function of the uncertainty in the time interval. Note the volatility function need not be a linear function in $(w_r - w_t)$. In a stochastic integral form, the evolution of states is:

$$x_t = x_s + \int_s^t f(r, x_r)\, \mathrm{d}r + \int_s^t \sigma(x_r, r)\, \mathrm{d}w_r$$

If we use in discrete time only a first-order (linear) approximation to the drift and the volatility, we have:

$$x_r \cong x_t + f(s, x_t)\, \Delta t + \sigma(t, x_t)\, \Delta w_t,$$
$$\Delta t = (r - t);\ \Delta w_t = (w_r - w_t),\ \alpha(.) = f(s, x_t)\, \Delta t;\ \beta(.) = \sigma(s, x_s)\, \Delta w_t$$

When we take a second-order approximation, an additional source of uncertainty is added. To see how this occurs, consider a more refined approximation based on a Taylor series expansion of the first two terms for the functions (f, σ). Then in the time interval $(t, s;\ t > s)$:

$$x_t = x_s +$$
$$+ \int_s^t \left[f(s, x_s) + \frac{\partial f(s, x_s)}{\partial t}(r - s) + \frac{\partial f(s, x_s)}{\partial x}(f(s, x_s)(r - s)\right.$$
$$\left. + \sigma(s, x_s)(w_r - w_s))\right]\mathrm{d}r + \int_s^t \left[\sigma(s, x_s) + \frac{\partial \sigma(s, x_s)}{\partial t}(r - s) + \frac{\partial \sigma(s, x_s)}{\partial x}\right.$$
$$\left.(f(s, x_s)(r - s) + \sigma(s, x_s)(w_r - w_s))\right]\mathrm{d}r$$

Since the stochastic integral is given by:

$$x_t = x_s + \int_s^t [w_r - w_s]\ \mathrm{d}w_r = \frac{1}{2}[(w_t - w_s)^2 - (t - s)]$$

Including terms of order $(t - s)$ only, we obtain:

$$x_t = x_s + f(s, x_s)(t - s) + \sigma(s, x_s)(w_t - w_s) + \frac{1}{2}\frac{\partial \sigma(s, x_s)}{\partial x}$$
$$[(w_t - w_s)^2 - (t - s)]$$

or:

$$x_t = x_s + f(s, x_s)\,\Delta t + \sigma(s, x_s)\,\Delta w_s + \frac{1}{2}\frac{\partial\sigma(s, x_s)}{\partial x}[(\Delta w_s)^2 - \Delta t]$$

Note that in this case, there are two sources of uncertainty. First we have a normal term $(w_t - w_s)$ of mean zero and variance $t - s$, while we also have a chi-square term defined by $(w_t - w_s)^2$. Their sum produces, of course, a nonlinear (stochastic) model.

An improvement of this approximation can further be reached if we take a higher-order Taylor series approximation. Although this is cumbersome, we summarize the final result here that uses the following stochastic integral relations (which are given without their development):

$$\int_s^t (r - s)\,\mathrm{d}r = \int_0^{t-s} u\,\mathrm{d}u = \frac{1}{2}(t - s)^2 = \frac{1}{2}(\Delta t)^2$$

$$\int_s^t (r - s)^2\,\mathrm{d}r = \int_0^{t-s} u^2\,\mathrm{d}u = \frac{(t - s)^3}{3} = \frac{(\Delta t)^3}{3}$$

$$\int_s^t (r - s)(w_r - w_s)\,\mathrm{d}r = \int_0^{t-s} \tau(w_r - w_s)\,\mathrm{d}\tau = \int_0^{t-s} \tau \int_0^{v} \mathrm{d}w(v)\,\mathrm{d}\tau$$

$$= \int_0^{t-s} \mathrm{d}w(v) \int_v^{t-s} \tau\,\mathrm{d}\tau = \frac{1}{2}\int_0^{t-s} (\Delta t^2 - v^2)\,\mathrm{d}w(v)$$

which has Normal probability distribution with mean zero and variance:

$$\mathrm{var}(x(t)) = \frac{1}{4}\int_0^{t-s} [\Delta t^2 - u^2]^2\,\mathrm{d}u = \frac{1}{4}\int_0^{t-s} [\Delta t^4 + u^4 - 2\Delta t^2 u^2]\,\mathrm{d}u = \frac{\Delta t^5}{6}$$

$$\int_s^t w(\tau)\,\mathrm{d}w(\tau) = \frac{1}{2}(w(t)^2 - w(s)^2) - \frac{1}{2}(t - s)$$

$$\int_s^t w(\tau)^2\,\mathrm{d}w(\tau) = \frac{1}{3}(w(t)^3 - w(s)^3) - \int_s^t w(\tau)\,\mathrm{d}\tau$$

$$= \frac{1}{3}(w(t)^3 - w(s)^3) - \frac{1}{2}\,(w(t)^2 - w(s)^2) + \frac{1}{2}(t - s)$$

And therefore we have at last:

$$
\begin{aligned}
x_t = {} & x_s + [f(s, x_s) + \sigma(s, x_s)]\,\Delta t + \\
& + \frac{1}{2}\left[\begin{array}{l} \dfrac{\partial f(s, x_s)}{\partial t} + \dfrac{\partial \sigma(s, x_s)}{\partial t} + \\ \dfrac{\partial f(s, x_s)}{\partial x} f(s, x_s) + \dfrac{\partial \sigma(s, x_s)}{\partial x} f(s, x_s) \end{array}\right] (\Delta t)^2 \\
& + \frac{1}{2.3}\left[\frac{\partial^2 f(s, x_s)}{\partial t^2} + \frac{\partial^2 \sigma(s, x_s)}{\partial t^2} + \frac{\partial^2 \sigma(s, x_s)}{\partial x^2} f(s, x_s)\right] (\Delta t)^3 \\
& + \frac{1}{2}\left[\frac{\partial f(s, x_s)}{\partial x} + \frac{\partial \sigma(s, x_s)}{\partial x}\right] \sigma(s, x_s)[(\Delta w_s)^2 - \Delta t] \\
& + \frac{1}{2.3}\left[\begin{array}{l} \dfrac{\partial^2 f(s, x_s)}{\partial x^2} \sigma^2(s, x_s) + \\ \dfrac{\partial^2 \sigma(s, x_s)}{\partial x^2} \sigma^2(s, x_s) \end{array}\right] \left\{\left(w_t^3 - w_s^3\right) - 3\left(w_t^2 - w_s^2\right) + 3\Delta t\right\} \\
& + \frac{\partial^2 \sigma(s, x_s)}{\partial x^2} f(s, x_s) \sigma(s, x_s) \frac{(\Delta t)^{5/2}}{\sqrt{6}} \Delta w_s
\end{aligned}
$$

This scheme allows the numerical approximation of the stochastic differential equation and clearly involves multiple sources of risk.

For the lognormal price process:

$$
\mathrm{d}x = \alpha x\,\mathrm{d}t + \beta x\,\mathrm{d}w
$$

The transformation of the Ito stochastic differential equation becomes:

$$
\mathrm{d}(\log x) = \left[\alpha - \frac{\beta^2}{2}\right] \mathrm{d}t + \beta\,\mathrm{d}w_s
$$

And therefore, a finite differencing, based on integration over a unit interval yields:

$$
\log x_t - \log x_{t-1} = \left[\alpha - \frac{\beta^2}{2}\right] + \beta \varepsilon_t;\ \varepsilon_t \equiv (W_t - W_{t-1})
$$

which can be used now to estimate the model parameters using standard statistical techniques. Interestingly, if we consider other intervals, such as smaller length intervals (as would be expected in intraday data), then the finite difference model would be instead:

$$
\Delta_\tau = \log x_t - \log x_{t-\tau} = \left[\alpha - \frac{\beta^2}{2}\right] + \beta\sqrt{\tau}\varepsilon_t^\tau;\ \varepsilon_t^\tau \sim N(0, 1)
$$

Of course, the estimators of the model parameters will be affected by this discretization. Higher-order discretization can be used as well to derive more precise results, albeit these results might not allow the application of standard fundamental finance results.

4.5.3 The Girsanov Theorem and martingales*

The Girsanov Theorem is important to many applications in finance. It defines a 'discounting process' which transforms a given price process into a martingale. A martingale essentially means, as we saw it earlier, that any trade or transaction will be 'fair' in the sense that the expected value of any such transaction is null. And therefore its 'transformed price' remains the same. In particular, define the transformation:

$$L = \exp\left\{\int_0^t \sigma(s)\,\mathrm{d}W(s) - \frac{1}{2}\int_0^t \sigma^2(s)\,\mathrm{d}s\right\}$$

where $\sigma(s),\ 0 \le s \le T$ is bounded and is the unique solution of:

$$\frac{\mathrm{d}L}{L} = \sigma\,\mathrm{d}W;\ L(0) = 1;\ E(L) = 1,\ \forall t \in [0, T]$$

and $L(t)$ is a martingale. The proof can be found, by an application of Ito's differential rule with $y = \log L$, or $\mathrm{d}y = \mathrm{d}L/L - (1/2L^2)(\mathrm{d}L)^2 = \sigma\,\mathrm{d}W - (1/2)\sigma^2\,\mathrm{d}t$ whose integration leads directly to:

$$y - y_0 = \ln(L) - \ln(L_0) = \int_0^t \sigma(s)\,\mathrm{d}W(s) - \frac{1}{2}\int_0^t \sigma^2(s)\,\mathrm{d}s$$

And therefore,

$$L(0) = L(t)\exp\left\{-\int_0^t \sigma(s)\,\mathrm{d}W(s) + \frac{1}{2}\int_0^t \sigma^2(s)\,\mathrm{d}s\right\},\quad L(0) = 1$$

In this case, note that y is a process with drift while that of L has no drift and thus L is a martingale. For example, for the lognormal stock price process:

$$\frac{\mathrm{d}S}{S} = \alpha\,\mathrm{d}t + \sigma\,\mathrm{d}W,\ S(0) = S_0$$

Its solution at time t is simply:

$$S(t) = S(0)\,\mathrm{e}^{\alpha t}\exp\left[-\frac{\sigma^2}{2}t + \sigma W(t)\right],\quad W(t) = \int_0^t \mathrm{d}W(t)$$

which we can write in terms of $L(t)$ as follows:

$$S(t) = S(0)\mathrm{e}^{\alpha t}\frac{L(t)}{L(0)}\ \text{or}\ \frac{S(t)}{L(t)} = \mathrm{e}^{\alpha t}\frac{S(0)}{L(0)}\ \text{or}\ \frac{S(0)}{L(0)} = \mathrm{e}^{-\alpha t}\frac{S(t)}{L(t)}$$

Note that the term $S(t)/L(t)$ is no longer stochastic and therefore no expectation is taken (alternatively, a financial manager would claim that, since it is deterministic, its future value ought to be at the risk-free rate). By the same token, consider both a stock price and a risk-free bond process given by:

$$\frac{\mathrm{d}S}{S} = \alpha\,\mathrm{d}t + \sigma\,\mathrm{d}W,\ S(0) = S_0;\ \frac{\mathrm{d}B}{B} = R_f\,\mathrm{d}t,\ B(0) = B_0$$

If the bond is defined as the numeraire, then we seek a transformation such that $S(t)/B(t)$ is a martingale. For this to be the case, we will see that the stock price is then given by:

$$\frac{\mathrm{d}S}{S} = R_f\,\mathrm{d}t + \sigma\,\mathrm{d}W^*,\ S(0) = S_0$$

where $\mathrm{d}W^*$ is the martingale measure which is explicitly given by:

$$\mathrm{d}W^* = \mathrm{d}W + \frac{\alpha - R_f}{\sigma}\,\mathrm{d}t$$

In this case, under such a transformation, we can apply the risk-neutral pricing formula:

$$S(0) = \mathrm{e}^{-R_f t} E^*\left(S(t)\right)$$

where E^* denotes expectation with respect to the martingale measure.

Problem

Let $y = x_1/x_2$ where $\{x_1(t), x_2(t); t \geq 0\}$ are two stochastic processes, each with known drift and known diffusion, then show that the stochastic total differential yields:

$$\mathrm{d}y = \frac{1}{x_2}\,\mathrm{d}x_1 - \frac{x_1}{x_2^2}\,\mathrm{d}x_2 + \frac{x_1}{x_2^3}(\mathrm{d}x_2)^2 - \frac{1}{x_2^2}(\mathrm{d}x_1\,\mathrm{d}x_2)$$

Further, show that if we use the stock price and the bond process, it is reduced to:

$$\mathrm{d}y = \frac{\mathrm{d}S}{B(t)} - \frac{S(t)}{B^2(t)}\,\mathrm{d}B + \frac{S(t)}{B^3(t)}(\mathrm{d}B)^2 - \frac{\mathrm{d}S\,\mathrm{d}B}{B^2(t)} = \frac{\mathrm{d}S}{B(t)} - \frac{S(t)}{B^2(t)}\,\mathrm{d}B$$

And as a result,

$$\frac{\mathrm{d}y}{y} = \left(\alpha - R_f\right)\,\mathrm{d}t + \sigma\,\mathrm{d}W$$

Finally, replace $\mathrm{d}W$ by

$$-\frac{\alpha - R_f}{\sigma}\,\mathrm{d}t + \mathrm{d}W^* = \mathrm{d}W \text{ and obtain: } \frac{\mathrm{d}y}{y} = \sigma\,\mathrm{d}W^*$$

which is of course a martingale under the transformed measure.

Martingale examples*

(1). Let the stock price process be defined in terms of a Bernoulli event where stock prices grow from period to period at rates $a > 1$ and $b < 1$ with probabilities defined below:

$$S_t = \begin{cases} aS_{t-1} & \text{w.p. } \dfrac{1-b}{a-b} \\ bS_{t-1} & \text{w.p. } \dfrac{a-1}{a-b} \end{cases}$$

This process is a martingale. First it can be summed and, further, we have to show that this a constant mean process with:

$$E\left(S_{t+1} \,|S_t, S_{t-1}, \ldots, S_0\right) = E\left(S_{t+1} \,|S_t\right) = S_t$$

Explicitly, we have:

$$E\left(S_{t+1} \,|S_t\right) = aS_t \frac{1-b}{a-b} + bS_t \frac{a-1}{a-b} = S_t \frac{b(a-1) + a(1-b)}{a-b} = S_t$$

By the same token, we can show that there are many other processes that have the martingale property. Consider the trinomial (birth–death) random walk defined by:

$$\begin{cases} S_{t+1} = S_t + \varepsilon_t, \ S_0 = 0 \\ \varepsilon_t = \begin{cases} +1 \text{ w.p. } p \\ 0 \text{ w.p. } r \\ -1 \text{ w.p. } q \end{cases} & ; p \geq 0, q \geq 0, r \geq 0, p+q+r = 1 \end{cases}$$

Then it is easy to show that $[S_t - t(p-q); t \geq 0]$ is a martingale. To verify this assertion, note that:

$$\begin{aligned} &E\left[S_{t+1} - (t+1)(p-q)/\Phi_t\right] = E\left[S_t + \varepsilon_t - (t+1)(p-q)/\Phi_t\right] \\ &= S_t - (t+1)(p-q) + E(\varepsilon_t) = S_t - (t+1)(p-q) + (p-q) \\ &= S_t - t(p-q) \end{aligned}$$

where $\Phi_t \equiv (S_0, S_1, S_2, \ldots, S_t)$ resumes the information set available at time t. Further:

$$\left[S_t^2 - t(p-q); t \geq 0\right], \left\{\left[S_t^2 - t(p-q)\right]^2 + t[(p-q)^2 - (p+q)]; t \geq 0\right\}$$

and $\{\lambda^{x_t}, \lambda = q/p, t \geq 0\}$, $p > 0$ are also martingales. The proof is straightforward and a few cases are treated below. By the same token, we can also consider processes that are not martingales and then find a transformation or another process that will render the original process a martingale.

*(2) The Wiener process is a martingale**. The Wiener process $\{w(t), t \geq 0\}$ is a Markov process and as we shall see below a martingale. Let $\mathbf{F}(t)$ be its filtration, in other words, it defines the information set available at time t on which a conditional expectation is calculated (and on the basis of which financial calculations are

assumed to be made). Then, we can state that the Wiener process is a martingale with respect to its filtration. The proof is straightforward since

$$E\{w(t+\Delta t)\,|\mathbf{F}(t)\} = E\{w(t+\Delta t) - w(t)\,|\mathbf{F}(t)\} + E\{w(t)\,|\mathbf{F}(t)\}$$

The first term is null while the independent conditional increments of the Wiener process imply:

$$E\{w(t+\Delta t) - w(t)\,|\mathbf{F}(t)\} = E\{w(t+\Delta t) - w(t)\} = 0$$

since $\{w(t+\Delta t) - w(t)\}$ is independent of $w(s)$ for $s \le t$. Thus by the law of conditional probabilities which are independent, we note that the Wiener process is a martingale.

(3) The process $x(t) = w(t)^2 - t$ is a martingale. This assertion can be proved by showing that $E[x(t+s)\,|\mathbf{F}(t)] = x(t)$. By definition we have:

$$\begin{aligned}
&E[w(t+s)\,|\mathbf{F}(t)] = E[w(t+s)^2\,|\mathbf{F}(t)] - (t+s)\\
&\quad = E[\{w(t+s) - w(t)\}^2 + 2w(t+s)w(t) - w(t)^2\,|\mathbf{F}(t)] - (t+s)\\
&\quad = E[\{w(t+s) - w(t)\}^2\,|\mathbf{F}(t)] + 2E[w(t+s)\,|\mathbf{F}(t)] - E[w(t)^2\,|\mathbf{F}(t)] - (t+s)
\end{aligned}$$

Independence of the non-overlapping increments implies that $w(t+s) - w(t)$ is independent of $\mathbf{F}(t)$ which makes it possible to write:

$$E[\{w(t+s) - w(t)\}^2\,|\mathbf{F}(t)] = E[\{w(t+s) - w(t)\}^2] = s$$

Conditional expectation and using the fact that $w(t)$ is a martingale with respect to its filtration imply:

$$E[\{w(t+s)w(t)\}\,|\mathbf{F}(t)] = w(t)^2$$

and since $E[w(t)^2\,|\mathbf{F}(t)] = w(t)^2$. We note therefore that:

$$E[x(t+s)\,|\mathbf{F}(t)] = s + 2w(t)^2 - w(t)^2 - (t+s) = w(t)^2 - t = x(t)$$

which proves that the process is a martingale.

(4) The process

$$x(t) = \exp\left(\alpha w(t) - \frac{\alpha^2 t}{2}\right)$$

where α is any real number is a martingale. The proof follows the procedure above. We have to show that $E[x(t+s)\,|\mathbf{F}(t)] = x(t)$, or

$$\begin{aligned}
E[x(t+s)\,|\mathbf{F}(t)] &= E\left[\exp\left(\alpha w(t+s) - \frac{\alpha^2(t+s)}{2}\right)|\mathbf{F}(t)\right]\\
&= E\left[x(t)\exp\left(\alpha[w(t+s) - w(t)] - \frac{\alpha^2 s}{2}\right)|\mathbf{F}(t)\right]
\end{aligned}$$

Independence and conditional expectations make it possible to write:

$$E\left[x(t+s)\,|\mathbf{F}(t)\right] = x(t)\mathrm{E}\left[x(t)\exp\left(\alpha[w(t+s)-w(t)]-\frac{\alpha^2 s}{2}\right)|\mathbf{F}(t)\right]$$
$$= x(t)\exp\left(-\frac{\alpha^2 s}{2}\right)\mathrm{E}[\exp\{\alpha[w(t+s)-w(t)]\}]$$

The term $\alpha\,[w(t+s)-w(t)]$ has a Normal probability distribution with zero mean and variance $\alpha^2 s$. Consequently, the term $\exp\{\alpha\,[w(t+s)-w(t)]\}$ has a lognormal probability distribution with expectation $\exp\left(\alpha^2 s/2\right)$ which leads to $E\left[x(t+s)\,|\mathbf{F}(t)\right] = x(t)$ and proves that the process is a martingale.

REFERENCES AND FURTHER READING

Arnold, L. (1974) *Stochastic Differential Equations*, John Wiley & Sons, Inc., New York.

Bachelier, L. (1900) *Théorie de la spéculation, thèse de mathématique*, Paris.

Barrois, T. (1834) Essai sur l'application du calcul des probabilités aux assurances contre l'incendie, *Mem. Soc. Sci. de Lille*, 85–282.

Bensoussan, A. (1982) *Stochastic Control by Functional Analysis Method*, North Holland, Amsterdam.

Bernstein, P.L. (1998) *Against the Gods*, John Wiley & Sons, Inc., New York.

Bibby, Martin, and M. Sorenson (1997) A hyperbolic diffusion model for stock prices, *Finance and Stochastics*, **1**, 25–41.

Bismut, J.M. (1976) Théorie Probabiliste du Controle des Diffusions, *Memoirs of the American Mathematical Society*, **4**, no. 167.

Born, M. (1954) Nobel Lecture, published in *Les Prix Nobel*, Nobel Foundation, Stockholm.

Brock, W.A., and P.J. de Lima (1996) Nonlinear time series, complexity theory and finance, in G. Maddala and C. Rao (Eds), *Handbook of Statistics*, Vol. 14, *Statistical Methods in Finance*, North Holland, Amsterdam.

Brock, W.A., D.A. Hsieh and D. LeBaron (1991) *Nonlinear Dynamics, Chaos and Instability: Statistical Theory and Economic Evidence*, MIT Press, Boston, MA.

Cinlar, E. (1975) *Introduction to Stochastic Processes*, Prentice Hall, Englewood Cliffs, NJ.

Cox, D.R., and H.D. Miller (1965) *The Theory of Stochastic Processes*, Chapman & Hall, London.

Cramer, H. (1955) *Collective Risk Theory*, Jubilee Volume, Skandia Insurance Company.

Doob, J.L. (1953) *Stochastic Processes*, John Wiley & Sons, Inc., New York.

Elliot, R.J. (1982) *Stochastic Calculus and Applications*, Springer Verlag, Berlin.

Feller, W. (1957) *An Introduction to Probability Theory and its Applications*, Vols. I and II, John Wiley & Sons, Inc., New York (second edition in 1966).

Gardiner, C.W. (1990) *Handbook of Stochastic Methods*, (2nd edn), Springer Verlag, Berlin.

Gerber, H.U. (1979) *An Introduction to Mathematical Risk Theory*, Monograph no. 8, Huebner Foundation, University of Pennsylvania, Philadelphia, PA.

Geske, R. and K. Shastri (1985) Valuation by approximation: A comparison of alternative option valuation techniques, *Journal of Financial and Quantitative Analysis*, **20**, 45–71.

Ghahgshaie S., W. Breymann, J. Peinke, P. Talkner and Y. Dodge (1996) Turbulent cascades in foreign exchange markets, *Nature*, **381**, 767.

Gihman, I.I., and A.V. Skorohod (1970) *Stochastic Differential Equations*, Springer Verlag, New York.

Harrison, J.M., and D.M. Kreps (1979) Martingales and arbitrage in multiperiod security markets, *Journal of Economic Theory*, **20**, 381–408.

Harrison, J.M., and S.R. Pliska (1981) Martingales and stochastic integrals with theory of continuous trading, *Stochastic Processes and Applications*, **11**, 261–271.

Iglehart, D.L. (1969) Diffusion approximations in collective risk theory, *Journal of Applied Probability*, **6**, 285–292.

Ito, K. (1961) *Lectures on Stochastic Processes,* Lecture Notes, Tata Institute of Fundamental Research, Bombay, India.

Ito, K., and H.P. McKean (1967) *Diffusion Processes and their Sample Paths*, Academic Press, New York.

Judd, K. (1998) *Numerical Methods in Economics*, MIT Press, Cambridge, MA.

Kalman, R.E. (1994) Randomness reexamined, *modeling, Identification and Control*, **15**(3), 141–151.

Karatzas, I., J. Lehocsky, S. Shreve and G.L. Xu (1991) Martingale and duality methods for utility maximization in an incomplete market, *SIAM Journal on Control and Optimization*, **29**, 702–730.

Karlin, S., and H.M. Taylor (1981) *A Second Course in Stochastic Processes*, Academic Press, San Diego, CA.

Kushner, H.J. (1990) Numerical methods for stochastic control problems in continuous time, *SIAM Journal on Control and Optimization*, **28**, 999–1948.

Levy, P. (1948) *Processus Stochastiques et Mouvement Brownien*, Paris.

Lorenz, E. (1966) Large-scale motions of the atmosphere circulation, in P.M. Hurley (Ed.), *Advances in Earth Science*, MIT Press, Cambridge, MA.

Lundberg, F. (1909) *Zur Theorie der Ruckversicherung Verdandlungskongress fur Versicherungsmathematik*, Vienna.

Malliaris, A.G., and W.A. Brock (1982) *Stochastic Methods in Economics and Finance*, North Holland, Amsterdam.

Mandelbrot, B. (1972) Statistical methodology for non-periodic cycles: From the covariance to R/S analysis, *Annals of Economic and Social Measurement*, **1**, 259–290.

Mandelbrot, B., and M. Taqqu (1979) Robust R/S analysis of long run serial correlation, *Bulletin of the International Statistical Institute*, **48**, book 2, 59–104.

Mandelbrot, B., and J. Van Ness (1968) Fractional Brownian motion, fractional noises and applications, *SIAM Review*, **10**, 422–437.

McKean, H.P. (1969) *Stochastic Integrals*, Academic Press, New York.

Milshtein, G.N. (1974) Approximate integration of stochastic differential equations, *Theory of Probability and Applications*, **19**, 557–562.

Milshtein, G.N. (1985) Weak approximation of solutions of systems of stochastic differential equations, *Theory of Probability and Applications*, **30**, 750–206.

Nelson, Daniel, and K. Ramaswamy (1990) Simple binomial processes as diffusion approximations in financial models, *The Review of Financial Studies*, **3**(3), 393–430.

Peter, Edgar E. (1995) *Chaos and Order in Capital Markets*, John Wiley & Sons, Inc., New York.

Pliska, S. (1986) A stochastic calculus model of continuous trading: Optimal portfolios, *Mathematics of Operations Research*, **11**, 371–382.

Révész, Pal (1994) *Random Walk in Random and Non-Random Environments*, World Scientific, Singapore.

Samuelson, P.A. (1965) Proof that properly anticipated prices fluctuate randomly, *Industrial Management Review*, **6**, 41–49.

Seal, H.L. (1969) *Stochastic Theory of a Risk Business*, John Wiley & Sons, Inc., New York.

Snyder, D.L. (1975) *Random Point Processes*, John Wiley & Sons, Inc., New York.

Spitzer, F. (1965) *Principles of Random Walk*, Van Nostrand, New York.

Stratonovich, R.L. (1968) *Conditional Markov Processes and their Applications to the Theory of Optimal Control*, American Elsevier, New York.

Sulem, A., and C.S. Tapiero (1994) Computational aspects in applied stochastic control, *Computational Economics*, **7**, 109–146.

Taylor, S. (1986) *modeling Financial Time Series*, John Wiley & Sons, Inc., New York.

CHAPTER 5

Derivatives Finance

5.1 EQUILIBRIUM VALUATION AND RATIONAL EXPECTATIONS

Fundamental notions such as rational expectations, risk-neutral pricing, complete and incomplete markets, underlie the market's valuation of risk and its pricing of derivatives assets. Both economics and mathematical finance use these concepts for the valuation of options and other financial instruments. Rational expectations presume that current prices reflect future uncertainties and that decision makers are rational, preferring more to less. It also means that current prices are based on the unbiased, minimum variance mean estimate of future prices. This property seems at first to be simple, but it turns out to be of great importance. It provides the means to 'value assets and securities', although, in this approach, bubbles are not possible, as they seem to imply a persistent error or bias in forecasting. This property also will not allow investors to earn above-average returns without taking above-average risks, leading to market efficiency and no arbitrage. In such circumstances, arbitrageurs, those 'smart investors' who use financial theory to identify returns that have no risk and yet provide a return, will not be able to profit without assuming risks.

The concept of rational expectation is due to John Muth (1961) who formulated it as a decision-making hypothesis in which agents are informed, constructing a model of the economic environment and using all the relevant and appropriate information at the time the decision is made (see also Magill and Quinzii, 1996, p. 23):

> I would like to suggest that expectations, since they are informed predictions of future events, are essentially the same as the predictions of the relevant economic theory... We call such expectations 'rational' ... The hypothesis can be rephrased a little more precisely as follows: that expectations ... (or more generally, the subjective probability distribution of outcomes) tend to be distributed, for the same information set, about the prediction of the theory (the objective probability of outcomes).

In other words, if investors are 'smart' and base their decisions on informed and calculated predictions, then, prices equal their discounted expectations. This hypothesis is essentially an equilibrium concept for the valuation of asset prices stating that under the 'subjective probability distribution' the asset price equals the

Risk and Financial Management: Mathematical and Computational Methods. C. Tapiero
 ISBN: 0-470-84908-8

expectation of the asset's future prices. In other words, it implies that investors' subjective beliefs are the same as those of the real world – they are neither pessimistic nor optimistic. When this is the case, and the 'rational expectations equilibrium' holds, we say that markets are complete or efficient. Samuelson had already pointed out this notion in 1965 as the martingale property, leading Fama (1970), Fama and Miller (1972) and Lucas (1972) to characterize markets with such properties as markets efficiency.

Lucas used a concept of rational expectations similar to Muth to confirm Milton Friedman's 1968 hypothesis of the long-run neutrality of the monetary policy. Specifically, Lucas (1972, 1978) and Sargent (1979) have shown that economic agents alter both their expectations and their decisions to neutralize the effects of monetary policy. From a practical point of view it means that an investor must take into account human reactions when making a decision since they will react in their best interest and not necessarily the investor's.

Martingales and the concept of market efficiency are intimately connected. If prices have the martingale property, then only the information available today is relevant to make a prediction on future prices. In other words, the present price has all relevant information embedding investors' expectations. This means that in practice (the weak form efficiency) past prices should be of no help in predicting future prices or, equivalently, prices have no memory. In this case, arbitrage is not possible and there is always a party to take on a risk, irrespective of how high it is. Hence, risk can be perfectly diversified away and made to disappear. In such a world without risk, all assets behave as if they are risk-free and therefore prices can be discounted at a risk-free rate. This is also what we have called risk-neutral pricing (RNP). It breaks down, however, if any of the previous hypotheses (martingale, rationality, no arbitrage, and absence of transaction costs) are invalid. In such cases, prices might not be valued uniquely, as we shall see subsequently.

There is a confrontation between economists, some of whom believe that markets are efficient and some of whom do not. Market efficiency is 'under siege' from both facts and new dogmas. Some of its critics claim that it fails to account for market anomalies such as bubbles and bursts, firms' performance and their relationship to size etc. As a result, an alternative 'behavioural finance' has sought to provide an alternative dogma (based on psychology) to explain the behaviour of financial markets and traders. Whether these dogmas will converge back together as classical and Keynesian economics have, remains yet to be seen. In summary, however, some believe that the current price imbeds all future information. And some presume that past prices and behaviour can be used (through technical analysis) to predict future prices. If the 'test is to make money', then the verdict is far from having been reached. Richard Roll, a financial economist and money manager argues:

> 'I have personally tried to invest money, my clients' and my own, in every single anomaly and predictive result that academics have dreamed up. And I have yet to make a nickel on these supposed market inefficiencies. An inefficiency ought to be an exploitable opportunity. If there is nothing that investors can exploit in a systematic way, time in and time out, then it's very hard to say that information is not being properly incorporated into stock prices. Real money investment strategies do not produce the results that academic papers say they

should' ... but there are some exceptions including long term performers that have over the years systematically beaten the market. (Burton Malkiel, *Wall Street Journal*, 28 December 2000)

Rational expectation models in finance may be applied wrongly. There are many situations where this is the case. Information asymmetries, insider trading and advantages of various sorts can provide an edge to individual investors, and thereby violate the basic tenets of market efficiency, and an opportunity for the lucky ones to make money. Further, the interaction of markets can lead to instabilities due to very rapid and positive feedback or to expectations that are becoming trader- and market-dependent. Such situations lead to a growth of volatility, instabilities and perhaps, in some special cases, to bubbles and chaos. Nonetheless, whether it is fully right or wrong, it seems to work sometimes. Thus, although rational expectations are an important hypothesis and an important equilibrium pillar of modern finance, they should be used carefully for making money. It is, however, undoubtedly in theoretical finance where it is used with simple models for the valuation of options and for valuing derivatives in general – albeit this valuation depends on a riskless interest rate, usually assumed known (i.e. mostly assumed exogenous). Thus, although the arbitrage-free hypothesis (or rational expectations) assumes that decision makers are acting intelligently and rationally, it still requires the risk-free rate to be supplied. In contrast, economic equilibrium theory, based on the clearing of markets by equating 'supply' to 'demand' for all financial assets, provides an equilibrium where interest rates are endogenous. It assumes, however, that beliefs are homogeneous, markets are frictionless (with no transaction costs, no taxes, no restriction on short sales and divisible assets) as well as competitive markets (in other words, investors are price takers) and finally it also assumes no arbitrage. Thus, general equilibrium is more elaborate than rational expectations (and arbitrage-free pricing) and provides more explicit results regarding market reactions and prices (Lucas, 1972). The problem is particularly acute when we turn to incomplete markets or markets where pricing cannot be uniquely defined under the rational expectations hypothesis. In this case, a decision makers' rationality is needed to determine asset prices. This was done in Chapter 3 when we introduced the SDF (stochastic discount factor) approach used to complement the no arbitrage hypothesis by a rationality that is sensitive to decision makers' utility of consumption. In Chapter 9, we shall return to this approach in an inter-temporal setting. For the present, we introduce the financial instruments that we will attempt to value in the next chapters.

5.2 FINANCIAL INSTRUMENTS[1]

There are a variety of financial instruments that may be used for multiple purposes, such as hedging, speculating, investing, and 'money multiplying' or leveraging. Their development and use require the ingenuity of financial engineers and the

[1] This section is partly based on a paper written by students at ESSEC, Bernardo Dominguez, Cédric Lespiau and Philippe Pages in the Master of Finance programme. Their help is gratefully acknowledged.

care of practising investors. Financial instruments are essentially contracts of various denominations and conditions on financial assets. Contracts by definition, however, are an agreement between two or more parties that involves an exchange. The terms of the contract depend on the purpose of contracting, the contractees, the environment and the information available to each of the parties. Examples of contracts abound in business, and more generally in society. For example, one theory holds that a firm is nothing more than a nexus of contracts both internal and external in nature.

Financial contracts establish the terms of exchange between parties mostly for the purpose of managing contractors' and contract holders' risks. Derivative assets or derivative contracts are special forms of contract that derive their value from an underlying asset. Such assets are also called contingent claim assets, as their price is dependent upon the state of the underlying asset. For example, warrants, convertible bonds, convertible preferred stocks, options and forward contracts, etc. are some well-known derivatives. They are not the only ones, however. The intrinsic value of these assets depends on the objectives and the needs of the buyer and the seller as well as the right and obligations these assets confer on each of the parties. When the number of buyers and (or) sellers is very large, these contingent assets are often standardized to allow their free trading on an open market. Many derivatives remain over-the-counter (OTC) and are either not traded on a secondary market or are in general less traded and hence less liquid than their market counterparts. The demand for such trades has led to the creation of special stock exchanges (such as the Chicago, London, and Philadelphia commodities and currency exchanges) that manage the transactions of such assets. A number of such contingent assets and financial instruments are defined next.

5.2.1 Forward and futures contracts

A futures contract gives one side, the holder of the contract, the obligation to buy or sell a commodity, a foreign currency etc. at some specified future time at a specified price, place, quantity, location and quality, according to the contract specification. The buyer or long side has at the end of the contract, called the maturity, the option to buy the underlying asset at a predetermined price and sell it back at the market price if he wishes to do so. The seller or short side (provider), however, has the obligation to sell the underlying asset at the predetermined price. In futures contracts, the exchange of the underlying asset at a predetermined price is between anonymous parties which is not the case in OTC forward contracts. Financial futures are used essentially for trading, hedging and arbitrage.

Futures contracts can be traded on the CBOT (the Chicago Board of Trade) and the CME (the Chicago Mercantile Exchange), as well as on many trading floors in the world. Further, many commodities, currencies, stocks etc. are traded daily in staggering amounts (hundreds of billions of dollars). A futures price at time t with delivery at time T can be written by $F(t, T)$. If $S(t)$ is the spot price, then clearly if $t = T$, we have by definition $F(t, t) = S(t)$ and $S(t) \geq F(t, T)$, $T \geq t$.

The difference between the spot assets to be pledged in a future contract and its futures price is often called the 'basis risk' and is given by $b(t, T) = S(t) - F(t, T)$. It is the risk one suffers when reversing a futures contracts position. Imagine we need to buy in 3-month pork for a food chain. We may buy futures contracts today that deliver the asset at a predetermined price in 6 months. After 3 months, we reverse or sell our futures position. The payoff is thus the change in the futures price less the price paid for the underlying asset, or $F(3, 6) - F(0, 6) - S(3)$. If we were at maturity, only $-F(0, 6)$ would remain. That is, the price of the underlying asset is set by a delayed physical transaction using futures contracts. However, if there remains a basis risk in the payoff, then $-F(0, 6) + b(3, 6)$ would remain. If the futures contract does not closely match the price of the underlying asset then the effectiveness of our hedging strategy will be reduced.

Futures contracts like forwards can be highly speculative instruments because they require no down payment since no financial exchange occurs before either maturity or the reversal of the position. Traders in the underlying assets can therefore use these markets to enhance their positions in the underlying asset either short or long. Unsurprisingly, a position in these contracts is considered levered or a borrowed position in the underlying asset, as the price of a forward and futures contract is nothing more than an arbitrage with the asset bought today using borrowed funds and delivered at maturity. There are differences between futures and forwards involving liquidity, marking to market, collaterals and delivery options, but these differences are generally glossed over.

The leverage implied in a futures contract explains why collaterals are required for forwards and marking to market for futures. In their absence, defaults would be much more likely to happen. For example, for a short futures contract, when prices fall, the investor is making a virtual loss since he would have to sell at a higher price than he started with (should he terminate his contract) and take an offsetting position by buying a futures contract. This is reflected in a 'futures market' when the bank adjusts the collateral account of the trader, called the margin. The margin starts at an initial level in, generally, the form of Treasury bills. It is adjusted every day to reflect the day's gains or losses. Should the margin fall below a maintenance level, the trader will ask the investor to add funds to meet margin requirements. If the investor fails to meet such requirements, the trader cuts his losses by reversing the position.

A *forward rate agreement (FRA)* is an agreement made between two parties seeking generally to protect or hedge themselves against a future interest-rate or price movement, for a specific hedging period, by fixing the future interest rate or price at which they will buy or sell for a specific principal sum in a specified currency. It requires that settlement be effected between the parties in accordance with an established formula. Typically, forward contracts, unlike futures contracts are not traded and can therefore be tailored to specific needs. This means that contracts tend to be much higher in size, far less liquid and less competitively priced, but suffer from no basis risk. The price at time t of a forward contract at time T in the future can be written by $p(t, T)$ or by $p(t, t + x)$, $x = T - t$ and is defined by the (delivery) price for which the contract value is null at delivery under risk neutral pricing.

$$\text{E(Future spot rate} - \text{Forward rate)} = 0$$

Of course, $p(T, T) = 1$ and therefore the derivative of the price with respect to T (or x) is necessarily negative, reflecting the lower value of the asset in the future compared to the same asset in the present.

The relationship between forward rates and spot prices is a matter of intensive research and theories. For example, the theory of rational expectations suggests that we equate the expected future spot rate to the current forward rate, that is (see also the next chapter):

$$\text{Forward rate} = \text{E (Future spot rate)}$$

For example, if s_t is the logarithm of the spot price of a currency at time t and f_t is the logarithm of the 1 month forward price, the expectation hypothesis means that:

$$f_t = E(s_{t+1})$$

Note that if S_t is the spot price at time t, $\Delta S_t = S_{t+1} - S_t$, then the rate of change, expressing the rates of return $\Delta S_t/S_t$ is given by $\Delta\,(\log S_t)$ with $s_t = \log S_t$. Empirical research has shown, at least for currencies forward, that it is misleading and therefore additional and alternative theories are often devised which introduce concepts of risk premium as well as the expected rate of depreciation to explain the incoherence between spot and forward market values and risk-neutral pricing.

Forward and futures contracts are not only used in financial and commodities markets. For example, a transport futures exchange has been set up on the Internet to help solve forward-planning problems faced by truckers and companies shipping around the world. The futures exchange enables companies to purchase transport futures, helping them to plan their freight requirements and shipments by road, rail and, possibly, barge. The exchange allows truckers and manufacturers to match transport capacity to their shipments and to match their spot requirements, buy and sell forward, and speculate on future movements of the market. This market completes other markets where one can buy and sell space on ocean-going ships. For example, London's Baltic Exchange handles spot trades in dry cargo carriers and tankers.

5.2.2 Options

Options are instruments that let the buyer of the option (the long side) the *right to exercise*, for a price, called the *premium*, the delivery of a commodity, a stock, a foreign currency etc. at a given price, called the *strike price*, at *(within)* a given time period, also called the *exercise date*. Such an option is called a *European (American) CALL* for the buyer. The seller of such an option (the short side), has by contrast the *obligation* to sell the option at the stated strike and exercise date. A *PUT* option (the long side) provides an option to sell while for the short seller this is an obligation to buy. There are many types of options, however. Below are a selected few (in the next two chapters we shall consider a far larger number of option contracts):

- *Call option* (long) (on foreign exchange (FX), deposit or futures etc.): an option contract that gives the holder the right to buy a specified amount of the commodity, stock or foreign currency for a premium on or before an expiration date as stated above. A call option (short), however, is an obligation to maintain the terms of this contract.
- *Put option* (long) (on FX, commodity etc.): gives the right to sell a specified amount at the strike price on (or before, for an American option) a specific expiration date. The short side of such a contract is an obligation, however, to meet the terms of this contract.
- *Swaps* (for interest rates, currency and cross-currency swaps, for example): transactions between two unrelated and independent borrowers, borrowing identical principal amounts for the same period from different lenders and with an interest rate calculated on a different basis. The borrowers agree to make payments to each other based on the interest cost of the other's borrowing. It is used both for arbitrage and to manage firm's liabilities. It can facilitate access of funding in a particular currency, provide export credits or other credits in a particular currency, provide access to various capital markets etc. These contracts are used intensively by banks and traders and will be discussed at length in the next chapter.
- *Caps*: a contract in which a seller pays a buyer predetermined payments at prespecified dates, with an interest (cap) rate calculated at later dates. If the rate of reference (the variable rate) is superior to a guaranteed rate, then the cap rate becomes effective, meaning that the largest interest rate is applied.
- *Floors*: products consisting in buying a cap and at the same time selling another product at a price compensating exactly the buying price of the cap. In this case, the floor is a contract in which the seller pays to the buyer for a predetermined period with a rate calculated at the fictive date. If the reference rate (the variable rate) is inferior to the guaranteed rate by the floor (rate), then the lower rate is applied.

Options again

Trading in options and other derivatives is not new. Derivative products were used by Japanese farmers and traders in the Middle Ages, who effectively bought and sold rice contracts. European financial markets have traded equity options since the seventeenth century. In the USA, derivative contracts initially started to trade in the CBOT (Chicago Board of Trade) in 1973. Derivatives were thus used for a long time without stirring up much controversy. It is not the idea that is new, it is the volume of trade, the large variety of instruments and the significant and growing number of users trading in financial markets that has made derivatives a topic that attracts permanent attention.

Today, the most active derivative market is the CBOT, while the CME (Mercantile Stock Exchange) ranks second. Other active exchanges are the CBOE,

PHLX, AMEX, NYSE and TSE (Toronto Stock Exchange). In Montreal a stock exchange devoted to derivatives was also started in 2001. In Europe the most active markets are LIFFE (London International Financial Futures Exchange), MATIF (Marché à Terme International de France), DTB (Deutsche Terminbrose), and the EOE (Amsterdam's European Options Exchange). The most voluminous markets in East Asia include TIFFE (Tokyo International Financial Futures Exchange), the Hong Kong Futures Exchange and SIMEX (Singapore International).

Options contracts in particular are traded on many trading floors and, mostly, they are defined in a standard manner. Nevertheless, there are also 'over-the-counter options' which are not traded in specific markets but are used in some contracts to fit specific needs. For example, there are 'Bermudan and Asian options'. The former option provides the right to exercise the option at several specific dates during the option lifetime while the latter defines the exercise price of the option as an average of the value attained over a certain time interval. Of course, each option, defined in a different way, will lead to alternative valuation formulas. More generally, there can be options on real assets, which are not traded but used to define a contract between two parties. For example, an airline company contracts the acquisition (or the option to acquire) a new (technology) plane at some future time. The contract may involve a stream or a lump sum payment to the contractor (Boeing or Airbus) in exchange for the delivery of the plane at a specified time. Since payments are often made prior to the delivery of the plane, a number of clauses are added in the contract to manage the risks sustained by each of the parties if any of the parties were to deviate from the contract stated terms (for example, late deliveries, technological obsolescence etc.). Similarly, a manufacturer can enter into binding bilateral agreements with a supplier by which agreed (contracted) exchange terms are used as a substitute for the free market mechanism. This can involve future contractual prices, delivery rates at specific times (to reduce inventory holding costs) and, of course, a set of clauses intended to protect each party against possible failures by the other in fulfilling the terms of the contract. Throughout the above cases the advantage resulting from negotiating a contract is to reduce, for one or both parties, the uncertainty concerning future exchange operating and financial conditions. In this manner, the manufacturer will be eager to secure long-term sources of supplies, and their timely availability, while the investor, buyer of options, would seek to avoid too large a loss implied by the acquisition of a risky asset, currency or commodity, etc. Since for each contract there, necessarily, needs to be one (or many) buyer and one (or many) seller, the price of the contract can be interpreted as the outcome of a negotiation process where both parties have an inducement to enter into a contractual agreement. For example, the buyer and the seller of an option can be conceived of as being involved in a game, the benefits of which for each of the players are deduced from premium and risk transfer. Note that the utility of entering into a contractual agreement is always positive ex-ante for all parties; otherwise there would not be any contractual agreement (unless such a contract were to be imposed on one of the parties!). When the number of buyers and sellers of such contracts becomes extremely large, transactions become 'impersonal' and it is the 'market price' that defines the value of the contract. Strategic behaviours tend to break down the larger the group and prices tend to become more efficient.

Making decisions with options

We shall see in Chapter 7, 'Options and Practice', some approaches using options in hedging and in speculating. Decisions involving options are numerous, e.g.:

- Buy and sell; on the basis of the stock price and the remaining time to its exercise.
- Buy and sell; on the basis of estimated volatility of the underlying or related statistics.
- Use options to hedge downside risk.
- Use stock options to motivate management and employees.
- Use options and stock options for tax purposes.
- Use options to raise money for investments.

These problems clearly require a competent understanding of options theory and financial markets and generally the ability to construct and compounds assets, options and other contracts into a portfolio of desirable characteristics. This is also called financial engineering and is also presented in the next chapter.

We shall use a theoretical valuation of options based on 'risk-neutral probabilities'. 'Uncertainty', defined by 'risk-neutral probabilities', unlike traditional (historical) probabilities, determined by interacting market forces, reflects the market resolution of demand and supply (equilibrium) for assets of various risks. This difference contrasts two cultures. It is due to economic and financial assumptions that current market prices 'endogenize' future prices (states and their best forecast based on available information). If this is the case, and it is so in markets we call *complete markets*, the current price ought to be determined by an appropriate discounting of expected future values. In other words, *it is the market that determines prices and not uncertainty.* We shall calculate explicitly these 'probabilities' in the next chapter when we turn to the technical valuation of options.

5.3 HEDGING AND INSTITUTIONS

Financial instruments are used in many ways to reduce risk (*hedging*), make money through *speculation* (which means that the trader takes a position, short or long, in the market) or through *arbitrage*. Arbitrage consists of taking positions in two or more markets so that a riskless profit is made (i.e. providing an infinite rate of return since money can be made without committing any investment). The number of ways to hedge is practically limitless. There are therefore many trading strategies financial managers and insurers can adopt to protect their wealth or to make money. Firms can use options on currency, commodities and other assets to protect their assets from unexpected variations. Financial institutions (such as banks and lending institutions) by contrast, use options to cover their risk exposure and immunize their investment portfolios. Insurance firms, however, use options to seek protection against excessive uncontrollable events, to diversify risks and to spread out risks with insured clients as we have outlined earlier. Generally, hedging strategies can be 'specialized services', tailored to individual and collective needs.

5.3.1 Hedging and hedge funds

Hedging is big business, with many financial firms providing a broad range of services for protecting investments and whatnot. The traditional approach, based on portfolio theory, optimizes a portfolio holding on the basis of risk return substitution (measured by the mean and variance of returns as we saw in Chapter 3). Hedging, however, proceeds to eliminate a particular risk in a portfolio through a trade or a series of trades (called hedges). While in portfolio management the investor seeks the largest returns given a risk level, in hedging – also used in the valuation of derivatives – a portfolio is constructed to eliminate completely the risk associated with the derivative (the option for example). In other words, a hedging portfolio is constructed, replicating the derivative security. If this can be done, then the derivative security and the replicating portfolio should have the same value (since they have exactly the same return properties). Otherwise, there would be a potential for arbitrage.

Hedge funds, however, may be a misnomer. They attract much attention because of the medias' fascination with their extremes – huge gains and losses. They were implicated in the 1992 crisis that led to major exchange rate realignments in the European Monetary System, and again in 1994 after a period of turbulence in international bond markets. Concerns mounted in 1997 in the wake of the financial upheavals in Asia. And they were amplified in 1998, with allegations of large hedge fund transactions in various Asian currency markets and with the near-collapse of Long-Term Capital Management (LTCM). Government officials, fearing this new threat to world financial markets, stepped in to coordinate a successful but controversial private–public sector rescue of LTCM. Yet, for all this attention, little concrete information is available about the extent of hedge funds' activities and how they operate. Despite a plethora of suggestions for reforms, no consensus exists on the implications of hedge fund activity for financial stability, or on how policy should be adapted.

The financial community defines a hedge fund as any limited partnership, exempted from certain laws (due to its legal location, shareholders features, etc.), whose main objective is to manage funds and profits. The term 'hedge fund' was coined in the 1960s when it was used to refer to investment partnerships that used sophisticated arbitrage techniques to invest in equity markets. Federal regulation of financial instruments and market participants in the USA is based on Acts of Congress seeking to protect individual market investors. However, by accepting investments only from institutional investors, companies, or high-net-worth individuals, hedge funds are exempt from most of investor protection and regulations. Consequently, hedge funds and their operators are generally not registered and are not required to publicly disclose data regarding their financial performance or transactions. Hence they have been accused of being speculative vehicles for financial institutions that are constrained by costly prudential regulations.

Hedge funds can also be eclectic investment pools, typically organized as private partnerships and often located offshore for tax and regulatory reasons. Their managers, who are paid on a fee-for-performance basis, may be free to use a variety of investment techniques, including short positions and leverage, to raise returns and cushion risk. While hedge funds are a rapidly growing part of the financial industry, the fact that they operate through private placements and restrict share ownership to rich individuals and institutions frees them from most disclosure and regulation requirements applied to mutual funds and banks. Further, funds legally domiciled outside the main financial markets and main trading countries are generally subject to even less regulation.

Hedge funds operate today as both speculators and hedgers, using a broad spectrum of risk-management tools. *Macro funds*, for example, base their investment strategies on the use of perceived discrepancies in the economic fundamentals of macroeconomic policies. Macro funds, may take large directional (unhedged) positions in national markets based on top-down analysis of macroeconomic and financial conditions, including current accounts, the inflation rate, the real exchange rate, etc. As a result, they necessarily are very sensitive to countries risk, global and national politics, economics and finance. Macro hedge funds may be classified into two essential categories:

- Arbitrage-based investment strategies
- Macro specific funds strategies

Arbitrage-based strategies seek to profit from current price discrepancies in two instruments (or portfolios) that will, at their maturity, have the same value. However, hedge funds that call themselves arbitrage-type use analytical models to profit from the discrepancy between their valuation 'model' and the actual market price. The arbitrage always involves two transactions: the purchase of an undervalued asset and the sale of an overvalued asset. Some commonly utilized arbitrage strategies include:

- Trade an instrument (cash instruments, index of equity securities, currency spot price, etc.) against its futures counterpart.
- Misalignments in prices of cash market fixed-income securities. A hedge fund might have a model for the levels of yield representing a number of bonds with various maturities. If the yield curve models differ from the yields of some bonds there is an opportunity for arbitrage.
- Misalignments because of the credit quality of two instruments. These sorts of trades are routinely executed by hedge funds examining the differences in the creditworthiness of various US corporate securities relative to the US Treasury bond yield spread.
- Convertible arbitrage involves purchasing convertible securities, mostly fixed-income bonds that can be converted into equity under certain circumstances. A portion of the equity risk embedded in the bond is hedged by selling short

the underlying equity. Sometimes the strategy will also involve an interest rate hedge to protect against general fluctuations in the yield curve. Thus, this trade would be designed to profit from mis-pricing of the equity associated with the convertible bond.

- Misalignments between options or other features imbedded in mortgage-backed securities. Often, complicated structures can be decomposed into various components that have market counterparts, permitting hedge funds to profit from deviations in prices of the underlying components and the structured product. The prepayment risk – the risk that the mortgage holder will prepay the mortgage prior to its maturity – is such an example.

Many of the determinants of a viable strategy are not specific to hedge funds, but are common to many types of investors. Virtually all hedge funds calculate whether the all-in return more than compensates for the risk undertaken. Three elements are taken into account in these calculations: (1) examining the market risk, which usually includes some type of 'stress test' to assess the downside risks of the proposed strategy; (2) examining the liquidity risk, that is, to see whether the hedge fund can enter and exit markets without extra costs in both normal times and in periods of market distress; (3) examining the timing and the cost of financing the position. If the expected duration of the trade is too long, with a prohibitive financing cost, the position will not be assumed.

Macro specific funds strategies are based on information regarding economic fundamentals. They seek incongruent relationships between the level of prices and the country's fundamentals – both economically and psychologically. Macro hedge funds are universally known for their 'top-down' global approach to investments, combining knowledge of economics, politics and history into a coherent view of things to come. In currency markets, a macro fund strategy might examine countries maintaining a pegged exchange rate to the dollar but having little economic reason for using the dollar for the peg. Some funds use rather detailed macroeconomic modelling techniques; others use less quantitative techniques, examining historical relationships among the various variables of interest. They may examine the safety and soundness of the banking sector and its connections to other parts of the financial sector. Excess liquidity and credit growth within the banking sector are often cited by funds as leading indicators of subsequent banking problems. Extensive use of unhedged foreign-currency-denominated debt of banks is also a tip-off for hedge funds. A pattern of high and fast appreciation of various assets is also used as a signal for a financial sector awaiting a downturn. Political risk and the probability that government's strategy may, or may not, be implemented are also used as signals on the basis of which positions may be taken. However, market funds are very sensitive to the potential for market exit (and thus liquidity) in the case where events are delayed or do not confirm expectations.

Risk management in a hedge fund is often planned and integrated across products and markets (related through correlation analysis). Scenario analysis and stress tests are common diagnostic techniques. Further, some trading risks are managed by limiting the types, the number and the market exposure of trades.

The criteria used are varied, such as the recent track record of the trader, the relative portfolio risk of the trade, and market liquidity.

5.3.2 Other hedge funds and investment strategies

There are of course, many strategies for hedge fund management and trading. We can only refer to a few.

Market hedge funds focus on either equity or debt markets of developing or emerging countries. In general, they are classified by geographical areas and combine arbitrage and macro hedge fund strategies. Since many emerging markets are underdeveloped and illiquid, we note three points. (1) The size of transactions is relatively small. (2) Pricing of various securities abounds. These are inefficient markets for a number of reasons such as a basic misunderstanding of their operation, due to selling agents behaviour governed by liquidity needs rather than by 'market rationality'. (3) Bets on political events may cause important differences in valuations. Political risk receives special attention for emerging market hedge funds compared to the Group of Ten leading countries where economic considerations are prominent.

Event-related funds focus on securities of firms undergoing a structural change (mergers, acquisitions, or reorganizations), seeking to profit from increases or decreases in both stock prices – before a merger or when valuation of the merged firm is altered appreciably. These funds may estimate the time to complete the merger and the annualized return on the investment if undertaken. Annualized returns includes the purchase and sale of the equity of the two merging companies and the cost of executing the short position, any dividends gained or lost and commissions. Using such returns calculations, the fund can assess the probability that a deal will be consumed. If annualized returns, including the probability that the merger will come through is greater than a 'baseline', the fund may execute the deal.

Value investing funds have a strategy close to mutual funds (portfolio) strategies seeking to profit from undervalued companies. Hedge funds are probably more likely to use hedging methodologies designed to offset industry risk and reduce market volatility, however.

Short-selling funds use short-selling strategies. They involve limited partnerships and offshore funds sponsored by wealthy individuals. In short sales, the investor sells short a stock at the current market price while the capital is invested in US Treasury securities with the same holding period. The amount of capital is then adjusted daily to reflect the change in the stock price. If the stock price decreases, free cash is released; when the stock price increases, the capital must be increased. Losses on a short position are unlimited since they must be paid in real time. As a result, the short seller may run out of capital, making the depth of the short sellers' pocket and the timing of trades important determinants of success of the fund.

Sector funds combine strategies described above but applied to the 'sector' the fund focuses on and in which it trades. A sector may have specific characteristics, recognized and capitalized on for making greater profits.

Hedge funds have raised concerns due to their often speculative and destabilizing character. For this reason, financial regulation agencies have devoted special attention to regulating funds. Further, hedge funds often use stabilizing strategies. Two such strategies are employed: 'counter' strategies and arbitrage strategies. Counter strategies involve buying when prices are thought to be too low and selling when they are thought to be high, countering current market movements. It is an obvious strategy when prices are naturally pushed back to their perceived fair value, thereby stabilizing prices. Arbitrage strategies are neither stabilizing nor destabilizing since the arbitrageur's action simply links one market to another. However, studies have shown that arbitrage activity on stock indices are in fact stabilizing, in the sense of reducing volatility of the underlying stocks.

In contrast, *destabilizing strategies* can be divided into two essential groups: (1) strategies that use existing prices and (2) strategies that use positions of other market participants for trading decisions. The first group is often called 'positive feedback trading'; if there are no offsetting forces, these participants can cause prices to 'overshoot' their equilibrium value, adding volatility relative to that determined by fundamental information. It can arise under a variety of circumstances, some of which are related to institutional features of markets. These include dynamic hedging, stop loss orders, and collateral or margin calls. On a simpler level, positive feedback strategies also incorporate general trend-following behaviour where investors use various technical rules to determine trends, reinforced by buying and selling on the trend. Among strategies inducing a positive feedback type behaviour, the most complex is dynamic hedging. Options sellers, for example (using a put protective strategy, see Chapter 7), sell the underlying asset as its price decreases in order to dynamically hedge to replicate put options. Thus, to hedge, they would be required to sell the underlying asset in a falling market to maintain a hedged position, potentially exacerbating the original movement. In general, hedge funds are typically buyers of options (not sellers) and do not need to hedge themselves; but dealers that sell those options to hedge funds do need to hedge. Other institutional features like collateral calls or margin calls can also lead to a positive feedback response. Collateral holders may require additional collateral from their customers when prices fall and losses are incurred. Often, the collateral is obtained by selling any number of instruments, causing further price declines and losses. Some intermediaries, providing margins to hedge funds can keep these funds on a very tight leash, requiring margin calls more than once a day if necessary.

A second group of trading behaviours that destabilize hedge funds results from herding – taking similar positions to other market participants, rather than basing decisions explicitly on prices. Positions can be mimicked directly by observing what other participants do or indirectly by using the same information, analysis and tools as other participants. Often, fund managers have an incentive to mimic other participants' behaviour to hide their own incompetence. There may be then a temptation to ignore private information and realign their performance on others. Since hedge fund managers have most of their wealth invested in the fund and are compensated on total absolute returns rather than on relative benchmarks, they are less inclined than other fund managers to 'herd', directly mimicking others.

However, many hedge funds probably hold the same analytical tools and have access to the same information, arriving necessarily, at similar assessments and at approximately the same time, creating an appearance of collusion. Further, even if hedge funds do not herd (directly or indirectly), other investors may herd with them or follow their lead into various markets.

Hedge funds, like other institutional investors, are potentially subject to three general types of prudential regulations: (1) those intended to protect investors, (2) those designed to ensure the integrity of markets, and (3) those meant to contain systemic risk. Investor protection regulation is employed when authorities perceive a lack of sophistication on the part of investors, for example, lacking the information needed to properly evaluate their investments. Then, regulations can either ensure that sufficient information is properly disclosed or exclude certain types of investors from participating in certain investments.

Regulation to protect market integrity seeks to ensure that markets are designed so that price discovery is reasonably efficient, that market power is not easily concentrated in ways that allow manipulation, and that pertinent information is available to potential investors.

Systemic risk is often the most visible element in the regulation of financial markets because it often requires coordination across markets and across regulatory and geographical boundaries. Regulation to protect market integrity and/or limit systemic risk, which includes capital requirements, exposure limits, and margin requirements, seeks to ensure that financial markets are sufficiently robust to withstand the failure of even the largest participants.

5.3.3 Investor protection rules

Shares in hedge funds are securities but, since they are issued through private placements,[2] they are exempt from making extensive disclosure and commitments in the detailed prospectuses required of registered investment funds. They must still provide investors with all material information about their securities and will generally do so in an offering memorandum. Non-accredited investors are generally not accepted by hedge funds, because they would have to be given essentially the same information that would have been provided as a registered offer. However, most hedge fund operators are likely to be subject to regulation under the Commodity Exchange Act, because of their activity as commodity pool operators and/or as large traders in the exchange-traded futures markets. Requirements for commodity pools and commodity pool operators (CPOs) are mainly relative to (1) personal records and exams to get registered, (2) disclosure and reporting on issues as risks relevant to the pool, historical performance, fees incurred by participants, business backgrounds of CPOs, any possible conflict of interest on the part of the CPOs, and (3) maintenance of detailed records at the head office.

[2] A private placement consists of an offering of securities made to investors on an individual (bilateral) basis rather than through broader advertising. It is not allowed to offer for sale the securities by any form of general solicitation or advertising.

Market integrity protection rules: Although hedge funds can opt out of many of the registration and disclosure requirements of the securities laws, they are subject to all the laws enacted to protect market integrity. The essential purpose of such laws is to minimize the potential of market manipulation by increasing transparency and limiting the size of positions that a single participant may establish in a particular market. Many of these regulations also help in containing the spillovers across markets and hence in mitigating systemic risks.

The Treasury monitors all 'large' participants in the derivatives markets. Weekly and monthly reports are required of large participants, defined as players with more than US$50 billion equivalent in contracts at the end of any quarter during the previous year. The Treasury puts out the aggregate data in its monthly bulletin but the desegregated data by participant are not published or revealed to the public. For government securities, the US Treasury is allowed to impose reporting requirements on entities having large positions in to-be-issued or recently issued Treasury securities. Such information is deemed necessary for monitoring large positions in Treasury securities and making sure that players are not squeezing other participants. The Security Exchange Act (SEA) also requires the reporting of sizeable investments in registered securities. It obliges any person who, directly or indirectly, acquires more than 5 % of the shares of a registered security to notify the SEC within 10 days of such acquisition. In overseeing the futures markets, the CFTC attempts to identify large traders in each market, their positions, interaction of related accounts, and, sometimes, even their trading intentions. Also, to reinforce the surveillance, each exchange is required to have its own system for identifying large traders. For example, the Chicago Mercantile Exchange requires position reports for all traders with more than 100 S&P 500 contracts. The regulators have the authority to take emergency action if they suspect manipulation, cornering of a market, or any hindrance to the operation of supply and demand forces.

Systemic risk reduction rules: The key systemic question is to what extent are large, and possibly leveraged, investors, including hedge funds, a source of risk to the financial institutions that provide them with credit and to the intermediaries, such as broker-dealers, who help them implement their investment strategies. Banks provide many services to hedge funds and accept hedge funds as profitable customers with associated risks controllable. They examine the structure of the collective investment vehicle, the disclosure documents submitted to regulators and those offered to clients, the financial statements, and the fund's performance history. Further, generally, a large proportion of the credit extended by banks to hedge funds is collateralized.

The SEC also monitors brokers' and dealers' credit risk exposure. The net capital rule fortifies a broker-dealer against defaults by setting minimum net capital standards and requiring it to deduct from its net worth the value of loans that have not been fully collateralized by liquid assets. Further, reporting rules enable a periodic assessment and, at times, continuous monitoring of the risks posed to broker-dealers by their material affiliates, including those involved in over-the-counter. Along with the bank and broker-dealer credit structures that protect against excessively large uncollateralized positions, the Treasury and CFTC large position and/or large trader reporting requirements, by automatically soliciting information, provide continuous monitoring of large players in key markets and hence allow early detection of stresses in the system.

Mutual funds regulation, however, is strict, protecting shareholders, by:

- *Regulatory requirements* to ensure that investors are provided with timely and accurate information about management, holdings, fees, and expenses and to protect the integrity of the fund's assets. Therefore, mutual fund holdings and strategies are also regulated. In contrast, hedge funds are free to choose the composition of their portfolios and the nature of their investment strategies.
- *Fees.* Federal law requires a detailed disclosure, a standardized reporting and imposes limits to mutual fund fees and expenses. Hedge fund fees need not be disclosed and there are no imposed limits, which generally are between 15 and 20 % of returns and between about 1 and 2 % of net assets.
- *Leverage* practices and derivative products are used to enhance returns or reduce risks and have a restricted usage in mutual funds, while hedge funds have no restrictions other than their own internal strategies or partnership agreements.
- *Pricing and liquidity.* Mutual funds are required to price their shares daily and to allow shareholders to redeem shares also on a daily basis. Hedge funds, however, have no rules about pricing their own shares and redemption of shares may be restricted by the partnership agreement if wanted.
- *Investors.* The minimum initial investment to enter a mutual fund is about US$1000–2500. To own shares in a hedge fund, it is commonly required to make a commitment of US$1 million. Such measures are designed to restrict share ownership and, in consequence, to fall in a much weaker investor protection rules environment.

REFERENCES AND ADDITIONAL READING

Asness, C., R. Krail and J. Liew (2001) Do hedge funds hedge?, *Journal of Portfolio Management*, Fall, 6–19.

Fama, E.F. (1970) Efficient capital markets: A review of theory and empirical work, *The Journal of Finance*, **25**, 383–417.

Fama, E.F., and M.H. Miller (1972) *The Theory of Finance*, Holt, Rinehart & Winston, New York.

Fothergill, M. and C. Coke (2001) Funds of hedge funds: An introduction to multi-manager funds, *Journal of Alternative Investments*, Fall, 7–16.

Henker, T. (1998) Naïve diversification for hedge funds, *Journal of Alternative Investments*, Winter, 32–42.

Liang, B. (2001) Hedge funds performance: 1990–1999, *Financial Analysts Journal*, Jan/Feb., **57**, 11–18.

Lucas, R.E. (1972), Expectations and the Neutrality of Money, *Journal of Economic Theory*, **4**(2), 103–124.

Lucas, R.E. (1978) Asset prices in an exchange economy, *Econometrica*, **46**, 1429–1446.

Magill, M., and M. Quinzii (1996) *Theory of Incomplete Markets*, Vol 1, MIT Press, Boston, MA.
Muth, J. (1961) Rational expectations and the theory of price movements, *Econometrica*, **29**, 315–335.
Sargent, T.J. (1979) *Macroeconomic Theory*, Academic Press, New York.

PART II

Mathematical and Computational Finance

CHAPTER 6

Options and Derivatives Finance Mathematics

6.1 INTRODUCTION TO CALL OPTIONS VALUATION

Options are some of the building blocks of modern corporate finance and financial economics. Their mathematical study is in general difficult, however. In this chapter and in the following one, we consider the valuation of options and their use in practice. Terms such as a trading strategy, risk-neutral pricing, rational expectations, etc. will be elucidated in simple mathematical terms. To value an option it is important to define first, and clearly, a number of terms. This is what we do next.

We begin by defining *wealth* at a given time t, $W(t)$. This is the amount of money an investor has either currently invested or available for investment. Investments can be made in a number of assets, some of which may be risky, providing uncertain returns, while others may provide a risk-free rate of return (as would be achieved by investing in a riskless bond) which we denote by R_f. A risky investment is assumed for simplicity to consist of an investment in securities. Let N_0 be the number of bonds we invest in, say zero coupon of \$1 denomination, bearing a risk-free rate of return R_f one period hence. Thus, at a given time, our investment in bonds equals $N_0 B(t, t+1)$ with $B(t+1, t+1) = 1$. This means that one period hence, this investment will be worth $B(t, t+1)N_0(1 + R_f) = N_0(1 + R_f)$ for sure. We can also invest in risky assets consisting of m securities each bearing a known price $S_i(t)$, $i = 1, \ldots, m$ at time t. The investment in securities is defined by the number of shares $N_1, N_2, \ldots, N_m$ bought of each security at time t. Thus, a trading strategy at this time is given by the portfolio composition $(N_0, N_1, N_2, \ldots, N_m)$. The total portfolio investment at time t, is thus given by:

$$W(t) = N_0 B(t, t+1) + N_1 S_1(t) + N_2 S_2(t) + \cdots + N_m S_m(t)$$

Risk and Financial Management: Mathematical and Computational Methods. C. Tapiero
 ISBN: 0-470-84908-8

For example, for a portfolio consisting of a bond and in a stock, we have:

$$W(t) = \begin{cases} \$N_0 & \text{invested in a riskless bond} \\ \$N_1 S_1(t) & \text{invested in a risky asset, a stock} \end{cases}$$

$$W(t) = N_0 + N_1 S_1(t)$$

A period later, the bond is cashed while security prices may change in an uncertain manner. That is to say, the price in the next period of a security i is a random variable that we specify by a 'tilde', or $\tilde{S}_i(t+1)$. The gain (loss) is thus the random variable:

$$\Delta S_i(t) = \tilde{S}_i(t+1) - S_i(t), \, i = 1$$

Usually, one attempts to predict the gain (loss) by constructing a stochastic process for $\Delta S_i(t)$. The wealth gain (loss) over one period is:

$$\Delta W(t) = \tilde{W}(t+1) - W(t)$$

where,

$$\tilde{W}(t+1) = N_0(1 + R_f) + N_1 \tilde{S}_1(t+1)$$

Thus, the net gain (loss) in the time interval $(t, t+1)$, is:

$$\Delta W(t) = N_0 R_f + N_1 \Delta S_1(t)$$

In general, a portfolio consists of multiple assets such as bonds of various denominations and maturities, stocks, options, contracts of various sorts and assets that may be more or less liquid (such as real estate or transaction-cost-prone assets). We restrict ourselves for the moment to an investment in a simple binomial stock and a bond.

Over two periods, future security prices assume two values only, one high S_H (the security price increases), the other low S_L (the security price decreases) with $0 < S_L < S_H$ as well as $S_L/S \leq 1 + R_f \leq S_H/S$. These conditions will exclude arbitrage opportunities as we shall see later on. Thus stock prices at t and at $t+1$ are (see Figure 6.1):

$$S(t) \quad \text{and} \quad \tilde{S}(t+1) = \begin{cases} S_H \\ S_L \end{cases}$$

This results in a portfolio that assumes two possible values at time $t+1$:

$$W(t) = \begin{cases} N_0 \\ N_1 S \end{cases} \quad \text{and} \quad \tilde{W}(t+1) = \begin{cases} N_0(1 + R_f) + N_1 S_H \\ N_0(1 + R_f) + N_1 S_L \end{cases}$$

In other words, at time t the current time, the price of a stock is known and given by $S = S(t)$. An instant of time later, at $(t+1)$, its price is uncertain and assumes the two values (S_H, S_L), with $S_H > S_L$. As a result, if at $t = 0$, wealth is invested in a bond and in a security, we have the investment process given by (see Figure 6.2):

$$W(0) = N_0 + N_1 S_1(0) \quad \text{and} \quad \tilde{W}(1) = N_0(1 + R_f) + N_1 \tilde{S}_1(1)$$

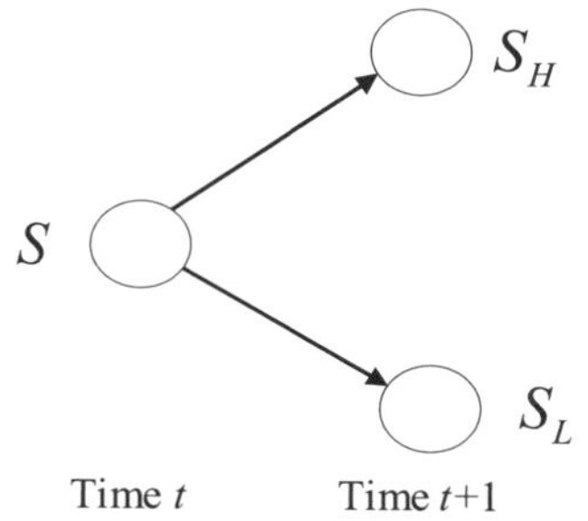

Figure 6.1

where in period 1, wealth can assume two values only since future prices are equal to either of (S_H, S_L), $S_H > S_L$ and the trading strategy is defined by (N_0, N_1). In this specific case, the price process is predictable, assuming two values only. This predictability is an essential assumption to obtain a unique value for the derivative asset, as we shall see subsequently.

For example, say that a stock has a current value of \$100 and say that a period hence (say a year), it can assume two possible values of \$140 and \$70. That is:

$$S(t) = S(=100) \quad \text{and} \quad \tilde{S}(t+1) = \begin{cases} S_H(=140) \\ S_L(=70) \end{cases}$$

The risk-free yearly interest rate is 12%, i.e. $R_f = 0.12$. Thus, if we construct a portfolio of N_0 units of a bond worth each \$1 and N_1 shares of the stock, then the portfolio investment and its future value one period hence are:

$$W(0) = N_0(1+0.12) + 100N_1 \quad \text{and} \quad \tilde{W}(1) = N_0(1+0.12) + \begin{cases} 140N_1 \\ 70N_1 \end{cases}$$

Now assume that we want to estimate the value of an option derived from such a security. Namely, consider a *call option* stating that at time $t = 1$, the strike time, the buyer of the option has the right to buy the security at a price of K, the exercise or strike price, with, for convenience, $S_H \geq K \geq S_L$. If the price is high, then the gain for the buyer of the option is $S_H - K > 0$ and the option is exercised

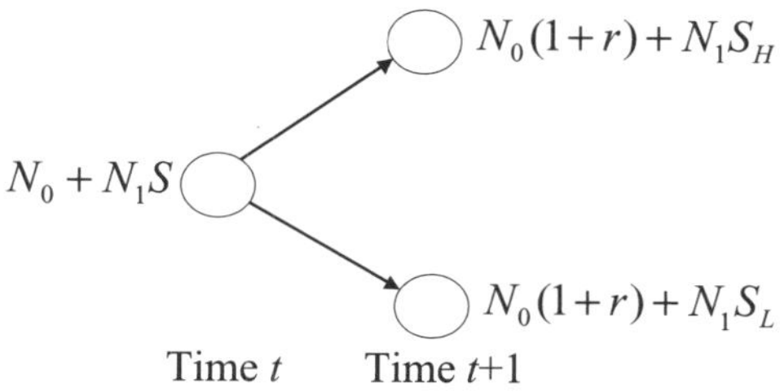

Figure 6.2

while the short seller of the option has a loss, which is $K - S_L$. If the price is low (below the strike price) then there is no gain and the only loss to the buyer of the call option is the premium paid for it initially. The problem we are faced with concerns the value/price of such a derived (option) contract. In other words, how much money would the (long) buyer of the option be willing to pay for this right. To find out, we proceed as follows. First, we note the possible payoffs of the option over one period and denote it by $\tilde{C}(1)$. Then we construct a portfolio replicating the exact cash flow associated to the option. Let the portfolio worth at the strike time be $\tilde{W}(1)$:

$$\tilde{W}(1) = N_0(1 + R_f) + N_1\tilde{S}(1) = \begin{cases} N_0(1 + R_f) + N_1 S_H \\ N_0(1 + R_f) + N_1 S_L \end{cases} \quad \text{and} \quad \tilde{W}(1) \equiv \tilde{C}(1)$$

To determine this equivalence, the portfolio composition N_0, N_1 has to be determined uniquely. If it were not possible to replicate the option cash flow uniquely by a portfolio, then we would not be able to determine a unique price for the option and we would be in a situation we call incomplete. This conclusion is based on the economic hypothesis that two equivalent and identical cash flows have necessarily the same economic value (or cost). If this were not the case, there may be more than one price or no price at all for the derivative asset. Our ability to replicate a risky asset by a portfolio uniquely underlies the notion of the 'no arbitrage' assumption, which implies in turn the 'law of the single price'. Thus, by constructing portfolios that have exactly the same returns with the same risks, their value ought to be the same. If this were not the case, then one of the two assets would be dominated and therefore their value could not be the same. Further, there would be an opportunity for profits that can be made with no investment – or equivalently, an opportunity for infinite rates of returns (assuming perfect liquidity of markets) that cannot be sustained (and therefore not maintain a state of equilibrium). Thus, to derive the option price, it is sufficient to estimate the replicating portfolio initial value. This is done next. Say that, for a call option, its value one period hence is:

$$\tilde{C}(1) = \begin{cases} S_H - K & \text{if the security price rises} \\ 0 & \text{if the security price decreases} \end{cases}$$

where $S_L < K < S_H$. A replicating portfolio investment equivalent to an option would thus be:

$$\tilde{W}(1) = \tilde{C}(1)$$

Or, equivalently,

$$\tilde{W}(1) = \tilde{C}(1) \Leftrightarrow \begin{cases} N_0(1 + R_f) + N_1 S_H = S_H - K \\ N_0(1 + R_f) + N_1 S_L = 0 \end{cases}$$

Note that these are two linear equations in two unknowns and have therefore a unique solution for the replicating portfolio:

$$N_1 = \frac{S_H - K}{(S_H - S_L)}, \quad N_0 = -\frac{S_L(S_H - K)}{(1 + R_f)(S_H - S_L)}$$

The procedure followed is summarized below.

$$\begin{array}{ccc} W(0) & \Leftarrow & \tilde{W}(1) \\ \Downarrow & & \Updownarrow \\ C(0) & & \tilde{C}(1) \end{array}$$

The call option's payoff is replicated by holding short bonds to invest in a stock ($N_0 < 0,\ N_1 > 0$). As the stock price increases, the portfolio is shifted from bonds to stocks. As a result, calling upon the 'no arbitrage' assumption, the option price and the replicating portfolio must be the same since they have identical cash flows. That is, as stated above:

$$\tilde{W}(1) = \tilde{C}(1) \Leftrightarrow W(0) = C(0) \quad \text{and} \quad \text{since: } W(0) = N_0 + N_1 S(0)$$

We insert the values for (N_0, N_1) calculated above and obtain the *call option price*:

$$C(0) = \frac{(S(1 + R_f) - S_L)(S_H - K)}{(1 + R_f)(S_H - S_L)}$$

Thus, if we return to our portfolio, and assume that the option has a strike price of \$120, then the replicating portfolio is:

$$N_1 = \frac{S_H - K}{(S_H - S_L)} = \frac{140 - 120}{140 - 70} = \frac{2}{7}$$

$$N_0 = -\frac{S_L(S_H - K)}{(1 + R_f)(S_H - S_L)} = -\frac{70(140 - 120)}{(1 + 0.12)(140 - 70)} = -\frac{20}{1.12}$$

and further, the option price is:

$$W(0) = N_0 + N_1 S(0) = -\frac{20}{1.12} + \frac{200}{7} = 10.72$$

which can be calculated directly from the formula above:

$$C(0) = \frac{[S(1 + R_f) - S_L](S_H - K)}{(1 + R_f)(S_H - S_L)} = \frac{(100(1.12) - 70)(140 - 120)}{(1 + 0.12)(140 - 70)} = 10.72$$

By the same token, say that the current price of a stock is $S = \$100$ while the price a period hence (at which time the option may be exercised) is either $S_H = \$120$ or $S_L = \$70$. The strike price is $K = \$110$ while the discount rate over the relevant period is 0.03. Thus, a call option taken for the period on such a stock has a price, which is given by:

$$C(0) = \frac{(100 - 70)(120 - 110)}{(1 + 0.03)(120 - 70)} = \$5.825$$

6.1.1 Option valuation and rational expectations

The rational expectations hypothesis claims that an expectation over 'future prices' determines current prices (see Figure 6.3). That is to say, assuming that

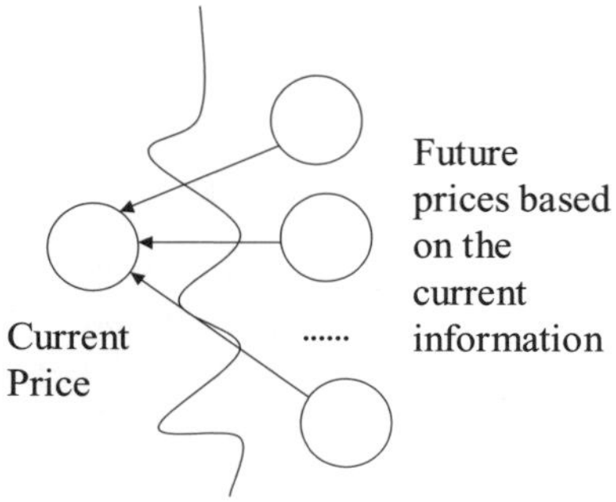

Figure 6.3

rational expectations hold, there is a probability measure that values the option in terms of its expected discounted value at the risk-free rate, or

$$C(0) = \frac{1}{1+R_f} E^* \tilde{C}(1)$$

where E^* is an expectation taken over the appropriate probability measure assumed to exist (in our current case it is given by $[p^{*,1} - p^*]$) and therefore:

$$C(0) = \frac{1}{1+R_f}[p^* C(1|S_H) + (1-p^*)C(1|S_L)]$$

where $C(1\,|S_H) = S_H - K,\ C(1\,|S_L) = 0$ are the option value at the exercise time and p^* denotes a 'risk-neutral probability'. *This probability is not, however, a historical probability of the stock moving up or down but a 'risk-neutral probability', making it possible to value the asset under a risk-neutrality assumption.* In this case, the option's price is the discounted (at a risk-free rate) expected value of the option,

$$C(0) = \frac{1}{1+R_f}[p^*(S_H - K) + (1-p^*)(0)]$$

And, using the value of the option found earlier, we have:

$$0 \le p^* = \frac{1}{S_H - S_L}[(1+R_f)S - S_L] \le 1$$

In our previous example, we have:

$$0 \le p^* = \frac{1}{120-70}[(1+0.03)100 - 70] = 0.66$$

By the same token, we can verify that:

$$S = \frac{1}{1+.03}\,[(0.66)(120) + (1-0.66)(70)] = 100$$

$$C(0) = \frac{1}{1+.03}\,[(0.66)(120-110)] = \frac{(0.66)(10)}{1.03} = 5.825$$

with $p^* = 0.66$. This 'risk-neutral probability' is determined in fact by traders in financial markets interacting with others in developing the financial market equilibrium – where profits without risk cannot be realized. For this reason, 'risk-neutral pricing' is 'determined by the market and provides the appropriate discount mechanism to value the asset in the following form (see also Chapter 3 and our discussion on the stochastic discount factor):

$$C(0) = E\{m_1\tilde{C}(1)\};\ m_1 = \frac{1}{1+R_f}$$

Risk-neutral probabilities, as we have just seen, allow a linear valuation of the option which hinges on the assumption of no arbitrage. Nonetheless, the existence of risk-neutral probabilities do not mean that we can use linear valuation, for to do so requires markets completeness (expressed by the fact in this section that we were able to replicate by portfolio the option value and derive a unique price of the option). In subsequent chapters, we shall be concerned with market incompleteness and see that this is not always the case. These situations will complicate the valuation of financial assets in general.

6.1.2 Risk-neutral pricing

The importance of risk-neutral pricing justifies our considering it in greater depth. In many instances, security prices can be conveniently measured with respect to a given process – in particular, a growing process called the *numeraire*, expressing the value of money (money market), a bond or some other asset. That is, allowing us to write (see also Chapter 3):

$$V(S(t)) = \frac{1}{1+R_f}E^*(V(\tilde{S}(t+1))) = \frac{1}{1+R_f}[p^*V(S_H) + (1-p^*)V(S_L)]$$

p^* is said to be a 'risk-neutral probability' and R_f is a risk-free discount rate. And for an option (since R_f has a fixed value):

$$C(t) = E^*\left[\frac{1}{1+R_f}\tilde{C}(t+1)\right] = \frac{1}{1+R_f}E^*[\tilde{C}(t+1)]$$

In general, for any value (whether it is an option or not) $\tilde{V}_i$ at time i with a risk-free rate R_f, we have, over one period: $V_0 = E^*\left(\frac{1}{1+R_f}\tilde{V}_1\right)$

By iterated expectations, we have as well:

$$V_1 = \mathrm{E}^*\left(\frac{1}{1+R_f}\tilde{V}_2\right) \quad \text{and}$$

$$V_0 = \frac{1}{(1+R_f)}\mathrm{E}^*\left(\mathrm{E}^*\left(\frac{1}{1+R_f}\tilde{V}_2\right)\right) = \mathrm{E}^*\left(\frac{1}{(1+R_f)^2}\tilde{V}_2\right)$$

and therefore, over n periods:

$$V_0 = \mathrm{E}^*\left(\frac{1}{(1+R_f)^n}\tilde{V}_n\right) = \frac{1}{(1+R_f)^n}\mathrm{E}^*(\tilde{V}_n)$$

If we set Φ_0, the information regarding the process initially, then we write

$$V_0(1+R_f)^n = \mathrm{E}^*(\tilde{V}_n\,|\Phi_0)$$

Further, application of iterated expectations has shown that this discounting process defines a martingale. Namely, we have:

$$V(S_0) = (1+R_f)^{-k}V(S_k) = \mathrm{E}^*\{(1+R_f)^{-(k+n)}V(S_{k+n})\,|\Phi_k\};$$
$$k = 0, 1, 2, \ldots \quad \text{and} \quad n = 1, 2, 3, \ldots$$

or, equivalently,

$$V(S_k) = \mathrm{E}^*\{(1+R_f)^{-n}V(S_{k+n})\,|\Phi_k\}; k = 0, 1, 2, \ldots \quad \text{and} \quad n = 1, 2, 3, \ldots$$

This result can be verified next using our binomial model. Set the unit one period risk-free bond, $B(t) = B(t, t+1)$ for notational convenience, then discounting a security price with respect to the risk-free bond yields:

$$S^*(t) = \frac{S(t)}{B(t)} \quad \text{or} \quad S^*(t) = \frac{S(t)}{(1+R_f)^t}$$

and $S^*(t)$ is a martingale. Generally, under the risk-neutral measure, P^* the discounted process

$$\{(1+R_f)^{-k}S_k\,|\Phi_k\},\ k = 0, 1, 2, \ldots$$

is, as we saw earlier, a martingale. Here again, the proof is simple since:

$$\mathrm{E}^*\left\{(1+R_f)^{-(k+1)}S_{k+1}\,|\Phi_k\right\} = (1+R_f)^{-k}S_k$$

and

$$\mathrm{E}^*\left\{(1+R_f)^{-(k+1)}S_{k+1}\,|\Phi_k\right\} = (1+R_f)^{-(k+1)}\left[p^*\frac{S_H}{S_k} + q^*\frac{S_L}{S_k}\right]S_k$$
$$= (1+R_f)^{-(k+1)}S_k[(1+R_f)] = (1+R_f)^{-k}S_k$$

This procedure remains valid if we consider a portfolio which consists of a bond and m stocks. In this case, dropping for simplicity the tilde over random variables, we have:

$$W^*(t) = N_0 + N_1S_1^*(t) + N_2S_2^*(t) + \cdots + N_mS_m^*(t)$$
$$W^*(t+1) = N_0 + N_1S_1^*(t+1) + N_2S_2^*(t+1) + \cdots + N_mS_m^*(t+1)$$

and

$$\Delta W^*(t) = N_1\Delta S_1^*(t) + N_2\Delta S_2^*(t) + \cdots + N_m\Delta S_m^*(t)$$

Equating these to the value of some derived asset, a period hence:

$$W^*(t+1) = C^*(t+1)$$

leads to a solution for $(N_0, N_1, N_2, \ldots, N_m)$ where $C^*(t+1)$ is a vector of assets we use to construct a riskless hedge and replicate the derivative product we wish to estimate (Pliska (1997) and Shreve *et al.* (1997) for example).

Example: Options and portfolios holding cost

Consider now the problem of valuing the price of a call option on a stock when the alternative portfolio consists in holding a risky asset (a stock) and a bond, for which there is a 'holding cost'. This cost is usually the charge a bank may require for maintaining in its books an investor's portfolio. In this case, the hedging portfolio is given by equating:

$$\tilde{W}(1) = \begin{cases} N_0(1 + R_f - c_B) + N_1(S_H - c_S) \\ N_0(1 + R_f - c_B) + N_1(S_L - c_S) \end{cases}$$

where c_B is the bond holding cost and c_S is the stock holding cost. The option's cash flow is:

$$\tilde{C}(1) = \begin{cases} S_H - K & \text{if the security price rises} \\ 0 & \text{if the security price decreases} \end{cases}$$

This leads to:

$$\begin{cases} N_0(1 + R_f - c_B) + N_1(S_H - c_S) = & S_H - K \\ N_0(1 + R_f - c_B) + N_1(S_L - c_S) = & 0 \end{cases}$$

and

$$N_1 = \frac{S_H - K}{S_H - S_L}, \; N_0 = -\frac{(S_H - K)(S_L - c_S)}{(S_H - S_L)(1 + R_f - c_B)}$$

Therefore, the option price is equal instead to:

$$C(0) = N_1 S + N_0 = S\frac{S_H - K}{S_H - S_L} + \frac{(S_H - K)(S_L - c_S)}{(S_H - S_L)(1 + R_f - c_B)}$$

For example, if we use the data used in the previous option's example with $S = 100$, $S_H = 140$, $S_L = 70$, $K = 120$, $R_f = 0.12$ and the 'holding costs' are: $c_S = 5$, $c_B = 0.02$, then

$$C(0) = 100\frac{140 - 120}{140 - 70} - \frac{(140 - 120)(70 - 5)}{(140 - 70)(1 + 0.12 - 0.02)} \quad \text{or}$$

$$C(0) = \frac{200}{7} - \frac{(20)(65)}{(70)(1.1)} = 28.57 - 16.88 = 11.68$$

which compares to a price of 10.64 without the holding cost. In this sense, holding costs will increase the price of acquiring the option. A general approach to this problem is treated by Bensoussan and Julien (2000) in continuous-time models. The costs of holding, denoted friction costs, are, however, far more complex, leading to incompleteness.

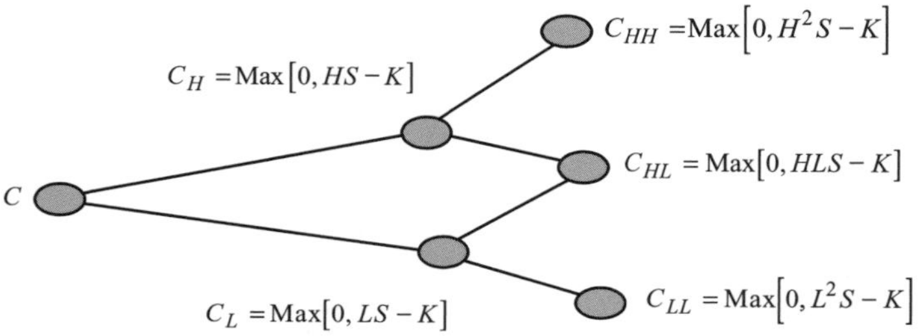

Figure 6.4 A two-period binomial tree.

6.1.3 Multiple periods with binomial trees

Over two or more periods, the problem remains the same. For one period, we saw that the price of a call option is: $C_H = \text{Max}\,[0, S_H - K]\,; C_L = \text{Max}\,[0, S_L - K]$ and by risk-neutral pricing,

$$C(0) = \frac{1}{1+R_f}E^*\tilde{C}(1) = \frac{1}{1+R_f}[p^*C_H + (1-p^*)C_L]$$

Over two periods, we have:

$$C(0) = \frac{1}{(1+R_f)^2}E^*\tilde{C}(2)$$

$$C_H = \frac{1}{1+R_f}[p^*C_{HH} + (1-p^*)C_{HL}];\ C_L = \frac{1}{1+R_f}[p^*C_{HL} + (1-p^*)C_{LL}]$$

which we insert in the previous equation, to obtain the option price for two periods (see Figure 6.4). Explicitly, we have the following calculations:

$$C(0) = \frac{1}{(1+R_f)}E^*\tilde{C}(1) = \frac{1}{(1+R_f)}E^*\left\{\frac{1}{(1+R_f)}E^*\tilde{C}(2)\right\} \text{ or}$$

$$C(0) = \left(\frac{1}{1+R_f}\right)^2[p^{*2}C_{HH} + 2p^*(1-p^*)C_{HL} + (1-p^{*2})C_{LL}]$$

Generally, the price of a call option at time t whose strike price is K at time T can be calculated recursively by:

$$C(t) = E^*\left(\frac{1}{1+R_f}\tilde{C}(t+1)\right); C(T) = Max\,[0, S(T) - K]$$

Explicitly, if we set, $S_H = HS,\ S_L = LS$, we have :

$$C(0) = \frac{1}{(1+R_f)^2}\left\{\begin{array}{l} p^{*2}(H^2S-K)^+ + \\ 2p^*(1-p^*)(HLS-K)^+ + \\ (1-p^*)^2\,(H^2S-K)^+ \end{array}\right\} = \frac{1}{(1+R_f)^2}\sum_{j=0}^{2}\binom{2}{j}$$

$$p^{*j}(1-p^*)^{2-j}\{H^jL^{2-j}S - K\}^+$$

We generalize to n periods and obtain by induction:

$$C(0) = \frac{1}{(1+R_f)^n}\left\{\sum_{j=0}^{n}\binom{n}{j}p^{*j}(1-p^*)^{n-j}(H^jL^{n-j}S-K)^+\right\}$$

We can write this expression in still another form:

$$C_n = \frac{1}{(1+R_f)^n}E\{(S_n-K)^+\} = \frac{1}{(1+R_f)^n}\sum_{j=0}^{n}P_j(\tilde{S}_j-K)^+$$

where

$$P_j = P(S_n = H^jL^{n-j}S) = \binom{n}{j}p^{*j}(1-p^*)^{n-j}$$

are the risk-neutral probabilities. This expression is of course valid only under the assumption of no arbitrage. This mechanism for pricing options is generally applicable to other types of options, however, such as American, Look-Back, Asiatic, esoteric and other options, as we shall see later on.

The option considered so far is European since exercise of the option is possible only at the option's maturity. American options, unlike European ones, give the buyer the right to exercise the option before maturity. The buyer must therefore take into account to optimal timing of his exercise. An option exercised too early may forgo future opportunities, while exercised too late it may lose past opportunities. The optimal exercise time will be that time that balances the live value of the option versus its 'dead' or exercise value. The recursive solution of the European call option can be easily modified for the exercise feature of the American option. Proceeding backward from maturity, the option will be exercised when its 'dead' value is larger than its 'live' one. Technically, the exercise time is a stopping time, as we shall see subsequently. Note that early exercise of the option is optimal only if the option value diminishes. For a call option (and in the absence of dividends), it does not diminish over time and therefore it will never pay to exercise an option early. For this reason we note that the price of a European and an American call are equal. For a put option, the present value of the payoff is a decreasing function of time hence, early exercise is possible irrespective of the existence of dividend payments.

6.2 FORWARD AND FUTURES CONTRACTS

A forward contract is an agreement to buy or sell an asset at a fixed date for a price determined today. The buyer agrees to buy the asset at the price F and sell it at the market price at maturity for a payoff $S - F$. The seller takes the opposite position and sells at the market price F and buys the asset at the market price S at maturity.

Forward contracts are thus an agreement between two parties or traders regarding the price, the *delivery price*, of a stock, a commodity or any another asset, settled at some future time – the *maturity*. Unlike options, forward contracts are

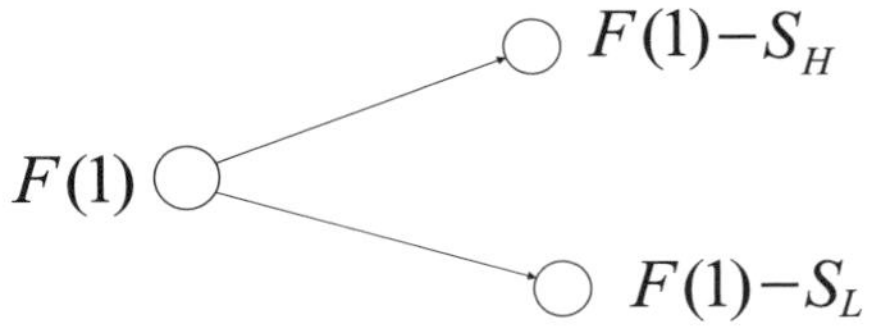

Figure 6.5 Forward contract valuation.

an obligation to be maintained by the buyer and the seller at maturity. The party that has agreed to buy the forward contract is said to assume a long position while the party that agrees to sell is said to assume the short position. Such contracts allow for the parties to exchange the price risk at maturity. For example, a wheat farmer may be exposed to a fall of the wheat price when he brings it to market. He can then enter in a forward contract to sell his wheat today at the fixed price F. At maturity, he may sell wheat at a predetermined price and buy it at the spot rate S from the buyer (say, the baker) of the forward for a payoff of $(F - S)$. The buyer (baker) takes the opposite position for a payoff of $(S - F)$. Both sell and buy are in the market and their position is $[(F - S) + S = F]$ and $[(S - F) - S = F]$ respectively. The parties have therefore perfectly eliminated their wheat price risk as their payoffs are determined at the initiation of the contract. In this example, we evolved into a world where risk can be completely shifted away, which is also the risk-neutral world that conveniently discounts risky payoffs at the risk-free rate (under an appropriately defined probability measure). This transformation to the 'risk-neutral world' breaks down when a seller cannot find a buyer with the exact opposite hedging needs and vice versa. In this case, speculators are needed to take on the risk and a risk neutral world will no longer exist. Depending on whether excess hedging is in long or short forwards, the pressure will be upward or downward compared to the risk-neutral price.

To calculate the forward price at times $t = 1$ and $t = 2$, say $F(1)$ and $F(2)$ we proceed as follows. Consider the first period only, at which the gain can be either $F(1) - S_H$ in case of a price increase or $F(1) - S_L$ in case of a price decrease (see Figure 6.5). Initially nothing is spent and therefore, initially we also get nothing. At present it is thus worth nothing. Assuming no arbitrage (otherwise we would not be able to use the risk-neutral probability), and proceeding as in the previous section, we have:

$$0 = \frac{1}{1 + R_f}[p^*(F(1) - S_H) + q^*(F(1) - S_L)];\ p^* + q^* = 1$$

which is an one equation in one unknown and where R_f is an effective risk-free annual rate. The forward price $F(1)$ resulting from the solution of the equation above is therefore:

$$F(1) = [p^* S_H + q^* S_L] = S(1 + R_f)$$

In other words, *the one period forward price equals the discounted current spot price.* For two periods we note equivalently that when the spot price is S_H or S_L, then (from period 1 to 2):

$$\tilde{F}(2) = \begin{cases} p^* S_{HH} + q^* S_{HL} = (1 + R_f) S_H \ w.p. \ p^* \\ p^* S_{HL} + q^* S_{LL} = (1 + R_f) S_L \ \ w.p. \ q^* \end{cases}$$

As a result, $F(2) = E^* \tilde{F}(2) = p^*(1 + R_f)S_H + q^*(1 + R_f)S_L$ and therefore $F(2) = (1 + R_f)^2 S$ and obviously:

$$F(n) = (1 + R_f)^n S$$

This means that the n periods forward price equals the n periods discounted current spot price (see also Figure 6.4). Of course, using the risk-neutral reasoning, since there is no initial expenditure at the time the forward contract is signed, while at time t, the profit realized equals the difference between the current price and the forward (agreed) on price at time zero which we write by $F(n)$, we have:

$$0 = \frac{1}{(1 + R_f)^n} E^* \left[S_n - F(n) \right] \quad \text{and} \quad F(n) = E^* \left[S_n \right]$$

Since under risk-neutral pricing,

$$S_0 = \frac{1}{(1 + R_f)^n} E^* (S_n) \rightarrow E^* (S_n) = S_0 (1 + R_f)^n$$

we obtain at last the general forward price:

$$F(n) = S_0 (1 + R_f)^n$$

In practice, there may be some problems because decision makers may use forward prices to revalue the spot price. Feedback between these markets can induce an opportunity for arbitrage. Further, it is also necessary to remember that we have assumed a risk-neutral world. As a result, when traders use historical data, there may again be some problems, leading to a potential for arbitrage since the fundamental assumption of rational expectations is violated. For example, if the spot price of silver is \$50, while the delivery price is \$53 with maturity in one year, while interest rates equal 0.08, then the no arbitrage price is: $50(1 + 0.08) = \$54$. This provides an arbitrage opportunity since in one year there is an arbitrage profit of \$1(=54 − 53) that can be realized.

A *futures contract* differs from a forward contract in that it is standardized, openly traded and marked to market. Marking to market involves adjusting an investor's initial margin deposit by the change in the futures contract price each day. If the investor's margin account falls below the maintenance margin, the trader asks the investor to fill the margin account back to the initial margin, posted in the form of interest-bearing T-bonds.

A *futures price* is determined as follows. The futures price one period hence $F(0, 1)$ at time $t = 1$ is set equal to the forward price for that time, since no cost is incurred. In other words, we have, $F(0, 1) = F(1)$. Now consider the futures price in two periods, $F(0, 2)$. If the spot price increases to S_H, the futures price turns out to equal the one-period forward price, or $F_H(1)$ (since only one more period is left till the exercise time). Similarly, if the spot price decreases to S_L,

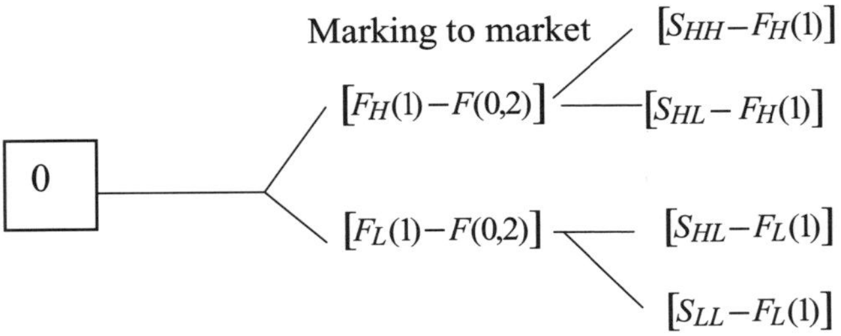

Figure 6.6 Future price valuation.

the future price is now $F_L(1)$. As a result, cash flow payments at the first and second periods are given by Figure 6.6. Initially, *the value of these flows is worth nothing, since nothing is spent and nothing is gained.* Thus, an expectation of futures flows is worth nothing today. That is,

$$0 = \frac{1}{1+R_f}\left[p^*\left(F_H(1) - F(0,2)\right) + q^*\left(F_L(1) - F(0,2)\right)\right]$$

$$+\frac{1}{(1+R_f)^2}\left[\begin{array}{l} p^{*2}\left(S_{HH} - F_H(1)\right) + p^*q^*\left(S_{HL} - F_H(1)\right) + \\ p^*q^*\left(S_{HL} - F_L(1)\right) + q^{*2}\left(S_{LL} - F_L(1)\right) \end{array}\right]$$

which is one equation and three unknowns. However, noting that for the one-period futures (forward) price, we have:

$$F_H(1) = (1+R_f)S_H, \ \ F_L(1) = (1+R_f)S_L$$

Inserting these results into our equation, we obtain the futures price: $F(0,2) = (1+R_f)^2 S_0$ which is equal to the forward price. This is the case, however, because the discount interest rate is deterministic. In a stochastic interest rate framework, this would not be the case. A generalization to n periods yields:

$$F(0,n) = (1+R_f)^n S_0$$

Futures contracts are stated often in terms of a *basis*, measuring the difference between the spot and the futures price. The basis may be mis-priced, however, because of mismatching of assets (cross-hedged), because of maturity (forward versus futures) and the quality of related assets (options). There are some fundamental differences between forward and futures contracts that we summarize in Table 6.1. These relate to the hedging quality of these financial products, their barriers to entry, etc. Further, although under risk-neutral pricing they have the same price, in practice (when interest rates are stochastic as stated above) they can differ appreciably. In many cases, futures contracts are preferred to forward contracts simply because they are more liquid and thereby more 'tradable'.

Example

We compare the consequences of forward and futures contracts on a volume of 100 Dax shares each worth 77E over say five periods. We obtained the following

Table 6.1 Forward and futures contracts: contrasts.

	Forward	Futures
Market	OTC (Private)	Exchange markets
Standard contract	No	Yes
Barrier to entry	Substantial	Weak
Security	Individual	Margin system
Daily controls	No	Yes
Flexibility	Inverse contract	Long–short
Hedge quality	Best	Problematic

results, pointing to differences in cash flow (Table 6.2). Calculations are performed as follows. The cash flows associated with a forward contract of five periods (denoted by *) and associated with a futures contract at period 2 (denoted by **) are given in Table 6.2.

Further, note that the sum of payments of mark to market are equal to the sum of payments of the forward since their initial prices (investment) were the same.

Table 6.2

T	1	2	3	4	5
DAX	7700	7770	7680	7650	7730
Price forward	7800	—	—	—	7730
Price future	7800	7845	7730	7675	7730
Cash flow forward	—	—	—	—	−7000(*)
Cash flow future	—	+4500(**)	−11500	−5500	+5500

(*) = (F[1;T] – F[0;T]) ∗ volume = (7730 – 7800) ∗ 100 = 7000
(**) = (7845–7800) ∗ 100 = 4500

Example
Say that K is the forward delivery price with maturity T while F is the current forward price. The value of the long forward contract is then equal to the present value of their difference at the risk-free rate R_f, or $P_L = (F - K)\,\mathrm{e}^{-R_f T}$. Similarly, the value of the short forward contract is $P_S = -P_L = (K - F)\,\mathrm{e}^{-R_f T}$.

Example: Futures on currencies
Let S be the dollar value of a euro and let $(R_\$,\ R_E)$ be the risk-free rate of the local (dollar) and the foreign currency (euro). Then the relative rate is $(R_\$ - R_E)$ and the future euro price T periods hence is: $F = S\mathrm{e}^{(R_\$ - R_E)T}$.

6.3 RISK-NEUTRAL PROBABILITIES AGAIN

Risk-neutral probabilities, conveniently, allow linear pricing measures. These probabilities are defined in terms of market parameters (although their existence hinges importantly on a risk-free rate, R_f, and rational traders) and differ markedly from historical probabilities. This difference, contrasting two cultures, is due to economic assumptions that the market price of a traded asset 'internalizes' all the past, future states and information that such an asset can be subjected to. If this is the case, and it is so in markets we call *complete markets*, then the current price ought to be determined by its future values as we have shown here. In other words, *the market determines the price and not historical (probability) uncertainty! If there is no unique set of risk-neutral pricing measures, then market prices are not unique and we are in a situation of market incompleteness, unable to value the asset price uniquely.*

It is therefore important to establish conditions for market completeness. Our ability to construct a unique set of risk neutral probabilities for the valuation of the stock at period 1 or the value of buying an option depends on a number of assumptions that are of critical importance in finance and must be maintained theoretically and practically. Pliska (1997) for example, emphasizes the importance of these assumptions and their implications for risk-neutral probabilities. Namely, there can be a linear pricing measure if and only if there are no dominating trading strategies. Further, if there are no dominant trading strategies, then the law of the single price holds, albeit the converse need not necessarily be true. And finally, if there were a dominating trading strategy, then there exists an arbitrage opportunity, but the converse is not necessarily true. Thus, risk-neutral pricing requires, as stated earlier:

- No arbitrage opportunities.
- No dominant trading strategies.
- The law of the single price.

When the assumption of market completeness is violated, it is no longer possible to obtain a unique set of *risk-neutral probabilities*. This means that one cannot duplicate the option with a portfolio or price it uniquely. In this case, an appropriate portfolio is optimized for the purpose of selecting risk-neutral probabilities. Such an optimization problem can be based on the best mean forecast as we shall outline below. These probabilities will, however, be a function of a number of parameters implied by the portfolio and decision makers' preferences and of course the information available to the decision maker. When this is not possible, we can, for a given set of parameters bound the relevant option prices.

6.3.1 Rational expectations and optimal forecasts

Rational expectations mean that economic agents can forecast the 'mean' price (since risk-neutral probabilities imply that an expected value is used to value the asset). In this case, a mean forecast can be selected by minimizing the forecast error (in which case the mean error is null). Explicitly, say that $\{x\} = \{x_1, x_2, \ldots, x_t\}$ stands for an information set (a time series, a stock price record, financial variables etc.). A forecast is thus an estimate based on the information set $\{x\}$ written for convenience by the function $f(.)$ such that $\bar{y} = f(x)$ whose error forecast is

$\varepsilon = y - \bar{y}$ where y is the actual record of the series investigated and its forecast is obtained by minimizing the least squares errors. Assume that the forecast is unbiased, that is, based on all the relevant information available, I; the forecast equals the conditional expectation, or $\bar{y} = E(y\,|I)$ whose error is $\varepsilon = y - \bar{y} = y - E(y\,|I)$. In this case, rational expectations exist when the expected errors are both null and uncorrelated with its forecast as well as with any observation in the information set. This is summarized by the following three conditions of rational expectations:

$$E(\varepsilon) = 0; E(\varepsilon\bar{y}) = E(\varepsilon E(y\,|I)) = 0\ ; E(\varepsilon x) = \mathrm{cov}(\varepsilon, x) = 0, \forall x \in I$$

Of course, there can be various information sets as well as various mechanisms that can be used to generate rational expectations. However, it is essential to note that the behaviour of forecast residual errors determine whether these forecasts are rational expectations forecasts or not.

6.4 THE BLACK–SCHOLES OPTION FORMULA

In continuous time and continuous state, the procedure for pricing options remains the same but its derivation is based on stochastic calculus. We shall demonstrate how to proceed by developing the Black–Scholes model for the valuation of a call option. Let $S(t)$ be a security-stock price at time t and let W be the value of an asset derived from this stock which we can write by the following function $W = f(S, t)$ assumed to be differentiable with respect to time and the security-stock $S(t)$. For simplicity, let the security price be given by a lognormal process:

$$\mathrm{d}S/S = \alpha\,\mathrm{d}t + \sigma\,\mathrm{d}w,\ \ S(0) = S_0$$

The procedure we follow consists in a number of steps:

(1) We calculate $\mathrm{d}W$ by applying Ito's differential rule to $W = f(S, t)$
(2) We construct a portfolio P that consists of a risk-free bond B and number 'a' of shares S. Thus, $P = B + aS$ or $B = P - aS$.
(3) A perfect hedge is constructed by setting: $\mathrm{d}B = \mathrm{d}P - a\,\mathrm{d}S$. Since, $\mathrm{d}B = R_f B\,\mathrm{d}t$, this allows determination of the stockholding 'a' in the replicating portfolio.
(4) We equate the portfolio and option value processes and apply the 'law of the single price to determine the option price. Thus, setting $\mathrm{d}P = \mathrm{d}W$, we obtain a second-order partial differential equation with appropriate boundary conditions and constraints, providing the solution to the option price.

Each of these steps is translated into mathematical manipulations. First, by an application of Ito's differential rule we obtain the option price:

$$\mathrm{d}W = \frac{\partial f}{\partial t}\,\mathrm{d}t + \frac{\partial f}{\partial S}\,\mathrm{d}S + \frac{1}{2}\frac{\partial^2 f}{\partial S^2}(\mathrm{d}S)^2 =$$
$$= \left(\frac{\partial f}{\partial t} + \alpha S\frac{\partial f}{\partial S} + \frac{\sigma^2 S^2}{2}\frac{\partial^2 f}{\partial S^2}\right)\mathrm{d}t + \left(\sigma S\frac{\partial f}{\partial S}\right)\mathrm{d}w$$

Next, we construct a replicating riskless portfolio by assuming that the same amount of money is invested in a bond whose riskless (continuously compounded) rate of return is R_f. In other words, instead of a riskless investment, say that we sell a units of the asset at its price S (at therefore at the cost of aS) and buy an option whose value is $\$W$. The return on such a transaction is $W - aS$. In a small time interval, this will be equal: $\mathrm{d}W - a\,\mathrm{d}S$. To replicate the bond's rate, R_f, we establish an equality between $\mathrm{d}B$ and $\mathrm{d}W - a\,\mathrm{d}S$, thus: $\mathrm{d}B \equiv \mathrm{d}W - a\,\mathrm{d}S$. This argument implies no arbitrage, i.e. the risk-free and the 'risky' market rates should yield an equivalent return, or

$$\mathrm{d}B = \mathrm{d}W - a\,\mathrm{d}S = \mathrm{d}W - a(\alpha S\,\mathrm{d}t + \sigma S\,\mathrm{d}w)$$

Inserting $\mathrm{d}W$, found above by application of Ito's Lemma, we obtain the following stochastic differential equation:

$$\mathrm{d}B = \mu\,\mathrm{d}t + \left(\frac{\partial f}{\partial S} - a\right) S\sigma\,\mathrm{d}w;\ \mu = \alpha S\frac{\partial f}{\partial S} - a\alpha S + \frac{\partial f}{\partial t} + \frac{\sigma^2 S^2}{2}\frac{\partial^2 f}{\partial S^2}$$

Since $\mathrm{d}B = R_f B\,\mathrm{d}t$, this is equivalent to: $\mu\,\mathrm{d}t = R_f B\,\mathrm{d}t$ and

$$\left(\frac{\partial f}{\partial S} - a\right) S\sigma = 0$$

These two equalities lead to the following conditions:

$$a = \frac{\partial f}{\partial S} \quad \text{and} \quad R_f S = \alpha S\frac{\partial f}{\partial S} - a\alpha S + \frac{\partial f}{\partial t} + \frac{\sigma^2 S^2}{2}\frac{\partial^2 f}{\partial S^2}$$

Inserting the value of a, we obtain:

$$R_f f = \frac{\partial f}{\partial t} + \frac{\sigma^2 S^2}{2}\frac{\partial^2 f}{\partial S^2}$$

Since $B = f - aS$, then inserting in the above equation we obtain the following second-order differential equation in $f(S, t)$, the price of the derived asset:

$$-\frac{\partial f}{\partial t} = R_f S\frac{\partial f}{\partial S} + \frac{\sigma^2 S^2}{2}\frac{\partial^2 f}{\partial S^2} - R_f f$$

To obtain an explicit solution, it is necessary to specify a boundary condition. If the derived asset is a European call option, then there are no cash flows arising from the European option until maturity. If T is the exercise date, then clearly,

$$f(0, t) = 0,\ \forall t \in [0, T]$$

At time T, the asset price is $S(T)$. If the strike (exercise price) is K, then if $S(T) > K$, the value of the call option at this time is $f(S, T) = S(T) - K$ (since the investor can exercise his option and sell back the asset at its market price at time T). If $S(T) \leq K$, the value of the option is null since it will not be worth exercising. In other words,

$$f(S, T) = \mathrm{Max}\,[0, S(T) - K]$$

This final condition, together with the asset price partial differential equation, can be solved providing thereby a valuation of the option, or the option price. Black and Scholes, in particular have shown that the solution is given by:

$$W = f(S, t) = S\Phi(d_1) - K\,\mathrm{e}^{-R_f t}\,\Phi(d_2)$$

where

$$\Phi(y) = (2\pi)^{-1/2}\int_{-\infty}^{y} \mathrm{e}^{-u^2/2}\,\mathrm{d}u;\quad d_1 = \left[\frac{\log(S/K) + (T-t)(R_f + \sigma^2/2)}{\sigma\sqrt{T-t}}\right];$$

$$d_2 = d_1 - \sigma\sqrt{T-t}$$

This result is remarkably robust and holds under very broad price processes. Further, it can be estimated by simulation very simply. There are many computer programs that compute these options prices as well as their sensitivities to a number of parameters.

The price of a put option is calculated in a similar manner and is therefore left as an exercise. For an American option, the value of the call equals that of a European call (as we have shown earlier). While for the American put, calculations are much more difficult, although we shall demonstrate at the end of this chapter how such calculations are made.

Properties of the European call are easily calculated using the explicit equation for the option value. It is simple to show that the option price has the following properties:

$$\frac{\partial W}{\partial S} \geq 0,\ \frac{\partial W}{\partial T} > 0,\ \frac{\partial W}{\partial K} < 0,\ \frac{\partial W}{\partial R_f} > 0$$

They express the option's sensitivity. Intuitively, the price of a call option is the discounted expected value (with risk-neutral probabilities) of the payoff $f(S, T) = \text{Max}\,[0, S(T) - K]$. The greater the stock price at maturity (or the lower the strike price), the greater the option price. The higher the interest rate, the greater the discounting of the terminal payoff and thus the stock price increases as it grows at the risk-free rate in the risk-neutral world. The net effect is an increase in the call option price. The longer the option's time to maturity the larger the chances of being in the money and therefore the greater the option price. The call option is therefore an increasing function of time. The higher the stock price volatility, the larger the stock option price. Because of the correspondence between the option price and the stock price volatility, traders often talk of volatility trading rather than options trading, trading upward on volatility with calls and downward with puts.

A numerical analysis of the Black–Scholes equation will reveal these relationships in fact. For example, if we take as a reference point a call option whose strike price is \$160, the expiration date is 5 months, stock current price is \$140, volatility is 0.5 and the compounded risk-free interest rate is 0.15, then the price of this option will turn out to be \$81.82. In other words, given the current parameters, an investor will be willing to pay \$81.82 for the right to buy the stock at a price of \$160 over the next 5 months. If we let the current price vary, we obviously see that as the stock price increases (at the time the call option is acquired), the price of the option increases and vice versa when the current stock prices decreases.

- Variation in the current stock price

Stock price	120	130	140	150	160
Option price	65.01	73.33	81.82	90.46	90.22

- Variation in the expiration date

Expiration T	3	4	5	6	7
Option price	60.53	72.14	81.82	89.98	96.92

- Variation in the strike price

Strike price	140	150	160	170	180
Option price	86.81	84.26	81.82	79.50	77.27

- Variation in volatility

Volatility	0.3	0.4	0.5	0.6	0.7
Option price	68.07	74.76	81.82	88.82	95.52

- Variation in the risk-free compounded interest rate

Interest rate	0.05	0.10	0.15	0.20	0.25
Option price	63.80	73.15	81.82	89.68	96.67

Variations in the strike price, the expiration date of the option, the stock volatility and the compounded risk-free interest rate are outlined below as well. Note that when the stock price, the expiration date, the volatility and the interest rate increase, the option price increases; while when the strike price increases, the option price declines. However, in practice, we note that beyond some level of the strike, the option price increases — this is called the smile and will be discussed in Chapter 8.

Call and put options are broadly used by fund managers for the leverage they provide or to cover a position. For example, the fund manager may buy a call option out of the money (OTM) in the hope that he will be in the money (ITM) and thus make an appreciable profit. If the fund manager owns an important number of shares of a given stock, he may then buy put options at a given price (generally OTM). In the case of a loss, the stock price decline might be compensated by exercising the put options. Unlike the fund manager, a trader can use call options to 'trade on volatility'. If the trader buys a call, the price paid will be associated to the implicit volatility (namely a price calculated based on Black–Scholes option value formula). If the volatility is in fact higher than the implicit volatility, the trader can probably realize a profit, and vice versa.

6.4.1 Options, their sensitivity and hedging parameters

Consider a derived asset as a function of its spot price S, time t, the standard deviation (volatility σ) and the riskless interest rate R_f. In other words, we set the derived asset price as a function $f \equiv f(S, t, \sigma, R_f)$ whose solution is known. Consider next small deviations in these parameters, then by Taylor series approximation, we can write:

$$\mathrm{d}f = \frac{\partial f}{\partial S}\,\mathrm{d}S + \frac{\partial f}{\partial t}\,\mathrm{d}t + \frac{\partial f}{\partial \sigma}\mathrm{d}\sigma + \frac{\partial f}{\partial R_f}\,\mathrm{d}R_f + \frac{1}{2}\frac{\partial^2 f}{\partial S^2}(\mathrm{d}S)^2 + \frac{1}{2}\frac{\partial^2 f}{\partial t^2}(\mathrm{d}t)^2 + \cdots$$

Terms with coefficients of order greater than $\mathrm{d}t$ are deemed negligible (for example $\partial^2 f/\partial t^2$). Each of the terms in the Taylor series expansion provides a measurement of local sensitivity with respect to the parameter defining the derivative price. In particular, in financial studies the following 'Greeks' are defined:

$$\mathit{DELTA} = \Delta = \frac{\partial f}{\partial S} \quad : \text{sensitivity to the spot price}$$

$$\mathit{THETA} = \Theta = \frac{\partial f}{\partial t} \quad : \text{sensitivity to time to expiration}$$

$$\mathit{VEGA} = \upsilon = \frac{\partial f}{\partial \sigma} \quad : \text{sensitivity to volatility}$$

$$\mathit{RHO} = \rho = \frac{\partial f}{\partial R_f} \quad : \text{sensitivity to the interest rate}$$

$$\mathit{GAMMA} = \Gamma = \frac{\partial^2 f}{\partial S^2}$$

Inserting into the derived asset differential equation, we have:

$$\mathrm{d}F = \Delta\,\mathrm{d}S + \Theta\,\mathrm{d}t + \nu\,\mathrm{d}\sigma + \rho\,\mathrm{d}R_f + \frac{1}{2}\Gamma(\mathrm{d}S)^2$$

For the sensitivity equations to make sense, however, the option price, a function of the spot price must assume certain mathematical relationships that imply convexity of the option price. Evidently, this equation will differ from one derived asset to another (for example, for a bond, a currency, a portfolio of securities etc. we will have an equation which expresses the parameters at hand and of course the underlying partial differential equation of the derived asset).The 'Greeks' can be calculated easily using widely available programs (such as MATLAB, MATHEMATICA etc.) that also provide graphical representations of 'Greeks' variations.

6.4.2 Option bounds and put–call parity

Bounds

An option is a right, not an obligation to buy or sell an asset at a predetermined (strike price) and at a given period in the future (maturity). A forward contract differs from the option in that it is an obligation and not a right to buy or sell. Hence, an option is inherently more valuable than a forward or futures contract for it can never lead to a loss at maturity. Explicitly, the value of the forward contract is the discounted payoff at maturity $F_T - F_0$ for a long futures contract and $F_0 - F_T$ for a short. The predetermined futures price F_0 is the strike price K in option terminology. Hence, the option price must obey the following inequalities that provide *lower bounds* on the option's call and put values (where we replaced the forward's price by its value derived previously):

$$c_E \geq \mathrm{e}^{-R_f(T-t)}\left(S\,\mathrm{e}^{R_f(T-t)} - F_0\right) = S - K\,\mathrm{e}^{-R_f(T-t)}$$

$$p_E \geq \mathrm{e}^{-R_f(T-t)}\left(F_0 - S\,\mathrm{e}^{R_f(T-t)}\right) = K\,\mathrm{e}^{-R_f(T-t)} - S$$

where c_E and p_E are the call and the put of a European option. Further, at the limit:

$$\operatorname*{Lim}_{t\to T} c_E = S - K$$

$$\operatorname*{Lim}_{t\to T} p_E = K - S$$

Similarly, we can construct option bounds on American options. Since these options have the additional right to exercise the option in the course of its lifetime, option writers are likely to ask for an additional premium to cover the additional risk transfer from the option buyer. Thus, as long as time is valuable to the investor the following bounds must also hold. Explicitly, let the price of an American and a European call option be C and c respectively while for put options we have also P and p. Then, for a non-paying dividend option it can be verified that (based on the equivalence of cash flows of two portfolios using European and American put and call options):

$$C = c, \; P > p \quad \text{when} \quad R_f > 0$$

Put–call parity

The put–call parity relationship establishes a relationship between p and c. It can be derived by a simple arbitrage between two equivalent portfolios, yielding the same payoff regardless of the stock price. As a result, their value must be the same. To do this, we construct the following two portfolios at time t:

$$\begin{bmatrix} & \text{Time } t & \text{Time } T \\ & S_T < K & S_T > K \\ c + K\,e^{-R_f(T-t)} & K & (S_T - K) + K = S_T \\ p + S_t & K = (K - S_T) + S_T & S_T \end{bmatrix}$$

We see that at time T, the two portfolios yield the same payoff $\text{Max}(S_T, X)$ which implies the same price at time t. Thus:

$$c + K\,e^{-R_f(T-t)} = p + S_t$$

If this is not the case, then there would be some arbitrage opportunity. In this sense, computing European options prices is simplified since, knowing one leads necessarily to knowing the other.

When we consider dividend-paying options, the put-call Parity relationships are slightly altered. Let D denote the present value of the dividend payments during the lifetime of the option (occurring at the time of its ex-dividend date), then:

$$c > S - D - K\,e^{-R_f(T-t)}$$

$$p > D + K\,e^{-R_f(T-t)} - S$$

Similarly, for put-call parity in a dividend-paying option, we have the following bounds:

$$S - D - K < C - P < S - K\,e^{-R_f(T-t)}$$

Upper bounds

An option's upper bound can be derived intuitively by considering the payoff irrespective of the options being American or European. The largest payoff for a put option $\text{Max}[K - S, 0]$ occurs when the stock price is null. The put option upper bound is thus,

$$p < P < K$$

For a European call option, a similar argument leads to the conclusion that the call price must be below the price of the stock at maturity. This is irrelevant to a trader who cannot predict the stock price. However, for an American option, $\text{Max}[S - K, 0]$, the largest payoff occurs when the strike price is set to zero and therefore, the American call upper bound is the stock price,

$$c < C < S$$

These relationships can be obtained also by using arbitrage arguments (see, for example, Merton 1973). Finally other bounds on options are considered in Chapter 8.

6.4.3 American put options

American options, unlike European options may be exercised prior to the expiration date. To value such options, we can proceed intuitively by noting that the valuation is defined in terms of exercise and continuation regions over the stock price. In a continuation region, the value of the option is larger than the value of its exercise and, therefore, it is optimal to wait. In the exercise region, it is optimal to exercise the option and cash in the profits. If the time to the option's expiration date is t, then the exercise of the option provides a profit $K - S(t)$. In this latter case, the exercise time is a 'stopping time', and the problem is terminated. Another way to express such a statement is:

$$f(S, t) = \text{Max}[K - S(t),\ e^{-R_f \text{dt}} Ef(S + \text{d}S, t + \text{d}t)]$$

where $f(S, t)$ is the option price at time t when the underlying stock price is S and one or the other of the two alternatives hold at equality. At the contracted strike time of the option, we have necessarily, $f(S, 0) = K - S(0)$. The solution of the option's exercise time is difficult, however, and has generated a large number of studies seeking to solve the problem analytically or numerically. Noting that the solution is of the barrier type, meaning that there is some barrier $X^*(t)$ that separates the exercise and continuation regions, we have:

$$\begin{cases} \text{If } K - S(t) \geq X^*(t) & \text{exercise region: stopping time} \\ \quad K - S(t) < X^*(t) & \text{continuation region} \end{cases}$$

The solution of the American put problem consists then in selecting the optimal exercise barrier (Bensoussan, 1982, 1985). A number of studies have attempted to do so, including Broadie and Detemple (1996), Carr *et al.* (1992) and Huang *et al.* (1996) as well as many other authors. Although the analytical solution of American put options is hard to achieve, we shall consider here some very simple and analytical problems. For most practical problems, numerical and simulation techniques are used.

Example: An American put option and dynamic programming*
American options, unlike European ones, provide the holder of the option with the option to exercise it whenever he may wish to do so within the relevant option's lifetime. For American call options the call price of the European equals the call price of the American. This is not the case for put options, however. Assume that an American put option derived from this stock is exercised at time $\tau < T$ where T is the option exercise period while the option exercise price is K. Let the underlying stock price be:

$$\frac{\text{d}S(t)}{S(t)} = R_f\,\text{d}t + \sigma\,\text{d}W(t),\ S(0) = S_0$$

Under risk-neutral pricing, the value of the option equals the discounted value (at the risk-free rate) at the optimal exercise time $\tau^* < T$, namely:

$$J(S, T) = \underset{\tau \leq T}{\text{Max}}\, E_S\, e^{-R_f \tau}\,(K - S(\tau), 0)$$

Thus,

$$J(S,t)=\begin{cases}K-S(t) & \text{exercise region: stopping time}\\ \mathrm{e}^{-R_f\,\mathrm{d}t}EJ(S+\mathrm{d}S,t-\mathrm{d}t) & \text{continuation region}\end{cases}$$

In the continuation region we have explicitly:

$$\begin{aligned}J(S,t)&=\mathrm{e}^{-R_f\,\mathrm{d}t}EJ(S+\mathrm{d}S,t+\mathrm{d}t)\\&=\left(1-R_f\,\mathrm{d}t\right)E\left(J(S,t)+\frac{\partial J}{\partial t}\,\mathrm{d}t+\frac{\partial J}{\partial S}\,\mathrm{d}S+\frac{1}{2}\frac{\partial^2 J}{\partial S^2}\,(\mathrm{d}S)^2\right)\end{aligned}$$

which is reduced to the following partial differential equation:

$$-\frac{\partial J}{\partial t}=-R_fJ(S,t)+\frac{\partial J}{\partial S}R_fS+\frac{1}{2}\frac{\partial^2 J}{\partial S^2}\sigma^2S^2$$

While in the exercise region:

$$J(S,t)=K-S(t)$$

For a perpetual option, note that the option price is not a function of time but of price only and therefore $\partial J/\partial t=0$ and the option price is:

$$0=-R_fJ(S)+\frac{\mathrm{d}J}{\mathrm{d}S}R_fS+\frac{1}{2}\frac{\mathrm{d}^2 J}{\mathrm{d}S^2}\sigma^2S^2$$

Here the partial differential equation is reduced to an ordinary differential equation of the second order. Assume that an interior solution exists, meaning that the option is exercised if its (optimal) exercise price is S^*. In this case, the option is exercised as soon as $S(t)\le S^*,\ S^*\le K$. These specify the two boundary conditions required to solve our equation.

- In the exercise region: $J(S^*)=K-S^*$
- For optimal exercise price

$$\frac{\mathrm{d}J(S)}{\mathrm{d}S}\,|_{S=S^*}=-1$$

Let the solution be of the type $J(S)=qS^{-\lambda}$. This reduces the differential equation to an equation we solve for λ:

$$\sigma^2\frac{\lambda(\lambda+1)}{2}-\lambda R_f-R_f=0\quad\text{and}\quad\lambda^*=\frac{2R_f}{\sigma^2}$$

At the exercise boundary S^*, however:

$$J(S^*)=qS^{*-\lambda*}=K-S^*;\frac{\mathrm{d}J(S^*)}{\mathrm{d}S^*}=-\lambda^*qS^{*-\lambda*-1}=-1$$

These two equations are solved for q and S^* leading to:

$$S^*=\frac{\lambda^*}{1+\lambda^*}K\quad\text{and}\quad q=\frac{(\lambda^*)^{\lambda*}K^{1+\lambda^*}}{(1+\lambda^*)^{1+\lambda^*}}$$

And the option price is:

$$J(S) = \left[\frac{(\lambda^*)^{\lambda^*} K^{1+\lambda^*}}{(1+\lambda^*)^{1+\lambda^*}} \right] S^{-\lambda*}, \ \lambda^* = \frac{2R_f}{\sigma^2}, \ S^* = \frac{\lambda^*}{1+\lambda^*} K$$

In other words, the solution of the American put will be:

$$\begin{cases} \text{sell if} & S \leq S^* \\ \text{hold if} & S > S^* \end{cases}$$

When the option time is finite, say T, the condition for optimality is reduced to one of the two equations equating zero:

$$0 = \begin{cases} J(S,t) - (K - S(t)) \\ -\dfrac{\partial J}{\partial t} + R_f J(S,t) + \dfrac{\partial J}{\partial S} R_f S + \dfrac{1}{2} \dfrac{\partial^2 J}{\partial S^2} \sigma^2 S^2 \end{cases}$$

This problem is much more difficult to solve, however. Below, we consider a paper that has in fact been solved analytically.

Example*: A solved case (Kim and Yu, 1993)
Let the underlying price process be a lognormal process:

$$\frac{\mathrm{d}S}{S} = \mu \, \mathrm{d}t + \sigma \, \mathrm{d}w, \ S(0) = S_0$$

As long as the option is kept, its price evolves following the (Black–Scholes) partial differential equation:

$$\frac{\partial f}{\partial t} + \mu S \frac{\partial f}{\partial S} + \frac{\sigma^2 S^2}{2} \frac{\partial^2 f}{\partial S^2} - R_f f = 0$$

In addition, we have the following boundaries:

$$f(S_T, T) = \text{Max}\,[0, K - S_T]$$

$$\lim_{S_t \to \infty} f(S_t, t) = 0$$

$$\lim_{S_t \to S_t^*} f(S_t, t) = K - S_t^*$$

The first boundary condition assumes that the option is exercised at its expiration date T, the second assumes that the value of the option is null if the stock price is infinite (in which case, it will never pay to sell the option) and finally, the third boundary condition measures the option's payoff at its exercise at time t. Let S_t^* be the optimal exercise price at time t, when the option is exercised prior to maturity, in which case (assuming that $f(S_t, t)$ admits first and second derivatives), we have:

$$\lim_{S_t \to S_t^*} \frac{\partial f(S_t, t)}{\partial S_t} = -1$$

Although the solution of such a problem is quite difficult, Carr *et al.* (1992) and Kim and Yu (1993) have shown that the solution can be written as the sum of the option price for the European part of the option plus another sum which accounts for the premium that the American option provides. This expression is explicitly given by:

$$P_0 = P(S_0, 0) = p_0 + \pi$$

$$\pi = R_f K \int_0^T \mathrm{e}^{-R_f t} \int_0^{S_t^*} \Psi(S_t, S_0)\, \mathrm{d}S_t\, \mathrm{d}t - (R_f - \mu) \int_0^T \mathrm{e}^{-R_f t} \int_0^{S_t^*} S_t \Psi(S_t, S_0)\, \mathrm{d}S_t\, \mathrm{d}t$$

where p_0 is the option price of a European put, the 'flexibility premium' associated with the American option is π, while $\Psi(S_t, S_0)$ is the transition probability density function to a price S_t at time t from a price S_0 at $t = 0$.

The analytical, as well as the numerical, solution of these problems is of course cumbersome. In the next chapter we shall consider a similar class of problems that seek to resolve simple problems of the type 'when to sell, when to buy, should we hold' both assets and options.

REFERENCES AND ADDITIONAL READING

Back, K. (1993) Asymmetric information and options, *Review of Financial Studies*, **6**, 435–472.

Beibel, M., and H.R. Lerche (1997) A new look at optimal stopping problems related to mathematical finance, *Statistica Sinica*, **7**, 93–108.

Bensoussan, A. (1982) *Stochastic Control by Functional Analytic Methods*, North Holland, Amsterdam.

Bensoussan, A. (1984) On the Theory of Option Pricing, *ACTA Applicandae Mathematicae*, **2**, 139–158.

Bensoussan, A., and H. Julien (2000) On the pricing of contingent claims with friction, *Mathematical Finance*, **10**, 89–108.

Bergman, Yaacov A. (1985) Time preference and capital asset pricing models, *Journal of Financial Economics*, **14**, 145–159.

Black, F., and M. Scholes (1973) The pricing of options and corporate liabilities, *Journal of Political Economy*, **81**, 637–659.

Boyle, P. P. (1992) *Options and the Management of Financial Risk*, Society of Actuaries, New York.

Brennan, M.J. (1979) The pricing of contingent claims in discrete time models, *The Journal of Finance*, **1**, 53–63.

Brennan, M.J., and E.S. Schwartz (1979) A Continuous Time Approach to the Pricing of Corporate Bonds, *Journal of Banking and Finance*, **3**, 133–155.

Brennan, M.J., and E.S. Schwartz (1989) Portfolio insurance and financial market equilibrium, *Journal of Business*, **62**(4), 455–472.

Briys, E., M. Crouhy and H. Schlesinger (1990) Optimal hedging under intertemporally dependent preferences, *The Journal of Finance*, **45**(4), 1315–1324.

Broadie, M., and J. Detemple (1996) American options valuation, new bounds, approximations and a comparison of existing methods, *Review of Financial Studies*, **9**, 1211–1250.

Brown, R.H., and S.M. Schaefer (1994) The term structure of real interest rates and the Cox, Ingersoll and Ross model, *Journal of Financial Economics*, **35**(1), 3–42.

Carr, P., R. Jarrow and R. Myneni (1992) Alternative characterizations of American Put options, *Mathematical Finance*, **2**, 87–106.

Cox, J.C., J.E. Ingersoll Jr and S. A. Ross (1981) The relation between forward prices and futures prices, *Journal of Financial Economics*, **9**(4), 321–346.
Cox, J.C., and S.A. Ross (1976) The valuation of options for alternative stochastic processes, *Journal of Financial Economics*, **3**, 145–166.
Cox, J.C., and S.A. Ross (1978) A survey of some new results in financial option pricing theory, *Journal of Finance*, **31**, 383–402.
Cox, J.C., S.A. Ross and M. Rubenstein (1979) Option pricing approach, *Journal of Financial Economics*, **7**, 229–263.
Cox, J., and M. Rubinstein (1985) *Options Markets*, Prentice Hall, Englewood Cliffs, NJ.
Davis, M.H.A., V.G. Panas and T. Zariphopoulou (1993) European option pricing with transaction costs, *SIAM Journal on Control and Optimization*, **31**, 470–493.
Duffie, D. (1988) *Security Markets: Stochastic Models*, Academic Press, New York.
Duffie, D. (1992) *Dynamic Asset Pricing Theory*, Princeton University Press, Princeton, N. J.
Geman, H., and M. Yor (1993) Bessel processes, Asian options and perpetuities, Mathematical Finance, **3**(4), 349–375.
Geske, R., and K. Shastri (1985) Valuation by approximation: A comparison of alternative option valuation techniques, *Journal of Financial and Quantitative Analysis*, **20**, 45–71.
Grabbe, J. O. (1991) *International Financial Markets* (2nd edn), Elsevier, New York.
Harrison, J.M., and D.M. Kreps (1979) Martingales and arbitrage in multiperiod security markets, *Journal of Economic Theory*, vol. 20, no. 3, 381–408.
Harrison, J.M., and S.R. Pliska (1981) Martingales and stochastic integrals with theory of continuous trading, *Stochastic Processes and Applications*, **11**, 261–271.
Haug, E.G. (1997) *The Complete Guide to Option Pricing Formulas*, McGraw-Hill, New York.
Henry, C. (1974) Investment decisions under uncertainty: The irreversibility effect, *American Economic Review*, **64**, 1006–1012.
Huang, C.F., and R. Litzenberger (1988) *Foundations for Financial Economics*, North Holland, Amsterdam.
Huang, J., M.G. Subrahmanyan and G. G. Yu (1996) Pricing and hedging Amercian options, *Review of Financial Studies*, **9**(3), pp. 277–300.
Hull, J. (1993) *Options, Futures and Other Derivatives Securities* (2nd edn), Prentice Hall, Englewood Cliffs, NJ.
Jacka, S.D. (1991) Optimal stopping and the American Put, *Journal of Mathematical Finance*, **1**, 1–14.
Jarrow, R.A. (1988) *Finance Theory*, Prentice Hall, Englewood Cliffs, NJ.
Karatzas, I., and S.E. Shreve (1998) *Methods of Mathematical Finance*, Springer, New York.
Kim, I.J., and G. Yu (1990) A simplified approach to the valuation of American options and its application, New York University, Working paper.
Kim, I.J. (1993) The analytic valuation of American options, *Review of Financial Studies*, **3**, 547–572.
Leroy, Stephen F. (1982) Expectation models of asset prices: A survey of theory, *Journal of Finance*, **37**, 185–217.
McKean, H.P. (1965) A free boundary problem for the heat equation arising from a problem in mathematical economics, *Industrial Management Review*, **6**, 32–39.
Merton, R. (1969) Lifetime portfolio selection under uncertainty: The continuous time case, *Review of Economics and Statistics*, **50**, 247–257.
Merton, R.C. (1973) Theory of rational option pricing, *Bell Journal of Economics and Management Science*, **4**, 141–183.
Merton, R.C. (1977) Optimum consumption and portfolio rules in a continuous time model, *Journal of Economic Theory*, **3**, 373–413.
Merton, R.C. (1992) *Continuous Time Finance*, Blackwell, Cambridge, MA.
Pliska, S.R. (1975) Controlled jump processes, *Stochastic Processes and Applications*, **3**, 25, 282.
Ross, S.A. (1976) Options and efficiency, *Quarterly Journal of Economics*, **90**.
Ross, S.A. (1976) The arbitrage theory of capital asset pricing, *Journal of Economic Theory, December*, **13**(3), 341–360.

Smith, C.W. (1976) Option pricing: A review, *Journal of Financial Economics*, **3**, 3–51.
Stoll, Hans, R. (1969) The relationship between put and call option prices, *Journal of Finance*, **24**, 802–824.
Wilmott, P. (2000) *Paul Wilmott on Quantitative Finance*, John Wiley & Sons, Ltd, Chichester.
Wilmott, P., J. Dewynne and S.D. Howison (1993) *Option Pricing: Mathematical Models and Computation*, Oxford Financial Press, Oxford.

CHAPTER 7

Options and Practice

7.1 INTRODUCTION

Option writers, are entrepreneurs in search of profits. As in any fight, fairness is not rewarded. In this spirit, option writers and their financial engineers seek to avoid fair competition by differentiating their products and fitting them to their clients' specific needs or responding to demands of new or seasoned hedgers and speculators. 'The best fight is the one that we cannot lose'! Profits may thus be realized when option writers create a market niche where competition is conspicuously lacking and where there may be some arbitrage profits. Of course, fees have to be set as a function of the writer's power which will depend on the risk of losing important clients, competition from other writers of the same and other products as well as the sophistication of large institutions with their own trading centres. Option writers, as other marketers in other areas, attempt to innovate by creating new products (or variants to currently marketed products), which is in fact a service of intangible characteristics catering to the attitude of investors, firms and individuals to uncertainty. The majority of investors, in fact, abhor uncertainty, while only few seek it or are willing to take positions that the majority will refuse. These participants in financial markets are 'human entities' and market gladiators are prospecting by providing services and trades that are sensitive to their 'psychological and economic' needs and profiles. For risky contracts (in times of crashes for example or very high volatility) speculators will be needed to provide liquidity. Hence, it is not surprising that complete markets and risk-neutral pricing breaks down when this is the case. When the supply of risk is overbearing and there may not be enough 'speculators' to assume it, markets will, at least, become incomplete. Market gladiators are neither risk-seeking nor pure hedgers, however. Management of conservative investment funds such as retirement funds also involve risk. Bonds, assumed generally safe investments, are also risky for they may default or, at least, their value may fluctuate as a function of interest rates, inflation and other economic variables. By the same token, as we have seen in Chapter 5, some hedge fund managers may share information regarding disparities between economic policies and economic fundamentals to generate a herd effect, or a potential run on a currency or an economic entity – bringing them back to alignment with a 'natural economic equilibrium'. Fortunes

Risk and Financial Management: Mathematical and Computational Methods. C. Tapiero
 ISBN: 0-470-84908-8

are made and lost on these 'casino runs' where money is made in an instant and lost in another.

Niche-seeking and product innovation responding to speculative and hedging needs are not the only tools available to market gladiators. A continuous concern for market participation, a concern to avoid regulatory interventions and the urge to avoid tax payments in a legally defensible manner, underpin another source of product innovation. An outstanding example is the creation of Eurodollars deposits of a domestic currency in a foreign country – just as currency swaps were started in the 1980s by the World Bank and have been used since then by CFOs of international firms and banks. Similarly, the concept of offshore funds was conceived to avoid tax payments. This fact underscores current government regulation seeking to limit the use of these funds.

For these reasons, option writers have produced as many tailored options as business imagination can construct. They can be used individually as well as in a combined manner. Financial engineers create and price the cost of products but it is only the market that prices these products. The more 'tailored' the products the less price-efficient the market is, compared to standard widely traded products.

Although most people believe that derivatives are a recent innovation, they date as far back as twelfth-century practices by Flemish traders. The first futures and options contracts resembling current option types were in fact implemented in the seventeenth century in Amsterdam, which was at that time the financial capital of the Western world, and in the rice market of Osaka. Practice in derivatives has truly expanded into global financial markets since mathematical finance and economic theory has made it possible to value such derivatives contracts. The result is an expansion of trades for both 'present and futures trades' are traded at the same time, providing broad flexibilities to financial managers and investors to select the time-risk profile substitutions they prefer.

Options products may be grouped in several categories, summarized by the following:

- Packaged options are usually expressed and valued in terms of plain vanilla options, combining them to generate desired risk properties and profiles. Options strategies such as covered call; protective put; bull and bear spread; calendar spread, butterfly spread, condors, laps and flex, warrants, and others, are such cases we shall consider in this chapter.
- Compound options are derived options based on exercise prices that may be uncertain (for example, warrants, stock options, options on corporate bonds etc.). In this case, it is a 'derived asset twice' – first on the underlying asset and then on some other variable on the basis of which the option is constructed.
- Forward starts are options with different states, awarding thereby the right to exercise the option at several times in the future.
- Path-dependent options depend on the price and the trajectory of other variables. Asian options, knock-out options and many other option types are of this kind, as we shall see in this chapter.

- Multiple assets options involve options on several and often correlated risky assets (such as quantos, exchange options etc.).

In addition, there are options in application areas such as currency options, commodity options, and options on futures, as well as climatic options that assume an increasingly important role in both insurance and energy-related contracts. Warrants are used by firms as call options on the firm's equity. When the warrant is exercised, firms usually issue new stock, thereby diluting current stockholders' equity holdings. There are in addition numerous contracts such as swaps, caps and floors, swaptions and captions etc. that we shall also elaborate on in this chapter. The number of options used in practice is therefore very large and this precludes a complete coverage. For this reason, we shall consider a few such options as examples, providing an opening for both the theoretically and applications minded investor and financial manager. Further study will be needed, however, to appreciate the mathematical intricacies and limitations of dealing with these problems and to augment the sensitivity to the economic rationale such options presume when they are used in practice and are valued by the available quantitative tools.

7.2 PACKAGED OPTIONS

Packaged options are varied. We consider first binary options. A payoff for binary options occurs if the value of the underlying asset $S(T)$ at maturity T is greater than a given strike price K. The amount paid may be constant or a function of the difference $S(T) - K$. The price of these options can be calculated easily if risk-neutral pricing is applicable (since, it equals the discounted value of the terminal payoff). When computations are cumbersome, it is still possible to apply standard (Monte Carlo) simulation techniques and calculate the expected discounted payoff (assuming again risk-neutral pricing, for otherwise simulation would be misleading). The variety of options that pay nothing or 'something' is large and therefore we can briefly summarize a few:

- *Cash or nothing*: Pays A if $S(T) > K$.
- *Asset or nothing*: Pays $S(T)$ if $S(T) \geq K$.
- *Gap*: Pays $S(T) - K$ if $S(T) \geq K$.
- *Supershare*: Pays $S(T)$ if $K_L \leq S(T) \leq K_H$.
- *Switch*: Pays a fixed amount for every day in $[0, T]$ that the stock trades above a given level K.
- *Corridor (or range notes)*: Pays a fixed amount for every day in $[0, T]$ that the stock trades above a level K and below a level L.
- *Lookback options*: Floating-strike lookback options that provide a payout based on a lookback period (say three months), equalling the difference between the largest value and the current price. There are Min and Max lookback options:

$$Min: \; V(T) = \text{Max}\,[0, S(T) - S_{\min}]; \; Max: \; V(T) = \text{Max}[0, S_{\max} - S(T)]$$

- *Asian options*: Asian options are calculated by replacing the strike price by the average stock price in the period. Let the average price be:

$$\bar{S} = \frac{1}{T}\int_0^T S(t)\,\mathrm{d}t;\, t \in [0,T]$$

Then the value of the call and put of an Asian option is simply:

$$\mathit{Put}:\ V(T) = \text{ Max }[0,\, \bar{S} - S(T)];\mathit{Call}:\ V(T) = \text{ Max}[0, S(T) - \bar{S}]$$

- *Exchange*: A multi-asset option that provides the option for a juxtaposition of two assets (S_1, S_2) and given by Max $[S_2(T) - S_1(T), 0]$. Such options can also be used to construct options on the maximum or minimum of two assets. For example, buying the option to exchange one currency (S_1) with another (S_2) leads to:

$$V(T) = \min[S_1(T),\, S_2(T)] = S_2 - \text{Max }[S_2(T) - S_1(T), 0]$$
$$V(T) = \max[S_1(T),\, S_2(T)] = S_1(T) + \text{Max }[S_2(T) - S_1(T), 0]$$

- *Chooser*: Provides the option to buy either a call or a put. Explicitly, say that (T_1, T_2) are the maturity dates of call and put options with strikes (K_1, K_2). Now assume that an option is bought on either of the options with strike $T \le (T_1, T_2)$. The payoff at maturity T is then equal to the max of a call $C[S(T), T_1 - T; K_1]$ and the put $P[S(T), T_2 - T; K_2]$:

$$\text{Max}\{C[S(T),\, T_1 - T;\, K_1],\, P[S(T),\, T_2 - T;\, K_2]\}$$

- *Barrier and other options*: Barrier options have a payoff contingent on the underlying assets reaching some specified level before expiry. These options have knock-in features (namely in barrier) as well as knock-out features (out-barrier). These options are solved in a manner similar to the Black–Scholes equation considered in the previous chapter, except for a specification of boundary conditions at the barriers. We can also consider barrier options with exotic and other features such as options on options, calls on puts, calls on calls, puts on calls etc., as well as calls on forwards and vice versa. These are compound options and are written using both the maturity dates and strike prices for both the assets involved. For example, consider a call option with maturity date and strike price given by (T_1, K_1). In this case, the payoff of a call on a call with maturity date T and strike K is a compound option given by:

$$C_c(T_1,\, K_1) = \text{Max}\{0,\, C[S(T),\, T_1 - T,\, K_1] - K_1\}$$

where $C[S(T), T_1 - T, K_1]$ is the value at time T of a European call option with maturity $T_1 - T$ and strike price K_1. By the same token, a compound put option on a call pays at maturity:

$$P_c(T_1,\, K_1) = \text{ Max }(0,\, K - C(S(T),\, T_1 - T,\, K_1))$$

Practically, the valuation of such options is straightforward under risk-neutral pricing since their value equals their present discounted terminal payoff (at the exercise time).

- *Passport options*: These are options that make it possible for the investor to engage in short/long (sell/buy) trading of his own choice while the option writer has the obligation to cover all net losses. For example, if the buyer of the option takes positions at times t_i, $i = 1, \ldots, n-1$, $t_0 = 0$, $t_n = T$ by buying or selling European calls on the stock, then the passport option provides the following payoff at time$T$ – the option exercise time:

$$\text{Max}\left(\sum_{i=0}^{n-1} u_i[S(t_{i+1}) - S(t_i)], 0\right)$$

where u_i is the number of shares (if bought, it is positive; if sold, it is negative) at time t_i and resolved at period t_{i+1}. In this case, the period profit or loss would be: $[S(t_{i+1}) - S(t_i)]$. Particular characteristic can be added such as the choice of the asset to trade, the number of trades allowed etc.

- *As you like it options*: These options allow the investor to chose after a specified period of time T, whether the option is a call or a put. If the option is European and the call and the put have the same strike price K, then put-call parity can be used. The value at exercise is $\text{Max}(c, p)$ and consists in selecting either the call or the put at the time the option exercise is made. Thus, put-call parity with continuous and compounded discounting and a dividend-paying stock at a rate of q implies (as we shall see later on):

$$c + \mathrm{e}^{-q(T-t)}\text{Max}[0, K\mathrm{e}^{-(R_f-q)(T-t)} - S(t)]$$

In other words, 'as you like it options' consist of a call option with strike K at T and $\mathrm{e}^{-q(T-t)}$ put options with a strike of $[K\mathrm{e}^{-(R_f-q)(T-t)}]$ at maturity T.

The finance trade and academic literature abounds with options that are tailored to clients' needs and to the market potential for such options. Therefore, we shall consider a mere few while the motivated reader should consult the numerous references at the end of the previous and the current chapter for further study and references to specific option types.

7.3 COMPOUND OPTIONS AND STOCK OPTIONS

Stocks are assets that represent equity shares issued by individual firms. They have various forms, granting various powers to stockholders. In general, stockholders are entitled to dividend payments made by the firm and to the right to vote at the firm's assembly. Stocks are also a claim to the value of the firm that they share with bondholders. For example, if the firm defaults on its interest payments, bondholders can force the firm into bankruptcy to recover the loans. A stockholder, a junior claimant in this case, has generally nothing left to claim. Hence, a bondholder has the right to sell the company at a given threshold or, equivalently, the bondholder holds a put on the value of the firm that the stockholder must hold short. Hence, a stock can be viewed as a claim or option on the value of the firm that is shared with bondholders. In practice, managers are often given stock options on their firm so they may align their welfare with those of the shareholders. The rationale of such compensation is that a manager whose income is heavily dependent on an upward move of the firm's stock price will be more likely to pursue an aggressive policy leading to a stock price rise as his payoff is a convex increasing function of the stock price. The shareholders will, of course, benefit from such a rise while it assumes some risk due to the call (stock) option's limited liability granted to the manager. This case illustrates some of the economic limits of risk-neutral pricing, which presumes that risk can be eliminated by trading it away. Further, this supposes the existence of another party willing to take the risk for no extra compensation. This can happen only if markets are perfectly liquid or there exists another investor willing to take on the exact opposite risk. Risk-neutrality presupposes therefore that there is always such an exact opposite. In reality, as is the case for executives' options, the strategy is set up so that the risk is not shifted away. For most applications, risk-neutrality may be used comfortably. But, the more out of the money options are, the less risk can be transferred and, thus, the more speculators are needed to take this risk. This means that in crash times or other extreme events, risk-neutral pricing tends to break down.

With these limitations in mind, we can apply risk-neutral pricing to value options or compound options (options on a stock option or some other underlying asset). Define a stock option (a claim) on the value of the firm (its stock price). To do so, say that a firm has N shares whose price is S and let the firm's debt be expressed by a pure discount bond B with maturity T. Initially, the value of the firm V can be written as $V = NS + B$. Assuming risk-neutral pricing, the stockprice (using an annual risk-free discount rate) over one and two periods is:

$$S = \frac{1}{(1+R_f)} E^* \tilde{S}(1) = \frac{1}{(1+R_f)^2} E^* \tilde{S}(2)$$

For a binomial process, shown in Figure 7.1, we have: $\tilde{S}(1) = (S_h, S_d)$ and $\tilde{S}(2) = (S_{hh}, S_{hd} S_{dd})$. By the same token, we compute recursively the value of the compound (stock) option by:

$$C^c = \frac{1}{(1+R_f)} E^* \tilde{C}^c(1) = \frac{1}{(1+R_f)^2} E^* \tilde{C}^c(2)$$

$$\text{with } \tilde{C}^c(1) = \left(\tilde{C}^c_h, \tilde{C}^c_d\right), \tilde{C}^c(2) = \left(\tilde{C}^c_{hh}, \tilde{C}^c_{hd}, \tilde{C}^c_{dd}\right)$$

$$C^c_{hh} = \text{Max}\,[0, h^2 V - B]; C^c_{dd} = \text{Max}\,[0, d^2 V - B]; C^c_{hd} = \text{Max}\,[0, h\,\text{d}V - B]$$

and therefore:

$$C^c = \left(\frac{1}{1+R_f}\right)^2 \begin{bmatrix} p^{*2}(\text{Max}[0, h^2 V - B]) + \\ 2(1-p^*)p^*\,(\text{Max}\,[0, h\,\text{d}V - B]) + \\ (1-p^*)^2 \text{Max}(0, \text{d}^2 V - B) \end{bmatrix}$$

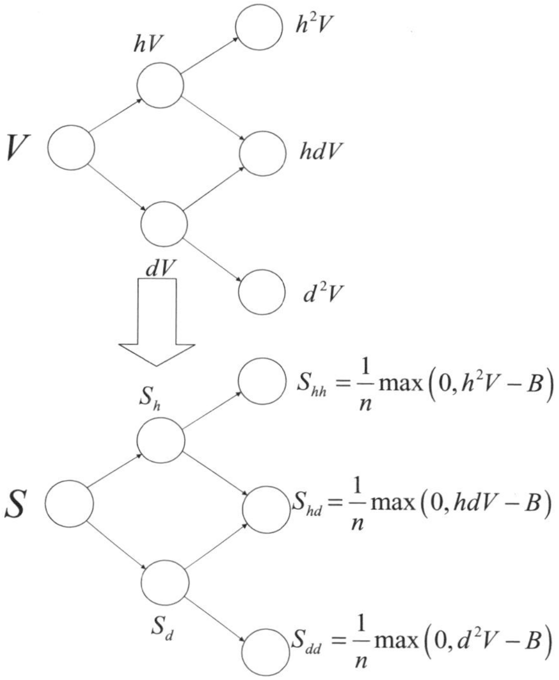

Figure 7.1 Compound option.

Here the risk neutral probability is:

$$p^* = \frac{1 + R_f - d}{h - d}$$

Note that this model differs from the simple plain vanilla model treated earlier, since in this case, $S_h \neq hS$; $S_d \neq \mathrm{d}S$. When the firm has no debt, the firm value is $V = NP$, a portion is invested in a risk-free asset and the other in a risky asset, similarly to the previous binomial case. For example, say that $u = 1.3$ while $d = 0.8$ and the risk free rate is $R_f = 0.1$. Thus the risk-neutral probability is:

$$p^* = \frac{1 + 0.1 - 0.8}{1.3 - 0.8} = \frac{0.3}{0.5} = 0.6,$$

$$C^c = \left(\frac{1}{1.1}\right)^2 \begin{bmatrix} 0.36\,[\text{Max}\,(0,\,1.69V - B)] + \\ 0.48\,[\text{Max}\,(0,\,0.8V - B)] + \\ 0.16\,[\text{Max}\,(0,\,0.64V - B)] \end{bmatrix}$$

Now, if bondholders have a claim on 40 % of the firm value, we have:

$$C^c = V\left(\frac{1}{1.1}\right)^2 [0.36\,(1.29) + 0.48\,(0.4) + 0.16\,(0.24)] = (0.57421)V$$

Problem
What are the effects of an increase of 5 % on bondholders' share of the firm on the option's price?

Problem
High-tech firms (and in particular start-ups) often offer their employees stock options instead of salary increases. When is it better to 'take the money' over the options and vice versa. Construct a model to justify your case.

7.3.1 Warrants

Warrants are compound options, used by corporations that issue call options with their stock as an underlying asset. When the option is exercised, new stock is issued, diluting other stockholders' holdings but adding capital to the corporation. A warrant is valued as follows. Say that V is the firm's value and let there be n warrants, providing the right to buy one share of stock at a price of x and assume no other source of financing. If all warrants are exercised, then the new value of the firm is $V + mx$ and thus, each warrant must at least be worth its price x, or:

$$\frac{V + mx}{N + m} > x$$

This means that a warrant is exercised only if:

$$V + mx > (N + m)x \quad \text{and} \quad V > Nx \quad \text{or} \quad x < V/N = S$$

If the value at time t is: $W(V, \tau)$, $\tau = T - t$ or at time t = 0,

$$W(V, 0) = \begin{cases} \dfrac{V + mx}{N + m} - x & V > Nx \\ 0 & V \leq Nx \end{cases}$$

Then, if we set: $\lambda = 1/[N + m]$ we have $\lambda(V + mx) - x = \lambda V - x(1 - m\lambda)$ and thereby the price $W(V, 0)$ can be written as follows:

$$W(V, 0) = \text{Max}\,[\lambda V - x(1 - m\lambda), 0] = \text{Max}\left[V - \frac{x(1 - m\lambda)}{\lambda}, 0\right]$$

This corresponds to an option whose price is V, the value of the firm, and whose strike (in a Black–Scholes model) is $[x(1 - m\lambda)]/\lambda$, thus, applying the Black–Scholes option pricing formula, we have at any one time:

$$W(V, \tau, \lambda, x) = \lambda W\left(V, \tau, \frac{x(1 - m\lambda)}{\lambda}\right) = W(\lambda V, \tau, x(1 - m\lambda))$$

Therefore, it is possible to value a warrant using the Black–Scholes option formula. For example, say that there are $m = 500$ warrants with a strike price $x = 100$, a time to maturity of $\tau = 0.25$ years, the yearly risk-free interest rate is $R_f = 10\,\%$, the stock price volatility is $\sigma = 20\,\%$ a year and let there be $N = 100\,000$ shares on the market while the current firm value is 150 000. Then, the warrant's price is calculated by:

$$\begin{aligned}W[\lambda V, \tau, x(1-m\lambda)] &= W\{1.5(10^5)\lambda, 0.25, 100[1-(500)\lambda]\}\\ &= W(14.285, 0.25, 95.23)\end{aligned}$$

where $\lambda = 1/(10\,500) = 0.095\,238$.

In a similar manner, other compound options such as *options on a call* (call on call, put on call) and *options on put* (put on put, put on call) etc. may be valued.

7.3.2 Other options

We consider next and briefly a number of other options in a continuous-time framework. Throughout, we assume that the underlying process is a lognormal process.

Options on dividend-paying stocks are options on stocks that pay dividends at a rate of D proportional to the stock price. Note that the underlying price process with dividends is then:

$$\frac{\mathrm{d}S}{S} = (\mu - D)\,\mathrm{d}t + \sigma\,\mathrm{d}W$$

Thus, applying risk-neutral pricing, the partial differential equation that values the option is given by:

$$-R_f V + \frac{\partial V}{\partial t} + (R_f - D)S\frac{\partial V}{\partial S} + \frac{1}{2}\sigma^2 S^2 \frac{\partial^2 V}{\partial S^2} = 0$$

and the boundary condition for a European call option is $V(S,T) = \mathrm{Max}\,(S(T) - K, 0)$. If we apply a no-arbitrage argument as we have in the previous chapter, we are left with $-D\,\mathrm{d}t$ which in essence deflates the price of the stock for the option holder (since the option holder, not owning the stock, does not benefit from dividend distribution). On this basis we obtain the option price deflated by dividends.

Options on foreign currencies are derived in the same manner. Instead of dividends, however, it is the foreign risk-free rate R_{for} that we use. In this case, the partial differential equation is:

$$-R_f V + \frac{\partial V}{\partial t} + (R_f - R_{for})S\frac{\partial V}{\partial S} + \frac{1}{2}\sigma^2 S^2 \frac{\partial^2 V}{\partial S^2} = 0$$

Again, by specifying the appropriate boundaries, we can estimate the value of the corresponding option.

Unlike options on dividend paying stocks, *options on commodities* involve a carrying charge of, say $qS\,\mathrm{d}t$, which is a fraction of the value of the commodity that goes toward paying the carrying charge. As a result, the corresponding differential equation is:

$$-R_f V + \frac{\partial V}{\partial t} + (R_f + q)S\frac{\partial V}{\partial S} + \frac{1}{2}\sigma^2 S^2 \frac{\partial^2 V}{\partial S^2} = 0$$

with an appropriate boundary condition, specified according to the type of option we consider (call, put, etc.)

Options on futures are defined by noting that (see also Chapter 8):

$$F = Se^{R_f(T_F - t)}$$

Thus, the value of an option on a stock and an option on its futures are inherently connected by the above relationship. However, futures differ from options on stock in that the underlying security is a futures contract. Upon exercise, the option holder obtains a position in the futures contract. If we apply Ito's differential rule to determine the value of the option on the futures, we have:

$$\mathrm{d}F = \frac{\partial F}{\partial t} + \frac{\partial F}{\partial S}\mathrm{d}S + \frac{1}{2}\frac{\partial^2 F}{\partial S^2}(\mathrm{d}S)^2 \quad \text{or} \quad \mathrm{d}F + R_f S\,\mathrm{d}t = \mathrm{e}^{R_f(T_F - t)}\,\mathrm{d}S$$

which is introduced in our partial differential equation to yield:

$$-R_f V_F + \frac{\partial V_F}{\partial t} + \frac{1}{2}\sigma^2 S^2 \frac{\partial^2 V_F}{\partial S^2} = 0$$

This is solved with the appropriate boundary constraint (determined by the contract we seek to value). Although options on futures have existed in Europe for some time, they have only recently become available in America. In 1982, the Commodity Futures Trading Commission allowed each commodity exchange to trade options on one of its futures contracts. In that year eight exchanges introduced options. These contracts included gold, heating oil, sugar, T-bonds and three market indices. Options on futures now trade on every major futures exchange. The underlying spot commodities include financial assets such as bonds, Eurodollars and stock indices, foreign currencies such as British pounds and euros, precious metals such as gold and silver, livestock commodities such as hogs and cattle and agricultural commodities such as corn and soybeans.

An *option on a futures price* for say, a commodity, can be related to the spot price by:

$$F = Se^{(R_f - q)(T_F - t)}$$

For a financial asset, q is the dividend yield on the asset, whereas for a commodity (which can be consumed), q must be modified to reflect the convenience yield less the carrying charge. Now in a risk-neutral economy the expected growth rate in the price of a stock which pays continuous dividends at a rate of q is $R_f - q$. In such an economy, the expected growth rate of a futures price should be zero, because trading a futures contract requires no initial investment. This means, that for pricing purposes, the value of q should be R_f. That is, for pricing an option on futures, the futures prices can be treated in the same way as a security paying a continuous dividend yield rate R_f. We substitute $G(t) = F(t)\,e^{-R_f(T_F - t)}$ into Merton's model, leading to the model above and whose solution for a European call option is (as established by Fisher–Black):

$$V_F(0) = \mathrm{e}^{-R_f T_F}[F(0)N(d_1^*) - XN(d_2^*)];\ d_1^* = \frac{1}{\sigma\sqrt{T_F}}\left[\ln\left(\frac{F(0)}{X}\right) + \frac{1}{2}\sigma^2 T_F\right]$$

and $d_2^* = d_1^* - \sigma\sqrt{T_F}$.

7.4 OPTIONS AND PRACTICE

Financial options and engineering is about making money, or, inversely, not losing it. To do so, pricing (valuation), forecasting, speculation and risk reduction through trading (hedging and risk trading management) are an essential activity of traders. For investors, hedgers, speculators and arbitageurs that consider the buying of options (call, put or of any other sort), it is important to understand the many statistics that abound and are provided by financial services and firms and how to apply such knowledge to questions such as:

- When to buy and sell (how long to hold on to an option or to a financial asset). In other words, what are the limits to buy/sell the stock on the asset.
- How to combine a portfolio of stocks, assets and options of various types and dates to obtain desirable (and feasible) investment risk profiles. In other words, how to structure an investment strategy.
- What are the risks and the profit potential that complex derivative products imply (and not only the price paid for them)?
- How to manage productively derivatives and trades.
- How to use derivatives to improve the firm positioning.
- How to integrate chart-trading strategies into a framework that takes into account financial theory, and how to value these strategies and so on.

The decision to buy (a long contract) and sell (a short contract, meaning that the contract is not necessarily owned by the investor) is not only based on current risk profiles, however. Prospective or expected changes in stock prices, in volatility, in interest rates and in related economic and financial markets (and statistics) are essential ingredients applied to solve the basic questions of 'what to do, when and where'. In practice, these questions are approached from two perspectives: the individual investor and the market valuation. The former has his own set of preferences and knowledge, while the latter results from demand and supply market forces interacting in setting up the asset's price. Trading and trading risks result from the diverging assessments of the market and the individual investor, and from external and environmental effects inducing market imperfections. These issues will be addressed here and in the subsequent chapters as well.

In practice, options and derivative products (forward, futures, their combinations etc.) are used for a broad set of purposes spanning hedging, credit and trading risk management, incentives for employees (serving often the dual purpose of an incentive to perform and a substitute for cash outlays in the form of salaries as we saw earlier) and as an essential tool for constructing financial packages (in mergers and acquisitions for example). Derivatives are also used to manage commodity trades, foreign exchange transactions, and interest risk (in bonds, in mortgage transactions etc.). The application of these financial products spans the simple 'buy'/'sell' decisions and complex trading strategies over multiple products, multiple markets and multiple periods of time. While there are many strategies, we shall focus our attention on a selected few. These strategies can be organized as follows:

- Strategies based on plain vanilla options, including:
  —strategies using call options only,
  —strategies using put options only,
  —strategies combining call and put options.
- strategies based on exotic options.
- Other speculating – buy–sell–hold-buy – strategies

These strategies are determined by constructing a portfolio combining options, futures, forwards, stocks etc. in order to obtain cash flows with prescribed and desirable risk properties. The calculation and the design of such portfolios is necessarily computation-intensive, except for some simple cases we shall use here.

7.4.1 Plain vanilla strategies

Call and put options

Plain vanilla options can be used simply and in a complex manner. A *long call* option consists in buying the option with a given exercise price and strike time specified. The portfolio implied by such a financial transaction is summarized in the table below and consists in a premium payment of c for an option, a function of K and T and the underlying process, whose payoff at time T is Max $(S_T - K, 0)$ (see also Figure 7.2):

Time $t = 0$	Final time T
c_0	Max $(S_T - K, 0)$

When the option is bought, the payoff is a random variable, a function of the future (at the expiration date T) market stock price $\tilde{S}_{T|0}$, reflecting the current information the investor has regarding the price process at time $t = 0$ and its strike K:

$$c_0 = E[\mathrm{e}^{-R_i(T-t)}\mathrm{Max}(\tilde{S}_{T|0} - K, 0)]$$

In this case, note that the discount rate R_i is the one applied by the investor. If markets are complete, we have risk-neutral pricing, then of course the option (market) price equals its expectation, and therefore:

$$c_0 = \mathrm{e}^{-R_f(T-t)}E^*[\mathrm{Max}(\tilde{S}_{T|0} - K, 0)]$$

where R_f is the risk-free rate and E^* is an expectation taken under the risk-neutral distribution. An individual investor may think otherwise, however, and his beliefs may of course be translated into (technical) decision rules where the individual attitudes to risk and beliefs as well as private and common knowledge combine to yield a decision to buy and sell (long or short) or hold on to the asset. In general, such call options are bought when the market (and/or the volatility) is bullish or when the investor expects the market to be bullish – in other words, when we expect that the market price may be larger than the strike price at its exercise time. The advantage in buying an asset long, is that it combines a limited downside risk (limited to the option's call price) while maintaining a profit potential if the price of the underlying asset rises above the strike price. For example, by buying a European call on a share of stock whose current price is \$110 at a premium of \$5 with a strike price of \$120 in six months, we limit our risk exposure to the premium while benefiting from any upside movement of the stock above \$120. Generally, for an option contract defined at time t by (c_t, K, T) and traded over the time interval $t \in [0, T]$, the profit resulting from such a transaction at time t is given by: $c_t - c_0$ where c_t is the option price traded at time t, reflecting the information available at this time. Using conditional estimates of this random variable, we have the following expectation:

$$c_t = \mathrm{e}^{-R_f(T-t)} \int_K^{\infty} (\tilde{S} - K)\,\mathrm{d}F_{T|t}(\tilde{S})$$

where $F_{T|t}(\tilde{S})$ is the (risk neutral) distribution function of the stock (asset) price at time T *based on the information at time t.*

There may also be some private information, resulting from the individual investors' analysis and access to information (information albeit commonly available but not commonly used), etc. Of course, as time changes, information will change as well, altering thereby the value of such a transaction. In other words, a learning process (such as filtering and forecasting of the underlying process) might be applied to alter and improve the individual investor's estimates of future prices and their probabilities. For example, say that we move from time t to time $t + \Delta t$. Assuming that the option is tradable (it can be bought or sold at any time), the expected value would be:

$$c_{t+\Delta t} = \mathrm{e}^{-R_f(T-t-\Delta t)} \int_K^{\infty} \left(\tilde{S} - K\right)\,\mathrm{d}F_{T|t+\Delta t}(\tilde{S})$$

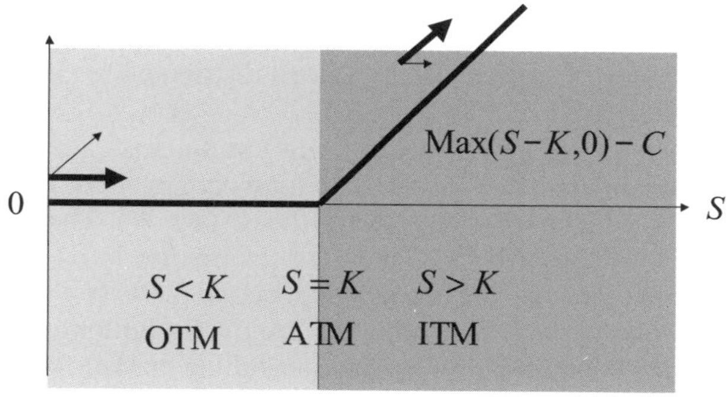

Figure 7.2 A plain call option.

where $F_{T|t+\Delta t}(\tilde{S})$ expresses the future stock price distribution reached at time $t + \Delta t$. Under risk-neutral pricing, the value of the option at any time equals its discounted expectation or $c_{t+\Delta t}$ changing over time and with incoming new information. As a result, if the option was bought at time t, it might be sold at time $t + \Delta t$ for a profit (or loss) of $(c_{t+\Delta t} - c_t)$, or it might be maintained with the accounting change in the value of the option registered if it is not sold. If the probabilities are 'objective' historical distributions, then of course, the profit – loss parameters are in fact random variables with moments we can calculate (theoretically or numerically). The decision to act one way or the other may be based on beliefs and the economic evaluation of the fundamentals or on technical analyses whose ultimate outcomes are: 'is the price of the stock rising or decreasing', 'will the volatility of the stock increase or decrease', 'are interest rates changing or not and in what direction' etc. The option's Greek sensitivity parameters (Delta, Vega etc.) provide an assessment of the effects of change in the respective parameters.

The same principle applies to other products and contracts that satisfy the same conditions, such as commodity trades, foreign exchange, industrial input factors, interest rates etc. In most cases, however, each contract type has its own specific characteristics that must be accounted for explicitly in our calculations. Further, for each contract bought there must be an investor (or speculator) supplying such a contract. In our case, in order to buy a long call, there must also be seller, who is buying the call short – in other words, collecting the premium c against which he will assume the loss of a profit in the case of the stock rising above the strike price at its exercise. Such transactions occur therefore because investors/speculators have varied preferences, allowing exchanges that lead to an equilibrium where demand and the supply for the specific contract are equal.

Example: Short selling

A short sell consists in the promise to sell a security at a given price at some future date. To do so, the broker 'borrows' the security from another client and sells it in the market in the usual way. The short seller must then buy back the security at some specified time to replace it in the client's portfolio. The short seller assumes then all costs and dividends distributed in the relevant period of the financial sell contract. For example, if we short sell 100 GM shares sold in January at \$25 while in March the contract is exercised when the price of the stock is \$20, and in February 1, a dividend of \$1 was distributed to GM shareholders, then, the short seller profit is: $100(25 - 20 - 1) = \$400$.

In a similar manner we may consider *long put options*. They consist in the option to sell a certain asset at a certain date for a certain strike price K or at the market price $S(T)$, whichever is largest, or Max $(S(T), K)$. The cost of such an option is denoted by p, the strike is K and the exercise time is T. For a speculator, such an option is bought when the investor expects the market to be bearish and/or the asset volatility to be bullish. Unlike the long call, the long put combines a limited upside exposure with a high gearing in a falling market. The costs/payoffs of a portfolio based on a single long put option contract is therefore given by the following.

A put option

Time $t = 0$	Final time T
$p + S_0$	Max (K, S_T)

As a result, in risk-neutral pricing, the price of the put is:

$$p = e^{-R_f T} E^*[\text{Max}\ (K, \tilde{S}_{T|0})] - S_0$$

Equivalently, it is possible at the time the put is bought, that a forward contract $F(0,T)$ for the security to be delivered at time T is taken (see also Chapter 8). In this case (assuming again risk-neutral pricing), $F(0,T) = e^{R_f T} S_0$ and therefore the value of a put option can be written equivalently as follows:

$$p = e^{-R_f T} E^*\{\text{Max}[K - F(0,T), \tilde{S}_{T|0} - F(0,T)]\}$$

Put–call parity can be proved from the two equations derived here as well. Note that the value of the call can be written as follows:

$$c = e^{-R_f T} E^*(\text{Max}[\tilde{S}_{T|0} - K, 0]) = e^{-R_f T} E^*(\text{Max}[\tilde{S}_{T|0}, K]) - e^{-R_f T} K$$

and therefore: $c + e^{-R_f T} K = e^{-R_f T} E^*(\text{Max}\ [\tilde{S}_{T|0}, K])$. Thus, we have the call-put parity seen in the previous chapter:

$$p + S_0 = c + e^{-R_f T} K$$

Since $p \geq 0$ and $c \geq 0$ the put–call parity implies trivially the following bounds: $p \geq e^{-R_f T} K - S_0$ and $c \geq S_0 - e^{-R_f T} K$. Thus, for a trader selling a put, the put writer, the maximum liability is the value of the underlying stock. These transactions are popular when they are combined with another (or several other) transactions. These strategies are used to both speculate and hedge. For example, say that a put option on CISCO is bought for a strike of \$42 per share whose premium is \$2.25 while the stock current price is \$46 per share. Then, an investor would be able to contain any loss due to stock decline, to the premium paid for the put (thereby hedging downside losses). Thus, if the stock falls below \$42, the maximum loss is: $(46 - 42) + 2.25 = \$6.25$ per share. While, if the stock increases to \$52, the gain would be: $-2.25 + \{52 - 46\} = 3.75$.

Some firms use put options as a means to accumulate information. For example, some investment firms buy puts (as warrants) in order to generate a signal from the firm they intend to invest in (or not). If a firm responds positively to a request to a put (warrant) contracts, then this may be interpreted as a signal of 'weakness' – the firm willing to sell because it believes it is overpriced – and vice versa, if it does not want to sell it might mean that the firm estimates that it is underpriced. Such information eventually becomes common knowledge, but for some investment firms, the signals they receive are private information which remains private for at least a certain amount of time and provides such firms with a competitive advantage (usually, less than four months) which is worth paying for and to speculate with.

7.4.2 Covered call strategies: selling a call and a share

Say that a pension fund holds 1000 GM shares with a current price of \$130 per share. A decision is reached to sell these shares at \$140 as well as a call expiring in 90 days with an exercise price of \$140 at a premium of \$5 per share. As a result, the fund picks up an immediate income of \$5000 while the fund would lose its profit share for the stock when it reaches levels higher than \$140. However, since it intended to sell its holdings at \$140 anyway, such a profit would have not been made. This strategy is called a *covered call.* It is based on a portfolio consisting in the purchase of a share of stock with the simultaneous sale of a (short) call on that stock in order to pick up an extra income (the call option price), on a transaction that is to be performed in any case. The price (per share) of such a transaction is the expectation of the following random variable written as follows:

$$S - c = \mathrm{e}^{-R_f T} E^*[\mathrm{Min}\,(K, \tilde{S}_{T|0})]$$

where c is the call premium received, K is the strike price of the option with an exercise at time T, while S_T is the stock price at the option exercise time. Under risk-neutral pricing, there is a gain for the seller of the call since he picks up the premium on a transaction that he is likely to perform anyway.

A covered call

Time t	Final time T
$S - c$	$\mathrm{Min}\,(K, S_T)$

The buyer of the (short) call, however, is willing to pay a premium because he needs the option to limit his potential losses. As a result, a market is created for buyers and sellers to benefit from such transaction. The terminal payoff of a covered call is given in the Figure 7.3.

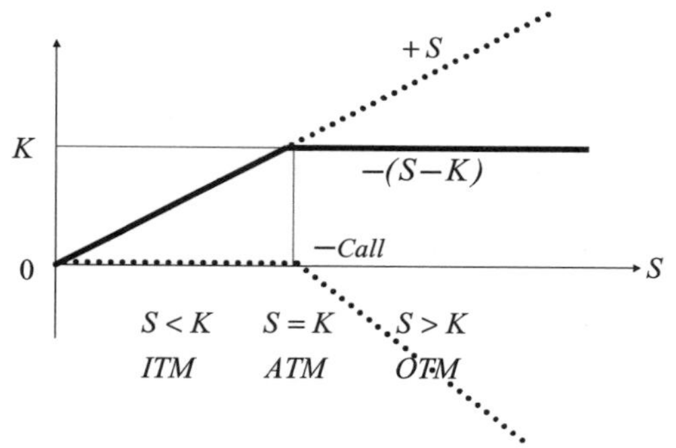

Figure 7.3 Terminal payoffs: covered call.

Problem
Write the payoff equation for a covered call when the seller of the stock is financed by a forward.

7.4.3 Put and protective put strategies: buying a put and a stock

The protective put is a portfolio that consists in buying a stock and a put on the stock. It is a strategy used when we seek protection from losses below the put option price. For this reason, it is often interpreted as an insurance against downside losses. Buying a put option on stock provides an investor with a limit on the downside risk while maintaining the potential for unlimited gains. Banks for example, use a protective put to protect their principal from interest rate increases. Similarly, say that a euro firm receives an income from the USA in dollars. If the dollar depreciates or the euro increases, it will of course be financially hurt. To protect the value of this income (in the local currency), the firm can buy a put option by selling dollars and obtain protection in case of a downward price movement. The protective put has therefore the following cash flow, summarized in the table below.

A protective put strategy

Time $t = 0$	Final time T
$p + S$	$\text{Max}(K, S_T)$

where p is the put premium and:

$$p + S = \mathrm{e}^{-R_f T} E^*[\text{Max}\ (K, \tilde{S}_{T|0})]$$

or, equivalently:

$$p = \mathrm{e}^{-R_f T} E^*[\text{Max}\ (K - \mathrm{e}^{-R_f T} S, \tilde{S}_{T|0} - \mathrm{e}^{-R_f T} S)]$$

This is shown graphically in Figure 7.4 and illustrated by exercise of the put.

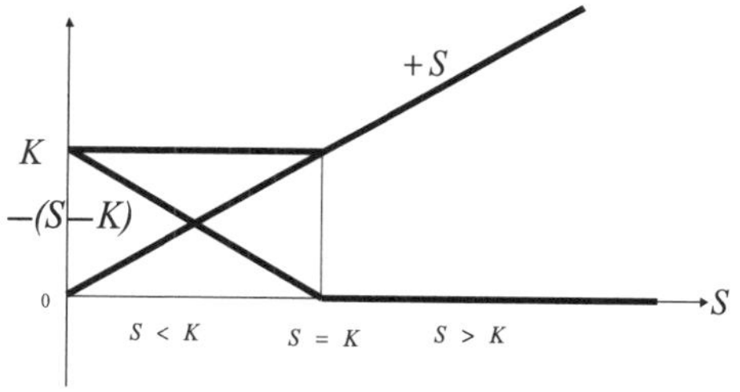

Figure 7.4 Terminal payoffs: Protective put.

7.4.4 Spread strategies

Spread strategy consists in constructing a portfolio by taking position in two or more options of the same type but with different strike prices. For example, a spread over two long call options can be written by $W = c_1 + c_2$ Where c_i, $i = 1, 2$ are call option prices associated with the strike prices K_i, $i = 1, 2$. The following table summarizes a spread strategy cash flow.

A call–call spread strategy

Time $t = 0$	Final time T
$c_1 + c_2$	Max $(K_1, S_T) +$ Max (K_2, S_T)

There are also long and short put spread versus call and, vice versa, long and short call versus put. In a short put spread versus call, we sell a put with strike price B and sell a put at a lower strike A and buy a call at any strike. The long call will generally be at a higher strike price, C, than both puts. The return profile turns out to be similar to that of a short put spread, but the long call provides an unlimited profit potential should the underlying asset rise above C. Such a transaction is performed when the investor expects the market and the volatility to be bullish. In a rising market the potential profit would be unlimited while in a falling market, losses are limited. This is represented graphically below. For example, in the expectation of a stock price increasing, a speculator will buy a call at a low strike price K_1 and sell another with a high strike price $K_2 > K_1$ (this is also called a *bull spread*). This will have the effect of delimiting the profit/loss potential of such a trade, as shown in the Figure 7.5. In this case, the value of such a spread is as follows.

A bull spread
The premium collected initially at time $t = 0$ is $c_1 - c_2$, while (under risk-neutral pricing):

$$c_1 - c_2 = e^{-R_f T} E^*\{\text{Max}[(\tilde{S}_{T|0} - K_2), 0]\} - e^{-R_f T} E^*\{\text{Max}[(\tilde{S}_{T|0} - K_1), 0]\}$$

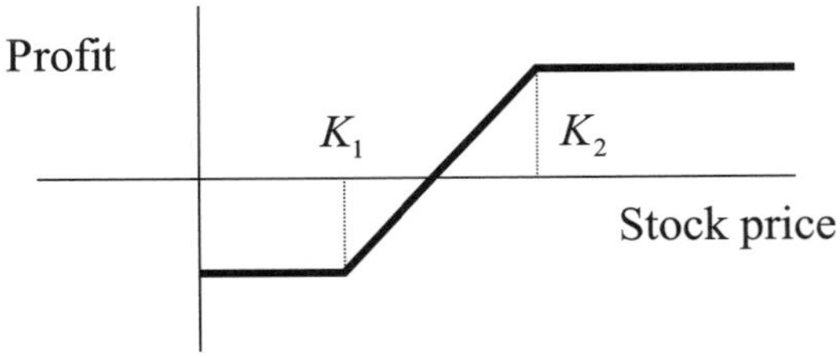

Figure 7.5 A bull spread.

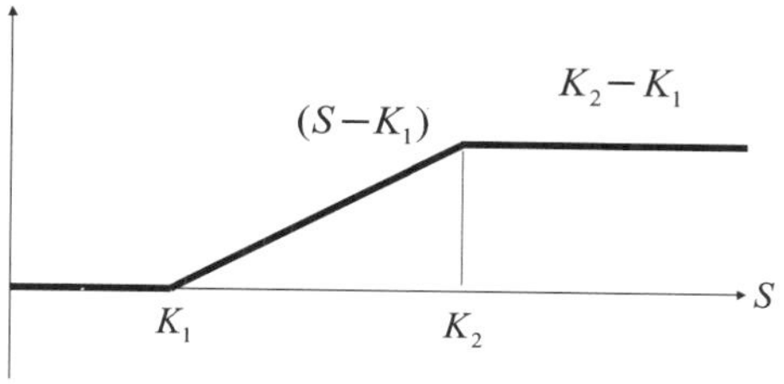

Figure 7.6 Terminal payoffs: bullish spread.

By contrast, in the expectation that stock prices will fall, the investor/speculator may buy a call option with a high strike price and sell a (short) call with a lower strike price. *This is also called a bear spread.* In this case, the initial cash inflow would be.

$$0 = c_1 - c_2 + e^{-R_f T} E^*\{\text{Max}[(\tilde{S}_{T|0} - K_2), 0]\} - e^{-R_f T} E^*\{\text{Max}[(\tilde{S}_{T|0} - K_1), 0]\}$$

For a bullish spread, however, we buy and sell:

$$\text{Long call } (K_1, T) + \text{ Short call } (K_2, T) = C(K_1, T) - C(K_2, T)$$

while the payoff at maturity is:

$S_T < K_1$	$K_1 < S_T < K_2$	$S_T > K_2$	Calls
0	$S_T - K_1$	$S_T - K_1$	$C(K_1, T)$
0	0	$-(S_T - K_2)$	$-C(K_2, T)$
0	$S_T - K_1$	$K_2 - K_1$	

This is given graphically in Figure 7.6.

7.4.5 Straddle and strangle strategies

A straddle consists in buying both a call and a put on a stock, each with the same strike price, K, at the exercise date, T, and selling a call at any strike. A straddle is used by investors who believe that the stock will be volatile (moving strongly but in an unpredictable direction). A straddle can be long and short. A long straddle versus call consists of buying both a call and a put at the same strike but in addition, selling a call at any strike. Similarly, expectation of a takeover or an important announcement by the firm is also a good reason for a straddle. The cash flows associated with a straddle based on buying a put and a call with the same exercise price and the same expiration date is thus (see Figure 7.7):

$$p + c = e^{-R_f T} E^* \text{Max}[\tilde{S}_{T|0} - K, K - \tilde{S}_{T|0}]$$

In contrast to a straddle, a *strangle* consists of a portfolio but with different strikes for the put and the call. The graph for such a strangle is given in Figure 7.8.

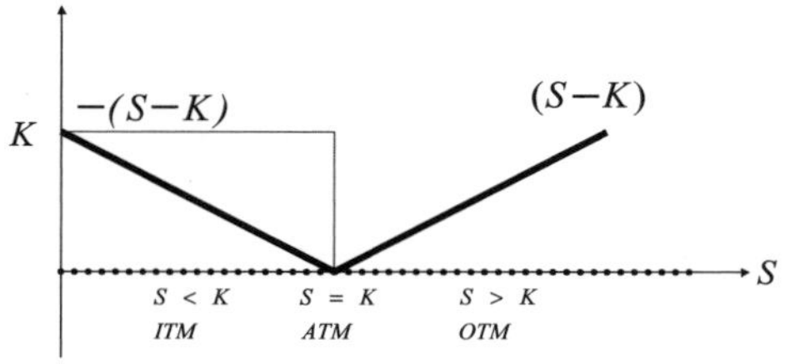

Figure 7.7 A Straddle strategy: terminal payoffs.

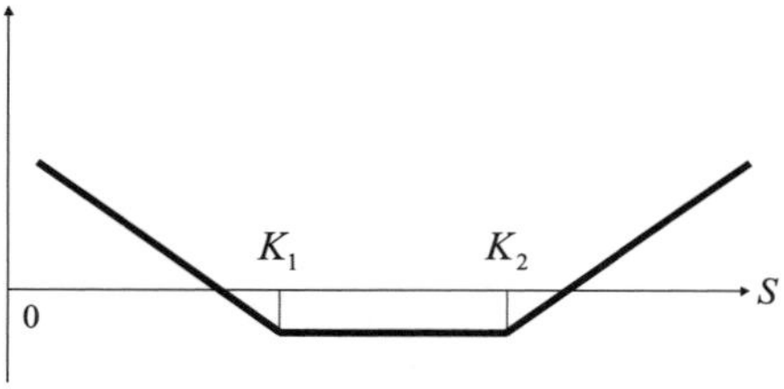

Figure 7.8 A strangle strategy.

7.4.6 Strip and strap strategies

If the investor believes that there is soon to be a strong stock price move, but with potentially a stronger probability of a downward move, then the investor can use a *strip*. This is similar to a straddle, but it is asymmetric. To implement a strip, the investor will take a long position in one call and two in puts. Inversely, if the investor believes that there is a stronger chance that the stock price will move upwards, then the investor will implement a *strap*, namely taking a long position on two calls and one on a put. In other words, the economic value of such strategies are given by the following payoff:

Strip strategies

	$S_T < K$	$S_T > K$
1 Long call	0	$S_T - K$
2 Long puts	$2(K - S_T)$	0
	$2(K - S_T)$	$S_T - K$

The graph of such a cash flow is given in Figure 7.9.

Strap strategies

In a strap (see Figure 7.10) we make an equal bet that the market will go up or down and thus a portfolio is constructed out of two calls and one put. The payoffs

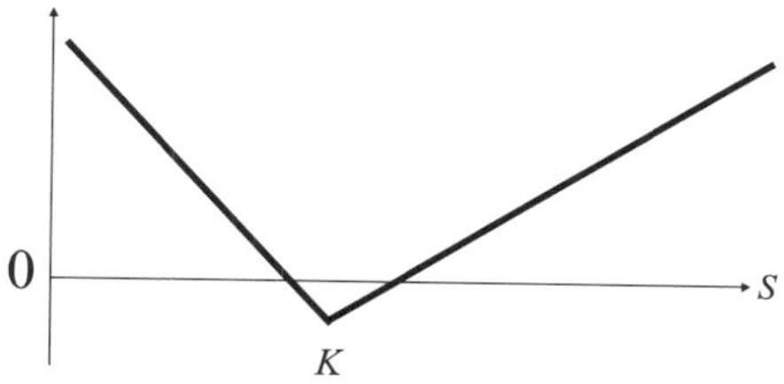

Figure 7.9 A strip strategy.

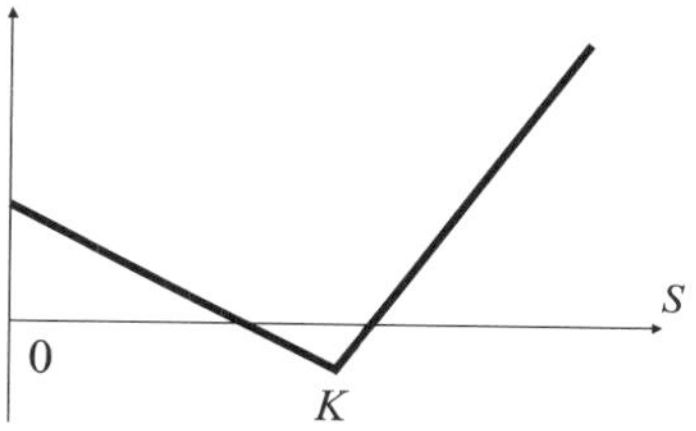

Figure 7.10 A Strap strategy.

are given by the following table.

	$S_T < K$	$S_T > K$
2 Long calls	0	$2(S_T - K)$
1 Long put	$(K - S_T)$	0
	$(K - S_T)$	$2(S_T - K)$

7.4.7 Butterfly and condor spread strategies

When investors believe that prices will remain the same, they may use *butterfly spread strategies* (see Figure 7.11). This will ensure that if prices move upward or downward, then losses will be limited, while if prices remain at the same level the investor will make money. In this case, the investor will buy two call options, one with a high strike price and the other with a low strike price and at the same time, will sell two calls with a strike price roughly halfway in between (roughly equalling the spot price). As a result, butterfly spreads merely involve options with three different strikes. *Condor* spreads are similar to butterfly spreads but involve options with four different strikes.

7.4.8 Dynamic strategies and the Greeks

In the previous chapter we have drawn attention to the 'Greeks', expressing the option's price sensitivity to the parameters used in calculating the option's price. These measures are summarized below with an interpretation of their signs

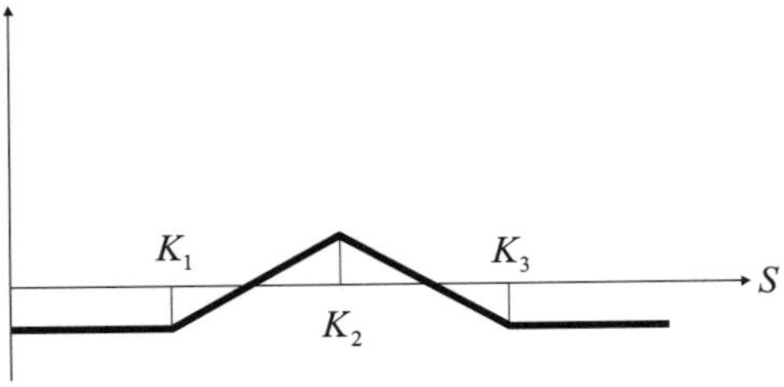

Figure 7.11 A butterfly spread strategy.

(Willmott, 2002). This sensitivity is used practically to take position on options and stocks:

Delta

$$\Delta = \left\{ \frac{\text{Exposure to direction of price changes, Dollar change in position value}}{\text{Dollar change in underlying security price}} \right.$$

When Delta is negative, it implies a bearish situation with a position that benefits from a price decline. When Delta is null, then there is little change in the position as a function of the price change. Finally, when the sign is positive, it implies a bullish situation with a position that benefits from a price increase.

Lambda

$$\Lambda = \left\{ \frac{\text{Leverage of position price elasticity, Percentage change in position value}}{\text{Percentage change in underlying security price}} \right.$$

The Lambda has the same implications as the Delta.

Gamma

$$\Gamma = \left\{ \frac{\begin{array}{l}\text{Exposure to price instability; `non - directional price change',}\\ \text{Change in position Delta}\end{array}}{\text{Dollar change in underlying security price}} \right.$$

When Gamma is positive or negative, the position benefits from price instability, while when it is null, it is not affected by price instability.

Theta

$$\Theta = \left\{ \frac{\text{Exposure to time decay, Dollar change in position value}}{\text{Decrease in time to expiration}} \right.$$

When Theta is negative the position value declines as a function of time and vice versa when Theta is positive. When Theta is null, the position is insensitive to time.

Kappa

$$\kappa = \left\{ \frac{\text{Exposure to changes in volatility of prices, Dollar change in position value}}{\text{One percent change in volatility of prices}} \right.$$

When Kappa is negative, the position benefits from a drop in volatility and vice versa. Of course, when Kappa is null, then the position is not affected by volatility.
The option strategies introduced above are often determined in practice by the 'Greeks'. That is to say, based on their value (positive, negative, null) a strategy is implemented. A summary is given in Table 7.1 (Wilmott, 2002).

Table 7.1 Common option strategies.

Delta	Gamma	Theta	Strategy	Implementation
Positive	Positive	Negative	Long call	Purchase long call option
Negative	Positive	Negative	Long put	Purchase long put option
Neutral	Positive	Negative	Straddle	Purchase call and put, both with same exercise and expiration date
Neutral	Positive	Negative	Strangle	Purchase call and put, each equally out of the money, and write a call and a put, each further out of the money, and each with the same expiration date
Neutral	Positive	Negative	Condor	Purchase call and put, each equally out of the money, and write a call and a put, each further out of the money than the call and put that were purchased. All options have the same expiration date
Neutral	Negative	Positive	Butterfly	Write two at-the-money calls, and buy two calls, one in the money and the other equally far out of the money
Positive	Neutral	Neutral	Vertical spread	Buy one call and write another call with a higher exercise price. Both options have the same time to expiration
Neutral	Negative	Positive	Time spread	Write one call and buy another call with a longer time to expiration. Both options have the same exercise price
Neutral	Positive	Negative	Back spread	Buy one call and write another call with a longer time to expiration. Both options have the same exercise price.
Neutral	Neutral	Neutral	Conversion	Buy the underlying security, write a call, and buy a put. The options have the same time to expiration and the same exercise price

Problem

Discuss the strategy buy a call and sell a put $c - p$. Discuss the strategy to buy a stock and borrow the present value $S - K/(1 + R_f)^T$. Verify that these two strategies are equivalent. In other words verify that: $c - p = S - K/(1 + R_f)^T$.

Problem: A portfolio with return guarantees

Consider an investor whose initial wealth is W_0 seeking a guaranteed wealth level $aS_T + b$ at time T where S_T is the stock price. This guarantee implies initially that $W_0 \geq aS_0 + b\,\mathrm{e}^{-R_f T}$ where R_f is the risk-free rate. Determine the optimal portfolio which consists of an investment in a zero coupon bond with nominal value, B_T at time T or $B_T = B_0 \mathrm{e}^{R_f T}$, an investment in a risky asset whose price S_t is a given by a lognormal process and finally, an investment in a European put option whose current price is P_0 with an exercise price K.

7.5 STOPPING TIME STRATEGIES*

7.5.1 Stopping time sell and buy strategies

Buy low, sell high is a sure prescription for profits that has withstood the test of time and markets (Connolly, 1977; Goldman *et al.*, 1979). Waiting too long for the high may lead to a loss, while waiting too little may induce insignificant gains and perhaps losses as well. In this sense, trade strategies of the type 'buy–hold–sell' involve necessarily both gains and losses, appropriately balanced between what an investor is willing to gamble and how much these gambles are worth to him.

Risk-neutral pricing (when it can be applied) has resolved the dilemma of what utility function to use to price uncertain payoffs. Simply none are needed since the asset price is given by the (rational) expectation of its future values discounted at the risk-free discount rate. Market efficiency thus implies that it makes no sense for an investor to 'learn', to 'seek an advantage' or even believe that he can 'beat the market'. In short, it denies the ability of an investor who believes that he can be cunning, perspicacious, intuition-prone or whatever else that may lead him to make profits by trading. In fact, in an efficient market, one makes money only if one is lucky, for in the short run, prices are utterly unpredictable. Yet, investors trade and invest trillions daily just because they think that they can make money. In other words, they may know something that we do not know, are plain gamblers or plain 'stupid'. Or perhaps, they understand something that makes the market incomplete and find potential arbitrage profits. This presumption, that traders and investors trade because they believe that markets are incomplete, results essentially in 'market' forces that will correct market incompleteness and thus lead eventually to efficient markets. For as long as there are arbitrage profits they will take advantage of the market inefficiency till it is no longer possible to do so and the market becomes efficient. In this sense a market may provide an opportunity for profits which cannot be maintained forever since the market will be self-correcting. The cunning investor is thus one who understands that an opportunity to profit from 'inefficiency' cannot last forever, knows when to detect it, identify and prospect on this opportunity and, importantly, knows as well when to get out. For at some time, this inefficiency will be obliterated by other investors who have realized that profits have been made and, wishing to share in it, will necessarily render the market efficient.

For these reasons, in practice, there may be problems in applying the risk-neutral framework. Numerous situations to be elaborated in Chapter 9, such as risk-sensitive individual investors expressing varied preferences, using discrete time and historical data, using private information etc. contribute to market inefficiencies. In such a framework, future price distributions possess risk properties that may be more or less desired by the individual investor/speculator who may either use a risk-adjusted discount factor or provide a risk qualification for the financial decisions he may wish to assume.

In this section we shall consider first a risk-neutral framework (essentially to maintain the risk-free discounting of risk-neutral pricing) and consider the decision to buy or sell an asset under the martingale probability measure. We shall show, using an example, that no profits can be made when the trade strategy is to sell as soon as a given price (greater than the current price) is reached. Of course, in practice we can use the actual probability measure (rather than the risk-neutral measure) but then it would be necessary to apply the individual investor's discount rate. This procedure is followed subsequently and we demonstrate through examples how the risk premium associated with a trading strategy for a risk-sensitive investor/trader is determined. Finally, we consider a quantile risk-sensitive (Value at Risk – VaR) investor and assess a number of trading strategies for such an investor. Both discrete and continuous-time price processes are considered and therefore some of the problems considered may be in an incomplete market situation. For practical purposes, when the risk-neutral framework can be applied, simulation can be used to evaluate a trading strategy, however complex it may be (since simulation merely applies an experimental approach and assess performance of the trading strategy based on frequency and average concepts). When this is not the case, simulation must be applied carefully for the discount rate applied to simulated cash streams must necessarily account for the premium payments to be incurred for the cash flow's random characteristics.

To demonstrate the technique used, we begin again with the lognormal stock price process:

$$\frac{\mathrm{d}S}{S} = \alpha \mathrm{d}t + \sigma \mathrm{d}W, \ S(0) = S_0$$

Its solution at time t is simply:

$$S(t) = S(0)\exp\left(\alpha - \frac{\sigma^2}{2}\right)t + \sigma W(t), \ W(t) = \int_0^t \mathrm{d}W(t)$$

where $W(t)$ is the Brownian motion. It is possible to rewrite this expression as follows (adding and subtracting in the exponential $R_f t$):

$$S(t) = S(0)\exp\left\{\left(R_f - \frac{\sigma^2}{2}\right)t + \sigma\left(W(t) + \frac{\alpha - R_f}{\sigma}t\right)\right\}$$

In order to apply the risk-neutral pricing framework, we define another probability measure or a numeraire with respect to which expectation is taken. Let

$$W^*(t) = \left(W(t) + \frac{\alpha - R_f}{\sigma}t\right),$$

then

$$S(t) = S(0)\exp\left\{\left(R_f - \frac{\sigma^2}{2}\right)t + \sigma W^*(t)\right\}$$

which corresponds to the (transformed) price process:

$$\frac{\mathrm{d}S}{S} = R_f \mathrm{d}t + \sigma \mathrm{d}W^*(t),\ S(0) = S_0$$

And the current price equals the expected future price under risk neutral pricing since:

$$\begin{aligned} S(0) &= \mathrm{e}^{-R_f t}E^*\left(S(t)\right) = \mathrm{e}^{-R_f t}E^*\left(S(0)\exp\left\{\left(R_f - \frac{\sigma^2}{2}\right)t + \sigma W^*(t)\right\}\right) \\ &= S(0)\,\mathrm{e}^{-\sigma^2 t/2}E^*\left(\mathrm{e}^{\sigma W^*(t)}\right) \end{aligned}$$

Since E^* is an expectation taken with respect to the risk neutral process, we have:

$$\mathrm{e}^{-\sigma^2 t/2}E^*\left(\mathrm{e}^{\sigma W^*(t)}\right) = \mathrm{e}^{-\sigma^2 t/2}\mathrm{e}^{\sigma^2 t/2} = 1$$

Thus, the current price equals an expectation of the future price. It is important to remember, however, that the proof of such a result is based on our ability to replicate such a process by a risk-free process (and thereby value it by the risk-free rate). In terms of the historical process, we have first:

$$\mathrm{d}W^*(t) = \mathrm{d}W(t) + \lambda \mathrm{d}t,\ \lambda = \frac{\alpha - R_f}{\sigma}$$

which we insert in the risk-neutral process and note that:

$$\frac{\mathrm{d}S}{S} = R_f \mathrm{d}t + \sigma[\mathrm{d}W(t) + \lambda \mathrm{d}t] = [R_f + \sigma\lambda]\,\mathrm{d}t + \sigma \mathrm{d}W(t)$$

where R_f is the return on a risk-free asset while $\lambda = (\alpha - R_f)/\sigma$ is the premium (per unit volatility) for an asset whose mean rate of return is α and its volatility is σ. In other words, risk-neutral pricing is reached by equating the stock price process whose return equals the risk-free rate plus a return of $\alpha - \sigma\lambda$ that compensates for the stock risk, or:

$$\begin{aligned} \frac{\mathrm{d}S}{S} &= (R_f + \alpha - \sigma\lambda)\,\mathrm{d}t + \sigma[\mathrm{d}W(t) + \lambda\,\mathrm{d}t] \\ &= R_f\,\mathrm{d}t + \sigma[\mathrm{d}W(t) + \lambda\,\mathrm{d}t] = R_f\,\mathrm{d}t + \sigma dW^*(t) \end{aligned}$$

Under such a transformation, risk-neutral pricing is applicable and therefore financial assets may be valued by expectations using the transformed risk-neutral process or adjusting the price process by its underlying risk premium (if it can be assessed of course by, say, regressions that can estimate risky stock betas using the CAPM). Consider next the risk-neutral framework constructed and evaluate the decision to sell an asset we own (whose current price is S_0) as soon as its price reaches a given level $S^* > S_0$. The profit of such a trade under risk-neutral pricing is:

$$\pi_0 = E^* \mathrm{e}^{-R_f \tau} S^* - S_0$$

Here the stopping (sell) time is random, defined by the first time the target sell price is reached:

$$\tau = \mathrm{Inf}\{t > 0,\ S(t) \geq S^*;\ S(0) = S_0\}$$

We shall prove that under the risk-neutral framework, there is an 'equivalence' to selling now or at a future date. Explicitly, we will show that $\pi_0 = 0$. Again, let the risk-neutral price process be:

$$\frac{\mathrm{d}S}{S} = R_f \mathrm{d}t + \sigma \mathrm{d}W^*(t)$$

and consider the equivalent return process $y = \ln S$

$$\mathrm{d}y = \left(R_f - \tfrac{\sigma^2}{2}\right)\mathrm{d}t + \sigma \mathrm{d}W^*(t),\ y(0) = \ln(S_0)$$
$$\tau = \mathrm{Inf}\{t > 0,\ y(t) \geq \ln(S^*);\ y(0) = \ln(S_0)\}$$

As a result, $E_S^*(\mathrm{e}^{-R_f\tau}) = E_y^*(\mathrm{e}^{-R_f\tau})$ which is the Laplace transform of the sell stopping time when the underlying process has a mean rate and volatility given respectively by $\mu = R_f - \sigma^2/2$, σ (see also the mathematical Appendix to this chapter):

$$g_{R_f}^*(S^*, \ln S_0) = \exp\left[\frac{\ln S_0 - \ln S^*}{\sigma^2}\left(-\mu + \sqrt{\mu^2 + 2R_f\mu\sigma^2}\right)\right],$$
$$\sigma > 0,\ -\infty < \ln S_0 \leq \ln S^* < \infty$$

The expected profit arising from such a transaction is thus:

$$\pi_0 = S^* E^*(\mathrm{e}^{-R_f\tau}) - S_0 = S^* g_{R_f}^*(\ln S^*, \ln S_0) - S_0$$
$$= S^* \exp\left[\frac{\ln S_0 - \ln S^*}{\sigma^2}\left(-\mu + \sqrt{\mu^2 + 2R_f\mu\sigma^2}\right)\right] - S_0$$

That is to say, such a strategy will in a risk-neutral world yield a positive return if $\pi_0 > 0$. Elementary manipulations show that this is to equivalent to:

$$\pi_0 > 0 \text{ if } \frac{\sigma^2}{2} > (R_f - 1)\mu \quad \text{or} \quad \frac{\sigma^2}{2} > (1 - R_f)\left(\frac{\sigma^2}{2} - R_f\right) \text{ if } R_f > \frac{\sigma^2}{2}$$

As a result,

$$\pi_0 = \begin{cases} > 0 & \text{if } R_f > \dfrac{\sigma^2}{2} \\ < 0 & \text{if } R_f < \dfrac{\sigma^2}{2} \end{cases}$$

The decision to sell or wait to sell at a future time is thus reduced to the simple condition stated above. An optimal selling price in these conditions can be found by optimizing the return of such a sell strategy which is found by noting that either it is optimal to have a selling price as large as possible (and thus never sell) or select the smallest price, implying selling now at the current (any) price. If the risk-free rate is 'small' compared to the volatility, then it is optimal to wait and, vice versa, a small volatility will induce the holder of the stock to sell. In other words,

$$\frac{d\pi_0}{dS^*} = \begin{cases} > 0 \ R_f < \sigma^2/2 \\ < 0 \ R_f > \sigma^2/2 \end{cases}$$

Combining this result with the profit condition of the trade, we note that:

$$\begin{cases} \dfrac{d\pi_0}{dS^*} > 0, \pi_0 < 0 \text{ if } R_f < \sigma^2/2 \\ \dfrac{d\pi_0}{dS^*} < 0, \pi_0 > 0 \text{ if } R_f > \sigma^2/2 \end{cases}$$

And therefore the only solution that can justify these conditions is $\pi_0 = 0$, implying that whether one keeps the asset or sells it is irrelevant, for under risk-neutral pricing, the profit realized from the trade or of maintaining the stock is equivalent. Say that $R_f < \sigma^2/2$ then a 'wait to sell' transaction induces an expected trade loss and therefore it is best to obtain the current price. When $R_f > \sigma^2/2$, the expected profit from the trade is positive but it is optimal to select the lowest selling price which is, of course, the current price and then, again, the profit transaction, $\pi_0 = 0$ will be null as our contention states.

Similar results are obtained when the time to exercise the sell strategy is finite. In this case,

$$\pi_{0,T} = S^* E^*(e^{-R_f \tau \wedge T}) - S_0 = 0$$

which turns out to have the same properties as above.

For a risk-sensitive investor (trader or speculator) whose utility for money is $u(.)$, a decision to sell or wait will be based on the following (using in this case the actual probability measure and an individual discount rate R_i):

$$\underset{S^* \geq S_0}{\text{Max}}\ Eu(\tilde{\pi}_0) = Eu(S^* e^{-R_i \tilde{\tau}} - S_0)$$

If, at the end of the period, we sell the asset anyway, then the optimal trading/sell condition is:

$$\underset{S^* \geq S_0}{\text{Max}} \int_0^T Eu(S^* e^{-R_i \tau} - S_0) e^{-R_i \tau} g(\tau)\, d\tau + E_{S(T)} u(S(T) e^{-R_i T} - S_0)[1 - G(T)]$$

where $g(.)$ and $G(.)$ are the inverse Gaussian probability and cumulative distributions respectively.

$$g(S^*, t; S_0) = \frac{|S_0 - S^*|}{\sqrt{2\pi\sigma^2 t^3}} \exp\left[-\frac{(S^* - S_0 - R_f t)}{2\sigma^2 t}\right],$$

$$1 - G(T) = \left[1 - 2\Phi\left(\frac{-(S^* - S_0)}{\sigma\sqrt{T}}\right)\right]$$

When we use historical data, the situation is different, as we discussed earlier. Say that the underlying asset has the following equation:

$$\mathrm{d}S/S = b\,\mathrm{d}t + \sigma \mathrm{d}W, \;\; S(0) = S_0$$

To determine the risk-sensitive rate associated with a trading strategy, we can proceed as follows. Say that a risk-free zero coupon bond pays S^* at time T whose current price is B^*. In other words, $B^* = \mathrm{e}^{-R_{f,T}T} S^*$. Thus, $S^* = B^* \mathrm{e}^{R_{f,T}T}$ where $R_{f,T}$ is a known discount rate applied to this bond. Since (see also the mathematical Appendix):

$$g^*_{R_f}(S^*, \ln S_0) = \exp\left[-\frac{\ln S^*/S_0}{\sigma^2}\left(-b + \sqrt{b^2 + 2R_f b\sigma^2}\right)\right], \text{ and}$$

$$S_0 = S^* \exp\left[-\frac{\ln S^*/S_0}{\sigma^2}\left(-b + \sqrt{b^2 + 2R_i b\sigma^2}\right)\right]$$

We obtain:

$$S_0 = B^* \exp\left[R_{f,T}T - \frac{\ln(B^* \mathrm{e}^{R_{f,T}T}/S_0)}{\sigma^2}\left(-b + \sqrt{b^2 + 2R_i b\sigma^2}\right)\right]$$

This is solved for the risk-sensitive discount rate:

$$R_i = \left[1 + \sigma^2 \frac{[R_{f,T}T - \ln(S_0/B^*)]}{2b\,\ln(B^* \mathrm{e}^{R_{f,T}T}/S_0)}\right] \frac{[R_{f,T}T - \ln(S_0/B^*)]}{\ln(B^* \mathrm{e}^{R_{f,T}T}/S_0)}$$

Example: Buying and selling on a random walk

The problem considered above can be similarly solved for a buy/sell strategy in a random walk. To do so, consider a binomial price process where price increase or decline by \$1 with probabilities p and q respectively. Set the initial price to $S_0 = 0$. When $p = q = 1/2$, the price process is a martingale. Assume that we own no stock initially but we construct the trading strategy: buy a stock as soon as it reaches the prices $S_n = -a$ and sell it soon thereafter as soon as it reaches the price $S_n = b$. Our problem is to assess the average profit (or loss) of such a trade. First set the first time that a buy order is made to:

$$T_a = \mathrm{Inf}\{n, S_n = -a\}, \;\; a > 0$$

Once the price '$-a$' is reached and a stock is bought, it is held until it reaches the price b. This time is:

$$T_b = \mathrm{Inf}\{n, S_n = b, S_0 = -a\}$$

At this time, the profit is $(a + b)$. In present value terms, the profit is a random variable given by:

$$\pi_{-a,b} = -aE(\rho^{-T_a}) + bE(\rho^{-(T_a+T_b)})$$

where ρ is the discount rate applied to the trade. Given the stoppin time generating function (to be calculated below), we can calculate the expected profit. Of course, initially, we put down no money and therefore, in equilibrium, it is also worth no money. That is $\pi_{-a,b} = 0$ and therefore $a/b = E(\rho^{-(T_a+T_b)})/E(\rho^{-T_a})$.

Problem

Calculate $\pi_{-a,b}$ and compare the results under a risk-free discount rate (when risk-neutral pricing can be applied and when it is not).

Now assume that we own a stock which we sell when the price decreases by a or when it increases by b, whichever comes first. If $-a$ is reached first, a loss '$-a$' is incurred, otherwise a profit b is realized. Let the probability of a loss be U_a and let the underlying price process be a symmetric random walk (and thereby a martingale), with $E(S_n) = E(S_T) = 0,\ T = \min(T_{-a}, T_b)$ while $E(S_T) = U_a(-a) - (1 - U_a)b$ which leads to:

$$U_a = \frac{b}{a+b}$$

As a result, in present value terms, we have a profit given by:

$$\theta_{-a,b} = -aU_aE(\rho^{-T_a}) + b(1 - U_a)E(\rho^{-T_b})$$

Some of the simple trading problems based on the martingale process can be studied using Wald's identity. Namely say that a stock price jump is $Y_i, i = 1, 2, \ldots$ assumed to be independent from other jumps and possessing a generating function:

$$\phi(\nu) = E(\mathrm{e}^{\nu Y_1}),\ S_n = \sum_{i=1}^{n} Y_i,\ S_0 = 0$$

Now set the first time T that the process is stopped by:

$$T = \mathrm{Inf}\,\{n, S_n \le -a, S_n \ge b\}$$

Wald's identity states that:

$$E[(\phi(\nu))^{-T}\mathrm{e}^{\nu S_n}] = 1 \text{ for any } \nu \text{ satisfying } \phi(\nu) \ge 1.$$

Explicitly, set a $\phi(\nu^*) = 1$ then by Wald's identity, we have:

$$E[(\phi(\nu^*))^{-T}\mathrm{e}^{\nu^* S_n}] = E[\mathrm{e}^{\nu^* S_n}] = 1$$

In expectation, we thus have:

$$1 = E[\mathrm{e}^{\nu^* S_T} | S_T \le -a]\ \Pr\{S_T \le -a\} + E[\mathrm{e}^{\nu^* S_T} | S_T \ge b]\ \Pr\{S_T \ge b\}$$

and therefore:

$$\Pr\{S_T \geq b\} = \frac{1 - E[e^{\nu^* S_T} | S_T \leq -a]}{E[e^{\nu^* S_T} | S_T \geq b] - E[e^{\nu^* S_T} | S_T \leq -a]}$$

As a result, the probability of 'making money' (b) is $\Pr\{S_T \geq b\}$ while the probability of losing money ($-a$) is $1 - \Pr\{S_T \geq b\}$.

Problem: Filter rule

Assess the probabilities of making or losing money in a filter rule which consists in the following. Suppose that at time '0' a sell decision has been generated. The trading rule generates the next buy signal (i.e. reaching price b first and then waiting for price $-a$ to be reached).

Problem: Trinomial models

Consider for simplicity the risk-neutral process (in fact, we could consider equally any other diffusion process):

$$\mathrm{d}S/S = R_f \,\mathrm{d}t + \sigma \,\mathrm{d}W, \;\; S(0) = S_0$$

and apply Ito's Lemma to the transformation $y = \ln(S)$ and obtain:

$$\mathrm{d}y = \left(R_f - \frac{1}{2}\sigma^2\right) \mathrm{d}t + \sigma \,\mathrm{d}W, \;\; y(0) = y_0$$

Given this normal (logarithmic) price process consider the trinomial random walk approximation:

$$Y_{t+1} = \begin{cases} Y_t + f_1 & \text{w.p.} \quad p \\ Y_t + f_2 & \text{w.p.} \quad 1 - p - q \\ Y_t + f_3 & \text{w.p.} \quad q \end{cases}$$

where:

$$E\left(Y_{t+1} - Y_t\right) = pf_1 + (1 - p - q)f_2 + qf_3 \approx \left(R_f - \frac{1}{2}\sigma^2\right)$$

$$E\left(Y_{t+1} - Y_t\right)^2 = pf_1^2 + (1 - p - q)f_2^2 + qf_2^2 \approx \sigma^2$$

First assume that $p + q = 1$ and calculate the stopping sell time for an asset we own. Apply also risk neutral valuation to calculate the price of a European call option derived from this price if the strike is K and if the exercise time is 2. If $p + q < 1$, explain why it is not possible to apply risk-neutral pricing? In such a case calculate the expected stopping time of a strategy which consists in buying the stock at $Y = -Y_a$ and selling at $Y = Y_b$.

Stopping times on random walks are often called the 'gambler's ruin' problem, inspired by a gambler playing till he loses a certain amount of capital or taking his winnings as soon as they reach a given level. For an asymmetric birth–death random walk with:

$$P(Y_i = +1) = p, \quad P(Y_i = -1) = q, \quad \text{and} \quad P(Y_i = 0) = r$$

It is well known (for example, Cox and Miller, 1965, p. 75) that the probability of reaching one or the other boundaries is given by,

$$P(Y_t = -a) = \begin{cases} \dfrac{1-(1/\lambda)^b}{1-(1/\lambda)^{a+b}} & \lambda \neq 1 \\ b/(a+b) & \lambda = 1 \end{cases} \quad P(Y_t = b) = \begin{cases} \dfrac{(1/\lambda)^b-(1/\lambda)^{a+b}}{1-(1/\lambda)^{a+b}} & \lambda \neq 1 \\ a/(a+b) & \lambda = 1 \end{cases}$$

where $\lambda = q/p$. Further,

$$E(T_{-a,b}) = \begin{cases} \left[\dfrac{1}{1-r}\right]\left[\dfrac{\lambda+1}{\lambda-1}\right]\left[\dfrac{a(\lambda^b-1)+b(\lambda^{-a}-1)}{\lambda^b-\lambda^{-a}}\right] & \lambda = q/p \\ \dfrac{ab}{1-r} & \lambda = 1 \end{cases}$$

Note that $(\lambda^{x_n}; n \geq 0)$ is a martingale which is used together with the stopping theorem to prove these results (see also Chapter 4). For example, as discussed earlier, at the first loss $-a$ or at the profit b the probability of 'making money' is $P(S_{T(-a,b)} = b)$ while the probability of losing it is $P(S_{T(-a,b)} = -a)$, as calculated above. The expected amount of time the trade will be active is also $E(T_{-a,b})$. For example, if a trader repeats such a process infinitely, the average profit of the trader strategy would be given by:

$$\bar{\pi}(-a, b) = \frac{bP(S_{T(-a,b)} = b) - aP(S_{T(-a,b)} = -a)}{E(T_{-a,b})}$$

Of course, the profit from such a trade is thus random and given by:

$$\tilde{\pi} = \begin{cases} -a & \text{w.p.} \begin{cases} \dfrac{1-(1/\lambda)^b}{1-(1/\lambda)^{a+b}} & \lambda \neq 1 \\ b/(a+b) & \lambda = 1 \end{cases} \\ b & \text{w.p.} \begin{cases} \dfrac{(1/\lambda)^b-(1/\lambda)^{a+b}}{1-(1/\lambda)^{a+b}} & \lambda \neq 1 \\ a/(a+b) & \lambda = 1 \end{cases} \end{cases}$$

And therefore, the expected profit, its higher moments and the average profit can be calculated. When $\lambda = 1$ in particular, the long run average profit is null and the variance equals *2ab*, or:

$$E(\tilde{\pi}) = \frac{(-ab+ba)}{a+b} = 0, \ \text{var}(\tilde{\pi}) = 2ab$$

Thus, a risk-averse investor applying this rule will be better off doing nothing, since there is no expected gain. When $\rho \neq 1$ while $r = 0$ (the random walk) we have (Cox and Miller, 1965, p. 31):

$$P(S_{T(-a,b)} = b) = \frac{\lambda^a - 1}{\lambda^{a+b}-1}; \ P(S_{T(-a,b)} = -a) = \frac{\lambda^{a+b}-\lambda^a}{\lambda^{a+b}-1}$$

and

$$E(T_{-a,b}) = \left[\frac{\lambda+1}{\lambda-1}\right]\left[\frac{a(\lambda^b-1)+b(\lambda^{-a}-1)}{\lambda^b-\lambda^{-a}}\right]$$

while the long run average profit is:

$$\bar{\pi}(-a,b) = \frac{[b(\lambda^a-1)-a(\lambda^{a+b}-\lambda^a)](\lambda-1)(\lambda^b-\lambda^{-a})}{[b(\lambda^{-a}-1)+a(\lambda^b-1)](\lambda+1)(\lambda^{a+b}-1)}$$

An optimization of the average profit over the parameters (a, b) when the underlying process is a historical process provides then an approach for selling and buying.

Problem
Consider the average profit above and optimize this profit with respect to a and b and as a function of $\lambda > 1$ and $\lambda < 1$.

Problem
Show that when $p > q$, then the mean time and its variance for a random walk to attain the value b is equal to:

$$E(T_b) = b/(p-q) \quad \text{and} \quad \operatorname{var}(T_b) = ([1-(p-q)^2]\,b)/(p-q)^3$$

Finally, show that when the boundary b becomes large that the standardized stopping time tends to a standard Normal distribution.

Example: Pricing a buy/sell strategy on a random walk*
Consider again an underlying random walk price (where the framework of risk-neutral pricing might not be applicable) (S_t), $t = 1, 2, \ldots$. The probability that the price increases is s while the probability that the stock price decreases is q. Assume that the current price is i_0 and let i be a target selling price $i = i_0 + \Delta i > i_0$, given by the binomial probability distribution:

$$P(S_n = i = i_0 + \Delta i) = \left(\begin{matrix} n \\ \left[\frac{\Delta i + n}{2}\right]\end{matrix}\right) s^{(n+\Delta i)/2} q^{(n-\Delta i)/2}$$

$$= \binom{\Delta i + 2\nu}{\Delta i + \nu} s^{\Delta i+\nu} q^{\nu};\ n - \Delta i = 2\nu,\ \ \nu = 0, 1, 2, 3, \ldots$$

where [ ] denotes the least integer. Since, prices i can be reached only at even values of $\Delta i + n$, it is convenient to rewrite the price process by:

$$P(S_{\Delta i+2\nu} = i) = \binom{\Delta i + 2\nu}{\Delta i + \nu} s^{\Delta i+\nu} q^{\nu} = \frac{s^{\Delta i+\nu} q^{\nu}}{\nu!} \prod_{k=1}^{\nu} (\Delta i + \nu + k);$$

$$n - \Delta i = 2\nu,\ \ \nu = 0, 1, 2, 3, \ldots$$

The amount of time $\tau(i) = n = 2\nu + \Delta i$ for an underlying process $(s > q)$ that can reach this price, however, is given by Feller (1957):

$$P(\tau(i) = 2\nu + \Delta i) = \frac{\Delta i}{2\nu + \Delta i} P(S_{2\nu+\Delta i} = i) = \frac{i}{(2\nu + i)} \frac{s^{i+\nu} q^{\nu}}{\nu!} \prod_{k=1}^{\nu} (i + \nu + k);$$
$$\nu = 0, 1, 2, 3, \ldots \; s > q$$

Thus, if a sell order for a stock is to be exercised at price i and if we use a risk-sensitive adjusted discount rate ρ, then the current expected value of this transaction is:

$$i_0 = E(\tilde{i}_0) = iE(\rho^{\tilde{\tau}(i)}) = i^2(\rho s)^i \sum_{\nu=0}^{\infty} \frac{\prod_{k=1}^{\nu} (i + \nu + k)}{(2\nu + i)\nu!} (\rho^2 sq)^{\nu}$$

Under risk-neutral pricing of course, the discount rate equals the risk free rate ρ_f, that is $\rho = \rho_f = 1/(1 + R_f)$ and for convenience:

$$i_0 = i^2(\rho s)^i \Lambda_i(\rho); \; \Lambda_i(\rho) = \sum_{\nu=0}^{\infty} \frac{\prod_{k=1}^{\nu} (i + \nu + k)}{(2\nu + i)\nu!} (\rho^2 sq)^{\nu}$$

We can also set $\rho = \rho_f + \pi$ where π is the risk premium associated with selling at a price i. A price i can also be obtained as well by buying a bond of nominal value i in m periods hence without risk. In this case, we will have:

$$B_i(m) = i(\rho_f)^m \quad \text{or} \quad i = B_i(m)(\rho_f)^{-m}$$

Replacing i, we have

$$i_0 = [B_i(m)(\rho_f)^{-m}]^2 (\rho s)^{[B_i(m)(\rho_f)^{-m}]} \Lambda_{B_i(m)(\rho_f)^{-m}}(\rho)$$

This provides a solution for the risk-sensitive discount rate and therefore its risk premium. Higher-order moments can be calculated as well. Set:

$$i_0 = i^2 \Lambda(\rho)$$
$$\text{var}(\tilde{i}_0) = i^3 \Lambda(\rho^2) - i^4 \Lambda^2(\rho)$$

The probability distribution $P(\tilde{i}_0 | i_0, \text{var}(\tilde{i}_0))$ of the current trade which is a function of the discount rate provides a risk specification for such a trade. If we have a quantile risk given by, $P(\tilde{i}_0 \leq i_0 - \Delta i | i_0, \text{var}(\tilde{i}_0)) \leq \xi$, then by inserting the mean variance parameters given above, and expressed in terms the discount rate ρ, we obtain an expression of the relationship between this discount rate and the VaR parameters $(\xi, i_0 - \Delta i)$.

Say that the problem is to sell or wait and let $i^* > i_0$ be the optimal future selling price (assumed to exist of course and calculated according to some criteria as we shall see below). If price i^* is reached for the first time at time n, then the price of such a trade is $i^* \rho^n$ which can be greater or smaller than the current price. If we sell now, then the probability that this decision is ill-taken is $P(i^* \rho^{\tau(i^*)} \geq i_0)$. As a result, the probability of a loss due to a future price increase has a quantile risk

$P(i^*\rho^{\tau(i^*)} - i_0 \geq V_s) \leq 1 - \theta_s$ where V_s is the value at risk for such a decision while $1 - \theta_s$ is the assigned probability associated with the risk of holding the stock. By the same token, if we wait and do not sell the stock, the probability of having made the wrong decision is now:

$$P(i_0 - i^*\rho^{\tau(i^*)} \geq V_h) = P(i^*\rho^{\tau(i^*)} - i_0 \leq V_h) \geq \theta_s$$

where V_h denotes the value at risk of holding the stock. If a transaction cost is associated with a trade and if we set c to be this cost (when there are no holding costs), then we have:

$$P[(i^* - c)\rho^{\tau(i^*)} - (i_0 - c) \geq V_s] \leq 1 - \theta_s$$

or

$$P\{i^*\rho^{\tau(i^*)} - [i_0 - c(1 - \rho^{\tau(i^*)})] \geq V_s\} \leq 1 - \theta_s$$

Therefore, a transaction cost has the net effect of depreciating the current price by $c(1 - \rho^{\tau(i^*)}) > 0$ and thereby favouring selling later rather than now in order to delay the cost of the transaction. In this sense, a transaction cost has the effect of reducing the number of trades! By the same token, an investor may seek to buy an asset believing that its current value is underpriced. In such a case, the buyer will compare the future discounted price with the current price and reach a decision accordingly. For example, the optimal buy price, based on the expected discounted future prices would be:

$$i_0^{**} = \operatorname*{Max}_{i>i_0} \left\{\mathrm{iE}\left(\rho^{\tau(i)}\right) - i_0 | i^{**} > i_0 + d\right\},\ \tau(i) = \mathrm{Inf}\{n, i > i_0,\}$$

where d is the buy transaction cost. An appropriate buy–sell–hold strategy is then defined by:

$$\begin{cases} \text{Do nothing} & i_0^* \leq i_0 \leq i_0^{**} \\ \text{Sell if} & i_0^* \leq i_0 \\ \text{Buy if} & i_0^{**} \geq i_0 \end{cases}$$

In this framework, the quantile risk approach provides in a simple and a uniform manner an approach to stopping as a function of the risks the investor is willing to sustain. In Chapter 10, this measure of risk will be considered in greater detail, however.

7.6 SPECIFIC APPLICATION AREAS

Foreign exchange is a fertile ground for the application of financial products, their pricing and their analyses. Basic transactions (through the interbank or the wholesale market) on spot, futures, forwards and swap and other products are applied extensively. FX trading is assuming greater importance. The Philadelphia Exchange trades, for example, options on the British pound, German mark, Japanese yen, Swiss franc and the Canadian dollar. The most heavily traded contracts are the Deutschemark and Japanese yen American-style options. The strike price for each foreign currency option is the US dollar price of a unit of foreign exchange. The expiration dates correspond to the delivery dates in futures. Specifically, the expiration dates correspond to the Saturday before the third Wednesday of the contract month. Contract months are March, June, September and December plus the two near-term contracts. The daily volume of contracts traded on the Philadelphia exchange has steadily increased to over 40 000 contracts per day.

Currency	Contract size	Strike price intervals	Premium quotations
Mark	62 500	1.0	Cents
Sterling	31 250	2.5	Cents
Swiss franc	62 500	1.0	Cents
Yen	6 250 000	0.01	Hundredth cent

Consider the British pound for example. Each contract is for 31 250 British pounds. Newspapers report the closing spot price, in cents per pound sterling. The strike prices are reported in cents per pounds, at 2.5-cent intervals. The call and put premiums are also in cents per pound. Consider the theoretical price of a European call option on the British pound that trades at the Philadelphia Exchange. The time to expiration is 6 months. The spot is \$1.60 per pound. An at-the-money call is to be valued where the exchange rate volatility is 10 % per year. The domestic interest rate and the foreign interest rate are both equal to 8 %. Using the theoretical price of the European call option given below, we find the option price to be \$0.0433. The equations for this problem are similar to the Black–Scholes model, as we saw earlier and in the previous chapter, and are summarized below.

Price of a European call option on foreign exchange

$$C^E(0) = G(0)N(d_1^*) - K\,\mathrm{e}^{-rT}N(d_2^*);\, d_1^* = \frac{1}{\sigma\sqrt{T}}\ln\left[G(0)/K\right]$$
$$+(r+\sigma^2/2)T;\; d_2^* = d_1^* - \sigma\sqrt{T}$$

with σ the volatility of foreign exchange and $G(0) = S(0)\,\mathrm{e}^{-r_F T}$ where r_F is the risk-free rate in the foreign currency.

FX swap contracts are made by drafting purchase–repurchase agreements by selling simultaneously one currency (say DM) in the spot and the forward market. Swap contracts are immensely popular contracts and will be treated in the next chapter in the context of interest-rate-related contracts.

In practice, financial products can be tailored to sources of risks and can respond to specific business, industrial or other needs. For example, financial products that meet firms' risks related to climatic risks and energy supplies. Climatic factors in particular account for a substantial part of insurance firms' costs. The December 1999 storm that hit France may have cost 44.5 billion francs! – hitting hard both the French insurance companies and reinsurance firms throughout Europe. Climatic risks also have an important effect on the US economy, accounting for approximately 20 % of GNP according to the Department of Energy. There are in fact few sectors that are immune to weather effects and thereby the importance of all risk-management activities related to meteorological forecasting, robust construction, tourism, fashion etc. Energy needs in particular, are determined by the intensity of summer heat and winter cold, generating fluctuations in demand and supply for energy sources. An expanding climatic volatility has only added to the management of these risks. For this reason, firms such as ENRON (now defunct), Koch, Aquila and Southern Energy have focused attention on the use of financial energy-related products that can protect sources of supplies and meet demands. As a result, since 1997, there have been energy products on the CME and, since 2000, on London's LIFFE, providing financial services to energy investors, speculators and firms. The underlying sources of risk of energy firms (as well as many supply contract) span:

(a) Price risk whose effects are given by $V \Delta p$ where V is the quantity of the energy commodity and Δp is the price change of the commodity.
(b) Quantity risk, $(\Delta V)p$.
(c) Correlation risk $\Delta V \Delta p$.

To manage these risks, derivative products on both the price and the supply contract are used. These contracts, over multiple sources of risks, are difficult to assess and are currently the subject of extensive research and applications.

7.7 OPTION MISSES

In the mid-nineties, media and regulators' attention was focused on option misses because of huge derivative-based losses that have affected significantly both large corporate firms and institutions. The belief that options are primarily instruments for hedging was severely shaken and the complexity and risks implied in trading with derivatives revealed. Management and boards, certain that derivatives were used only to hedge and reduce price risk, were astounded by the consequences of positions taken in the futures and options markets – for the better and for the worse. According to the *Wall Street Journal* for 12 April, 1996, J.P. Morgan earnings in the first quarter jumped by 72 % from the previous year, helped by an unexpectedly strong derivatives business that more than doubled the bank's overall trading revenue! By the same token, firms were driven to bankruptcy due to derivative losses. Business managers also discovered that managing risk with derivatives can be tempting, often only understood by a few mathematically inclined financial academics. At the same time there is a profusion of derivatives contracts, having a broad set of characteristics and responding in different ways to the many factors that beset firms and individuals, which have invaded financial

Table 7.2 Derivatives losses of industries and organizations.

Name	Losses ($ million)	Main cause
AIG	90	Derivatives revaluation
Air Product	113	Leveraged and currency swaps
Arco (Pension Fund)	25	Structured notes
Askine Securities	605	MBS model
Bank of America	68	Fraud
Procter & Gamble	450	Leveraged and currency swaps
Barings PLC	1400	Futures trading
Barnnet Banks	100	Leveraged swaps
Cargil	100	Mortgage derivatives
Codelco (Chile)	210	Futures trading–copper
Community Bankers	20	Leveraged swaps
Dell Computers	35	Leveraged swaps
Gestetner	10	Leveraged swaps
Glaxo	200	Derivatives and swaps
Harris Trust	52	Mortgage derivatives
Kashima Oil	1450	Currency derivatives
Kidder Peabody	350	Fraud trading
Mead	12	Leveraged swaps
Metalgesellschaft	1300	Futures trading
Granite Partners	600	Leveraged CMOs
Nippon Steel	30	Currency derivatives
Orange County	1700	Mortgage derivatives
Pacific Horizon	70	Structured notes
Piper Jaffray	700	Leveraged CMOs
Sandoz	80	Derivatives transactions
Showa/Shell Sekiyu	1400	Forward contracts
Salomon Brothers	1000	Fraud (cornering)
United Services	95	Leveraged swaps
Estimated losses	12 265	

markets. In many cases too, derivatives hype has ignited investors' imagination, for they provided a response to many practical and real problems hitherto not dealt with. An opportunity to manage risks, enhance yields, delay debt records to some future time, exploit arbitrage opportunities, provide corporate liquidity, leverage portfolios and do whatever might be needed are just a very few such instances.

A derivative mania has generated at the same time their misuse, leading to large losses, as many companies and individuals have experienced. Derivatives became the culprit for many losses, even if derivatives could not intelligently be blamed. For example, the continuous increase in interest rates during 1993–94 pushed down T-bond prices so that the market lost hundreds of billions in US dollars. Additionally, the sharp drop in the IBM stock price in 1992–94 from 175 to 50 created a market loss of approximately $70 billion for one firm only! These losses could not surely be blamed on derivatives trading! Table 7.2

summarizes some of the losses, assembled from various sources and sustained by corporate America (Meir Amikam, 1996). These derivative losses were found to be due to a number of reasons including: (1) a failure to understand and identify firms' sensitivities to different types of risk and calculating risk exposure; (2) over-trusting – trusting traders with strong personality led to huge losses in Barings, megalomania in Orange County, Gestetner etc.; (3) miscalculation of risks – overly large positions undertaken which turned out sour; (4) information asymmetry – the lack of internal control systems and audits of trading activities has led traders to assume unreasonable positions in hope; (5) poor technology – lack of computer-aided tools to follow up trading activity for example; (6) applying real-time trading techniques, responding to volatility rather than to fundamental economic analysis, that have also contributed to ignoring risks.

Of courser, a number of risk management tools and models have been suggested to institutional and individual investors and traders to prevent these risks wherever possible. For example, value at risk, extremes loss distributions, mark to market and using the Delta of model-based risk, to be considered in Chapter 10, have been found useful. These models have their own limitations, however, and cannot replace the expectations of qualified professional judgement. By the same token, these expectations introduce a *systematic risk* that can lead to unexpected volatility and cause severe losses, not only to speculators, but also to hedgers. Some of the great failures in 1993–94, for example, were incurred because users were caught by a surprise interest rate hike. A similar scenario occurred when the price of oil dropped. Thus, the use of derivatives for speculation purposes can cause large losses if traders turn out to be on the wrong side of their market expectations.

There are some resounding losses, however, that have been the subject of intense scrutiny. Below we outline a few such losses.

Bankers Trust/Procter & Gamble/Gibson Greeting: 'Our policy calls for plain vanilla type swaps', Erick Nelson, CFO, Procter & Gamble.

Procter & Gamble incurred a loss of $157 million loss from interest swaps in both US and German markets, swapping fixed for floating rates. In effect they had a put option given to Bankers Trust. P&G's strategic error was the belief that exchange rates would continue to fall both in the USA and in Germany. Swaps were thus made for the purpose of reducing interest costs. The actual state of interest rates turned out to be quite different, however. In the USA, the expectation of lower interest rates meant that the value of bonds would increase, rendering the put option worthless to Bankers Trust. In fact interest rates did not fall and therefore, Treasury bonds increased in value, forcing P&G to purchase the bonds from Bankers Trust at a higher price. This resulted in a first substantial loss. By the same token, the expectations of a decline in interest rates in Germany meant that the German Bund would decline rendering again the put option worthless to Bankers Trust. Therefore P&G would have lower interest costs as well. Rates increased instead, and therefore, the value of the German Bund decreased, forcing thereby a purchase of bonds from Banker's Trust at higher prices then the current market price – inducing again a loss. The combined losses reached $157 million.

In other words, P&G took an interest-rate gamble instead of protecting itself in case the 'bet' turned out to be wrong.

In April 1994, Procter & Gamble and Gibson Greetings claimed that Bankers Trust, had sold them high-risk, leveraged derivatives. The companies claim that those instruments, on which profits and losses can multiply sharply in certain circumstances, had been bought without giving the companies adequate warning of the potential risks. Bankers Trust countered that the firms were trying to escape loss-making contracts. P&G sued Bankers Trust in October 1994, and again in February 1995, for additional damages on the leveraged interest-rate swap tied to the yields on Treasury bonds. P&G has also claimed damages on a leveraged swap tied to DM interest rates for $195 million, insisting that the status of the first swap was not fully and accurately disclosed. The case was settled out of court in January 1995. It seemed in the course of the court deliberations that Bankers Trust may have not accurately disclosed losses and Bankers Trust may have hoped that market movements would turn to match their positions, so they did not notify Gibson immediately of the true magnitude of the intrinsic risk of the derivatives bought. Had Gibson been aware of the risk, the losses could have been minimized. In response the bank has fired one manager, reassigned others, and shacked up the leveraged derivatives unit. It also agreed to pay a $10 million fine to regulators and entered into a written agreement with the New York Federal Reserve Bank that allows regulators unprecedented oversight of the bank's leveraged derivatives business. The agreement is open-ended and highly embarrassed Bankers Trust. In addition, the lawsuits have tarnished the reputation of Bankers Trust. Outsiders have wondered if it was more the deal-making culture that was to blame, rather than the official tale of a few rogue employees. Other banks, such as Merrill Lynch, First Boston and J.P. Morgan have also run into trouble selling leveraged derivatives and other high-risk financial instruments.

The *Orange County (California)* case was a cause célèbre amplified by the media and regulation agencies warning of the dangers of financial markets speculations by public authorities. The Orange County strategy is called 'On the street, a kind of leveraged reverse repo strategy, also coined the *death spiral* because one significant market move can blow down the strategy in one puff'. This can occur if managers fail to understand properly the firm's sensitivities to different sorts of risks or do not regard financial risks as an integral part of the institution or corporate strategy. Speculation by the Orange County treasurer, who initially generated huge profits, ultimately bankrupted it. As the treasurer supervisor stated: 'This is a person who has gotten us all millions of dollars. I don't care how the hell he does it, but it makes us all look good.'

The county lost $2.5 billion. Out of a $7.8 billion portfolio! The loss was faulted on borrowing short to invest long in risk-structured notes. In other words, the county treasurer leveraged the (public) portfolio by borrowing $2 for each dollar in the portfolio, equivalent to investing on margin. The county then used the repo (reverse purchase agreement) market to borrow short in order to purchase long (term) government bonds. In the repo agreement, the county pledged the long-term bonds it was purchasing as collateral for secured loans. These loans were then rolled over every three to six months. As interest rates began to rise, the cost of borrowing increased while the value of long-term bonds decreased. This situation resulted in a substantial loss. In effect, Orange County was betting that interest rates would remain low or even decrease some and that spreads between long- and short-term rates would remain high. Something else happened. When interest rates rose, the cost of short-term borrowing increased, the value of the long-term bonds purchased decreased, the rates on the inverse floaters (consisting for an initial period of a fixed rate and then of a variable rate) fell and the rates on the spread bonds (consisting of a fixed percentage plus a long-term interest minus a short-term rate) fell as the yield curve flattened. This generated a huge loss for Orange County.

Metalgesellschaft: 'Pride in integrity takes a blow' (Cooke and Cramb, *Financial Times*). Metalgesellschaft, a \$15 billion sales commodity and engineering conglomerate, blamed its near collapse on reckless speculation in energy derivatives by its New York subsidiary. To save the firm that employed 46 000 people, banks and shareholders provided a \$2.1 billion bail-out. The subsidiary MGRM (MG Refining & Marketing) negotiated long-term, fixed-price contracts to sell fuel to gas stations and other small businesses in 1992. The fixed price was slightly higher than the prevailing spot price. To lock in this profit and hedge against rising fuel prices, the company hedged itself by buying futures on the New York Mercantile Exchange (NYMEX). Maintaining such a 'stacked rollover hedge' when prices are falling could require large amounts of liquidity. To hedge the long-term contracts, MGRM was obligated to buy short-term futures contracts to cover its delivery commitment since matching its supply obligations with contracts of the same maturity was impossible. That strategy was based on rolling over the short-term futures just before they expired. The hedging depended on the assumption that oil markets, which were in backwardation over two-thirds of the time over the past decade, would remain in that state for most of the time. By entering into futures contracts, MGRM would be able to hedge their short positions in the forward sales contracts. This assumption was predicated on the fact that, as expiration approaches, the future price MGRM paid for the contracts would be less than the spot price. That trading and hedging strategy had some inherent risks that happened to be crucial:

- *Market fluctuations risk*: oil prices, futures and spot price, that did not meet expectations.
- *Proper hedge risk*: Resulting from mismatched timing of contracts and entering into a speculative hedge.
- *Funding risk*: futures contracts require marking to market; this margin call caused by futures losses is not offset by forward contract gains, which are unrealized until delivery.

By September 1993, MGRM's obligation was equivalent to 160 million barrels. In November 1993, oil prices dropped by \$5 to \$14.5 as a reaction to OPEC's decision to cut production. That drop wiped out 20 % of MGRM short-term futures contracts and led to a cumulative trading loss of \$660 million. German GAAP does not allow the offset of gains or losses on hedging positions using futures against corresponding gains or losses on the underlying hedge asset, however. Further, Deutsche Bank, advising MG and apprehensive about the short-term losses as they mounted, convinced MG to close out its positions, thus causing the loss. The risk MG was taking was using a short-term instrument hedging strategy for a long-term exposure, creating thereby a mismatch that could be considered a bet that turned out to be wrong. The paper loss converted into a real heavy loss as the futures positions were closed. In 1994 the group announced that the potential losses of unwinding its positions could bring the total losses up to \$1.9 billion over the next three years. It is argued, though, among academics that the strategy could have worked had MGRM not unwound their futures position. Thus, liquidity was essential for this strategy to work. Alternatively MGRM could have reversed its futures position when oil prices dropped. It is important to look at current market conditions in reassessing the merits of one's strategy. MGRM traders did not (and, in fact, could not) properly hedge. They were speculating on the correlation between the underlying and the cash market. They ignored the risks of a speculative hedge, trusting that they could predict the relationship and changes in prices from month to month.

Barings: 'Ultimately, if you want to cover something up, it's not that difficult . . . Derivative positions change all the time and balance sheets don't give a proper picture of what's going on. For anyone on the outside to keep track is virtually impossible' (SIMEX trader, quoted by the *Financial Times*, February 1995).

The much-publicized *Barings* loss of \$1.3 billion was incurred by its branch in Singapore. It was incurred in three weeks by trading on the Nikkei Index. Leeson the Singapore Office Head of Trading was speculating that the Nikkei Index would rally after the Kobe earthquake, so he amassed a \$27 billion long position in Nikkei Index futures. The Nikkei Index fell, however, and Leeson was forced to sell put and call options to cover the margin calls. In an effort to recoup losses, Leeson increased the size of his exposure and held 61 039 long contracts on the Nikkei 225 and 26 000 short contracts on Japanese bonds. When he decided to flee, the Nikkei dropped to 17 885. Leeson was betting that the Index would trade in a range and he would therefore earn the premium from the contracts (to pay the margins). No one was aware of such trades and the risk exposure it created for Barings (as a result, it generated a much-needed and heated discussion regarding the needs for controls. The main office 'seemed' to focus far more on the potential gains rather than on the potential losses! This loss induced the demise of Barings, a venerable and longstanding English institution, which was sold to ABN AMRO.

Initially Leeson was responsible for settlement. In a short time he turned out to be a successful trader whose main job was to arbitrage variations among the prices of futures and options on the Nikkei 225, having the unique advantage that Barings had seats both on SIMEX and on OSE (Osaka Stock Exchange). Contracts on the Nikkei 225 and Nikkei 300 were OSE's only futures and options and accounted for 30 % of SIMEX business. As a member he enjoyed the privilege of seeing the orders ahead of non-members and of taking suitable positions with low risk. His strategy was mainly based on small spreads in which he invested large amount of money. Later on he was promoted to be responsible for trading and for settlement.

Granite partners: Granite Partners lost $600 million in mortgage derivatives in the mid-nineties. Fund managers promised their investors little risk in their investment policy since they used derivatives mainly for hedging purposes. By using CMO derivatives they expected to take advantage of market movements. But the disclosure emphasized that Granite had the option to wait, if need be, until market conditions suited the funds' position. Leveraging with CMO derivatives was much more than what was promised. The portfolio was leveraged based on the assumption of some in-house models. To their detriment, the bond market took a direction that went against the funds' positions. Since the portfolio was highly leveraged, the losses grew tremendously and Granite was shut down.

Freddie Mac: In January 2003, PriceWaterhouseCoopers, Freddie Mac's auditor for less than a year, revealed that the company might have misreported some of its derivatives trades. As a result, Freddie Mac later said that some earnings that should have been reported in 2001 and 2002 were improperly shifted into the future. It is not clear that the top executives were not attempting to distort the company's books. But recent corporate crises suggest that if someone wants to hide something, derivatives can help (*New York Times*, 12 June 2003).

Lessons from these loss cases, as well as many others, are summarized in Table 7.3. Generally, the most common cause was speculation – the market moving in directions other than presumed, the trade strategy collapsed – causing unexpected losses. There is little information relating to internal and external audit and control, however, implying perhaps that in most cases management does not realize the risk exposures they take on and thus controls end up being very poor. There were no reports of written policies that were supposed to limit positions and losses. Had there been such policies, traders could have easily ignored them, trusting their strategic 'cunning and assessments'. Although these figures were assembled from various sources, including the media, and the actual reasons for such losses were varied, a distribution of the main causes for the losses were: management, 18; poor audit or no controls, 20; wrong methods and trade strategies applied, 21; market fluctuation (poor forecasts), 17; and, finally, frauds and traders' megalomania, 5. Problems associated with audit and controls are resurging in various contexts today. For example, in the wake of multibillion-dollar accounting scandals, companies are under intense pressure to make sure that their financial results do not paint a misleading, rosy picture. Insurance firms for example, are swamped with billions of dollars in corporate bonds that they bought years ago and that are still maintained at their original value in their

Table 7.3 Main reasons which led to losses.

Firm	Management	Audit/ control	Methods/ strategy	Market fluctuations	Frauds/ megalomania
AIG		+	+	+	
Air Product	+	+	+		
Arco (Pension Fund)	+	+		+	
Askine Securities			+	+	
Bank of America		+	+		
Bankers Trust/ Procter and Gamble			+	+	+
Barings PLC	+	+	+		+
Barnnet Banks		+	+		
Cargil	+	+	+	+	
Codelco (Chile)	+	+	+	+	
Community Bankers	+	+	+	+	
Dell Computers	+	+	+		
Gestetner	+	+	+		+
Glaxo	+	+	+	+	
Harris Trust	+		+		
Kashima Oil		+	+	+	
Kidder Peabody	+	+	+	+	
Mead		+	+		
Metalgesellschaft	+	+		+	
Granite Partners			+	+	
Nippon Steel	+	+			
Orange County	+	+		+	+
Pacific Horizon	+	+		+	
Piper Jaffray	+		+	+	
Sandoz			+	+	
Showa/Shell Sekiyu	+	+			
Salomon Brothers				+	+
United Services		+	+		

books! Now, financial regulators are suggesting that they should be accounted at their true value, which could lead many insurance firms to the brink, or at the least to reporting huge losses and to borrowing large amounts of money to meet their capital requirements (International Herald Tribune, 17 June 2003, Business Section).

REFERENCES AND ADDITIONAL READING

Albizzati, M.O., and H. Geman (1994) Interest rate risk management and valuation of surrender option in life insurance policies, *Journal of Risk and Insurance*, **61**, 616–637.

Amikam, H., (1996), Private Communication.

Arndt, K. (1980) Asymptotic properties of the distribution of the supremum of a random walk on a Markov chain, *Probability Theory and Applications*, **46**, 139–159.

Barone-Adesi, G., and R.E. Whaley (1987) Efficient analytical approximation of American option values, *Journal of Finance*, **42**, 301–320.

Barone-Adesi, G., W. Allegretto and R. Elliott (1995) Numerical evaluation of the critical price and American options, *The European Journal of Finance*, **1**, 69–78.

Basak, S., and A. Shapiro (2001) Value at risk based risk management: Optimal policies and asset prices, 2001, *Review of Financial Studies*, **14**, 371–405.

Beibel, M., and H.R. Lerche (1997) A new look at warrant pricing and related optimal stopping problems. Empirical Bayes, sequential analysis and related topics in statistics and probability, *Statistica Sinica*, **7**, 93–108.

Benninga, S. (1989) *Numerical Methods in Finance*, MIT Press, Cambridge, MA.

Boyle, P. (1977) Options: A Monte Carlo approach, *Journal of Financial Economics*, **4**, 323–338.

Boyle, P., and Y. Tse (1990) An algorithm for computing values of options on the maximum or minimum of several assets, *Journal of Financial and Quantitative Analysis*, **25**, 215–227.

Brennan, M., and E. Schwartz (1977) The valuation of American put options, *Journal of Finance*, **32**, 449–462.

Capocelli, R.M., and L.M. Ricciardi (1972) On the inverse of the first passage time probability problem, *Journal of Applied Probability*, **9**, 270–287.

Caraux, G., and O. Gascuel (1992) Bounds on distribution functions of order statistics for dependent variates, *Statistical Letters*, **14**, 103–105.

Carr, P., R. Jarrow and R. Myeneni (1992) Alternative characterizations of American put options, *Journal of Mathematical Finance*, **2**, 87–106.

Cho, H., and K. Lee (1995) An extension of the three jump process models for contingent claim valuation, *Journal of Derivatives*, **3**, 102–108.

Chow, Y.S., H. Robbins and D. Siegmund (1971) *The Theory of Optimal Stopping*, Dover Publications, New York.

Chow, Y.S., H. Robbins and D. Siegmund (1971) *Great Expectations: The Theory of Optimal Stopping*, Houghton Mifflin, Boston, MA.

Coffman, E.G., P. Flajolet, L. Flatto and M. Hofri (1997) The max of a random walk and its application to rectangle packing, Research Report, INRIA (France), July.

Connoly, K.B. (1977) *Buying and Selling Volatility*, John Wiley & Sons, Inc., New York.

Cox, D.R., and H.D. Miller (1965) *The Theory of Stochastic Processes*, Chapman & Hall, London.

Darling, D.A., and A.J.F. Siegert (1953) The first passage time for a continuous Markov process, *Annals of Math. Stat.*, **24**, 624–639.

Duffie, D., and H.R. Richardson (1991) Mean-variance hedging in continuous time, *Annals of Applied Probability*, **1**, 1–15.

Durbin, J. (1992) The first passage time of the Brownian motion process to a curved boundary, *Journal of Applied Probability*, **29**, 291–304.

Embrechts, P., C. Kluppelberg and T. Mikosch (1997) *Modelling Extremal Events*, Springer Verlag, Berlin & New York.

Feller, W. (1957) *An Introduction to Probability Theory and its Applications*, John Wiley & Sons, Inc., New York.

Galambos, J. (1978) *The Asymptotic Theory of Extreme Order Statistics*, John Wiley and Sons, Inc., New York.

Garman, M.B., and S.W. Kohlhagen (1983) Foreign currencies option values, *Journal of International Money and Finance*, **2**, 231–237.

Gerber, H.U., and E.S.W. Shiu (1994a) Martingale approach to pricing perpetual American options, *ASTIN Bulletin*, **24**, 195–220.

Gerber, H.U., and E.S.W. Shiu (1994b) Pricing financial contracts with indexed homogeneous payoff, *Bulletin of the Swiss Association of Actuaries*, **94**, 143–166.

Gerber, H.U., and E.S.W. Shiu (1996) Martingale approach to pricing perpetual American options on two stocks, *Mathematical Finance*, **6**, 303–322.

Gerber, H.U., and Shiu, E.S.W. (1996). Actuarial bridges to dynamic hedging and option pricing, *Insurance: Mathematics and Economics*, **18**, 183–218.

Geske, R., (1979) The valuation of compound options, *Journal of Financial Economics*, **7**, 63–81.

Geske, R., and H.E. Johnson (1984) The American put option valued analytically, *Journal of Finance*, **39**, 1511–1524.

Geske, R., and K. Shastri (1985) Valuation by approximation: A comparison of alternative option valuation techniques, *Journal of Financial and Quantitative Analysis*, **20**, 45–71.

Goldman, M.B., H. Sosin and M. Gatto (1979) Path-dependent options: buy at the low, sell at the high, *Journal of Finance*, **34**, 1111–1128.

Graversen, S.E., G. Peskir and A.N. Shiryaev (2001) Stopping Brownian motion without anticipation as close as possible to its ultimate maximum, *Theory Probability and Applications*, **45**(1), 41–50.

He, H. (1990) Convergence from discrete to continuous time contingent claim prices, *The Review of Financial Studies*, **3**, 523–546.

Hull, J., and A. White (1990) Valuing derivative securities using the explicit finite difference method, *Journal of Financial and Quantitative Analysis*, **25**, 87–100.

Jacka, S.D. (1991) Optimal stopping and the American put, *Journal of Mathematical Finance*, **1**, 1–14.

Johnson, N.L., and S. Kotz (1969) *Discrete Distributions*, Houghton Mifflin, New York.

Johnson, N.L., and S. Kotz (1970a) *Continuous Univariate Distributions* – 1, Houghton Mifflin, New York.

Johnson, N.L., and S. Kotz (1970b) *Continuous Univariate Distributions* – 2, Houghton Mifflin, New York.

Karatzas, I. (1989) Optimization problems in the theory of continuous trading, *SIAM Journal on Control and Optimization*, **27**, 1221–1259.

Kijima, M. and M. Ohnishi (1999) Stochastic orders and their applications to financial optimization, *Mathematical Methods of Operation Research*, **50**, 351–372.

Kim, I.J. and G. Yu (1996) An alternative approach to the valuation of American options and applications, *Review of Derivative Research*, **1**, 61–85.

Korczak, J., and P. Roger (2002) Stock timing and genetic algorithms, *Applied Stochastic Models in Business and Industry*, **18**, 121–134.

Korshunov, D.A. (1997) On distribution tail of the maximum of a random walk, *Stochastic Processes and Applications*, **72**(1), 97–103.

Korshunov, D.A. (2001) Large-deviation probabilities for maxima of sums of independent random variables with negative mean and subexponential distribution. *Theory Probab. Appl.*, **46**(2) 387–397. (In Russian.)

Lamberton, D. (2002) Brownian optimal stopping and random walks, *Applied Mathematics and Optimization*, **45**, 283–324.

Leadbetter, M.R., G. Lindgren and H. Rootzen (1983) *Extremes and Related Properties of Random Sequences and Processes*, Springer Verlag, New York.

Murphy, J. (1998) *Technical Analysis of the Financial Markets*, New York Institute of Finance, New York.

Peksir, G. (1998) Optimal stopping of the maximum process: The maximality principle, *Annal. Prob.*, **26**, 1614–1640.

Révész, Pal (1994) *Random Walk in Random and Non-Random Environments*, World Scientific, Singapore.

Ritchken, P., and R. Trevor (1999) Pricing options under generalized GARCH and stochastic volatility process, *Journal of Finance*, **54**, 377–402.

Rychlik, T. (1992) Stochastically extremal distribution order statistics for dependent samples, *Statistical Probability Letters,* **13**, 337–341.

Rychlik, T. (2001) Mean-variance bounds for order statistics from dependent DFR, IFR, DFRA and IFRA samples, *Journal of Statistical Planning and Inference*, **92**, 21–38.

Schweizer, M. (1995) Varian-optimal hedging in discrete time, *Mathematics of Operations Research*, February (1), 1–32.

Shaked, M., and J.G. Shantikumar (1994) *Stochastic Orders and their Applications*, Academic Press, San Diego, CA.

Shepp, L.A., and A.N. Shiryaev (1993) The Russian option: Reduced regret, *Annals of Applied Probability*, **3**, 631–640.
Shepp, L. A., and A.N. Shiryaev (1994) A new look at the Russian option, *Theory Prob. Appl.*, **39**, 103–119.
Shirayayev, A.N. (1978) *Optimal Stopping Rules*, Springer-Verlag, New York.
Shiryaev, A.N. (1999) *Essentials of Stochastic Finance*, World Scientific, Singapore.
Tapiero, C.S. (1977) *Managerial Planning: An Optimum and Stochastic Control Approach*, Gordon & Breach, New York.
Tapiero, C.S. (1988) *Applied Stochastic Models and Control in Management*, North Holland, New York.
Tapiero, C.S. (1996) *The Management of Quality and its Control*, Chapman & Hall, London.
Wilmott P., (2000) Paul Wilmott on Quantitative Finance, John Wiley & Sons Ltd., Chichester.
Zhang, Q. (2001) Stock trading and optimal selling rule, *SIAM Journal on Control*, **40**(1), 64–87.

APPENDIX: FIRST PASSAGE TIME*

A first time to some state, say S (a given stock price, an exercise option price, a given interest rate level and so on), may be defined by:

$$T(x_0) = \text{Inf}\,\{t > 0; x0) = x_0, x(t) \geq S\}$$

where x_0 is the initial state (at time $t = 0$). The 'target state' can be thought of as an *absorbing state*. Let $f(x,t)$ be the probability of state x at time t of a Markov process. Thus, the probability that the passage time exceeds the current time is:

$$\Pr\{T(x_0) > t\} = \int_{-\infty}^{S} f(x, t/x_0)\,\mathrm{d}x$$

As a result, the passage time probability can be written by deriving: the $\Pr\{T(x_0) \leq t\} = 1 - \Pr\{T(x_0) > t\}$, leading to the distribution function $g(S, t/x_0),\ 0 \leq t < \infty$:

$$g(t) = -\frac{\partial}{\partial t}\int_{-\infty}^{S} f(x, t/x_0)\,\mathrm{d}x$$

with the additional (existence) conditions:

$$g(S, t/x_0) \geq 0,\ \forall S, t, x_0;\ 0 < \int_{0}^{\infty} g(S, t/x_0)\,\mathrm{d}t \leq 1,\ \forall S, t, x_0;$$

$$\times \lim_{x_0 \to S} g(S, t/x_0) = \delta(t)$$

Of course, if the probability distribution $f(.\,,.)$ can be found analytically, then the stopping time distribution can be calculated explicitly in some cases. An example to this effect is considered below, which clearly points out to some mathematical difficulties when analytical solutions are sought. Consider a forward Kolmogorov (Fokker–Plank) equation (corresponding to the stochastic differential equation with drift $b(x)$ and diffusion $a(x)$:

$$\frac{\partial f}{\partial t} = -\frac{\partial}{\partial x}[b(x)f] + \frac{\partial^2}{\partial x^2}[a(x)f]$$

which we write for convenience by the operator:

$$\frac{\partial f}{\partial t} = \mathbf{L}f,\ \mathbf{L} = -\frac{\partial}{\partial x}[b(x)f] + \frac{\partial^2}{\partial x^2}[a(x)f]$$

Using the fact that state S is absorbing, an expectation of the passage time can be obtained by defining a simpler differential operator (expressed as a function of the initial condition x_0 and not of time and as we can see by the application of Ito's differential rule). That is to say, the Laplace transform of the passage time distribution, defined in the terms of the initial state and the target (absorbing) state, is defined by:

$$g_\lambda^*(S, x_0) = \int_0^\infty e^{-\lambda t} g(S, t; x_0)\,dt,\ 0 < g_0^*(S, x_0) \le 1,\ \operatorname*{Lim}_{x_0 \to S} g_\lambda^*(S, x_0) = 1$$

An application of Ito's differential rule yields the second-order differential equation:

$$a(x_0)\frac{d^2 g_\lambda^*}{dx_0^2} + b(x_0)\frac{dg_\lambda^*}{dx_0} - \lambda g_\lambda^* = 0$$

which we write in terms of an adjoint operator $\mathbf{L}^+$ by:

$$\mathbf{L}^+ g_\lambda = \lambda g_\lambda,\ \mathbf{L}^+ = \frac{\partial}{\partial x}[b(x_0)f] + \frac{\partial^2}{\partial x^2}[a(x_0)f]$$

If $\lambda > 0$, then the solution for g_λ is necessarily bounded and is the Laplace transform of a passage time distribution for an Ito stochastic differential equation which is given by:

$$dx = b(x)\,dt + a(x)\,dw,\ x(0) = x_0$$

For (a, b) constants, we have as a special case:

$$g_\lambda^*(S, x_0) = \exp\left[\frac{x_0 - S}{a^2}\left(-b + \sqrt{b^2 + 2\lambda b a^2}\right)\right],\ a > 0, -\infty < x_0 \le S < \infty$$

whose inverse transform yields the inverse Gaussian distribution:

$$g(S, t; x_0) = \frac{(x_0 - S)}{\sqrt{2\pi a^2 t^3}} \exp\left[-\frac{(S - x_0 - bt)^2}{2a^2 t}\right]$$

In other words, if the decision is to sell a stock at a price S, then the probability distribution of the time at which the stock is sold is given by $g(S, t; x_0)$. The current discounted value of such a policy, however, is given by: $V(S) = E(S\,\mathrm{e}^{-R\tau})$ where $E\left(\mathrm{e}^{-R\tau}\right)$ is the stopping time Laplace transform with the risk-free rate replacing the transform's variable. As a result, we have:

$$V(S) = S\ \exp\left[\frac{x_0 - S}{a^2}\left(-b + \sqrt{b^2 + 2R_f ba^2}\right)\right]$$

For a study of first passage time problems the reader should refer to Darling and Siegert (1953) as well as Capocelli and Ricciardi (1972) who provides the first passage time distribution for a lognormal process as well.

CHAPTER 8

Fixed Income, Bonds and Interest Rates

8.1 BONDS AND YIELD CURVE MATHEMATICS

Bonds are binding obligations by a bond issuer to pay the holder of the bond pre-agreed amounts of money at future and given dates. Thus, unlike stocks, bonds have payouts of known quantities and at known dates. Bonds are important instruments that make it possible for governments and firms to raise funds now against future payments. They are considered mostly safe investments, although they can be subject to default and their dependence on interest rates affect their price. As a result, although the nominal values of bonds are known, their price is derived from underlying interest rates. There are as well many types of bonds, designed to meet investors' needs, firms and governments' needs and payment potential when raising capital and funds. For example, there are zero-coupon bonds, coupon-bearing bonds paid at discrete irregular and regular time intervals, there are floating rate bonds, fixed rate bonds, repos (involving a repurchase agreement at some future date and at an agreed-on price). There are also strips bonds (meaning Separate Trading of Registered Interest and Principal of Securities) in which the coupon and the principal of normal bonds are split up, creating an artificial zero-coupon bond of longer maturity. There are options on bonds, bonds with call provisions allowing their recall prior to redemption etc.

Bond values express investors' 'impatience' measured by the rate of interest (discount) used in determining their value. When a bond is totally risk-free, the risk-free rate (usually the Treasury Bills rate of the US Government) is used. When a bond is also subject to various sources of uncertainties (due to interest-rate processes, due to defaults, inflation etc.) then a risk-sensitive discount rate will be applied reflecting an attitude toward these uncertainties and interactions between 'impatience and risk'. We shall see later on that these 'risk-sensitive discount rates' can also be determined in terms of the ongoing risk-free rates and the rates term (of payments and time) structure.

Bond market sizes and trades dwarf all other financial markets and provide therefore a most important source and fundamental information for the valuation

Risk and Financial Management: Mathematical and Computational Methods. C. Tapiero
 ISBN: 0-470-84908-8

of financial assets in general and the economic health of nations and firms. Rated bonds made by financial agencies, such as Standard and Poors, Moody's and their like, are closely watched indicators that have a most important impact on both firm's equity value and governments' liquidity.

In this chapter, essential elements of bond valuation and bond-derived contracts will be elaborated. Further we shall also provide an introduction to interest-rate modelling which is equivalent to bond modelling for one reflects the other and vice versa. It is also a topic of immense economic practical and research interest. Irving Fisher in his work on interest (1906, 1907, 1930) gave the first modern insight into the market interest rate as a balance between agents' impatience (and attitude towards time) and the productivity (returns) of capital (investments). These studies were performed in the spirit of a general equilibrium theory whose foundations were posed by Walras in his *Elements d'Economie Politique Pure* in 1874. Subsequent economic studies (Arrow, 1953) have introduced uncertainty in equilibrium theory based on this approach. A concise review can be found for example in Magill and Quinzzi (1996). Subsequent studies have formalized both the theory of interest rates and its relation to time (the term structure of interest rates) as well as model the exogenous and endogenous sources of uncertainty in interest rates evolution. These studies are of course of paramount importance and interest for bond pricing, whether they are risk-free or default-prone (as it is the case for some corporate bonds). When a bond is risk-free then of course we use the risk-free rate associated with the time of payment. However, since interest rates may vary over time, the bond 'productivity or yield' may shift in various ways (according to the uncertain evolution of interest rates as well as the demand by borrowers and the supply by lenders), which renders the risk-free rate time-varying. When bonds are subject to default of various types, risks are compounded, affecting thereby the discount rate applied to the payment of bonds (and thus the bond price). In this sense, the study and the valuation of bonds is imbued with uncertainty and the risks it generates.

In this chapter we introduce some basic notions for the valuation of bonds. We consider rated bonds, with and without default, with reliable and unreliable rating, for which a number of results and examples will be treated. These results are kept simple except in some cases where over-simplification can hide some important aspects in bond valuation. In such cases we 'star' the appropriate section. In addition a number of results regarding options on bonds, the use of bonds to value the cash flows of corporate rated firms (such as computing 'net present values' of investment projects by such firms etc.) are derived.

Bond markets are, as stated above, both extremely large and active. By far, the most-traded bonds are Treasury bills. These are zero-coupon bonds with a maturity of less than one year. Treasury bills are issued in increments of \$5 000 above a minimum amount of \$10 000. In economic journals, T-bills are quoted by their maturity, followed by a price expressed by the bank discount yield. Below some simple examples are treated to appreciate both the simplicity and the complexity of bond valuation. At the same time, we shall elaborate on a broad number of transactions that can be valued using the bond terminology.

8.1.1 The zero-coupon, default-free bond

A zero coupon bond consists in an obligation to pay at a given future date T (the maturity date), a certain amount of money (the bond nominal value). For simplicity, let this amount be \$1. The price of such a bond at a given time t, $B(t, T)$, denotes the current price of a dollar payment at time T. The value of this bond is essentially a function of:

(i) the time to payment or $\tau = T - t$ and
(ii) the discount factor used at t for a payment at T or, equivalently, it is expressed in terms of the bond yield, denoted by $y(t, T)$.

In other words, the value of a bond can be written in terms of these variables by:

$$B(t, T) = V(y(t, T), \tau);\ \tau = T - t;\ \frac{\partial V(y(t, T), \tau)}{\partial y} < 0,\ \frac{\partial V(y(t, T), \tau)}{\partial \tau} < 0$$

Note that the larger the amount of time left to payment (the bond redemption time) the smaller the bond price (explaining its negative derivative). Further, the larger the discount factor-yield at time t, the smaller the value of the bond (see Figure 8.1).

The definition of a bond's price and its estimation is essential for financial management and mathematics. The behaviour of such a value and its properties underlies the process of interest-rate formation and vice versa, interest-rate processes define the value of bonds. Some obvious properties for bond values are:

$$B(t, t) = 1;\ \lim_{(T-t)\to\infty} B(t, T) = 0;\ B(s', s) > B(s'', s) \text{ if } s'' > s'$$

In other words, a bond paid instantly equals its nominal value, while a bond redeemed at infinity is null. Finally, two similar bond payouts, with one bond due before the other, imply that the one is worth more than the other. Thus, to value a bond, we need to express the time preference for money by the yield, representing

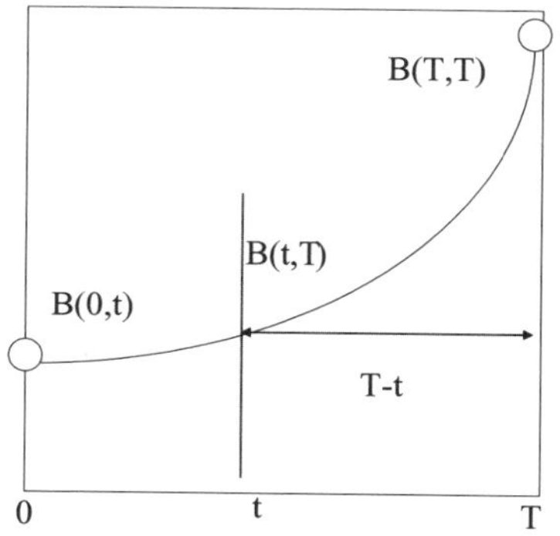

Figure 8.1

the effective bond discount rate $y(t, T)$ at time t associated with a payment $T - t$ periods later. The yield is one of the important functions investors and speculators alike seek to define. It is used by economists to capture the overall movement of interest rates (which are known as 'yields' in Wall Street parlance). There are various interest rates moving up and down, not necessarily in unison. Bonds of various maturities may move independently with short-term rates and long-term rates often moving in opposite directions simultaneously. The overall pattern of interest-rate movement – and what it means about the future of the economy and Wall Street – are the important issues to reckon with. They are thus like tea leaves, only much more reliable if one knows how to read them. Ordinarily, short-term bonds carry lower yields to reflect the fact that an investor's money has less risk. The longer our cash is tied up, the theory goes, the more we should be rewarded for the risk taken. A normal yield curve, therefore, slopes gently upward as maturities lengthen and yields rise. From time to time, however, the curve twists itself into a few recognizable shapes, each of which signals a crucial, but different, turning point in the economy. When those shapes appear, it is often time to alter one's assumptions about economic growth.

In discrete time, the value of a bond is given by discounting the future payout over the remaining period of time using the yield associated to the payment to be received. Of course, we can also calculate the yield as a function of the bond price and its time to maturity. This is done when data regarding bond values are more readily available than yields. For a discrete time bond, it would be written as follows:

$$B(t, T) = [1 + y(t, T)]^{-(T-t)} \quad \text{or} \quad y(t, T) = 1 - [B(t, T)]^{-1/T-t}$$

while in continuous time, it is written as follows:

$$B(t, T) = e^{-y(t,T)(T-t)} \quad \text{or} \quad y(t, T) = -\frac{\ln(B(t, T))}{(T-t)}$$

In other words, the yield and the bond are priced uniquely – one reflecting the other and vice versa. If this were not the case, then markets would be incomplete and 'bond arbitrageurs', for example, would identify such situations and profit from the 'mis-pricing' of bonds. For example, say that a bond paying \$1 in 2 years has a current market value of 0.85. Thus, the yield is found by solving the following equation:

$$0.85 = [1 + y(0, 2)]^{-2} \text{ and therefore the yield is } y(0, 2)$$

$$= \sqrt{\left[\frac{1}{0.85}\right]} - 1 = 0.08455$$

In this case, the bond has a return of 8.455 %. If risk-free interest rates for the same period are 9 %, then clearly it is economically appealing to use the difference in interest rates to make money.

8.1.2 Coupon-bearing bonds

Pure discount bonds such as the above are one of the 'building blocks' of finance and can be used to evaluate a variety of financial instruments. For example, if a default-free bond pays a periodic payment of \$$c$ (the coupon payment) as well as a terminal nominal payment F at time T(the bond face value at maturity), then its price can be expressed in terms of zero-coupon bonds. Its value would be:

$$B_c(t, T) = c \sum_{k=t+1}^{T-1} B(t, k) + FB(t, T)$$

The same value expressed in terms of the yield will be, of course:

$$B_c(t, T) = c \sum_{k=t+1}^{T} \left[\frac{1}{1 + y(t, k)}\right]^{k-t} + F\frac{1}{[1 + y(t, T)]^{T-t}}$$

For example, a bond whose face value is \$1000, with a coupon payout of \$50 yearly with a 9 % interest rate has a current value of \$713.57. By the same token, another bond whose current price is \$800 and has the same properties (payout and face value) has a yield which is necessarily smaller than putting money in the bank and collecting money at the risk-free rate of 9 %.

Valuation in continuous time yields the following equation:

$$B_c(t, T) = c \int_t^T B(t, \tau)\,\mathrm{d}\tau + FB(t, T) \text{ or as yields}$$

$$B_c(t, T) = c \int_t^T e^{-y(t,\tau-t)(\tau-t)}\,\mathrm{d}\tau + F\,\mathrm{e}^{-y(t,T)(T-t)}$$

Mortgage payments, debt payment of various forms and sorts, investments yielding a fixed income etc. can be written in terms of bonds (assuming all payments to be default-free). In some cases, such as reverse mortgage, one may have to be careful in using bonds for the valuation of a financial contract. For example, in reverse mortgage, the bank would assume the responsibility of paying a fixed amount, say c, every month to a homeowner, as long as he lives. At death (which is a random time) the bank would 'at last' take ownership of the home. The value of the bond, thus, equals a coupon payout made for a random amount of time while receiving at the final random time (when the homeowner passes away) an amount equalling the home (random) value. These situations, of course, render the valuation of such contracts more difficult. What may seem at first a profitable contract may turn out to be disastrous subsequently. For this reason, considerable attention is devoted to these situations so that an appropriate pricing procedure and protection (hedging) may be structured. If sources of uncertainty can be determined in a fairly reliable manner, we can at least write the value of the bond equation in terms of these uncertain ingredients and proceed to numerical or simulation techniques to obtain a solution, providing that we can equate these

Table 8.1 Term structure interest rates (source: ECB, 2000).

	1 year	2 years	3 years	4 years	5 years
$y(0, t)$	0.0527	0.053	0.0537	0.0543	0.0551
$y(1, t)$	—	0.0533	0.0542	0.0548	0.0557
$y(2, t)$	—	—	0.0551	0.0556	0.0565
$y(3, t)$	—	—	—	0.0561	0.0572
$y(4, t)$	—	—	—	—	0.0583

values to some replicating risk-free portfolio that would allow calculation of the appropriate discount rate. Additional problems are met when we introduce rated bonds, default bonds, junk bonds etc., as we shall see subsequently.

Example

Consider a coupon-paying bond with a payout of \$100 a year for 4 years at the end of which the principal of \$1000 is redeemed. The current yield is found and given in Table 8.1. On the basis of this information, we are able to calculate the current bond price. Namely, assuming that this is a default-free bond, the bond price is:

$$B_{100}(0, 5) = 100 \sum_{k=1}^{4} \left[\frac{1}{1 + y(0, k)} \right]^k + 1000 \frac{1}{[1 + y(0, 5)]^5}$$

Table 8.1 provides the yields at and for various periods. Yields are calculated by noting that if there is no arbitrage then a dollar invested at time '0' for t periods should have the same value as a dollar invested for s periods and then reinvested for the remaining $t - s$ periods. In other words, in complete markets, when there can be no arbitrage profit, we have:

$$[1 + y(0, t)]^t = [1 + y(0, s)]^s \, [1 + y(s, t)]^{t-s} \quad \text{and}$$

$$y(s, t) = \left(\frac{[1 + y(0, t)]^t}{[1 + y(0, s)]^s} \right)^{1/(t-s)} - 1$$

Thus, yields $y(0, t)$ provide all the information needed to calculate the bond current price. Say that we are currently in the year 2000. This means that we have to insert the term structure rates of year 2000 in our equation in order to calculate the current bond price, or:

$$B_{100}(0, 5) = 100 \left[\begin{array}{l} \dfrac{1}{(1 + 0.0527)} + \dfrac{1}{(1 + 0.053)^2} + \\ + \dfrac{1}{(1 + 0.0537)^3} + \dfrac{1}{(1 + 0.0543)^4} \end{array} \right]$$

$$+ 1100 \frac{1}{(1 + 0.0551)^5} = 1192.84$$

To determine the price a period hence, the appropriate table for the rates term structure will have to be used. If we assume no changes in rates, then the bond value is calculated by:

$$B_{100}(1,5) = 100\sum_{k=1}^{4}\left[\frac{1}{1+y(1,k)}\right]^k + 1000\frac{1}{[1+y(1,5)]^5} = 1155.72$$

which is a decline in the bond value of 1192.84 − 1155.72 = 37.12 dollars.

8.1.3 Net present values (NPV)

The NPV of an investment providing a stream of known-for-sure payments over a given time span can be also written in terms of zero-coupon bonds. The traditional NPV of a payment stream $C_0, C_1, C_2, C_3, \ldots, C_n$ with a fixed risk-free discount rate R_f is:

$$NPV = C_0 + \frac{C_1}{1+R_f} + \frac{C_2}{(1+R_f)^2} + \frac{C_3}{(1+R_f)^3} + \cdots + \frac{C_n}{(1+R_f)^n}$$

There are some problems with this formula, however, for it is not market-sensitive, ignoring the rates term structure and the uncertainty associated with future payouts. If the payout is risk-free, it is possible to write each payment C_i, $i = 0, 1, 2, \ldots, n$ in terms of zero-coupon (risk-free) bonds. At time $t = 0$,

$$NPV_0 = C_0 + C_1B(0,1) + C_2B(0,2) + C_3B(0,3) + \cdots + C_nB(0,n)$$

While a period later, we have:

$$NPV_1 = C_1 + C_2B(1,2) + C_3B(1,3) + \cdots + C_nB(1,n)$$

with each bond valued according to its maturity. When a zero-coupon bond is rated or subject to default (which has not been considered so far), applying a constant discount rate to evaluate the NPV can be misleading since it might not account for changes in interest rates over time, their uncertainty as well as the risks associated with the bond payouts and the ability of the bond issuer to redeem it as planned. If a bond yield is time-varying, deterministically or in a random manner, then the value of the bond will change commensurately, altering over time the NPV. Corporate bonds (rated by financial agencies such as Standard and Poors, Moody's, Fitch), the value of corporations' cash flows must similarly reflect the corporate rating and their associated risks. In section 8.3, we consider these bonds and thereby provide an approach to valuing cash flows of rated corporations as well. The net present value at time t of a corporate cash flow is thus a random variable reflecting interest-rate uncertainty and the corporate rate and its reliability. An appropriate and equivalent way to write the NPV (assuming that cash payments are made for sure) using the yield $y(0,i)$ associated with each zero-coupon bond with maturity i), is:

$$NPV(0\,|\Re) = C_0 + \frac{C_1}{1+y(0,1)} + \frac{C_2}{(1+y(0,2))^2} + \frac{C_3}{(1+y(0,3))^3} + \cdots + \frac{C_n}{(1+y(0,n))^n}$$

Thus, generally, we can write a net present value at time t by:

$$NPV(t\,|\Re) = \sum_{i=t}^{n} C_i B(t,i) = \sum_{i=t}^{n} \frac{C_i}{[1+y(t,i)]^{i-t}}$$

In a similar manner, a wide variety of cash flows and expenses may be valued. The implication of this discussion is that all cash flows, their timing and the uncertainty associated with these flows may also be valued using 'bond mathematics'. When coupon payments are subject to default, we can represent the NPV as a sum of default-prone bonds, as will be discussed later on. Similarly, if the NPV we calculate is associated to a corporation whose debt (bond) is rated, then such rating also affects the value of the bond and thereby the corporation's cash flow. Generally, bonds are used in many ways to measure asset values, to measure risks and to provide an estimate of many contracts that can be decomposed into bonds that can be, or are, traded.

8.1.4 Duration and convexity

'Duration' is a measure for exposure to risk. It expresses the sensitivity of the bond price to (small) variations in interest rates. In other words, the duration at time t of a bond maturing at time T, written by $D(t,T)$, measures the return per unit for a move Δy in the yield, or

$$D(t,T) = -\frac{1}{B(t,T)}\frac{[\Delta B(t,T)]}{\Delta y(t,T)}$$

For small intervals of time, we can rewrite this expression as follows:

$$D(t,T) = -\frac{[\Delta \log B(t,T)]}{\Delta y(t,T)} \approx -\frac{\mathrm{d}[\log B(t,T)]}{\mathrm{d}y(t,T)}$$

Since the bond rate of return is $R(t,T) = \Delta B(t,T)/B(t,T) \approx \Delta \log B(t,T)$ and $\Delta y(t,T)$ is a rate move at time t, we can write:

$$R(t,T) = -D(t,T)\,[\Delta y(t,T)]$$

In words:

$$\text{Rate of returns on bonds} = -(\text{Duration}) * (\text{Yield rate move})$$

At time t, a zero-coupon bond maturing at time T has, of course, a duration of $T-t$. For a coupon bond with payments of C_i at times t_i, $i=1,\ldots,n$ and a bond price yield denoted by $B(0, n)$ and $y(0, n)$, then (in continuous-time discounting):

$$B(0, n) = \sum_{i=1}^{n} C_i \, \mathrm{e}^{-y(0,n)t_i}$$

and the duration is measured by time-weighted average of the bond prices:

$$D(0, n) = \frac{\sum_{i=1}^{n} C_i t_i \, \mathrm{e}^{-y(0,n)t_i}}{\sum_{i=1}^{n} C_i \, \mathrm{e}^{-y(0,n)t_i}}$$

This result can be proved by simple mathematical manipulations since:

$$-\frac{\mathrm{d}(\log B)}{\mathrm{d}y} = D \text{ implies } -\frac{\mathrm{d}\left(\log \sum_{i=1}^{n} C_i \, \mathrm{e}^{-yt_i}\right)}{\mathrm{d}y} = \frac{\sum_{i=1}^{n} C_i t_i \, \mathrm{e}^{-yt_i}}{\sum_{i=1}^{n} C_i \mathrm{e}^{-yt_i}} = D$$

While duration reflects a first-order change of the bond return with respect to its yield, convexity captures second-order effects in yield variations. Explicitly, let us take a second-order approximation to a bond whose value is a function of the yield. Informally, let us write the first three terms of a Taylor series expansion of the bond value:

$$B(t, y + \Delta y) = B(t, y) + \frac{\partial B(t, y)}{\partial y} \Delta y + \frac{1}{2} \frac{\partial^2 B(t, y)}{\partial y^2} (\Delta y)^2$$

Dividing by the bond value, we have:

$$\frac{\Delta B(t, y)}{B} = \frac{1}{B} \frac{\partial B(t, y)}{\partial y} \Delta y + \frac{1}{2} \frac{1}{B} \frac{\partial^2 B(t, y)}{\partial y^2} (\Delta y)^2$$

And, approximately, for a small variation in the yield of Δy, we have (replacing partial differentiation by differences):

$$\frac{\Delta B}{B} = \frac{1}{B} \frac{\Delta B}{\Delta y} \Delta y + \frac{1}{2} \frac{1}{B} \frac{\Delta^2 B}{\Delta y^2} (\Delta y)^2$$

If we define convexity by:

$$\Upsilon(t, T) = \frac{1}{B} \frac{\Delta^2 B}{\Delta y^2}$$

then, an expression for the bond rate of return in terms of the duration and the convexity is:

$$\frac{\Delta B}{B} = -D(t, T) \Delta y + \frac{1}{2} \Upsilon(t, T) (\Delta y)^2,$$

or

$$\begin{pmatrix} \text{Rate of} \\ \text{returns on bonds} \end{pmatrix} = -(\text{Duration}) * \begin{pmatrix} \text{Yield} \\ \text{rate move} \end{pmatrix} + \frac{1}{2}(\text{Convexity}) * \begin{pmatrix} \text{Yield} \\ \text{rate move} \end{pmatrix}^2$$

Thus, a fixed-income bond will lose value as the interest rate (i.e. $\Delta y > 0$) increases and, conversely, it loses value when the interest decreases (i.e. $\Delta y < 0$). For example, say that a coupon-bearing bond at time t_i, $i = 1, 2, 3, \ldots$ with yield y is given at time t by:

$$B(t, T) = K\mathrm{e}^{-y(T-t)} + \sum_{i=1}^{C_i} \mathrm{e}^{-y(t_i-t)}$$

Note that:

$$\frac{\mathrm{d}B}{\mathrm{d}y} = -K(T-t)\,\mathrm{e}^{-y(T-t)} - \sum_{i=1} C_i(t_i - t)\,\mathrm{e}^{-y(t_i-t)}$$

$$\frac{\mathrm{d}^2 B}{\mathrm{d}y^2} = -K(T-t)^2\,\mathrm{e}^{-y(T-t)} - \sum_{i=1} C_i(t_i - t)^2\,\mathrm{e}^{-y(t_i-t)}$$

And therefore the duration and the convexity express explicitly first- and second-order effects of yield variation, or:

$$D(t, T) = \frac{\mathrm{d}B}{B\mathrm{d}y};\ \Upsilon(t, T) = \frac{\mathrm{d}^2 B}{B\mathrm{d}y^2}$$

Example
We consider the following bond and calculate its duration:

Actual price:	100
Nominal interest rate:	10 % (p.a.)
Buy back value:	100
Years remaining:	4
The actual market interest rate:	10 %

The duration is defined by:

$$\text{Macaulay duration} = -\frac{PV'(Y)}{PV(Y)} = \frac{\sum_{i=1}^{n} t_i * c_i * (1+Y)^{-t_i}}{\sum_{i=1}^{n-1} c_i * (1+Y)^{-t_i} + (100 + c_i) * (1+Y)^{-t_i}}$$

where PV′ is the derivative of $PV(Y)$. This means that the duration of the bond equals:

$$D_M^C = \frac{[1*10\,000*(1.1)^{-1} + 2*10\,000*(1.1)^{-2} + 3*10\,000*(1.1)^{-3} + 4*110\,000*(1.1)^{-4}]}{100\,000} = 3.5$$

If one invests in a bond at a given time and for a given period, the yield does not represent the rate of return of such an investment. This is due to the fact that coupon payments are reinvested at the same yield, which is not precise since yields are changing over time and coupon payments are reinvested at the prevailing yields when coupons are distributed. As a result, changing yield has two opposite effects on the investor rate of return. On the one hand, an increase in the yield decreases the bond value, as we saw earlier, while it increases the rate of return on the coupon. These two effects cancel out exactly when the investor holds the bond for a time period equal to its duration. Thus, by doing so, the rate of return will be exactly the yield at the time he acquired the bond and thus his investment is immune to changing yields. This strategy is called immunization. This strategy is in fact true only for small changes in the yield.

Explicitly, let $B(t, y) = B(t, y : T)$ be the bond price at time t when the yield is y and the maturity T. Consider another instant of time $t + \Delta t$ and let the yield at this time be equal $y + \Delta y$. In the (continuous) time interval $[t, t + \Delta t]$ the coupon payment c is reinvested continuously at the new yield and therefore the bond values at time t and $t + \Delta t$ are given by:

Time t: $B(t, y)$

Time $t + \Delta t$: $B(t + \Delta t, y + \Delta y) + \int_t^{t+\Delta t} c\,e^{-(y+\Delta y)(t+\Delta t - z)} dz$

Thus, for immunization we require that the bond rate of return equals its current yield, or:

$$y = \frac{1}{\Delta t}\left[\frac{B(t + \Delta t, y + \Delta y) + \int_t^{t+\Delta t} c\,e^{-(y+\Delta y)(t+\Delta t - z)} dz - B(t, y)}{B(t, y)}\right]$$

Since

$$\int_t^{t+\Delta t} c\,e^{-(y+\Delta y)(t+\Delta t - z)}\, dz = \frac{c}{(y + \Delta y)}\left[1 - e^{-(y+\Delta y)\,\Delta t}\right] \approx c\,\Delta t$$

and for small Δt,

$$B(t + \Delta t, y + \Delta y) \approx B(t, y + \Delta y) + \frac{\partial B(t, y + \Delta y)}{\partial t}\,\Delta t$$

Inserting in our equation, we have:

$$1 + y\,\Delta t = \left[\frac{B(t, y+\Delta y) + \dfrac{\partial B(t, y+\Delta y)}{\partial t}\Delta t + c\,\Delta t}{B(t, y)}\right]$$

and

$$\frac{\partial B(t+\Delta y)}{\partial t} = \frac{\partial}{\partial t}\left(\int_t^T c\,e^{-(y+\Delta y)(z-t)}\,dz + e^{-(y+\Delta y)(T-t)}\right)$$

$$= c\,(y+\Delta y)\int_t^T e^{-(y+\Delta y)(z-t)}\,dz - c + (y+\Delta y)\,e^{-(y+\Delta y)(T-t)}$$

$$= (y+\Delta y)\,B - c$$

Replacing these terms in the previous equation, we have:

$$1 + y\Delta t = \left[\frac{B(t, y+\Delta y) + (y+\Delta y)\,B\,\Delta t - c\,\Delta t + c\,\Delta t}{B(t, y)}\right]$$

This is reduced to:

$$1 + y\Delta t = \left[\frac{B(t, y+\Delta y)}{B(t, y)}\right][1 + (y+\Delta y)\,\Delta t]$$

$$= \left[1 + \frac{B(t, y+\Delta y) - B(t, y)}{B(t, y)}\right][1 + (y+\Delta y)\,\Delta t]$$

or

$$1 + y\,\Delta t = \left[1 + \frac{\Delta B}{B\Delta y}\,\Delta y\right][1 + (y+\Delta y)\,\Delta t]$$

$$= [1 - D(t, y)\Delta y]\,[1 + (y+\Delta y)\,\Delta t]$$

Additional manipulations lead to the condition for immunization, namely that Δt equals the duration or $\Delta t = D(t, y) - D(t, y)(y+\Delta y)\,\Delta t \approx D(t, y)$ and finally for very small Δt

$$\Delta t = D(t, y)$$

8.2 BONDS AND FORWARD RATES

A forward rate is denoted by $F(t, t_1, t_2)$ and is agreed on at time t, but for payments starting to take effect at a future time t_1 and for a certain amount of time $t_2 - t_1$. In Figure 8.2, these times are specified.

A relationship between forward rates and spot rates hinges on an arbitrage argument. Roughly, this argument states (as we saw earlier), that two equivalent

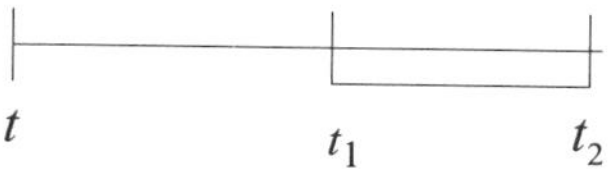

Figure 8.2

investments (from all points of view) have necessarily the same returns. Say that at time t we invest \$1 for a given amount of time $t_2 - t$ at the available spot rate (its yield). The price of such an investment using a bond is then: $B(t, t_2)$. Alternatively, we could invest \$1 for a certain amount of time, say $t_1 - t$, $t_1 \leq t_2$ at which time the moneys available will be reinvested at a forward rate for the remaining time interval: $t_2 - t_1$. The price of such an investment will then be $B(t, t_1)B_f(t_1, t_2)$ where $B_f(t_1, t_2) = [1 + F(t, t_1, t_2)]^{-(t_2-t_1)}$ is the value of the bond at time t_1 paying \$1 at time t_2 using the agreed-on (at time t) forward rate $F(t, t_1, t_2)$. Since both payments result in \$1 both received at time t_2 they have the same value, for otherwise there will be an opportunity for arbitrage. For this reason, assuming no arbitrage, the following relationship must hold (and see Figure 8.3):

$$B(t, t_2) = B(t, t_1)B_f(t_1, t_2) \text{ implying } B_f(t_1, t_2) = \frac{B(t, t_2)}{B(t, t_1)}$$

In discrete and continuous time, assuming no arbitrage, this leads to the following forward rates:

$$[1 + F(t, t_1, t_2)]^{t_2-t_1} = \frac{[1 + y(t, t_2)]^{t_2-t}}{[1 + y(t, t_1)]^{t_1-t}} \text{ (discrete time)}$$

$$F(t, t_1, t_2) = \frac{y(t, t_2)(t_2 - t) - y(t, t_1)(t_1 - t)}{(t_2 - t_1)} \text{ (continuous time)}$$

In practice, arbitrageurs can make money by using inconsistent valuations by bond and forward rate prices. For complete markets (where no arbitrage is possible), the spot rate (yield) contains all the information regarding the forward market rate and, vice versa, the forward market contains all the information regarding the spot market rate, and thus it will not be possible to derive arbitrage profits. In practice, however, some pricing differences may be observed, as stated above, opening up arbitrage opportunities.

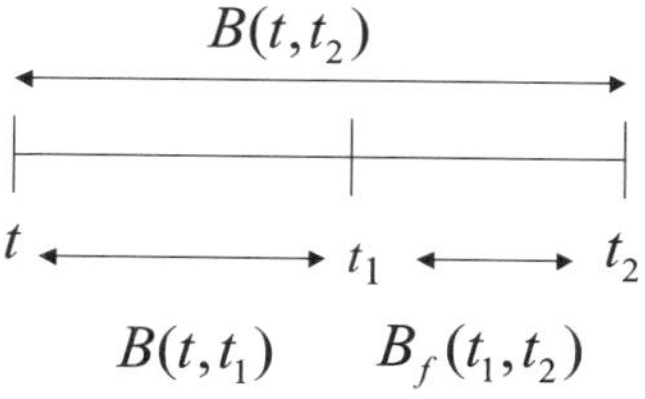

Figure 8.3

Problem
An annuity pays the holder a scheduled payment over a given amount of time (finite or infinite). Determine the value of such an annuity using bond values at the current time. What would this value be in two years using the current observed rates?

Problem
What will be the value of an annuity that starts in T years and will be paid for M years afterwards? How would you write this annuity it is terminated at the time the annuity holder passes away (assuming that all payments are then stopped)?

Problem
Say that we have an obligation whose nominal value is \$1000 at the fixed rate of 10 % with a maturity of 3 years, reimbursed in fine. In other words, the firm obtains a capital of \$1000 whose cost is 10 %. What is the financial value of the obligation? Now, assume that just after the obligation is issued the interest rate falls from 10 to 8 %. The firm's cost of finance could have been smaller. What is the value of the obligation (after the change in interest rates) and what is the 'loss' to the firm.

8.3 DEFAULT BONDS AND RISKY DEBT

Bonds are rated to qualify their standard risks. Standard and Poors, Moody's and other rating agencies use for example, AAA, AA, A, BB, etc. to rate bonds as more or less risky. We shall see in section 8.4 that these rating agencies also provide Markov chains, expressing the probabilities that rated firms switch from one rating to another, periodically adapted to reflect market environment and the conditions particularly affecting the rated firm (for example, the rise and fall of the technology sector, war and peace, and their likes).

Consider a portfolio of B-rated bonds yielding 14 %; typically, these are bonds which currently are paying their coupons, but have a high likelihood of defaulting or have done so in the recent past. A Treasury bond of similar duration yields 5.5 %. Thus, in this example, the Junk–Treasury Spread (JTS) is 8.5 %. Now, let us take a look at the spread's history over the past 13 years (Jay Diamond, *Grant's Interest Rate Observer* data).

The spread depicted in Figure 8.4 corresponds roughly to a B-rated debt. Note the very wide range of spreads, from just below 3 % to almost 10 %. What does a JTS of 3 % mean? Very bad news for the junk buyer, because he or she will have been better off in Treasuries if the loss rate exceeds 3 %. And even if the loss rate is only half of that, a 1.5 % return premium does not seem adequate to compensate for this risk. There is a wealth of data on the bankruptcy/default rate, allowing us to evaluate whether the prevailing risk premium amounts to adequate compensation.

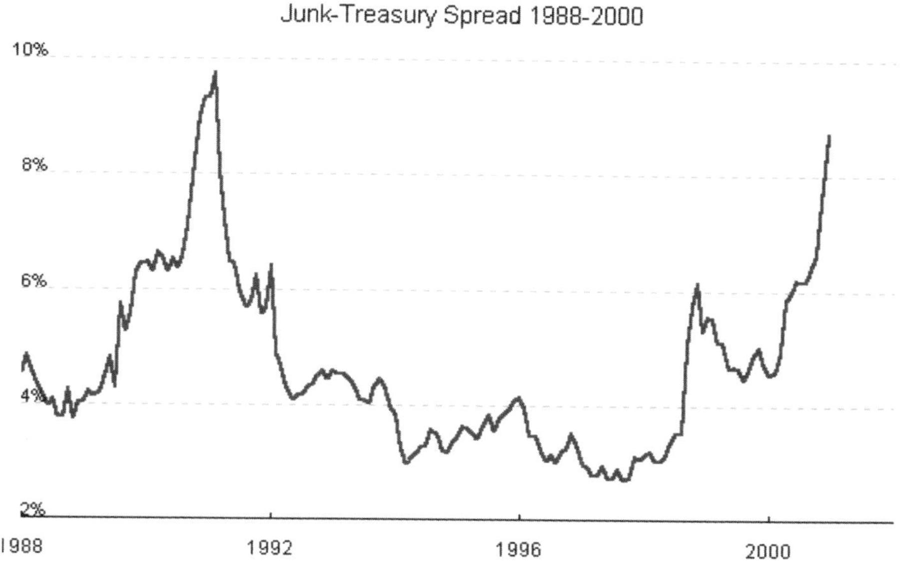

Figure 8.4 Junk–Treasury spread 1988–2000 (Jay Diamond, *Grant's Interest Rate Observer* data).

Rating agencies often use terms such as default rate and loss rate which are important to understand. The former defines the proportion of companies defaulting per year. But not all companies that default go bankrupt. The recovery rate is the proportion of defaulting companies that do not eventually go bankrupt. So a portfolio's reduction in return is calculated as the default rate times one minus the recovery rate: if the default rate is 4 % and the recovery rate is 40 %, then the portfolio's total return has been reduced by 2.4 %. The loss rate, how much of the portfolio actually disappears, is simply the default rate minus the absolute percentage of companies which recover. According to Moody's, the annual long-term default rate of bonds rated BBB/Baa (the lowest 'investment grade') is about 0.3 %; for BB/Ba, about 1.5 %; and for B, about 7 %. But in any given year, the default rate varies widely. Further, because of the changes in the high-yield market that occurred 15 years ago, the pre-1985 experience may not be of great relevance to high-yield investing today.

Prior to the use of junk bonds the overwhelming majority of speculative issues were 'fallen angels', former investment-grade debt which had fallen on hard times. But, after 1985, most high-yield securities were speculative right from their initial offering. Once relegated to bank loans, poorly rated companies were for the first time able to issue debt themselves. This was not a change for the better. Similar to speculative stock IPOs, these new high-yield bond issues tended to have less secure 'coverage' (based on an accounting term defined as the ratio of earnings-before-taxes-and-interest to total interest charges) than the fallen angels of yore, and their default rates were correspondingly higher.

Many financial institutions hold large amounts of default-prone risky bonds and securities of various degrees of complexity in their portfolios that require a reliable estimate of the credit exposure associated with these holdings. Models of default-prone bonds fall into one of two categories: *structural models* and *reduced-form models*. Structural models specify that default occurs when the firm value falls below some explicit threshold (for example, when the debt to equity ratio crosses a given threshold). In this sense, default is a 'stopping time' defined by the evolution of a representative stochastic process. Merton (1974) first considered such a problem; it was studied further by many researchers including Black and Cox (1976), Leland (1994), and Longstaff and Schwartz (1995). These models determine both equity and debt prices in a self-consistent manner via arbitrage, or contingent-claims pricing. Equity is assumed to possess characteristics similar to a call option, while debt claims have features analogous to claims on the firm's value. This interpretation is useful for predicting the determinants of credit-spread changes, for example.

Some models assume as well that debt-holders get back a fraction of the debt, called the recovery ratio. This ratio is mostly specified a priori, however. While this is quite unrealistic, such an assumption removes problems associated to the debt seniority structure, which is a drawback of Merton's (1974) model. Some authors, for example, Longstaff and Schwartz (1995), argue that, by looking at the history of defaults and recovery ratios for various classes of debt of comparable firms, one can find a reliable estimate of the recovery ratio. Structural models are, however, difficult to use in valuing default-prone debt, due to difficulties associated with determining the parameters of the firm's value process needed to value bonds. But one may argue that parameters could always be retrieved from market prices of the firm's traded bonds. Further, they cannot incorporate credit-rating changes that occur frequently for default-prone (risky) corporate debts.

Many corporate bonds undergo credit downgrades by credit-rating agencies before they actually default, and bond prices react to these changes (often brutally) either in anticipation or when they occur. Thus, any valuation model should take into account the uncertainty associated with credit-rating changes as well as the uncertainty surrounding default and the market's reactions to such changes. These shortcomings make it necessary to look at other models for the valuation of defaultable bonds and securities that are not predicated on the value of the firm and that take into account credit-rating changes. For example, a meltdown of financial markets, wars, political events of economic importance are such cases, where the risk is exogenous (rather than endogenous). This leads to reduced-form models.

The problem of rating the credit of bonds and credit markets is in fact more difficult than presumed by analytical models. Information asymmetries compound these difficulties. Akerlof in his 2001 Nobel allocution pointed to these effects further.

A bank granting a credit has less information than the borrower, on his actual default risk. . . . On the same token, banks expanding into new, unknown markets are at a particular risk. On the one hand, due to their imperfect market knowledge, they must rely on the equilibrium between supply and demand to a large extent. On the other hand, under asymmetric information, it is very

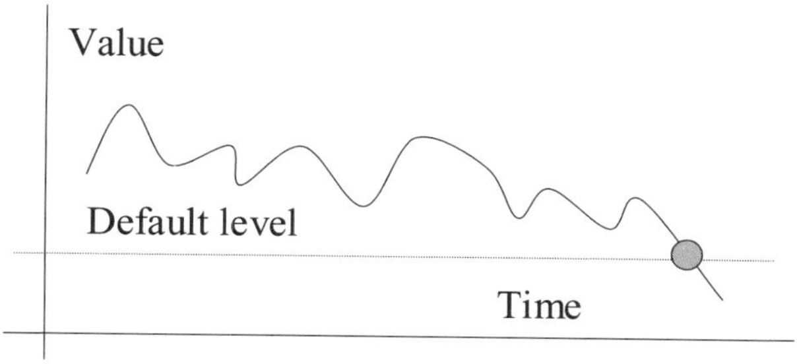

Figure 8.5 Structural models of default.

easy for clients to hide risks and to give too optimistic profit estimates, possibly approaching fraud in extreme cases. Adverse selection then implies a markedly increased default risk for such banks. Banks can use interest rates and additional security as instruments for screening the creditworthiness of clients when they estimate that their information is insufficient. Credit risk and pricing models, of course, are complementary tools. Based on information provided by the client, they produce risk-adjusted credit spreads and thus may set limits to the principle of supply and demand. On the other hand, borrowers with a credit rating may use this rating to signal the otherwise private information on their solvency, to the bank. In exchange, they expect to receive better credit conditions than they would if the bank could only use information on sample averages.

Technically, the value process is defined in terms of a stochastic process $\{x, t \geq 0\}$ while default is defined by the first time τ (the stopping time) the process reaches a predefined threshold-default level. In other words, let the threshold space be $\Re$, then:

$$\tau = \text{Inf}\,\{t > 0, x(t) \notin \Re\}$$

where $\Re$ is used to specify the set of feasible states for an operating firm. As soon as the firm's value is out of these states, default occurs.

Reduced-form models specify the default process explicitly, interpreting it as an exogenously motivated jump process, usually expressed as a function of the firm value. This class of models has been investigated, for example by Jarrow and Turnbull (1995), Jarrow *et al.* (1997) and others. Although these models are useful for fitting default to observed credit spreads, they mostly neglect the underlying value process of the firm and thus they can be less useful when it is necessary to determine credit spread variations. Jarrow *et al.* (1997) in particular have adopted the rating matrix used by financial institutions such as Moody's, Standard and Poors and others as a model of credit rating (as we too shall do in the next section).

Technically, default is defined exogenously by a random variable $\tilde{T}$ where $t < \tilde{T} < T$, with T, the bond expiry date. The conditional probability of default is assumed given by:

$$P(\tilde{T} \in (t + \mathrm{d}t) \,\big|\, t < \tilde{T} < T) = q(x)\,\mathrm{d}t + 0(\mathrm{d}t)$$

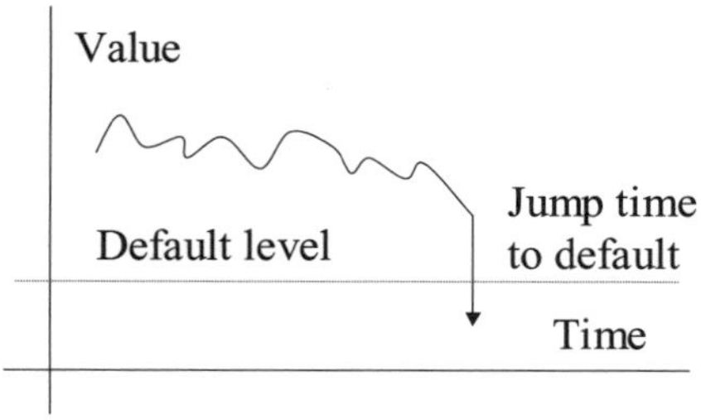

Figure 8.6 Reduced-form models default.

This means that the conditional probability of default q in a small time interval $(t + \mathrm{d}t)$, given that no default has occurred previously, is a function of an underlying stochastic process $\{x, t \geq 0\}$. If the probability q is independent of the process $\{x, t \geq 0\}$, this implies that the probability of default is of the exponential type. That is to say, it implies that at each instant of time, the probability of default is time-independent and independent of the underlying economic fundamentals. These are very strong assumptions and therefore, in practice, one should be very careful in using these models.

A comparison between structural and reduced-form models (see Figure 8.6) is outlined in Table 8.2. Selecting one model or the other is limited by the underlying risk considered and the mathematical and statistical tractability in applying such a model. These problems are extensively studied, as the references at the end of the chapter indicate.

A general technical formulation, combining both structural and reduced-form models leads to a time to default we can write by $\mathrm{Min}(\tau, \tilde{T}, T)$ where T is the maturity reached if no default occurs, while exogenous and endogenous default are given by the random variables $(\tau, \tilde{T})$. If the yield of such bonds at time t for a payout at s is given by, $Y(t, s) \equiv y(\tau, \tilde{T}, s)$, the value of a pure default-prone bond paying \$1 at redemption is then $E \exp(-Y(t, T)\,\mathrm{Min}(\tau, \tilde{T}, T))$. Of course if there was no default, the yield would be $y(t, s)$ and therefore $Y(t, s) > y(t, s)$ in order to compensate for the default risk. The essential difficulty of these problems is to determine the appropriate yield which accounts for such risks, however. For example, consider the current value of a bond retired at $\mathrm{Min}(\tau, \tilde{T}, T)$ and paying an indexed coupon payout indexed to some economic variable or economic index (inflation, interest rate etc.). Uncertainty regarding the coupon payment, its nominal value and the bond default must then be appropriately valued through the bond yield.

When a bond is freely traded, the coupon payment can also be interpreted as a 'bribe' paid to maintain bond holding. For example, when a firm has coupon payments that are too large, it might redeem the bond (provided it incurs the costs associated with such redemption). By the same token, given an investor with other opportunities, deemed better than holding bonds, it might lead the investor to forgo future payouts and principal redemption, and sell the bond at its current

Table 8.2 A comparison of selected models.

Model	Advantages	Drawbacks
Merton (1974)	Simple to implement.	(a) Requires inputs about the firm value. (b) Default occurs only at debt maturity. (c) Information about default and credit-rating changes cannot be used.
Longstaff and Schwartz (1995)	(a) Simple to implement. (b) Allows for stochastic term structure and correlation between defaults and interest rates.	(a) Requires inputs related to the firm value. (b) Information in the history of defaults and credit-rating changes cannot be used.
Jarrow, Lando, and Turnbull (1997)	(a) Simple to implement. (b) Can match exactly existing prices of default-risky bonds and thus infer risk-neutral probabilities for default and credit-rating changes. (c) Uses the history of default and credit-rating change.	(a) Correlation not allowed between default probabilities and the level of interest rates. (b) Credit spreads change only when credit ratings change.
Lando (1998)	(a) Allows correlation between default probabilities and interest rates. (b) Allows many existing term-structure models to be easily embedded in the valuation framework.	Historical probabilities of defaults and credit-rating changes are used assuming that the risk premiums due to defaults and rating changes, is null.
Duffie and Singleton (1997)	(a) Allows correlation between default probabilities and the level of interest rates. (b) Recovery ratio can be random and depend on the pre-default value of the security. (c) Any default-free term-structure model can be accommodated, and existing valuation results for default-free term-structure models can be readily used.	Information regarding credit-rating history and defaults cannot be used.
Duffie and Huang (1996) (swaps)	(a) Has all the advantages of Duffie and Singleton. (b) Asymmetry in credit qualities is easily accommodated. (c) ISDA guidelines for settlement upon swap default can be incorporated.	(a) Information regarding credit-rating history and defaults cannot be used. (b) Computationally difficult to implement for some swaps, such as cross-currency swaps, if domestic and foreign interest rates are assumed to be random.

market value. The number of cases we might consider is very large indeed, but only a few such cases will be considered explicitly here.

Structural and reduced-form models for valuing default-prone debt do not incorporate financial restructuring (and potential recovery) that often follows default. Actions such as renegotiating the terms of a debt by extending the maturity or lowering/postponing promised payments, exchanging debt for other forms of security, or some combination of the above (often being the case after default), are not considered. Similarly, institutional and reorganization features (such as bankruptcy) cannot be incorporated in any of these models simply. Further, anticipated debt restructurings by the market is priced in the value of a defaultable bond in ways that none of these models captures. In fact, many default-prone securities are also thinly traded. Thus, a liquidity premium is usually incorporated into these bond prices, hiding their risk of default. Finally, empirical evidence for these models is rather thin. Duffie and Singleton (1997, 1999) find that reduced-form models have problems explaining the observed term structure of credit spreads across firms of different credit qualities. Such problems could arise from incorrect statistical specifications of default probabilities and interest rates or from models' inability to incorporate some of the features of default/bankruptcy mentioned above. Bond research, just like finance in general, remains therefore a domain of study with many avenues to explore and questions that are still far from resolved.

8.4 RATED BONDS AND DEFAULT

The potential default of bonds and changes in rating are common and outstanding issues to price and reckon with in bonds trading and investment. They can occur for a number of reasons, including some of the following:

(1) Default of the payout or default on the redemption of the principal.
(2) Purchasing power risk arising because inflationary forces can alter the value of the bond. For example, a bond which is not indexed to a cost of living index may in fact generate a loss to the borrower in favour of the lender should inflation be lower than anticipated thereby increasing the real or inflation-deflated payments.
(3) Interest-rate risk, resulting from predictable and unpredictable variations in market interest rates and therefore the bond yield.
(4) Delayed payment risk, and many other situations associated with the financial health of the bond issuer and its credit reliability.

These situations are difficult to analyse but rating agencies specializing in the analysis and the valuation of financial assets provide ratings to nations and corporate entities that are used to price bonds. These agencies provide explicit matrices that associate to various bond classes (AAA, AA, B etc.) probabilities (a Markov chain) of remaining in a given class or switching to another (higher or lower) risk (rating) class. Table 8.3 shows a scale of ratings assigned to bonds by financial

Table 8.3 Ratings.

Moody's	S&P	Definition
Aaa	AAA	Highest rating available
Aa	AA	Very high quality
A	A	High quality
Baa	BBB	Minimum investment grade
Ba	BB	Low grade
B	B	Very speculative
Caa	CCC	Substantial risk
Ca	CC	Very poor quality
C	D	Imminent default or in default

firms services (Moody's, Standard and Poors etc.) that start from the best quality to the lowest. In addition to these ratings, Moody's adds a '1' to indicate a slightly higher credit quality; for instance, a rating of 'A1' is slightly higher than a rating of 'A' whereas 'A3' is slightly lower. S&P ratings may be modified by the addition of a '+' or '−' (plus or minus). 'A+' is a slightly higher grade than 'A' and 'A−' is slightly lower. Occasionally one may see some bonds with an 'NR' in either Moody's or S&P. This means 'not rated'; it does not necessarily mean that the bonds are of low quality. It basically means that the issuer did not apply to either Moody's or S&P for a rating. Government agencies are a good example of very high quality bonds that are not rated by S&P. Other things being equal, the lower the rating, the higher the yield one can expect. Insured bonds have the highest degree of safety of all non-government bonds. Bond insurance agencies guarantee the payment of principal and interest on the bonds they have insured (since insurance reduces the bond's risk). When bonds are insured by one of the major insurance agencies, they automatically attain 'AAA' rating, identifying the bond as one of the highest quality one can buy. Some of the major bond insurers are AMBAC, MBIA, FGIC and FSA. In such circumstances, bonds have almost no default risk.

The Moody's rating matrix shown in Table 8.4 is an example. For AAA bonds, the probability that it maintains such a rating is .9193 while there is a .0746 probability that the bond rating is downgraded to a AA bond and so on for remaining values. These matrices are updated and changed from time to time as business conditions change. Given these matrices we observe that even a triple AAA bond is 'risky' since there is a probability that it be downgraded and its price reduced to reflect such added risk.

In some cases, the price of downgrading the credit rating of a firm can be much larger than presumed. For example, buried in Dynegy Inc.'s. annual report for 2001 is a '$301 million paragraph'. The provision is listed on page 28 of a 114-page document is the only published disclosure showing that Dynegy will have to post that much collateral if the ratings of its Dynegy Holdings Inc. unit are cut to junk status, or below investment grade. Debtors like Dynegy, WorldCom Inc. and Vivendi Universal SA are obligated to pay back billions of dollars if their

Table 8.4 A typical Moody's rating matrix.

	AAA	AA	A	BBB	BB	B	CCC	D	NR
AAA	91.93 %	7.46 %	0.48 %	0.08 %	0.04 %	0.00 %	0.00 %	0.00 %	—
AA	0.64 %	91.81 %	6.76 %	0.60 %	0.06 %	0.12 %	0.03 %	0.00 %	—
A	0.07 %	2.27 %	91.69 %	5.12 %	0.56 %	0.25 %	0.01 %	0.04 %	—
BBB	0.04 %	0.27 %	5.56 %	87.88 %	4.83 %	1.02 %	0.17 %	0.24 %	—
BB	0.04 %	0.10 %	0.61 %	7.75 %	81.48 %	7.90 %	1.11 %	1.01 %	—
B	0.00 %	0.10 %	0.28 %	0.46 %	6.95 %	82.80 %	3.96 %	5.45 %	—
CCC	0.19 %	0.00 %	0.37 %	0.75 %	2.43 %	12.13 %	60.45 %	23.69 %	—
D	0.00 %	0.00 %	0.00 %	0.00 %	0.00 %	0.00 %	0.00 %	100.00 %	—

credit ratings fall, their stock drops or they fail to meet financial targets. Half of these so-called triggers have not been disclosed publicly, according to Moody's Investors Service Inc., which has been investigating the presence of such clauses since the collapse of the Enron Corp. (International Herald Tribune, May 9 May 2002, p. 15). In other words, in addition to a corporation's rating, there are other sources of information, some revealed and some hidden, that differentiate the value of debt for such corporations, even if they are equally rated. In other words, their yield may not be the same even if they are equally rated. The rating of bonds is thus problematic, although there is an extensive insurance market for bonds that index premiums to the bond rating. A dealer's quotes in Moody's provides, for example, an estimate of insurance costs for certain bonds (determined by the swap market), some of which are reproduced in Table 8.5. The premium paid varies widely, however, based on both the rating and the perceived viability of the company whose bond is insured.

Table 8.5 Date-Moody's, 20 May 2002.

Company	Premium cost per \$M for 5 years	Moody's Senior Debt Rating
Merrill Lynch	10 000	Aa3
Lehman Brothers	9 500	A2
American Express	8 000	A1
Bear Stearns	7 500	A2
Goldman Sachs	6 500	A1
GE Capital	6 500	Aaa
Morgan Stanley	6 500	Aa3
JP Morgan Chase	6 500	Aa3*
AIG	5 300	Aaa
Citigroup	4 000	Aa1*
Bank of America	4 000	Aa2*
Bank One	3 500	Aaa

The valuation of rated bonds is treated next by making some simplifying assumptions to maintain an analytical and computational tractability, and by solving some problems that highlight approaches to valuing rated bonds. Rating can only serve as a first indicator to future default risk. Good accounting, information (statistical and otherwise) and economic analyses are still necessary.

8.4.1 A Markov chain and rating

Consider first a universe of artificial (and in fact non-existing) coupon-bearing rated bonds with a payment of a dollar fraction ℓ_i at maturity T, depending on the rating of the bond at maturity. Risk is thus induced only by the fraction $1 - \ell_i$ lost at maturity. Further, define the bond m-ratings matrix by a Markov chain $[p_{ij}]$ where

$$0 \leq p_{ij} \leq 1, \quad \sum_{j=1}^{m} p_{ij} = 1$$

denotes the objective transition probability that a bond rated i in a given period will be rated j in the following one. Discount factors are a function of the rating states, thus a bond rated i has a spot (one period) yield R_{it}, $R_{it} \leq R_{jt}$ for $i < j$ at time t. As a result, a bond rated i at time t and paying a coupon c_{it} at this time has, under the usual conditions, a value given by:

$$B_{i,t} = c_{it} + \sum_{j=1}^{m} \frac{p_{ij}}{1 + R_{jt}} B_{j,t+1}; \; B_{i,T} = \ell_i, \; i = 1, 2, 3, \ldots, m$$

Note that the discount rate R_{jt} is applied to a bond rated j in the next period. For example, for an 'imaginary' rated bond A and D only, each with (short-term) yields (R_{At}, R_{Dt}) at time t and a rating matrix specified by the transition probabilities, $[p_{ij}]$; $i, j = A, D$, we have:

$$B_{A,t} = c_{A,t} + \left[\frac{p_{AA}}{1 + R_{At}} B_{A,t+1} + \frac{p_{AD}}{1 + R_{Dt}} B_{D,t+1}\right]; B_{A,T} = \ell_A = 1$$

$$B_{D,t} = c_{D,t} + \left[\frac{p_{DA}}{1 + R_{At}} B_{A,t+1} + \frac{p_{DD}}{1 + R_{Dt}} B_{D,t+1}\right]; B_{D,T} = \ell_D$$

where $(c_{A,t}, c_{D,t})$ are the payouts associated with the bond rating and (ℓ_A, ℓ_D) are the bond redemption values at maturity T, both a function of the bond rating. In other words, the current value of a rated bond equals current payout plus the expected discount value of the bond rated at all classes, using the corresponding yield for each class at time $t + 1$. Of course, at the terminal time, when the bond is due (since there is not yet any default), the bond value equals its nominal value. If at maturity the bond is rated A, it will pay the nominal value of one dollar ($\ell_A = 1$) while if it is rated D, it will imply a loss of $1 - \ell_D$ for a bond rated initially A. If we define D as a default state (i.e. where the bondholder cannot recuperate the bond nominal value), the ℓ_D can be interpreted as the recuperation ratio. Of course, we can assume as well $\ell_D = 0$ as will be the case in a number of examples below. In vector notation we have:

$$\begin{bmatrix} B_{A,t} \\ B_{D,t} \end{bmatrix} = \begin{bmatrix} c_{A,t} \\ c_{D,t} \end{bmatrix} + \begin{bmatrix} \dfrac{p_{AA}}{1+R_{At}} & \dfrac{p_{AD}}{1+R_{Dt}} \\ \dfrac{p_{DA}}{1+R_{At}} & \dfrac{p_{DD}}{1+R_{Dt}} \end{bmatrix} \begin{bmatrix} B_{A,t+1} \\ B_{D,t+1} \end{bmatrix}; \begin{bmatrix} B_{A,T} \\ B_{D,T} \end{bmatrix} = \begin{bmatrix} \ell_A \\ \ell_D \end{bmatrix};$$

where ℓ_i is the nominal value of a bond rated i at maturity. And generally, for an m-rated bond,

$$\mathbf{B}_t = \mathbf{c}_t + \mathbf{F}_t \mathbf{B}_{t+1}, \mathbf{B_T} = \mathbf{L}$$

Note that the matrix $\mathbf{F}_t$ has entries $[p_{ij}/(1+R_{jt})]$ and $\mathbf{L}$ is a diagonal matrix of entries ℓ_i, $i = 1, 2, \ldots, m$. For a zero-coupon bond, we have:

$$\mathbf{B}_t = \prod_{k=t}^{T} \mathbf{F}_k \mathbf{L}.$$

By the same token, rated bonds discounts $q_{it} = 1/(1+R_{it})$ are found by solving the matrix equation:

$$\begin{bmatrix} q_{1t} \\ q_{2t} \\ \cdots \\ \cdots \\ q_{mt} \end{bmatrix} = \begin{bmatrix} p_{11}B_{1,t+1} & p_{12}B_{2,t+1} & \cdots \cdots & p_{1m}B_{m,t+1} \\ p_{21}B_{1,t+1} & p_{22}B_{2,t+1} & & p_{2m}B_{m,t+1} \\ \cdots & & & \\ \cdots & & & \\ p_{m1}B_{1,t+1} & p_{m2}B_{2,t+1} & & p_{mm}B_{m,t+1} \end{bmatrix}^{-1} \begin{bmatrix} B_{1,t} - c_{1t} \\ B_{2,t} - c_{2t} \\ \cdots \\ \cdots \\ B_{m,t} - c_{mt} \end{bmatrix}$$

where at maturity T, $B_{i,T} = \ell_i$. Thus, in matrix notation, we have:

$$\bar{\mathbf{q}}_t = \Gamma_{t+1}^{-\mathbf{1}} (\mathbf{B}_t - \mathbf{c}_t)$$

Note that one period prior to maturity, we have: $\bar{\mathbf{q}}_{T-1} = \Gamma_T^{-\mathbf{1}} (\mathbf{B}_{T-1} - \mathbf{c}_{T-1})$ where Γ_T is a matrix with entries $p_{ij}B_{j,T} = p_{ij}\ell_j$. For example, for the two-ratings bond, we have:

$$B_{1,t} = c_{1t} + q_{1t}p_{11}B_{1,t+1} + q_{2t}p_{12}B_{2,t+1}; \quad B_{1,T} = \ell_1$$
$$B_{2,t} = c_{2t} + q_{1t}p_{21}B_{1,t+1} + q_{2t}p_{22}B_{2,t+1}; \quad B_{2,T} = \ell_2$$

Equivalently, in matrix notation, this is given by:

$$\begin{bmatrix} q_{1t} \\ q_{2t} \end{bmatrix} = \begin{bmatrix} p_{11}B_{1,t+1} & p_{12}B_{2,t+1} \\ p_{21}B_{1,t+1} & p_{22}B_{2,t+1} \end{bmatrix}^{-1} \begin{bmatrix} B_{1,t} - c_{1t} \\ B_{2,t} - c_{2t} \end{bmatrix}$$

In this sense the forward bond price can be calculated by the rated bond discount rate and vice versa.

Example

We consider the matrix representing a rated bond supplied by Moody's (Table 8.6). The discount rates R_i, $i = AAA, \ldots, D$ for each class and the corresponding coupon payments are given in Table 8.7. For example, the discount rate of an AAA bond is 0.06 yearly while that of a BBB bond is 0.1. In addition, the AAA-rated bond has a coupon paying \$1, while if it were rated BB its coupon payment

Table 8.6

	AAA	AA	A	BBB	BB	B	CCC	D
AAA	0.9193	0.0746	0.0048	0.0008	0.0004	0	0	0
AA	0.0064	0.9181	0.0676	0.006	0.0006	0.0012	0.0003	0
A	0.0007	0.0227	0.9169	0.0512	0.0056	0.0025	0.0001	0.0004
BBB	0.0004	0.0027	0.0556	0.8788	0.0483	0.0102	0.0017	0.0024
BB	0.0004	0.001	0.0061	0.0775	0.8148	0.079	0.0111	0.0101
B	0	0.001	0.0028	0.0046	0.0695	0.828	0.0396	0.0545
CCC	0.0019	0	0.0037	0.0075	0.0243	0.1213	0.6045	0.2369
D	0	0	0	0	0	0	0	1

would have to be \$1.4. In this sense, both the size of the coupon and the discount applied to the rated bond are used to pay for the risk associated with the bond. The bond nominal value is \$100 with a lifetime of ten years. An elementary program will yield then the following bond value, shown above for each class at each year till the bond's redemption. For example, initially, the premium paid for a AAA bond compared to a AA one is (63.10–58.81) = \$4.29. At the end of the fifth year, however, the AAA–AA bond price differential is only 82.59–80.11 = \$2.48. In fact, we note that the smaller the amount of time left to bond redemption, the smaller the premium.

8.4.2 Bond sensitivity to rates – Duration

For the artificial rated bond considered above, we can calculate the duration of a rated bond through the rated bond sensitivity to the yields of each rating. For simplicity, assume that short yields are constants and calculate the partial derivatives for a bond rated A or D only. In this case, we seek to calculate the partials:

$$\left[\frac{\partial B_{A,t}}{\partial R_A}, \frac{\partial B_{A,t}}{\partial R_D}, \frac{\partial B_{D,t}}{\partial R_A}, \frac{\partial B_{D,t}}{\partial R_D}\right],$$

$$\frac{\partial B_{A,t}}{\partial R_A} = -\frac{p_{AA}}{(1+R_A)^2}B_{A,t+1} + \frac{p_{AA}}{1+R_A}\frac{\partial B_{A,t+1}}{\partial R_A} + \frac{p_{AD}}{1+R_D}\frac{\partial B_{D,t+1}}{\partial R_A}; \frac{\partial B_{A,T}}{\partial R_A} = 0$$

$$\frac{\partial B_{A,t}}{\partial R_D} = \left[\frac{p_{AA}}{1+R_A}\frac{\partial B_{A,t+1}}{\partial R_D} - \frac{p_{AD}}{(1+R_D)^2}B_{D,t+1} + \frac{p_{AD}}{1+R_D}\frac{\partial B_{D,t+1}}{\partial R_D}\right]; \frac{\partial B_{A,T}}{\partial R_D} = 0$$

$$\frac{\partial B_{D,t}}{\partial R_A} = \left[-\frac{p_{DA}}{(1+R_A)^2}B_{A,t+1} + \frac{p_{DA}}{1+R_A}\frac{\partial B_{A,t+1}}{\partial R_A} + \frac{p_{DD}}{1+R_D}\frac{\partial B_{D,t+1}}{\partial R_A}\right]; \frac{\partial B_{D,T}}{\partial R_A} = 0$$

$$\frac{\partial B_{D,t}}{\partial R_D} = \left[\frac{p_{DA}}{1+R_A}\frac{\partial B_{A,t+1}}{\partial R_D} - \frac{p_{DD}}{(1+R_D)^2}B_{D,t+1} + \frac{p_{DD}}{1+R_D}\frac{\partial B_{D,t+1}}{\partial R_D}\right]; \frac{\partial B_{D,T}}{\partial R_D} = 0$$

Table 8.7

AAA	1
AA	1.1
A	1.2
BBB	1.3
BB	1.4
B	1.5
CCC	1.6
D	1.7

AAA	0.06
AA	0.07
A	0.08
BBB	0.09
BB	0.1
B	0.11
CCC	0.12
D	0.13

Results	AAA	AA	A	BBB	BB	B	CCC	D
T=0	63.10	58.81	54.85	51.30	47.89	44.84	42.32	39.88
T=1	65.89	61.78	57.96	54.50	51.15	48.13	45.61	43.15
T=2	68.84	64.96	61.32	58.00	54.74	51.78	49.29	46.83
T=3	71.98	68.37	64.95	61.80	58.70	55.84	53.41	51.00
T=4	75.31	72.02	68.88	65.95	63.04	60.34	58.03	55.71
T=5	78.84	75.92	73.12	70.48	67.83	65.35	63.20	61.03
T=6	82.59	80.11	77.70	75.41	73.10	70.91	68.99	67.05
T=7	86.56	84.58	82.64	80.79	78.89	77.08	75.48	73.84
T=8	90.78	89.38	87.99	86.65	85.27	83.93	82.75	81.52
T=9	95.25	94.51	93.76	93.04	92.28	91.55	90.88	90.20
T=10	100	100	100	100	100	100	100	100

We can write in vector notation a system of six simultaneous equations given by:

$$\mathbf{\Gamma}_t = \left[B_{A,t}, B_{D,t}, \frac{\partial B_{A,t}}{\partial R_A}, \frac{\partial B_{A,t}}{\partial R_D}, \frac{\partial B_{D,t}}{\partial R_A}, \frac{\partial B_{D,t}}{\partial R_D} \right];$$

where $\mathbf{\Gamma}_t = \mathbf{C}_t + \mathbf{\Phi}\mathbf{\Gamma}_{t+1}$, $\mathbf{\Gamma}_T = [\ell_A, \ell_D, 0, 0, 0, 0]$; $\mathbf{C}_t = [c_A, c_D, 0, 0, 0, 0]$

$$\mathbf{\Phi} = \begin{bmatrix} \frac{p_{AA}}{1+R_A} & \frac{p_{AD}}{1+R_D} & 0 & 0 & 0 & 0 \\ \frac{p_{DA}}{1+R_A} & \frac{p_{DD}}{1+R_D} & 0 & 0 & 0 & 0 \\ -\frac{p_{AA}}{(1+R_A)^2} & 0 & \frac{p_{AA}}{1+R_A} & 0 & \frac{p_{AD}}{1+R_D} & 0 \\ 0 & -\frac{p_{AD}}{(1+R_D)^2} & 0 & \frac{p_{AA}}{1+R_A} & 0 & \frac{p_{AD}}{1+R_D} \\ -\frac{p_{DA}}{(1+R_A)^2} & 0 & \frac{p_{DA}}{1+R_A} & 0 & \frac{p_{DD}}{1+R_D} & 0 \\ 0 & -\frac{p_{DD}}{(1+R_D)^2} & 0 & \frac{p_{DA}}{1+R_A} & 0 & \frac{p_{DD}}{1+R_D} \end{bmatrix}$$

A solution to this system of equations is found similarly by backward recursion. Namely, for a time-invariant coupon payout, we have:

$$\mathbf{\Gamma}_{T-n} = \sum_{\mathbf{j=1}}^{n} [\mathbf{\Phi}]^{\mathbf{j-1}} \mathbf{C} + [\mathbf{\Phi}]^n \mathbf{\Gamma}_T \mathbf{L},\ n = \mathbf{0}, \mathbf{1}, \mathbf{2}, \ldots$$

while for a nonpaying coupon bond, we have:

$$\mathbf{\Gamma}_{T-n} = [\mathbf{\Phi}]^n \mathbf{\Gamma}_T,\ n = \mathbf{0}, \mathbf{1}, \mathbf{2}, \ldots$$

These equations can be solved numerically providing thereby a combined estimate of rated bond prices and their yield sensitivity.

Generally, we can also calculate rated bonds' duration and their 'cross-duration'. Bond duration is now defined in terms of partial durations, expressing the effects of all yields rates. Explicitly, say that a bond is rated i at time t and for simplicity let the yields be time-invariant. The duration of a bond rated i with respect to its yield is denoted by $D_{ii}(t, T)$, $\mathrm{i} = 1, 2, \ldots, m$, with,

$$D_{ii}(t, T) = -\frac{1}{B_{i,t}} \frac{\partial B_{i,t}}{\partial R_i}$$

while the partial duration, of the bond rated i with respect to any other yield, R_j, $i \neq j$ is:

$$D_{ij}(t, T) = -\frac{1}{B_{i,t}} \frac{\partial B_{i,t}}{\partial R_j} = -\frac{\partial \left(\log B_{i,t}\right)}{\partial R_j};\ i \neq j$$

By the same token, for convexity we have:

$$\Upsilon_{ii}(t,T) = \frac{1}{B_{i,t}}\frac{\partial^2 B_{i,t}}{\partial R_i^2} \quad \text{and} \quad \Upsilon_{ij}(t,T) = \frac{1}{B_{i,t}}\frac{\partial^2 B_{i,t}}{\partial R_i \partial R_j}$$

The partial durations and convexities express the sensitivity of a bond rated i is thus:

$$\frac{\mathrm{d}B_{i,t}}{B_{i,t}} = \frac{1}{B_{i,t}}\frac{\partial B_{i,t}}{\partial t}\,\mathrm{d}t + \frac{1}{B_{i,t}}\sum_{j=1}^{m}\frac{\partial B_{i,t}}{\partial R_j}\,\mathrm{d}R_j + \frac{1}{2}\frac{1}{B_{i,t}}\sum_{j=1}^{m}\frac{\partial^2 B_{i,t}}{\partial R_j \partial R_i}\,\mathrm{d}R_i\mathrm{d}R_j$$

Or:

$$\frac{\mathrm{d}B_{i,t}}{B_{i,t}} - \frac{\partial \log B_{i,t}}{\partial t}\,\mathrm{d}t = -\sum_{j=1}^{m} D_{ij}\,\mathrm{d}R_j + \frac{1}{2}\sum_{j=1}^{m}\Upsilon_{ij}\,\mathrm{d}R_i\,\mathrm{d}R_j$$

Example

Consider a bond with three ratings (1, 2 and 3) and assume constant yields for each given by $(R_1, R_2, R_3) = (0.05; 0.07; 0.10)$. Let the ratings transition matrix be:

$$\mathbf{P} = \begin{bmatrix} 0.9 & 0.1 & 0 \\ 0.05 & 0.8 & 0.15 \\ 0.00 & 0.05 & 0.95 \end{bmatrix}$$

Then the bond recursive equation for a pure bond paying \$1 in two periods is:

$$B_{i,t} = \sum_{j=1}^{m}\frac{p_{ij}}{1+R_j}B_{j,t+1};\ B_{i,2} = \ell_i,\ i = 1,2,3,\ldots,m,\ t = 0,1,2$$

For $m = 3$, this is reduced to:

$$B_{1,0} = (0.81q_{11} + 0.005q_{12})\ell_1 + (0.09q_{12} + q_{22}0.08)\ell_2 + 0.015q_{23}\ell_3$$

$$B_{2,0} = (0.045q_{11} + 0.04q_{12})\ell_1 + (0.005q_{12} + 0.64q_{22} + 0.0075q_{23})\,\ell_2 + (0.12q_{23} + 0.1425q_{33})\,\ell_3$$

$$B_{3,0} = (0.0025q_{12})\,\ell_1 + (0.04q_{22} + 0.0475q_{23})\ell_2 + (0.0075q_{23} + 0.9025q_{33})\ell_3$$

with the notation:

$$q_{ij} = \frac{1}{1+R_i}\frac{1}{1+R_j}$$

and,

$$q_{11} = .9068,\ q_{12} = 0.8899,\ q_{13} = 0.8656,$$
$$q_{22} = 0,8732,\ q_{23} = 0.8494,\ q_{33} = 0.8262$$

Thus,

$$B_{1,0} = 0.7389\ell_1 + 0.097\,88\ell_2 + 0.0127\ell_3$$
$$B_{2,0} = 0.0763\ell_1 + 0.5695\ell_2 + 0.2196\ell_3$$
$$B_{3,0} = 0.002\,22\ell_1 + 0.0752\ell_2 + 0.7456\ell_3$$

Therefore, a bond rated '1' is worth more than a bond rated '2' and a '2' is worth more than a '3' if $B_{1,0} > B_{2,0} > B_{3,0}$. Their difference accounts for the yield differential associated with each bond rating. Of course, if the rated bond is secured throughout the two periods (i.e. it does not switch from class to class), we have:

$$B_{1,0} = \frac{1}{(1+R_1)^2} = 0.907; \; B_{2,0} = \frac{1}{(1+R_2)^2} = 0.8734;$$
$$B_{3,0} = \frac{1}{(1+R_3)^2} = 0.8264$$

The difference between these numbers accounts for a premium implied by the ratings switching matrix. For the bond rated '3', we note that the secured 'rate 3' bond is worth less than the rated bond, accounting for the potential gain in yield if the bond credit quality is improved. Alternatively, we can use the risk-free discount rate $R_f = 0.04$ assumed for simplicity to equal 4 % yearly (since the bond has no default risk). In this case,

$$B_{f,0} = \frac{1}{(1+R_f)^2} = \frac{1}{(1+0.04)^2} = 0.9245$$

The premium for such a risk-free bond compared to a secured bond rated '1', is 0.9245 – 0.907l.

8.4.3 Pricing rated bonds and the term structure risk-free rates*

When the risk-free term structure is available, and assuming no arbitrage, we can construct a portfolio replicating the bond, thereby valuing the rated bond yields for each bond class. Explicitly, consider a portfolio of rated bonds consisting of $N_i,\ i = 1, 2, 3, \ldots, m$ bonds rated i, each providing ℓ_i dollars at maturity. Let the portfolio value at maturity be equal to \$1. That is to say

$$\sum_{i=1}^{m} N_i B_{i,T} = \sum_{i=1}^{m} N_i \ell_i = 1$$

One period (year) prior to maturity, such a portfolio would be worth

$$\sum_{i=1}^{m} N_i B_{i,T-1}$$

dollars. By the same token, if we denote by $R_{f,T-1}$ the risk-free discount rate for one year, then assuming no arbitrage, one period prior to maturity, we have:

$$\sum_{i=1}^{m} N_i B_{i,T-1} = \frac{1}{1+R_{f,T-1}}; B_{i,T-1} = c_{it} + \sum_{j=1}^{m} q_{j,T-1} p_{ij} B_{j,T};$$
$$B_{i,T} = \ell_i, \ i = 1, 2, \ldots, m$$

with $q_{jt} = 1/(1 + R_{j,t})$ and $R_{j,t}$ is the one-period discount rate applied to a j-rated bond. This system of equations provides $2m$ unknown rates and the portfolio composition with only one equation is therefore under-determined. For two periods we have an additional equation:

$$\sum_{i=1}^{m} N_i B_{i,T-2} = \frac{1}{(1+R_{f,T-1})^2}$$

While the bond price is given by:

$$B_{i,T-2} = c_{i,T-2} + \sum_{j=1}^{m} q_{j,T-2} p_{ij} B_{j,T-1}; B_{i,T} = \ell_i, i = 1, 2, \ldots, m$$

as well as:

$$B_{i,T-2} = c_{i,T-2} + \sum_{j=1}^{m} q_{j,T-2,2} p_{ij}^{(2)} B_{j,T}; B_{i,T} = \ell_i, i = 1, 2, \ldots, m$$

where $p_{ij}^{(2)}$ is the probability that the bond is rated j two periods hence while $q_{j,T-2,2}$ is the discount rate for a j-rated bond for two periods forward (that might differ from the rate $q_{j,T-2,1}$ applied for one period only). Here again, we see that there are $2m$ rates while there are only two equations. For three periods we will have three equations per rating and so on. Generally, k periods prior to maturity, assuming no arbitrage, we have the following conditions for no arbitrage:

$$\sum_{i=1}^{m} N_i B_{i,T-k} = \frac{1}{(1+R_{f,T-k})^k} \ k = 0, 12, 3, \ldots, T$$
$$B_{i,T-k} = c_{i,T-k} + \sum_{j=1}^{m} q_{j,T-k,h} p_{ij}^{(h)} B_{j,T-(k-h)}; B_{i,T} = \ell_i,$$
$$i = 1, 2, \ldots, m; \ k = 1, 2, 3, \ldots, T; \ h = 1, 2, 3, \ldots, k$$

where $R_{f,T-k}$, $k = 1, 2, 3, \ldots,$ is the risk-free rate term structure and $p_{ij}^{(h)}$ is the ij entry of the h-power of the rating matrix. These provide a system of $T + 1$ simultaneous equations spanning the bond life. In matrix notation this is given by:

$$\mathbf{NB}_{T-k} = \frac{1}{(1+R_{f,T-k})^k} \ k = 0, 1, 2, \ldots, T; \ \mathbf{N} = (N_1, \ldots, N_m);$$
$$\mathbf{B}_{T-k} = (B_{1,T-k}, \ldots, B_{m,T-k})$$

as well as:

$$\mathbf{B}_{t-k} = \mathbf{c}_{t-k} + \mathbf{F}_t^{(h)}\mathbf{B}_{t+(k-h)}; \mathbf{B}_T = \mathbf{L},$$

$$\mathbf{F}_t^{(h)} = \left[q_{j,T-k,h}\, p_{ij}^{(h)}\right]; \; h = 1, 2, \ldots, k; \; k = 1, 2, \ldots, T$$

This renders the estimation of the term structure of ratings discount grossly under-determined. However, some approximations can be made which may be acceptable practically. Such an approximation consists in assuming that the rates at a given time are assumed time-invariant and the term structure of risk-free rates is known and we only estimate the short ratings discount. We assume first the case of a maturity larger than the number of rating classes.

Case $T \geq 2m$

When the bond maturity is larger than the number of ratings $T \geq 2m$, and $q_{j,T-k,h} = q_{j,h}$, $h = 1$ and $q_{j,1} = q_j$, the hedging portfolio of rated bonds is found by a solution of the system of linear equations above (with $h = 1$), leading to the unique solution:

$$\mathbf{N}^* = \Im^{-1}\mathbf{\Omega}$$

where $\Im$ is the matrix transpose of $[B_{i,T-j+1}]$ and $\mathbf{\Omega}$ is a column vector with entries $[1/(1 + R_{f,T-s})^s]$, $s = 0, 1, 2, \ldots, m - 1$. Explicitly, we have:

$$\begin{bmatrix} N_1 \\ N_2 \\ \cdots \\ \cdots \\ N_m \end{bmatrix} = \begin{bmatrix} B_{1,T} & B_{2,T} & B_{3,T} & \ldots & B_{m,T} \\ B_{1,T-1} & B_{2,T-1} & B_{3,T-1} & \ldots & B_{m,T-1} \\ \cdots & & & & \cdots \\ \cdots & & & & \cdots \\ B_{1,T-m} & B_{2,T-m} & B_{3,T-m} & \ldots & B_{m,T-m} \end{bmatrix}^{-1} \begin{bmatrix} 1 \\ 1/(1+R_{f,T-1})^1 \\ \cdots \\ \cdots \\ 1/(1+R_{f,T-m})^m \end{bmatrix}$$

Thus, the *condition for no arbitrage* is reduced to satisfying a system of system of nonlinear equations:

$$\Im^{-1}\mathbf{\Omega}\mathbf{B}_{T-k} = \frac{1}{(1 + R_{f,T-k})^k}; \; k = m, m + 1, \ldots, T$$

For example, for a zero-coupon rated bond and stationary short discounts, we have $\mathbf{B}_{t-k} = (\mathbf{F})^k\,\mathbf{L}$ and therefore, the no-arbitrage condition becomes:

$$\Im^{-1}\mathbf{\Omega}\,(\mathbf{F})^k\,\mathbf{L} = \frac{1}{(1 + R_{f,T-k})^k}; \; k = m, m + 1, \ldots, T$$

where $\mathbf{F}$ has entries $q_j p_{ij}$. This provides, therefore, $T + 1 - m$ equations applied to determining the bond ratings short (one period) discount rates q_j.

Our system of equations may be over- or under-identified for determining the ratings discount rates under our no-arbitrage condition. Of course, if $T + 1 - m = m$, we have exactly m additional equations we can use to solve the ratings discount rates uniquely (albeit, these are nonlinear equations and can be solved only numerically). If $(T \geq 2m + 1)$ we can use the remaining equations to calculate some of the term structure discounts of bond ratings as well. For example, for a bond with maturity three times the number of ratings, $T = 3m$, we have the following no-arbitrage condition:

$$\Im^{-1}\mathbf{\Omega}\mathrm{B}_{T-k} = \frac{1}{(1 + R_{f,T-k})^k}; \; k = m, m+1, \ldots, T$$

and

$$B_{i,T-k} = c_{i,T-k} + \sum_{j=1}^{m} q_{j,1} p_{ij}^{(1)} B_{j,T-(k-1)}; \; B_{i,T} = \ell_i,$$
$$B_{i,T-k} = c_{i,T-k} + \sum_{j=1}^{m} q_{j,2} p_{ij}^{(2)} B_{j,T-(k-2)}; \; B_{i,T} = \ell_i,$$
$$i = 1, 2, \ldots, m; k = 1, 2, 3, \ldots, T$$

Thus, when the bond maturity is very large (or if we consider a continuous-time bond), an infinite number of equations is generated which justifies the condition for no arbitrage stated by Jarrow *et al.* (1997).

When data regarding the risk-free term structure is limited, or for short bonds, we have, $m \leq T < 2m$ and the Markov model is incomplete. We must, therefore, proceed to an approach that can, nevertheless, provide an estimate of the ratings discount rates. We use for convenience a sum of squared deviations from the rated bond arbitrage condition, in which case we minimize the following expression (for estimating the short discount rates only):

$$\underset{0 \leq q_1, q_2, \ldots, q_{m-1}, q_m \leq 1}{\text{Minimize}} \sum_{k=m}^{T} \left(\Im^{-1}\mathbf{\Omega}\mathbf{B}_{T-k} - \frac{1}{(1 + R_{f,T-k})^k} \right)^2$$

subject to a number of equalities used in selecting the portfolio, namely:

$$B_{i,T-k} = c_{i,T-k} + \sum_{j=1}^{m} q_{j,1} p_{ij}^{(1)} B_{j,T-(k-1)}; \; B_{i,T} = \ell_i, k = 1, 2, \ldots, T$$

Additional constraints, reflecting expected and economic rationales of the ratings discounts q_j might be added, such as:

$$0 \leq q_j \leq 1 \text{ as well as } 0 \leq q_m \leq q_{m-1} \leq q_{m-2} \leq q_{m-3}, \ldots, \leq q_2 \leq q_1 \leq 1$$

These are typically nonlinear optimization problems, however. A simple two-ratings problem and other examples are considered to highlight the complexities in determining both the hedging portfolio and the ratings discounts provided the risk-free term structure is available.

Example: Valuation of a two-rates rated bond

For a portfolio of two-rates bonds over one period where $\ell_1 = 1$, $\ell_2 = 0.2$ we have the following two equations that can be used to calculate the risk-free portfolio composition:

$$N_1 + 0.2N_2 = 1 \quad \text{or} \quad \mathrm{N}_2 = 5(1 - N_1)$$

$$N_1 B_{1,T-1} + 5(1 - N_1)B_{2,T-1} = \frac{1}{1 + R_{f,T-1}}$$

and,

$$N_1 = \frac{B_{2,T-1}(1 + R_{f,T-1}) - 0.2}{(1 + R_{f,T-1})(B_{2,T-1} - 0.2B_{1,T-1})}$$

If we assume a bond of maturity of three periods only, then only two additional equations are available ($T - 2$, $T - 3$) providing a no-arbitrage estimate for the rated bond discounts and given by:

$$\left(\frac{B_{2,T-1}(1 + R_{f,T-1}) - 0.2}{(1 + R_{f,T-1})(B_{2,T-1} - 0.2B_{1,T-1})}\right) \times (B_{1,T-2} - 5B_{2,T-2}) + 5B_{2,T-2} = \frac{1}{(1 + R_{f,T-2})^2}$$

$$\left(\frac{B_{2,T-1}(1 + R_{f,T-1}) - 0.2}{(1 + R_{f,T-1})(B_{2,T-1} - 0.2B_{1,T-1})}\right) \times (B_{1,T-3} - 5B_{2,T-3}) + 5B_{2,T-3} = \frac{1}{(1 + R_{f,T-3})^3}$$

$$B_{1,T-1} = c_{1,T-1} + q_1 p_{11} + 0.2q_2 p_{12};$$

$$B_{2,T-1} = c_{2,T-1} + q_1 p_{21} + q_2 p_{22} 0.2$$

$$B_{1,T-2} = c_{1,T-2} + q_1 p_{11} B_{1,T-1} + q_2 p_{12} B_{2,T-1};$$

$$B_{2,T-2} = c_{2,T-2} + q_1 p_{21} B_{1,T-1} + q_2 p_{22} B_{2,T-1}$$

$$B_{1,T-3} = [c_{1,T-3} + q_1 p_{11} c_{1,T-2} + q_2 p_{12} c_{2,T-2}] + [q_1^2 p_{11}^2 + q_1 q_2 p_{12} p_{21}]B_{1,T-1} + [q_2 q_1 p_{11} p_{12} + q_2^2 p_{12} p_{22}]B_{2,T-1}$$

$$B_{2,T-3} = [c_{2,T-3} + q_1 p_{21} c_{1,T-2} + q_2 p_{22} c_{2,T-2}] + [q_1 q_1 p_{21} p_{11} + q_1 q_2 p_{12} p_{21}]B_{1,T-1} + [q_2 q_1 p_{21} p_{22} + q_2^2 p_{22}^2]B_{2,T-1}$$

where, $B_{1,T-1}$, $B_{1,T-2}$, $B_{1,T-3}$ are functions of the bond redemption values and the ratings discount rates q_1 and q_2. That is to say, we have a system of two independent equations in two unknowns only that we can solve by standard numerical analysis. For example, consider a zero-coupon bond with a rating matrix given by: $p_{11} = 0.8$, $p_{12} = 0.2$, $p_{21} = 0.1$, $p_{22} = 0.9$. In addition, set $\ell_1 = 1$, $\ell_2 = 0.6$ (and therefore a recuperation rate of 60 % on bond default) and $R_{f,T-1} = 0.07$, $R_{f,T-2} = 0.08$ thus:

$$B_{1,T-1} = 0.8q_1 + 0.12q_2; B_{2,T-1} = 0.1q_1 + 0.54q_2$$

$$B_{1,T-2} = 0.8q_1 B_{1,T-1} + 0.2q_2 B_{2,T-1}; \quad B_{2,T-2} = 0.1q_1 B_{1,T-1} + 0.9q_2 B_{2,T-1}$$

$$B_{1,T-3} = \left[0.64q_1^2 + 0.02q_1q_2\right] B_{1,T-1} + \left[0.16q_2q_1 + 0.18q_2^2\right] B_{2,T-1}$$

$$B_{2,T-3} = \left[0.08q_1^2 + 0.02q_1q_2\right] B_{1,T-1} + \left[0.09q_2q_1 + 0.81q_2^2\right] B_{2,T-1}$$

and therefore

$$\left(\frac{1.07B_{2,T-1} - 0.6}{1.07((0.8q_1 - 0.6)B_{1,T-1} + 0.2q_2 B_{2,T-1})}\right)$$
$$\times (0.6334q_1 B_{1,T-1} - 1.2294.2q_2 B_{2,T-1})$$
$$+ (0.1666q_1 B_{1,T-1} + 1.4994q_2 B_{2,T-1}) = 0.8733$$

$$\left(\frac{1.07B_{2,T-1} - 0.6}{1.07((0.8q_1 - 0.6)B_{1,T-1} + 0.2q_2 B_{2,T-1})}\right)$$
$$\times \begin{pmatrix} \left[0.5072q_1^2 - 0.013\,32q_1q_2\right] B_{1,T-1} + \\ + \left[0.010\,06q_2q_1 - 1.169\,46q_2^2\right] B_{2,T-1} \end{pmatrix}$$
$$+ \left(\left[0.133\,28q_1^2 + 0.033\,32q_1q_2\right] B_{1,T-1} + \left[0.14994q_2q_1 + 1.3446q_2^2\right] B_{2,T-1}\right)$$
$$= 0.7937$$

This is a system of six equations six unknowns that can be solved numerically by the usual methods.

8.4.4 Valuation of default-prone rated bonds*

We consider next the more real and practical case consisting in a bonds defaulting prior to maturity and generally we consider the first time n, a bond rated initially i, is rated j and let the probability of such an event be denoted by, $f_{ij}(n)$. This probability equals the probability of not having gone through a jth rating in prior transitions and be rated j at time n. For transition in one period, this is equal to the transition bond rating matrix (S&P or Moody's matrix), while for a transition in two periods it equals the probability of transition in two periods conditional on not having reached rating j in the first period. In other words, we have:

$$f_{ij}(1) = p_{ij}(1) = p_{ij}; f_{ij}(2) = p_{ij}(2) - f_{ij}(1)p_{jj}$$

By recursion, we can calculate these probabilities:

$$f_{ij}(n) = p_{ij}(n) - \sum_{k=1}^{n-1} f_{ij}(k)p_{jj}(n-k)$$

The probability of a bond defaulting prior to time n is thus,

$$F_{km}(n-1) = \sum_{j=1}^{n-1} f_{km}(j)$$

while the probability that such a bond does not default is:

$$\bar{F}_{km}(n-1) = 1 - F_{km}(n-1)$$

At present, denote by $\Phi_i(n)$ the probability that the bond is rated i at time n. In vector notation we write, $\bar{\Phi}(n)$. Thus, given the rating matrix, $[\mathbf{P}]$ we have:

$$\bar{\Phi}(n) = [\mathbf{P}]' \, \bar{\Phi}(n-1), \; n = 1, 2, 3, \ldots, \quad \text{and} \quad \bar{\Phi}(0) \text{ given}$$

with $[\mathbf{P}]'$, the matrix transpose. Thus, at time n, $\bar{\Phi}(n) = [\mathbf{P}']^n \bar{\Phi}(0)$ and the present value of a coupon payment (given that there was no default at this time) is therefore discounted at $R_{j,n}$, $q_{j,n} = 1/(1 + R_{j,n})$ if the bond is rated j. In other words, its present value is:

$$\sum_{j=1}^{m-1} c_{j,n} q_{j,n}^{n} \Phi_{j,n}; \Phi_{j,n} = \sum_{i=1}^{m-1} \Phi_{i,0} p_{ij}^{(n)}$$

where $p_{ij}^{(n)}$ is the *ijth* entry of the transpose power matrix $[\mathbf{P}']^n$ and $\Phi_{i,0}$ is the probability that initially the bond is rated i.

When a coupon-bearing default bond rated i at time s, defaults at time, $s+1$, $T-(s+1)$ periods before maturity with probability $f_{im}(s+1-s) = f_{im}(1)$, we have a value:

$$V_{s,i} = \left(c_{i,T-s} + q_i^{\ell_{m,T-(s+1)}}\right) \text{ w.p. } f_{im}(1)$$

If such an even occurs at time, $s+2$, with probability $f_{im}(s+2-s) = f_{im}(2)$, we have:

$$V_{s,i} = \left(c_{i,T-s} + \sum_{k=1}^{m-1} q_k c_{k,T-(s+1)} \Phi_{k,(s+1)-s} + q_i^2 \ell_{m,T-(s+2)}\right) \text{ w.p. } f_{im}(2)$$

where

$$\Phi_{k,1} = \sum_{i=1}^{m-1} \Phi_{i,0} p_{ik}^{(1)}$$

and $\Phi_{i,0}$ is a vector whose entries are all zero except at i (since at s we conditioned the bond value at being rated i). By the same token three periods hence and prior to maturity, we have:

$$V_{s,i} = \left(c_{i,T-s} + \sum_{k=1}^{m-1} q_k c_{k,T-(s+1)} \Phi_{i,0} p_{ik}^{(1)} \right.$$
$$\left. + \sum_{k=1}^{m-1} q_k^2 c_{k,T-(s+2)} \Phi_{i,0} p_{ik}^{(2)} + q_i^3 \ell_{m,T-(s+3)}\right) \text{ w.p. } f_{im}(3)$$

and, generally, for any period prior to maturity,

$$V_{s,i} = \left(c_{i,T-s} + \sum_{\theta=1}^{\tau-1} \sum_{k=1}^{m-1} q_k^{\theta} c_{k,T-(s+\theta)} \Phi_{i,0} p_{ik}^{(\theta)} + q_i^{\tau} \ell_{m,T-(s+\tau)} \right) \text{ w.p. } f_{im}(\tau)$$

In expectation, if the bond defaults prior to its maturity, its expected price at time s is,

$$EB_{i,D}(s,T) = c_{i,T-s} + \sum_{\tau=1}^{T-s} \left(q_i^{\tau} \ell_{m,T-(s+\tau)} + \sum_{\theta=1}^{\tau-1} \sum_{k=1}^{m-1} q_k^{\theta} c_{k,T-(s+\theta)} \Phi_{i,0} p_{ik}^{(\theta)} \right) f_{im}(\tau)$$

With $\ell_{m,T-j}$, the bond recovery when the bond defaults, assumed to be a function of the time remaining for the faultless bond to be redeemed. And therefore, the price of such a bond is:

$$B_{i,ND}(s,T) = \left(c_{i,T-s} + \sum_{k=1}^{m-1} q_k^{T-s} \ell_k \Phi_{i,0} p_{ik}^{(T-s)} \right.$$
$$\left. + \sum_{\theta=1}^{T-s-1} \sum_{k=1}^{m-1} q_k^{\theta} c_{k,T-(s+\theta)} \Phi_{i,0} p_{ik}^{(\theta)} \right) \left[1 - \sum_{u=1}^{T-s} f_{im}(u) \right]$$

where ℓ_i denotes the bond nominal value at redemption when it is rated i. Combining these sums, we obtain the price of a default prone bond rated i at time s:

$$B_i(s,T) = c_{i,T-s} + \left(c_{i,T-s} + \sum_{k=1}^{m-1} q_k^{T-s} \ell_k \Phi_{i,0} p_{ik}^{(T-s)} \right.$$
$$\left. + \sum_{\theta=1}^{T-s-1} \sum_{k=1}^{m-1} q_k^{\theta} c_{k,T-(s+\theta)} \Phi_{i,0} p_{ik}^{(\theta)} \right) \left[1 - \sum_{u=1}^{T-s} f_{im}(u) \right]$$
$$+ \sum_{\tau=1}^{T-s} \left(q_i^{\tau} \ell_{m,T-(s+\tau)} + \sum_{\theta=1}^{\tau-1} \sum_{k=1}^{m-1} q_k^{\theta} c_{k,T-(s+\theta)} \Phi_{i,0} p_{ik}^{(\theta)} \right) f_{im}(\tau)$$

For a zero-coupon bond, this is reduced to:

$$B_i(s,T) = \left(\sum_{k=1}^{m-1} q_k^{T-s} \ell_k \Phi_{i,0} p_{ik}^{(T-s)} \right) \left[1 - \sum_{u=1}^{T-s} f_{im}(u) \right]$$
$$+ \sum_{\tau=1}^{T-s} \left(q_i^{\tau} \ell_{m,T-(s+\tau)} \right) f_{im}(\tau)$$

To determine the (short) price discounts rates for a default-prone rated bond we can proceed as we have before by constructing a hedged portfolio consisting of $N_1, N_2, \ldots, N_{m-1}$ shares of bonds rated, $i = 1, 2, \ldots, m-1$. Again, let $R_{f,T-u}$ be the risk-free rate when there are u periods left to maturity. Then, assuming no arbitrage and given the term structure risk-free rate, we have:

$$\sum_{i=1}^{m-1} N_i B_i(s,T) = \frac{1}{\left(1+R_{f,T-s}\right)^s},\quad s=0,1,2,\ldots$$

with $B_i(s,T)$ defined above. Note that the portfolio consists of only $m-1$ rated bonds and therefore, we have in fact $2m-1$ variables to be determined based on the risk-free term structure. When the system is over-identified (i.e. there are more terms in the risk-free term structure than there are short ratings discount to estimate), additional equations based on the ratings discount term structure can be added so that we obtain a sufficient number of equations. Assuming that our system is under-determined (which is the usual case), i.e. $T \leq 2m-1$, we are reduced to solving the following minimum squared deviations problem:

$$\underset{\substack{0\leq q_1\leq q_2\leq \ldots \leq q_{m-1}\leq 1;\\ N_1,N_2,N_3\ldots\ldots,N_{m-1}}}{\text{Minimize}} \quad \sum_{s=0}^{T}\left(\sum_{k=1}^{m-1} N_k B_k(s,T) - \frac{1}{\left(1+R_{f,T-s}\right)^s}\right)^2$$

subject to:

$$\begin{aligned} B_i(s,T) = c_{i,T-s} &+ \left(c_{i,T-s} + \sum_{k=1}^{m-1} q_k^{T-s}\ell_k \Phi_{i,0} p_{ik}^{(T-s)}\right. \\ &\left. + \sum_{\theta=1}^{T-s-1}\sum_{k=1}^{m-1} q_k^{\theta} c_{k,T-(s+\theta)}\Phi_{i,0} p_{ik}^{(\theta)}\right)\left[1-\sum_{u=1}^{T-s} f_{im}(u)\right] \\ &+ \sum_{\tau=1}^{T-s}\left(q_i^{\tau}\ell_{m,T-(s+\tau)} + \sum_{\theta=1}^{\tau-1}\sum_{k=1}^{m-1} q_k^{\theta} c_{k,T-(s+\theta)}\Phi_{i,0} p_{ik}^{(\theta)}\right) f_{im}(\tau) \end{aligned}$$

This is, of course, a linear problem in N_k and a nonlinear one in the rated discounts which can be solved analytically with respect to the hedged portfolio (and using the remaining equations to calculate the ratings discount rates). Explicitly, we have:

$$\sum_{s=0}^{T}\sum_{k=1}^{m-1} N_k B_j(s,T)\, B_k(s,T) = \sum_{s=0}^{T}\frac{B_j(s,T)}{\left(1+R_{f,T-s}\right)^s}$$

This is a system of linear equation we can solve by:

$$\sum_{k=1}^{m-1} N_k A_{jk} = D_j;\ j=1,2,3,\ldots,m-1;\ D_j = \sum_{s=0}^{T}\frac{B_j(s,T)}{\left(1+R_{f,T-s}\right)^s};$$

$$A_{jk} = \sum_{s=0}^{T} B_j(s,T)\, B_k(s,T)$$

and, therefore, in matrix notation

$$\mathbf{N}^*\mathbf{A} = \mathbf{D} \rightarrow \mathbf{N}^* = \mathbf{A}^{-1}\mathbf{D}$$

and obtain the replicating portfolio for a risk-free investment. This solution can be inserted in our system of equations to obtain the reduced set of equations for the ratings discount rates of the default bond. A solution can be found numerically.

Example: A two-rated default bond

Consider a two-rated zero-coupon bond and define the transition matrix:

$$\mathbf{P} = \begin{bmatrix} p & 1-p \\ 0 & 1 \end{bmatrix} \quad \text{with} \quad \mathbf{P}^n = \begin{bmatrix} p^n & 1-p^n \\ 0 & 1 \end{bmatrix}$$

The probability of being in one of two states after n periods is $(p^n, 1-p^n)$. Further,

$$f_{12}(1) = 1-p;\ f_{12}(2) = p_2^{(2)} - (1) f_{12}(1) = 1 - p^2 - (1-p) = p(1-p)$$

Thus, for a no-coupon paying bond, we have:

$$B_i(s,T) = c_{i,T-s} + q^{T-s\ell}\left[1 - \sum_{u=1}^{T-s} f_{12}(u)\right] + \sum_{\tau=1}^{T-s} (q^{\tau}\ell_{2,T-(s+\tau)})\, f_{12}(\tau)$$

In particular,

$$\begin{aligned}
B_1(T,T) &= \ell \\
B_1(T-1,T) &= q\ell\,[1 - f_{12}(1)] + q\ell_{2,0}\, f_{12}(1) \\
B_1(T-2,T) &= q^2\ell\,[1 - f_{12}(1) - f_{12}(2)] + q\ell_{2,1} f_{12}(\tau) + q^2\ell_{m,0} f_{12}(2) \\
B_1(T-3,T) &= q^3\ell\,[1 - f_{12}(1) - f_{12}(2) - f_{12}(3)] + \left(q\ell_{2,2}\right)\, f_{12}(1) \\
&\quad + q^2\ell_{2,1}\, f_{12}(2) + q^3\ell_{2,0}\, f_{12}(3)
\end{aligned}$$

If we have a two-year bond, then the condition for no arbitrage is:

$$NB_1(T,T) = N\ell = 1 \quad \text{and} \quad N = 1/\ell$$

$$NB_1(T-1,T) = \frac{1}{1+R_{f,T-1}} \Rightarrow 1 + R_{1,T-1} = \frac{1+R_{f,T-1}}{1 - \left(1 - \ell_{2,0}/\ell\right)\, f_{12}(1)}$$

If we set, $\ell_{2,0} = 0$, implying that when default occurs at maturity, the bond is a total loss, then:

$$1 + R_{1,T-1} = \frac{1+R_{f,T-1}}{1 - f_{12}(1)} = 1 + \frac{(1-p) + R_{f,T-1}}{p} \quad \text{or}$$

$$R_{1,T-1} = \frac{(1-p) + R_{f,T-1}}{p}$$

This provides an explicit determination of the rated '1' bond in terms of the risk-free rate. If there is no loss ($p = 1$), then, $R_{1,T-1} = R_{f,T-1}$. If we have a two-year bond, then the least quadratic deviation cost rating can be applied. Thus,

$$\begin{aligned}
\underset{0 \le q \le 1}{\text{Minimize}}\ \mathbb{Q} &= ((1/\ell)B(T-1,T) - (q_{f,T-1}))^2 \\
&\quad + ((1/\ell)B(T-2,T) - (q_{f,T-2}))^2
\end{aligned}$$

Subject to:

$$B_1(T-1,T) = q\ell\,[1 - f_{12}(1)] + q\ell_{2,0}\, f_{12}(1)$$
$$B_1(T-2,T) = q^2\ell\,[1 - f_{12}(1) - f_{12}(2)] + q\ell_{2,1} f_{12}(1) + q^2\ell_{2,0} f_{12}(2)$$

leading to a cubic equation in q we can solve by the usual methods. Rewriting the quadratic deviation in terms of the discount rate yields:

$$\underset{0 \le q \le 1}{\text{Minimize}}\ (q[1 - f_{12}(1)(1 - (\ell_{2,0} / \ell))] - (q_{f,T-1}))^2$$
$$+ (q^2[1 - f_{12}(1) - (1 - \ell_{2,0}/\ell) f_{12}(2)] + q(\ell_{2,1}/\ell) f_{12}(1) - (q_{f,T-2}))^2$$

Set

$$a = [1 - f_{12}(1)(1 - (\ell_{2,0} / \ell))];\ b = [1 - f_{12}(1) - (1 - \ell_{2,0}/\ell) f_{12}(2)];$$
$$c = (\ell_{2,1}/\ell) f_{12}(1)$$

Then an optimal q is found by solving the equation:

$$2q^3b^2 + 3q^2bc + q(a^2 - 2bq_{f,T-2} + c^2) - (aq_{f,T-1} + cq_{f,T-2}) = 0$$

Assume the following parameters,

$$R_{f,T-1} = 0.07;\ R_{f,T-2} = 0.08,\ \ p = 0.8,\ \ell = 1,\ \ \ell_{2,0} = 0.6,\ \ell_{2,1} = 0.4$$

In this case,

$$f_{12}(1) = 1 - p = 0.2 \quad \text{and} \quad f_{12}(2) = p(1-p) = 0.16$$

For a one-period bond, we have:

$$1 + R_{1,T-1} = \frac{1 + 0.07}{1 - (0.084)} = 1.168$$

and, therefore, we have a 16.8 % discount, $R_{1,T-1} = 0.168$.

Problem

An AAA-rated bond has a clause that if it is downgraded to an AA bond, it must increase its coupon payment by 12 % while if it is downgraded to a B bond, then the firm has to redeem the bond in its entirety. For simplicity, say that the firm has only credit-rating classes (AAA, AA, B) and that it cannot default. In this case, how would you value the bond if initially it were an AAA bond?

Problem

By how much should a coupon payment be compensated when the bond class is downgraded? Should we sell a bond when it is downgraded? What are the considerations to keep in mind and how can they be justified?

Example

For simplicity, consider a coupon-paying bond with two credit ratings, A and D. ‘D’ denotes default and at redemption at time T, \$1 is paid. Let $q = 1 - p$ be a constant probability of default. Thus, if default occurs for the first time at $n \le T$, the probability that default occurs at any time prior to redemption is given by the geometric distribution $p^{n-1}(1-p)$ and therefore the conditional probability of default at $n \le T$ is given by:

$$f(n\,|n \le T) = \frac{p^{n-1}(1-p)}{1-F(T)}, n = 1, 2, \ldots, T$$

where $F(T)$ is the probability of default before or at bond redemption, or

$$F(T) = (1-p)\left(\frac{1-p^{T+1}}{1-p} - 1\right) = p(1-p^T)$$

As a result, if the yield of this bond is y, the bond price is given by:

$$B(0) = \left\{c\sum_{j=1}^{T}\left(\frac{1}{1+y}\right)^j\left(\frac{p^{j-1}(1-p)}{1-p(1-p^T)}\right)\right\} + \left(\frac{1}{1+y}\right)^T[1-p(1-p^T)]$$

While for a risk-free bond we have at the risk-free rate:

$$B_f(0) = \left\{c\sum_{j=1}^{T-1}\left(\frac{1}{1+R_f}\right)^j\right\} + 1\left(\frac{1}{1+R_f}\right)^T$$

The difference between the two thus measures the premium paid for a rated bond. These expressions can be simplified, however. This is left as an exercise.

Problem

Following Enron's collapse, Standard and Poors Corp. has said that it plans to rank the companies in the S&P500 stock index for the quality of their public disclosures as investors criticize the rating agency for failing to identify recent bankruptcies. As a result, a complex set of criteria will be established to construct a 'reliability rating' of S&P's own rating. Say that the credit rating of a bond is now given both by its class (AAA, AA, etc.) and by a reliability index, meaning that to each rating, there is an associated probability with the complement probability associated to a bond with lower rating. How would you proceed to integrate this reliability in bond valuation?

Example: Cash valuation of a rated firm

Earlier we noted that the cash flow of a firm can be measured by a synthetic sum of zero-coupon bonds. If these bonds are rated, then of course it is necessary for cash flow valuation to recognize this rating. Let $k,\ k = 1, 2, 3, \ldots, m$ be the m rates a firm assumes and let $q_k(t)$ be the probability that the bond is rated k at time t. In vector notation we write $\bar{q}(t)$ which is given in terms of the rating matrix transpose $\mathbf{P}'$. In other words, $\bar{q}(t) = [\mathbf{P}']^{\mathbf{t}}\bar{q}(0), \quad k = 1, 2, 3, \ldots$ and $\bar{q}(0)$ given. Thus, the NPV of a rated firm is given by:

$$NPV(t\,|\Re) = \sum_{s=t}^{n}\sum_{k=1}^{m} C_s B_k(t, s) q_k(s), \ \sum_{k=1}^{m} q_k(s) = 1$$

Our analysis can be misleading, however. Bonds entering in a given state may remain there for a certain amount of time before they switch to another state. Unstable countries and firms transit across rated states more often than say 'stable countries' and 'firms'. Further, they will usually switch to adjacent states or directly to a default state rather than to 'distant states'. For example, an AAA bond may be rated after some time to an AA bond while it is unlikely that it would transit directly to rating C. It is possible, however, that for some (usually external) reason the bond defaults, even if initially it is highly rated. These possibilities extend the Markov models considered and are topic for further empirical and theoretical study.

8.5 INTEREST-RATE PROCESSES, YIELDS AND BOND VALUATION*

Bonds, derivative securities and most economic time series depend intimately on the interest-rate process. It is therefore not surprising that much effort has been devoted to constructing models that can replicate and predict reliably the evolution of interest rates. There are, of course, a number of such models, each expressing some economic rationale for the evolution of interest rates. So far we have mostly assumed known risk-free interest rates. In fact, these risk-free (discounting) interest rates vary over time following some stochastic process and as a function of the discount period applied. Generally, and mostly for convenience, an interest-rate process $\{r(t), t \geq 0\}$ is represented by an Ito stochastic differential equation:

$$\mathrm{d}r = \mu(r,t)\,\mathrm{d}t + \sigma(r,t)\,\mathrm{d}w$$

where μ and σ are the drift and the diffusion function of the process, which may or may not be stationary. Table 8.8 summarizes a number of interest rates models. Note that while Merton's model is nonstationary (letting the

Table 8.8

Author	Drift	Diffusion	Stationary
Merton (1973)	β	σ	no
Cox (1975)	0	$\sigma r^{3/2}$	yes
Vasicek (1977)	$\beta(\alpha - r)$	σ	yes
Dothan (1978)	0	σr	yes
Brennan–Schwartz (1979)	$\beta r[\alpha - \ln(r)]$	σr	yes
Courtadon (1982)	$\beta(\alpha - r)$	σr	yes
March–Rosenfeld (1983)	$\alpha r^{-(1-\delta)} + \beta r$	$\sigma r^{\delta/2}$	yes
Cox–Ingersoll–Ross (1985)	$\beta(\alpha - r)$	$\sigma r^{1/2}$	yes
Chan et al. (1992)	$\beta(\alpha - r)$	σr^{λ}	yes
Constantinidis (1992)	$\alpha + \beta r + \gamma r^2$	$\sigma + \gamma r$	yes
Duffie–Kan (1996)	$\beta(\alpha - r)$	$\sqrt{\sigma + \gamma r}$	yes

diffusion-volatility be time-variant), other models have attempted to model this diffusion coefficient. Of course, to the extent that such a coefficient can be modelled appropriately, the technical difficulties encountered when the coefficients are time-variant can be avoided and the model parameters estimated (even though with difficulty, since these are mostly nonlinear stochastic differential equations). Further, note that the greater part of these interest rate models are of the 'mean reversion' type. In other words, over time short-term interest rates are pulled back to some long-run average level. Thus when the short rate is larger than the average long rate, the drift coefficient is negative and vice versa. Black and Karasinski (1991) (see also Sandmann and Sonderman, 1993) have also suggested that interest models can be modelled as well as a lognormal process. Explicitly, let the annual effective interest rate be given by the nonstationary lognormal model:

$$\frac{\mathrm{d}r_a(t)}{r_a(t)} = \beta(t)\,\mathrm{d}t + \sigma(t)\,\mathrm{d}W; r_a(0) = r_{a,0}$$

and consider the continuously compounded rate $R(t) = \ln(1 + r_a(t))$. An application of Ito's Lemma to this transformation yields also a diffusion process:

$$\mathrm{d}R(t) = (1 - \mathrm{e}^{-R(t)})\left[\theta(t) - \frac{1}{2}(1 - \mathrm{e}^{-R(t)})\sigma^2\right]\mathrm{d}t + \sigma\mathrm{d}W(t)$$

Another model suggested, and covering a broad range of distributional assumptions, includes the following (Hogan and Weintraub, 1993):

$$\mathrm{d}R(t) = R(t)\left[\theta(t) - a\ln R(t) + \frac{1}{2}\sigma^2\right]\mathrm{d}t + R(t)\sigma\,\mathrm{d}W(t)$$

The valuation of a bond when interest rates are stochastic is difficult because we cannot replicate the bond value by a risk-free rate. In other words, when rates are stochastic there is no unique way to price the bond. Mathematically this means that there are 'many' martingales we can use for pricing the bond and determine its yield (the integral of the spot-rate process). The problem we are faced with is, therefore, to determine a procedure which we can use to select the 'appropriate martingale' which can replicate observed bond prices. Specifically, say that the interest-rate model is defined by a stochastic process which is a function of a vector parameters Λ. In other words, we write the stochastic process:

$$\mathrm{d}r = \mu(r, t, \Lambda)\,\mathrm{d}t + \sigma(r, t, \Lambda)\,\mathrm{d}w$$

If this were the case, the theoretical price of a zero-coupon bond paying \$1 at time T is:

$$B_{Th}(0, T; \Lambda) = E^{*}\exp\left(-\int_0^T r(u, \Lambda)\mathrm{d}u\right) = E^{*}\left(\mathrm{e}^{-y(0,T;\Lambda)T}\right)$$

where $y(0, T; \Lambda)$ is the yield, a function of the vector parameters Λ. Now assume that these bond prices can be observed at time zero for a whole set of future times T and denote these observed values by, $B_{obs}(0, T)$. In order to determine the parameters set Λ we must find therefore some mathematical mechanism that would minimize in some manner some function of the 'error'

$$\Delta_B = B_{obs}(0, T) - B_{Th}(0, T; \Lambda)$$

There are several alternatives to doing so, as well as numerous mathematical techniques we can apply to solving this problem. This is essentially a computational problem (see, for example, Nelson and Siegel, 1987; Wets *et al.*, 2002; Kortanek and Medvedev, 2001; Kortanek, 2003; Delbaen and Lorimier, 1992; Filipovic, 1999, 2000, 2001).

The Nelson and Siegel approach is applied by many banks and consists in estimating the zero-coupon yield curve by fitting for all available bonds data in a sector credit combination the yield curve:

$$\begin{aligned} B_{Th}(0, T; \Lambda) &= E\left(\mathrm{e}^{-r(0,T;\Lambda)T}\right);\ r[0, T; \Lambda(\beta_i), i = 1, \ldots, 4] \\ &= \beta_0 + (\beta_1 + \beta_2)\left(\frac{1 - \mathrm{e}^{-\beta_3 T}}{\beta_3 T}\right) - \beta_2\,\mathrm{e}^{-\beta_3 T} \end{aligned}$$

where $r\,(0, T; \Lambda\,(\beta_i)\,, i = 1, \ldots, 4)$ is the spot rate and $\Lambda\,(\beta_i)$ are the model parameters. The Roger Wets approach (www.episolutions.com) is based upon a Taylor series approximation of the discount function in integral form. It is based on an approximation, and in this sense it shares properties with purely spline methods. Kortanek and Medvedev (2001), however, use a dynamical systems approach for modelling the term structure of interest rates based on a stochastic linear differential equation by constructing perturbation functions on either the unobservable spot interest rate or its integral (the yield) as unknown functions. Functional parameters are then estimated by minimizing a norm of the error comparing computed yields against observed yields over an observation period, in contrast to using the expectation operator for a stochastic process. When applied to a future period, the solved-for spot-rate function becomes the forecast of the unobservable function, while its integral approximates the yield function to the desired accuracy.

Some prevalent methods for computing (extracting) the zeros, curve-fitting procedures, equating the yield curve to observed data in the central bank include, among others: in Canada using the Svensson procedure and David Bolder (Bank of Canada); in Finland the Nelson–Siegel procedure; in France, the Nelson–Siegel, Svensson procedures; in Japan and the USA the banks use smoothing splines etc. (see Kortanek and Medvedev, 2001; Filipovic, 1999, 2000, 2001). Explicit solutions can be found for selected models, as we shall see below when a number of examples are solved. In particular, we shall show that approaches based on the optimal control of selected models can also be used.

8.5.1 The Vasicek interest-rate model

The Vasicek model has attracted much attention and is used in many theoretical and empirical studies. Its validity is of course, subject to empirical verification. An analytical study of the Vasicek model is straightforward since it is a classical model used in stochastic analysis (also called the Ornstein–Uhlenbeck process, as we saw in Chapter 4). In Vasicek's model the interest-rate change fluctuates around a long-run rate, α. This fluctuation is subjected to random and normal perturbations of mean zero and variance $\sigma^2 \mathrm{d}t$ however.

$$\mathrm{d}r = \beta(\alpha - r)\,\mathrm{d}t + \sigma\,\mathrm{d}w$$

This model's solution at time t when the interest rate is $r(t)$ is $r(u;t)$:

$$r(u;t) = \alpha + \mathrm{e}^{-\beta(u-t)}(r(t) - \alpha) + \sigma \int_t^u \mathrm{e}^{-\beta(u-\tau)}\,\mathrm{d}w(\tau)$$

In this theoretical model we might consider the parameters set $\Lambda \equiv (\alpha, \beta, \sigma)$ as determining a number of martingales (or bond prices) that obey the model above, namely bond prices at time $t = 0$ can theoretically equal the following:

$$B_{th}(0, T; \alpha, \beta, \sigma) = E^* \exp\left(-\int_0^T r(u; \alpha, \beta, \sigma)\,\mathrm{d}u\right)$$

In this simple case, interest rates have a normal distribution with a known mean and variance (volatility) evolution. Therefore

$$\int_0^T r(u, \alpha, \beta, \sigma)\,\mathrm{d}u$$

has also a normal probability distribution with mean and variance given by:

$$m(r(0), T) = \alpha T + (1 - \mathrm{e}^{-\beta T})\frac{r(0) - \alpha}{\beta}$$
$$v(r(0), T) = v(T) = \frac{\sigma^2}{2\beta^3}(4\,\mathrm{e}^{-\beta T} - \mathrm{e}^{-2\beta T} + 2\beta T - 3)$$

In these equations the variance is independent of the interest rate while the mean is a linear function of the interest which we write by:

$$m(r(0), T) = \alpha\left[T - \frac{(1 - \mathrm{e}^{-\beta T})}{\beta}\right] + r(0)\frac{(1 - \mathrm{e}^{-\beta T})}{\beta}$$

This property is called an affine structure and is of course computationally desirable for it will allow a simpler calculation of the desired martingale. Thus, the theoretical zero-coupon bond price paying \$1 T periods hence can be written by:

$$B_{th}(0, T; \alpha, \beta, \sigma) = E \exp\left(\int_0^T r(u, \alpha, \beta, \sigma)\, du\right)$$

$$= e^{-m(r(0),T)+v(T)/2} = e^{A(T)-r_0 D(T)}$$

$$A(T) = -\alpha\left[T - \frac{(1 - e^{-\beta T})}{\beta}\right] + \frac{\sigma^2}{4\beta^3}(4\,e^{-\beta T} - e^{-2\beta T} + 2\beta T - 3);$$

$$D(T) = \frac{(1 - e^{-\beta T})}{\beta}$$

Now assume that a continuous series of bond values are observed and given by $B_{obs}(0, T)$ which we write for convenience by, $B_{obs}(0, T) = e^{-R_T T}$. Without loss of generality we can consider the yield error term given by:

$$\Delta_T = R_T - (A(T) - r_0 D(T))$$

and thus select the parameters (i.e. select the martingale) that is closest in some sense to observed values. For example, a least squares solution of n observed bond values yields the following optimization problem:

$$\underset{\alpha,\beta,\sigma}{\text{Min}} \sum_{i=1}^{n} (\Delta_i)^2$$

When the model has time-varying parameters, the problem we faced above turns out to have an infinite number of unknown parameters and therefore the yield curve estimation problem we considered above might be grossly underspecified. Explicitly, let the interest rate model be defined by:

$$dr(t) = \beta\,[\alpha(t) - r(t)]\, dt + \sigma\, dw$$

The theoretical bond value has still an affine structure and therefore we can write:

$$B_{th}[t, T; \alpha(t), \beta, \sigma] = E^* \exp\left(-\int_0^T r(u; \alpha, \beta, \sigma)\, du\right) = e^{A(t,T)-r(t)D(t,T)}$$

The integral interest-rate process is still normal with mean and variance leading to:

$$A(t, T) = \int_t^T \left\{\frac{1}{2}\sigma^2 D^2(s, T) - \beta\alpha(s)D(s, T)\right\} ds;\ D(t, T) = \frac{1}{\beta}\left\{1 - e^{-\beta(T-t)}\right\}$$

or:

$$\frac{dA(t, T)}{dt} = \alpha(t)\left\{1 - e^{-\beta(T-t)}\right\} - \frac{\sigma^2}{2\beta^2}\left\{1 - e^{-\beta(T-t)}\right\}^2,\ A(T, T) = 0$$

in which $\alpha(t), \beta, \sigma$ are unspecified. If we equate this equation to the available bond data we will obviously have far more unknown variables than data points and therefore the yield curve estimate will depend again on the optimization technique we use to generate the best fit parameters β^*, σ^* and the function, $\alpha^*(t)$. Such problems can be formulated as standard problems in the calculus of variations (or optimal control theory). For example, if we consider the observed prices $B_{obs}(t, T),\ t \leq T < \infty$, for a specific time, T, and minimize the following squared error in continuous time, we obtain the following singular control problem:

$$\underset{\alpha(u)}{\text{Min}} = \int_0^t [A(u, T) - c(u, T)]^2 \mathrm{d}u$$

subject to:

$$\frac{\mathrm{d}A(u, T)}{du} = \alpha(u)a(u, T) - b(u, T),\ \ A(T, T) = 0$$

with

$$c(u, t) = y_{obs}(u, T) + r(u)\left[\frac{1}{\beta}\left\{1 - \mathrm{e}^{-\beta(T-u)}\right\}\right],$$

$$a(u, t) = \left\{1 - \mathrm{e}^{-\beta(T-u)}\right\};\ b(u, t) = \frac{\sigma^2}{2\beta^2}\left\{1 - \mathrm{e}^{-\beta(T-u)}\right\}^2$$

and $\alpha(u)$ is the control and $A(u, T)$ is the state which can be solved by the usual techniques in optimal control. The solution of this problem leads either to a bang-bang solution, or to a singular solution. Using the deterministic dynamic programming framework, the long-run (estimated) rate is given by solving:

$$-\frac{\partial J}{\partial u} = \underset{\alpha(u)}{\text{Min}}\left\{[A(u, T) - c(u, T)]^2 + \frac{\partial J}{\partial A}[\alpha(u)a(u, T) - b(u, T)]\right\}$$

On a singular strip, $\partial J/\partial A = 0$ where $a(u, t) \neq 0$ and therefore in order to calculate $\alpha(u)$, we can proceed by a change of variables and transform the original control problem into a linear quadratic control problem which can be solved by the standard optimal control methods. Explicitly, set:

$$y(u) = [A(u, T) - c(u, T)]$$

$$\text{with } \frac{\mathrm{d}w(u)}{du} = \alpha(u) \quad \text{and} \quad z(u) = y(u) - a(u, T)w(u)$$

Thus, the problem is reduced to:

$$\underset{w(u)}{\text{Min}} = \int_0^t [z(u) + a(u, T)w(u)]^2\, \mathrm{d}u$$

subject to:

$$\frac{dz(u)}{du} = -\dot{a}(u, T)w(u) - b(u, T) - c(u, T); \quad \dot{a}(u, T) = da(u, T)/du$$

and at time T,

$$z(T) = -c(T, T) - a(T, T)w(T)$$

This is a linear control problem whose objective is quadratic in both the state and the control. As a result, the problem solution of this standard control problem is the linear feedback form:

$$w(u) = Q(u) + S(u)z(u) \quad \text{or} \quad \alpha(t) = \int_0^t w(u)\,du$$

The functions $Q(u)$, $S(u)$ can be found by inserting in the problem's conditions for optimality. This problem is left for self-study, however (see also Tapiero, 2003).

Problem: The cox–ingersoll–Ross (CIR) model

By changing the interest-rate model, we change naturally the results obtained. Cox, Ingersoll and Ross (1985), for example, suggested a model, called the *square root process*, which has a volatility given as a function of interest rates as well, namely, they assume that:

$$dr = \beta(\alpha - r)\,dt + \sigma\sqrt{r}dw$$

First show that the interest rate process is not normal but its mean and variance are given by:

$$E(r(t)\,|r_0) = c(t)\left[\frac{4\beta\alpha}{\sigma^2} + \xi\right]; \mathrm{Var}(r(t)\,|r_0) = c(t)^2\left[\frac{8\beta\alpha}{\sigma^2} + 4\xi\right]$$

where

$$c(t) = \frac{\sigma^2}{4\beta}[1 - e^{-\beta t}]; \; \xi = \frac{4r_0\beta}{\sigma^2\,[\exp(\beta t) - 1]}$$

Demonstrate then that this process has an affine structure as well by verifying that:

$$B(r, t, T) = E\exp\left(-\int_t^T r(T - u)\,du\right) = e^{(A(t,T) - rD(t,T))}; \; B(r, T, T) = 1$$

and at the boundary $A(T, T) = 0$, $D(T, T) = 0$. Finally, calculate both $A(t, T)$ and $D(t, T)$ and formulate the numerical problem which has to be solved in order to determine the bond yield curve based on available bond prices.

Problem: The nonstationary Vasicek model
Show for the nonstationary model $\mathrm{d}r = \mu(t)(m(t) - r)\,\mathrm{d}t + \sigma r\,\mathrm{d}w$ that its solution is:

$$r(t) = \exp[-A(t)]\left[y + \int_0^t \mu(s)m(s)\exp[-A(s)]\right]$$

$$A(t) = M(t) + \sigma^2 t/2 - \sigma\int_0^t \mathrm{d}w \quad \text{and} \quad M(t) = \int_0^t \mu(s)\,\mathrm{d}s$$

8.5.2 Stochastic volatility interest-rate models

Cotton, Fouque, Papanicolaou and Sircar (2000) have shown that a single factor model (i.e. with one source of uncertainty) driven by Brownian motion implies perfect correlation between returns on bonds for all maturities T, which is not seen in empirical analysis. They suggest, therefore, that the volatility in the Vasicek model ought to be stochastic as well. Their derivation, based on a mean reverting model in the short rate, shows an exponential decay in the short-term, (two weeks). This is small compared to bonds with maturities of several years. Denote the variance in an interest model by $V = \sigma^2(r, t)$, then an interest-rate 'stochastic volatility model' consists of two stochastic differential equations, with two sources of risk (w_1, w_2) which may be correlated or not. An example would be:

$$\mathrm{d}r = \mu(r, t)\,\mathrm{d}t + \sqrt{V(r, t)}\,\mathrm{d}w_1$$

$$\mathrm{d}V = \nu(V, r, t)\,\mathrm{d}t + \gamma(V, r)\,\mathrm{d}w_2$$

where the variance V appears in both equations. Hull and White (1988) for example suggest that we use a square root model with a mean reverting variance model given by:

$$\frac{\mathrm{d}r}{r} = \mu\,\mathrm{d}t + \sqrt{V}\mathrm{d}w_1;\ \mathrm{d}V = \alpha(\beta - V)\,\mathrm{d}t + \gamma r V^{\lambda}\,\mathrm{d}w_2,\ \ \rho\,\mathrm{d}t = E\,\mathrm{d}w_1\,\mathrm{d}w_2$$

In this case, note that when stock prices increase, volatility increases. Further when volatility increases, interest rates (or the underlying asset we are modeling) increase as well. Cotton *et al.* (2000), in contrast, suggested that, in a CIR-type model such as $\mathrm{d}r = (\mu - r)\,\mathrm{d}t + \sigma r^{\gamma}\mathrm{d}W$, γ is not equal to a half but rather is equal to one and half and thereby certainly greater than one. The model they suggest turns out:

$$\mathrm{d}r = \theta_r\,(\mu_r - r)\,\mathrm{d}t + \sqrt{\alpha_r + \beta_r V}\mathrm{d}W^*$$

$$\mathrm{d}V = \theta_V\,(\mu_V - V)\,\mathrm{d}t + \sqrt{\alpha_V + \beta_V V}\mathrm{d}Z^*$$

where $(\mathrm{d}W^*, \mathrm{d}Z^*)$ are Brownian motion under the pricing measure. Note here that the volatility is a mean reverting driving process. The advantage in using such a model is that it also leads to an affine structure where the time-dependent coefficients are given by the solutions of differential equations. In this case, estimation of the yield curve can be reached, as we have stated above, by the solution of an optimal control problem. In other words, once a theoretical estimate of the bond price is found, and observed bond prices are available, we can calculate the parameters of the model by solving the appropriate optimization problem.

8.5.3 Term structure and interest rates

Interest rates applied for known periods of time, say T, change necessarily over time. In other words, if $r(t, T)$ is the interest rate applied at t for T, then at $t+1$, the relevant rate for this period T would be $r(t+1, T-1)$, while the going interest for the same period would be $r(t+1, T)$. If these interest rates are not equal, there may be an opportunity for refinancing. As a result, the evolution of interest rates for different maturity dates is important. For example, if a model is constructed for interest rates of maturity T, then we may write:

$$\mathrm{d}r(t, T) = \mu(r, T)\,\mathrm{d}t + \sigma(r, T)\,\mathrm{d}w$$

The price of a zero-coupon bond is a function of such interest rates and is given by $B(t, T) = \exp[-r(t, T)(T-t)]$ whose differential equation (see the mathematical Appendix to this chapter)

$$0 = \frac{\partial B}{\partial t} + \frac{\partial B}{\partial r}[\mu(r, T) - \lambda(r, t)] + \frac{1}{2}\frac{\partial^2 B}{\partial r^2}\sigma^2(r, T) - rB$$

$$B(r, T, T) = 1$$

where the price of risk, a known function of r and time t, is proportional to the returns standard deviation and given by:

$$\alpha(r, t, T) = r + \lambda(r, t)\frac{1}{B}\frac{\partial B}{\partial r}$$

The solution of this equation, although cumbersome, can in some cases be determined analytically, and in others it can be solved numerically. For example, if we set $(\mu(r, T) - \lambda(r, t)) = \theta; \sigma^2(r, T) = \rho^2$ where (θ, ρ) are constant then a solution of the partial differential equation of the bond price (see the Mathematical Appendix), we obtain an affine structure type:

$$B(r, t, T) = \exp\left\{-r(T-t) - \frac{1}{2}\theta(T-t)^2 + \frac{1}{6}\rho^2(T-t)^3\right\}$$

Problem
Set the following equalities: $\mu(r, T) - \lambda(r, t) = k(\theta - r)$; $\sigma^2(r, T) = \rho^2 r$ (which is the CIR model seen earlier) and show that the solution for the bond price equation is of the following form:

$$B(r, t, T) = \exp\{A(T-t) + rD(T-t)\}$$

A solution for the function $A(.)$ and $D(.)$ can be found by substitution.

8.6 OPTIONS ON BONDS*

Options on bonds are compound options, traded popularly in financial markets. The valuation of these options requires both an interest-rate model and the valuation of term structure bond prices (which depend on the interest rates for various maturities of the bond). For instance, say that there is a T bond call option, which confers the right to exercise it at time $S < T$. The procedure we adopt in valuing a call option on a bond consists then in two steps. First we evaluate the term structure for a T and an S bond. Then we can proceed to value the call on the T bond with exercise at time S (used to replace the spot price at time S in the plain vanilla option model of Black–Scholes). The procedure is explicitly given by the following. First we construct a hedging portfolio consisting of the two bonds maturities S and T ($S < T$). Such a portfolio can generate a synthetic rate, equated to the spot interest rate so that no arbitrage is possible. In this manner, we value the option on the bond uniquely. An extended development is considered in the Mathematical Appendix while here we summarize essential results. Let, for example, the interest process:

$$\mathrm{d}r = \mu(r,t)\,\mathrm{d}t + \sigma(r,t)\,\mathrm{d}w$$

A portfolio (n_S, n_T) of these two bonds has a value and a rate of return given by:

$$V = n_S B(t,S) + n_T B(t,T) \quad \text{and} \quad \frac{\mathrm{d}V}{V} = n_S \frac{\mathrm{d}B(t,S)}{B(t,S)} + n_T \frac{\mathrm{d}B(t,T)}{B(t,T)}$$

The rates of return on T and S bonds are assumed given as in the previous section. Each bond with maturity T and S has at its exercise time a \$1 denomination, thus the value of each of these (S and T) bonds is given by $B_T(t,r)$ and $B_S(t,r)$. Given these two bonds, we define the option value of a call on a T bond with $S < T$ and strike price K, to be:

$$X = \text{Max}\,[B(S,T) - K, 0]$$

with $B(S,T)$ the price of the T bond at time S. The bond value $B(S,T)$ is of course found by solving for the term structure equation and equating $B(r,S,T) = B(S,T)$. To simplify matters, say that the solution (value at time t) for the T bond is given by $F(t,r,T)$, then at time S, this value is: $F(S,r,T)$ to which we equate $B(S,T)$. In other words,

$$X = \text{Max}\,[F(S,r,T) - K, 0]$$

Now, if the option price is $B(.)$, then, as we have seen in the plain vanilla model in Chapter 6, the value of the bond is found by solving for $P(.)$ in the following partial differential equation:

$$0 = \frac{\partial B}{\partial t} + \mu(r,t)\frac{\partial B}{\partial r} + \sigma^2 \frac{1}{2}\frac{\partial^2 B}{\partial r^2} - rB, \; B(S,r) = \text{Max}\,[F(S,r,T) - K, 0]$$

A special case of interest consists again in using an affine term structure (ATS) model as shown above in which case:

$$F(t, r, T) = \mathrm{e}^{A(t,T)-rD(t,T)}$$

where *A(.)* and *D(.)* are calculated by the term structure model. The price of an option of the bond is thus given by the solution of the bond partial differential equation, for which a number of special cases have been solved analytically. When this is not the case, we must turn to numerical or simulation techniques.

8.6.1 Convertible bonds

Convertible bonds confer the right to the bond issuer to convert the bond into stock or into certain amounts of money that include the conversion cost. For example, if the bond can be converted against m shares of stock, whose price dynamics is:

$$\mathrm{d}S = \mu S\,\mathrm{d}t + \sigma S\,\mathrm{d}w$$

Then, the bond price is necessarily a function of the stock price and given by $V(S, t)$. To value such a bond we proceed 'as usual' by constructing an equivalent risk-free and replicating portfolio. Let this risk-free portfolio be:

$$\pi = V + \alpha S \quad \text{and} \quad \text{therefore } \mathrm{d}\pi = \mathrm{d}V + \alpha\,\mathrm{d}S$$

For this portfolio to be risk-free, we equate it to a portfolio whose rate of return is the risk-free rate R_f. Thus, $\mathrm{d}\pi = \mathrm{d}V + \alpha\,\mathrm{d}S = R_f \pi\,\mathrm{d}t = R_f\,(V + \alpha S)\,\mathrm{d}t$ and $\mathrm{d}V = R_f\,(V + \alpha S)\,\mathrm{d}t - \alpha\,\mathrm{d}S$. Using Ito's Lemma, we calculate $\mathrm{d}V$ leading to:

$$\frac{\partial V}{\partial t}\,\mathrm{d}t + \left(\frac{\partial V}{\partial S}\right)\mathrm{d}S + \frac{1}{2}\frac{\partial^2 V}{\partial S^2}\,(\mathrm{d}S)^2 = R_f\,(V + \alpha S)\,\mathrm{d}t - \alpha\,\mathrm{d}S$$

which can be rearranged to:

$$\left[\frac{\partial V}{\partial t} + \mu S\frac{\partial V}{\partial S} + \frac{\sigma^2 S^2}{2}\frac{\partial^2 V}{\partial S^2} + \alpha\mu S - R_f V - \alpha R_f S\right]\mathrm{d}t$$
$$+\sigma S\left(\frac{\partial V}{\partial S} + \alpha\right)\mathrm{d}w = 0$$

A risk-free portfolio has no volatility and therefore we require:

$$\sigma S\left(\frac{\partial V}{\partial S} + \alpha\right)\mathrm{d}w = 0 \quad \text{or} \quad \alpha = -\frac{\partial V}{\partial S}$$

Inserting $\alpha = -\,\partial V/\partial S$ yields the following partial differential equation:

$$\frac{\partial V}{\partial t} + \frac{\sigma^2 S^2}{2}\frac{\partial^2 V}{\partial S^2} + R_f S\frac{\partial V}{\partial S} - R_f V = 0,\ \ V(S, T) = 1$$

where at redemption the bond equals \$1. If the conversion cost is $C(S, t) = mS$, the least cost is Min $\{V(S, t), C(S, t)\}$. Therefore, in the continuation region (i.e. as long as we do not convert the bonds into stocks), we have: $V(S, t) \geq C(S, t) = mS$ while in the stopping region (i.e. at conversion) we have: $V(S, t) \leq mS$. In other words, the convertible option has the value of an American option which we solve as indicated in Chapter 6. It can also be formulated as a stopping time problem, but this is left as an exercise for the motivated reader.

8.6.2 Caps, floors, collars and range notes

A *cap* is a contract guaranteeing that a floating interest rate is capped. For example, let r_ℓ be a floating rate and let r_c be an interest rate cap. If we assume that the floating rate equals approximately the spot rate, $r \approx r_\ell$ then a simple *caplet* is priced by:

$$\frac{\partial V}{\partial t} + [\mu(r,t) - \lambda\sigma]\frac{\partial P}{\partial r} + \sigma^2 \frac{1}{2}\frac{\partial^2 V}{\partial r^2} - rV, \;\; V(r,T) = \text{Max}\,[r - r_c, 0]$$

with the cap being a series of caplets. By the same token, a *floor* ensures that the interest rate is bounded below by the rate floor: r_f. Thus, the rate at which a cash flow is valued is: Max $(r_f - r_\ell, 0), \; r_\ell \geq r_c$. Again, if we assume that the floating rate equals the short rate, we have a *floorlet* price given by:

$$\frac{\partial V}{\partial t} + [\mu(r,t) - \lambda\sigma]\frac{\partial P}{\partial r} + \sigma^2 \frac{1}{2}\frac{\partial^2 V}{\partial r^2} - rV, \;\; V(r,T) = \text{Max}\left[r_f - r, 0\right]$$

while the floor is a series of floorlets. A *collar*, places both an upper and a lower bound on interest payments, however. A collar can thus be viewed as a long position on a cap, with a given strike r_c and a short position on a floor with a lower strike r_f. If the interest rate falls below r_f, the holder is forced into paying the higher rate of r_f. The strike price of the call is often set up so that the cost of the cap is exactly subsidized by the revenue from the sale of the floor. When the interest rate on a notional principal is bounded above and below, then we have a *range* note. In this case, the value of the range note can be solved by using the differential equations framework as follows:

$$\frac{\partial V}{\partial t} + [\mu(r,t) - \lambda\sigma]\frac{\partial P}{\partial r} + \sigma^2 \frac{1}{2}\frac{\partial^2 V}{\partial r^2} - rV + \Xi(r)$$

$$= 0; \; \Xi(r) = \begin{cases} r & \text{if } r < r < \bar{r} \\ 0 & \text{otherwise} \end{cases}$$

This is only an approximation since, in practice, the relevant interest rate will have a finite maturity (Wilmott, 2000).

8.6.3 Swaps

An *interest rate swap* is a private agreement between two parties to exchange one stream of cash flow for another on a specific amount of principal for a specific period of time. Investors use swaps to exchange fixed-rate liabilities/assets into floating-rate liabilities/assets and vice versa.

Interest rate swaps are most important in practice. They emerged in the 1980s and their growth has been spectacular ever since. They are essentially customized commodity exchange agreements between two parties to make periodic payments to each other according to well-defined rules. In the simplest of interest rate swaps, one part periodically pays a cash flow determined by a fixed interest rate and receives a cash flow determined by a floating interest rate (Ritchken, lecture notes, 2002).

For example, consider Company A with \$50 000 000 of floating-rate debt outstanding on which it is paying LIBOR plus 150 bps (basis points), i.e. if LIBOR is 4 %, the interest rate would be 5.5 %. The company thinks that interest rates will rise, i.e. company's interest expense will rise, and the company decides to convert its debt from floating-rate into fixed-rate debt. Now consider Company B which has \$50 000 000 of fixed-rate 6 % debt. The company thinks that interest rates will fall, which would benefit the company if it has floating-rate debt instead of fixed-rate debt, since its interest expense will be reduced. By entering into an interest rate swap with Company A, both parties can effectively convert their existing liabilities into the ones they truly want. In this swap, Company A might agree to pay Company B fixed-rate interest payments of 5 % and Company B might agree to pay Company A floating-rate interest payments of LIBOR. Therefore Company A will pay LIBOR + 150 to its original lender and 5 % in the swap, giving a total of LIBOR + 6.5 %; it receives LIBOR in the swap. This leaves an all-in cost of funds of 6.5 %, a fixed rate. In the case of Company B, it pays 6 % to its original lender and LIBOR in the swap, giving a total of LIBOR + 6 %. In return it receives 5 % in the swap, leaving an all-in cost of LIBOR + 1 %, a floating rate (see Figure 8.7).

There are four major components to a swap: the notional principal amount, the interest rate for each party, the frequency of cash exchange and the duration of the swap. A typical swap in swap jargon might be \$20m, two-year, pay fixed, receive variable, semi. Translated, this swap would be for \$20m notional principal, where one party would pay a fixed interest-rate payment for every 6 months based on the \$20m and the counterparty would pay a variable rate payment every 6 months based on the \$20m. The variable-rate payment is usually based on a specific short-term interest rate index such as the 6-months LIBOR. The time period specified by the variable rate index usually coincides with the frequency of swap payments. For example, a swap that is fixed versus 6 months LIBOR would have semiannual payments. Of course there can be exceptions to this rule.

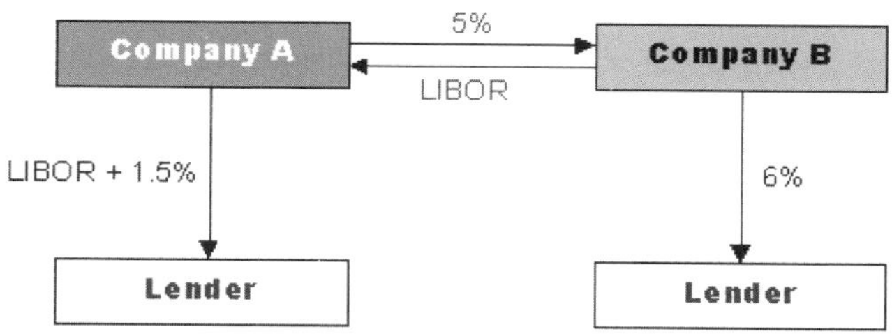

Figure 8.7 A swap contract.

For example, the variable-rate payment could be linked to the average of all T-bill auction rates during the time period between settlements.

Most interest rate swaps have payment date arrears. That is, the net cash flow between parties is established at the beginning of the period, but is actually paid out at the end of the period. The fixed rate for a generic swap is usually quoted as some spread over benchmark US treasuries. For example a quote of '20 over' for a 5-year swap implies that the fixed rate on a 5-year swap will be set at the 5-year Treasury yield that exists at the time of pricing plus 20 basis point. Usually, swap spreads are quoted against the two, three, five, seven and ten benchmark maturities. The yield used for other swaps (such as a 4-year swap) is then obtained by averaging the surrounding yields.

Finally, a 'swaption' is an option to swap. It confers the right to enter into a swap contract at a predetermined future date at a fixed-rate and be a payer at the fixed rate. 'Captions' and 'floortions' are similarly options on caps and floors respectively.

A swap price can be defined by a cap–floor parity, where

$$\text{Cap} = \text{Floor} + \text{Swap}$$

In other words, the price of a swap equals the price difference between the cap and the floor. A caplet can be shown to be equivalent to the price of a put option expiring at a time t_{i-1} prior to a bond's maturity at time t_i. Set the payoff $\text{Max}(r_\ell - r_c, 0)$ a period hence, which is discounted to t_{i-1} and yielding:

$$\frac{1}{1+r_\ell}\text{Max }(r_\ell - r_c, 0) = \text{Max}\left(\frac{r_\ell - r_c}{1+r_\ell}, 0\right) = \text{Max}\left(1 - \frac{1+r_c}{1+r_\ell}, 0\right)$$

However $(1 + r_c)/(1 + r_\ell)$ is the price of paying $1 + r_c$ a period hence. Thus a caplet is equivalent to a put option expiring at t_{i-1} on a bond with maturity t_i.

REFERENCES AND ADDITIONAL READING

Arrow, K.J. (1953) Le role des valeurs boursières pour la repartition la meilleur des risques, in econometric, *Colloquia International du CNRS*, **40**, 41–47. In English, The role of securities in the optimal allocation of risk bearing, *Review of Economic Studies*, **31**, 91–96, 1963.

Augros, Jean-Claude (1989) *Les Options sur Taux d'Intérêt*, Economica, Paris.

Belkin, B., S. Suchower and L. Forest Jr (1998) A one-parameter representation of credit risk and transition matrices, *CreditMetrics® Monitor*, third quarter, 1998. (http://www.riskmetrics.com/cm/pubs/index.cgi)

Bingham, N.H. (1991) Fluctuation theory of the Ehrenfest urn, *Advances in Applied Probability*, **23**, 598–611.

Black F. and J.C. Cox, (1976), Valuing corporate securities: Some effects of bond indenture provisions, *Journal of Finance*, **31**(2), 351–367.

Black, F., E. Derman and W. Toy (1990) A one-factor model of interest rates and its application to treasury bond options, *Financial Analysts Journal*, January–February, 133–139.

Black, F., and P. Karasinski (1991) Bond and options pricing when short rates are lognormal, *Financial Analysts Journal*, **47**, 52–59.

Black, F., and M. Scholes (1973) The pricing of options and corporate liabilities, *Journal of Political Economy*, **81**(3), 637–654.

Brennan, M.J., and E.S. Schwartz (1977) Convertible bonds: valuation and optimal strategies for call and conversion , *Journal of Finance*, **32**, 1699–1715.

Brennan, M.J., and E.S. Schwartz (1979) A continuous time approach to the pricing of corporate bonds, *Journal of Banking and Finance*, **3**, 133–155.

Chan, K.C., G.A. Karolyi, F.A. Longstaff and A.B. Sanders (1992) An empirical comparison of alternative models of the short term interest rate, *Journal of Finance*, **47**, 1209–1227.

Chance, D. (1990) Default risk and the duration of zero-coupon bonds, *Journal of Finance*, **45**(1), 265–274.

Cotton, Peter, Jean-Pierre Fouque, George Papanicolaou and K. Ronnie Sircar (2000) Stochastic volatility corrections for interest rate derivatives, Working Paper, Stanford University, 8 May.

Courtadon, G. (1982) The pricing of options on default free bonds, *Journal of Financial and Quantitative Analysis*, **17**, 75–100.

Cox, J.C., J.E. Ingersoll and S.A. Ross (1985) A theory of the term structure of interest rates, *Econometrica*, **53**, 385–407.

Delbaen, F., and S. Lorimier (1992) Estimation of the yield curve and forward rate curve starting from a finite number of observations. *Insurance: Mathematics and Economics*, **11**, 249–258.

Duffie, D., and D. Lando (2001) Term structures of credit spreads with incomplete accounting information, *Econometrica*, **69**, 633–664.

Duffie, D. and K.J. Singleton (1997) Modeling term structures of defaultable bonds, *Review of Financial Studies*, **12**(4), 687–720.

Duffie, D. and K.J. Singleton (1999) Modelling term structures of defaultable bonds, *Review of Financial Studies*, **12**(4), 687–720.

Duffie, G.R. (1998) The relation between Treasury yields and corporate bond yield spreads, *Journal of Finance*, **53**, 2225–2242.

Duffie, G.R. (1999) Estimating the price of default risk, *Review of Financial Studies*, **12**, 197–226.

Duffie, J.D., and R. Kan (1996) A yield-factor model of interest rates, *Mathematical Finance*, **6**, 379–406.

Fabozzi, F.J. (1996) *Bond Markets: Strategies and Analysis*, Prentice Hall, Englewood Cliffs, NJ.

Fama, E., and K. French (1993) Common risk factors in the returns on stocks and bonds, *Journal of Financial Economics*, **33**(1), 3–56.

Filipovic, D. (1999) A note on the Nelson–Siegel family, *Mathematical Finance*, **9**, 349–359.

Filipovic, D. (2000) Exponential-polynomial families and the term structure of interest rates, *Bernoulli*, **6**, 1–27.

Filipovic, D. (2001) Consistency problems for Heath–Jarrow–Morton interest rate models, Number 1760 in *Lecture Notes in Mathematics*, Series Editors J.-M. Morel, F. Takens and B. Teissier. Springer Verlag, New York.

Fisher, I. (1906) *The Nature of Capital and Income*, Sentry Press, New York. Reprinted by Augustus M. Kelly, New York, 1965.

Fisher, I. (1907) , *The Rate of Interest*, Macmillan, New York.

Fisher, I. (1930) *The Theory of Interest*, Macmillan, New York. Reprinted by Augustus M. Kelly, New York, 1960.

Flajollet, F., and F. Guillemin (2000) The formal theory of birth–death processes, lattice path combinatorics and continued fractions, *Advances Applied in Probability*, **32**, 750–778.

Geske, R. (1977) The valuation of corporate liabilities as compound options, *Journal of Financial and Quantitative Analysis*, **12**(4), 541–552.

Heath, D., R. Jarrow and A. Morton (1990) Contingent claim valuation with a random evolution of interest rates, *The Review of Future Markets*, 54–76.

Heath, D., R. Jarrow and A. Morton (1992) Bond pricing and the term structure of interest rates: A new methodology for contingent claims evaluation, *Econometrica*, **60**, 77–105.

Ho, T.S.Y., and S.B. Lee (1986) Term structure movements and pricing of interest rate contingent claims, *Journal of Finance*, **41**, December, 1011–1029.

Hogan, M., and K. Weintraub (1993) The lognormal interest model and Eurodollars futures, Working Paper, Citibank, New York.

Hull J. and A. White (1988) An analysis of the bias in option pricing caused by a stochastic volatility, *Advances in Futures and Options Research*, **3**(1), 29–61

Hull J. and A. White (1993) Bond Option pricing based on a model for the evolution of bond prices, *Advances in Futures and Options Research*, **6**, 1–13.

Jarrow, Robert A. (1996) *Modelling Fixed Income Securities and Interest Rate Options*, McGraw Hill, New York.

Jarrow, R., and S. Turnbull (1995) Pricing derivatives on financial securities subject to credit risk, *Journal of Finance*, **50**, 53–86.

Jarrow, R.A., D. Lando and S. Turnbull (1997) A Markov model for the term structure of credit spreads, *Review of Financial Studies*, **10**, 481–523.

Karlin, S., and J.L. McGregor (1957) The differential equation of birth–death processes, and the Stieltjes moment problem, *Transactions of the American Mathematical Society*, **85**, 489–546.

Karlin, S., and H.M. Taylor (1975) *A First Course in Stochastic Processes*, 2nd edn, Academic Press, New York.

Karlin, S., and H.M. Taylor (1981) *A Second Course in Stochastic Processes*, 2nd edn, Academic Press, New York.

Kim, J. (1999) Conditioning the transition matrix, *Credit Risk*, a special report by Risk, October.

Kortanek, K.O. (2003) Comparing the Kortanek & Medvedev GP approach with the recent Wets approach for extracting the zeros, April 26 (Internet paper).

Kortanek, K.O., and V. G. Medvedev (2001) *Building and Using Dynamic Interest Rate Models*, John Wiley & Sons, Ltd., Chichester.

Lando D. (1998) Oncox processes and credit risky securities, *Review of Derivatives Research*, **2**(2, 3), 99–120.

Lando, D. (2000) Some elements of rating-based credit risk modeling, in: N. Jegadeesh and B. Tuckman (Eds), *Advanced Fixed-Income Valuation Tools*, John Wiley & Sons, Inc., New York.

Leland H. E. (1994) Corporate debt value, bond covenants and optimal capital structure, *Journal of Finance*, **49**(4), 1213–1252.

Longstaff, F., and E. Schwartz (1995) A simple approach to valuing risky fixed and floating rate debt, *Journal of Finance*, **50**, 789–819.

Merton, R. (1974) On the pricing of corporate debt: the risk structure of interest rates, *Journal of Finance*, **29**, 449–470.

Moody's Special Report (1992) Corporate bond defaults and default rates, *Moody's Investors Services*, New York.

Nelson, C.R., and A. F. Siegel (1987) Parsimonious modeling of yield curves, *Journal of Business*, **60**, 473–489.

Nickell, P., W. Perraudin and S. Varotto (2000) Stability of rating transitions, *Journal of Banking and Finance*, **24** (1–2), 203–227.

Rebonato, R. (1998) *Interest-rate option models, 2nd edn*, John Wiley & Sons, Ltd, Chichester.

Sandmann, K., and D. Sondermann (1993) A term structure model and the pricing of interest rates derivatives, *The Review of Future Markets*, **12**(2), 391–423.

Sandmann, K., and D. Sondermann (1993) In the stability of lognormal interest rate models, SFB 303, Universität Bonn, Working Paper B-263.

Standard & Poors Special Report (1998) Corporate defaults rise sharply in 1998, Standard & Poors, New York.

Standard & Poors (1999) Rating performance 1998, stability and transition, Standard & Poors, New York.

Tapiero, C.S. (2003) Selecting the optimal yield curve: An optimal control approach, Working Paper, ESSEC, France.
Vasicek, O. (1977) An equilibrium characterization of the term structure, *Journal of Financial Economics*, **5**, 177–188.
Walras, L. (1874) *Elements d'Economie Politique Pure*, Corbaz, Lausanne. (English translation, *Elements of Pure Economy*, Irwin, Homewood, IL, 1954.
Wets, R.J.B., S.W. Bianchi and L. Yang (2002) Serious zero curves. Technical report, EpiSolutions, Inc, El Cerrito, California,
Wilmott, P., (2000), Paul Wilmott on Quantitative Finance, John Wiley & Sons Ltd., Chichester.

MATHEMATICAL APPENDIX

A.1: Term structure and interest rates

Let the interest-rate process:

$$dr(t, T) = \mu(r, T)\,dt + \sigma(r, T)\,dw$$

The price of a zero-coupon bond is a function of such interest rates which we can assume to be:

$$\frac{dB(t, T)}{B(t, T)} = \alpha(r, t, T)\,dt + \beta(r, t, T)\,dw$$

with $(\alpha(r, t, T), \beta(r, t, T))$ the parameters denoting the drift and diffusion of the bond's return. To determine these parameters, we apply Ito's Lemma's to:

$$B(t, T) = \exp[-r(t, T)(T - t)]$$

leading to:

$$dB(t, T) = \left[\frac{\partial B}{\partial t} + \frac{\partial B}{\partial r}\mu(r, T) + \frac{1}{2}\frac{\partial^2 B}{\partial r^2}\sigma^2(r, T)\right]dt + \frac{\partial B}{\partial r}\sigma(r, T)\,dw$$

and therefore, by equating the two equations, the bond and the term structure interest-rate model, we have:

$$\alpha(r, t, T)B = \left[\frac{\partial B}{\partial t} + \frac{\partial B}{\partial r}\mu(r, T) + \frac{1}{2}\frac{\partial^2 B}{\partial r^2}\sigma^2(r, T)\right];\ \beta(r, t, T)B = \frac{\partial B}{\partial r}\sigma(r, T)$$

Now assume that the risk premium is proportional to the returns standard deviation. Assuming that the price of risk is a known function of r and time t, we have thus:

$$\alpha(r, t, T) = r + \lambda(r, t)\frac{1}{B}\frac{\partial B}{\partial r}$$

which we insert in the Bond equation derived above leading to:

$$rB + \lambda(r, t)\frac{\partial B}{\partial r} = \left[\frac{\partial B}{\partial t} + \frac{\partial B}{\partial r}\mu(r, T) + \frac{1}{2}\frac{\partial^2 B}{\partial r^2}\sigma^2(r, T)\right]$$

and finally we obtain a partial differential equation whose solution provides the price of a zero-coupon bond with maturity T,

$$0 = \frac{\partial B}{\partial t} + \frac{\partial B}{\partial r}(\mu(r,T) - \lambda(r,t)) + \frac{1}{2}\frac{\partial^2 B}{\partial r^2}\sigma^2(r,T) - rB$$

$$B(r,T,T) = 1$$

A.2: Options on bonds

Let for example, the interest process:

$$\mathrm{d}r = \mu(r,t)\,\mathrm{d}t + \sigma(r,t)\,\mathrm{d}w$$

A synthetic portfolio (n_S, n_T) of two S and T bonds has a value and a rate of return given by:

$$V = n_S B(t,S) + n_T B(t,T) \quad \text{and} \quad \frac{\mathrm{d}V}{V} = n_S \frac{\mathrm{d}B(t,S)}{B(t,S)} + n_T \frac{\mathrm{d}B(t,T)}{B(t,T)}$$

The rates of return on the T and S bonds are (as shown previously):

$$\frac{\mathrm{d}B(t,T)}{B(t,T)} = \alpha_T(r,t)\,\mathrm{d}t + \beta_T(r,t)\,\mathrm{d}w$$

$$\alpha_T(r,t) = \frac{1}{B(t,T)}\left[\frac{\partial B}{\partial t} + \frac{\partial B}{\partial r}\mu(r,T) + \frac{1}{2}\frac{\partial^2 B}{\partial r^2}\sigma^2(r,T)\right];$$

$$\beta_T(r,t) = \frac{1}{B(t,T)}\frac{\partial B}{\partial r}\sigma(r,T)$$

$$\frac{\mathrm{d}B(t,S)}{B(t,S)} = \alpha_S(r,t)\,\mathrm{d}t + \beta_S(r,t)\,\mathrm{d}w$$

$$\alpha_S(r,t) = \frac{1}{B(t,S)}\left[\frac{\partial B}{\partial t} + \frac{\partial B}{\partial r}\mu(r,S) + \frac{1}{2}\frac{\partial^2 B}{\partial r^2}\sigma^2(r,S)\right];$$

$$\beta_S(r,t) = \frac{1}{B(t,S)}\frac{\partial B}{\partial r}\sigma(r,S)$$

We replace these terms in the synthetic bond portfolio leading to:

$$\frac{\mathrm{d}V}{V} = (n_S\alpha_S + n_T\alpha_T)\,\mathrm{d}t + (n_S\beta_S + n_T\beta_T)\,\mathrm{d}w$$

A risk-free portfolio has no volatility, however. If the portfolio initial value is one dollar ($V = 1$), we can then specify two equations in the two unknown portfolio parameters (n_S, n_T) which we can solve simply. Explicitly, these equations are:

$$\begin{cases} n_S\beta_S + n_T\beta_T = 0 \\ n_S + n_T = 1 \end{cases} \quad \text{and} \quad \begin{cases} n_S = \dfrac{\beta_T}{\beta_T - \beta_S} \\ n_T = -\dfrac{\beta_S}{\beta_T - \beta_S} \end{cases}$$

The risk-free (synthetic) portfolio has thus a rate of growth, called the synthetic rate $k(t)$, explicitly given by:

$$\frac{\mathrm{d}V}{V} = \left(\frac{\beta_T\alpha_S - \beta_S\alpha_T}{\beta_T - \beta_S}\right)\mathrm{d}t = k(t)\,\mathrm{d}t$$

This rate is equated to the spot rate $r(t)$, providing thereby the following equality:

$$r(t) = \frac{\beta_T\alpha_S - \beta_S\alpha_T}{\beta_T - \beta_S} \quad \text{or} \quad \lambda(t) = \frac{r(t) - \alpha_S}{\beta_S} = \frac{r(t) - \alpha_T}{\beta_T}$$

where $\lambda(t)$ denotes the price of risk per unit volatility. Each bond with maturity T and S has at its exercise time a \$1 denomination, thus the value of each of these (S and T) bonds is:

$$0 = \frac{\partial B_T}{\partial t} + \frac{\partial B_T}{\partial r}[\mu(r, T) - \lambda\beta_T] + \frac{1}{2}\frac{\partial^2 B}{\partial r^2}\beta_T^2 - rB_T, \; B(r, T) = 1$$

$$0 = \frac{\partial B_S}{\partial t} + \frac{\partial B_S}{\partial r}[\mu(r, S) - \lambda\beta_S] + \frac{1}{2}\frac{\partial^2 B}{\partial r^2}\beta_S^2 - rB_S, \; B(r, S) = 1$$

Given a solution to these two equations, we define the option value of a call on a T bond with $S < T$ and strike price K, to be: $X = \text{Max}\,[B(S, T) - K, 0]$ with $B(S, T)$ the price of the T bond at time S. The bond value $B(S, T)$ is of course found by solving for the term structure equation and equating $B(r, S, T) = B(S, T)$. To simplify matters, say that the solution (value at time t) for the T bond is given by $F(t, r, T)$, then at time S, this value is: $F(S, r, T)$ to which we equate $B(S, T)$. In other words,

$$X = \text{Max}\,[F(S, r, T) - K, 0]$$

Now, if the option price is $B(.)$, then as we have seen in the plain vanilla model in the previous chapter, the value of the bond is found by solving for $B(.)$ in the following partial differential equation:

$$0 = \frac{\partial B}{\partial t} + \mu(r, t)\frac{\partial B}{\partial r} + \sigma^2\frac{1}{2}\frac{\partial^2 B}{\partial r^2} - rB, \; B(S, r) = \text{Max}\,[F(S, r, T) - K, 0]$$

as indicated in the text.

CHAPTER 9

Incomplete Markets and Stochastic Volatility

9.1 VOLATILITY DEFINED

Volatility pricing, estimation and analysis are topics of considerable interest in finance. The value of an option, for example, depends on the volatility, which cannot be observed directly but must be estimated or guessed – the larger the volatility the larger the value of an option. Thus, trading in options requires that volatility be predicted and positions taken to profit from forthcoming high volatility and vice versa from forthcoming low volatility. In many instances, attempts are also made to manage volatility, either by using derivative-based strategies or by some other creative means, such as 'certification'. In a past issue of *The Economist* (18 August 2001, p. 56), an article on 'Fishy Math' pointed out that salmon certification may stabilize prices and thereby profit Alaska's fishermen. To do so, options were used by the MSC (the Marine Stewardship Council, a not-for-profit agency that campaigns for sustaining fishing), to value the certification of Alaska salmon, claimed to ensure a certain standard of fishery and environmental management which customers are said to value. For fishermen, a long-term benefit would be to reduce the volatility of salmon prices and thereby increase the value of their catch. The valuation of such profits was found by the MSC using Black–Scholes options. That is to say, the options prices implied by those two levels of volatility – what a reasonable person would expect to pay to hedge the price risk before and after certification – were calculated and compared, indicating a profit for fishermen, a profit sufficient to cover the cost of certification. Choosing a model of volatility is critical in the valuation of derivatives, however. In a stable economic environment it makes sense to use plain vanilla models. However, there is ample historical evidence that this may not be the case and therefore volatility, and in particular stochastic volatility, can be the cause of market incompleteness and create appreciable difficulties in pricing assets and their derivatives. The study of volatility is thus important, for both these and many other reasons. For example, the validation of fundamental financial theory presumes both the 'predictability' of future prices and interest rates, as well as other relevant

Risk and Financial Management: Mathematical and Computational Methods. C. Tapiero
 ISBN: 0-470-84908-8

time series. Financial markets and processes where the underlying uncertainty is modelled by 'random walks' are such an instance, since they can provide future predictions, albeit characterized by a known probability distribution. The random walk hypothesis further implies, as we saw earlier; independent increments, independently and identically distributed Gaussian random variables with mean zero and a linear growth of variance. Statistically independent increments imply in fact, 'a linear growth of uncertainty'. Technically, this is shown by noting that the functional relationship, implying independence, $f(t+s) = f(t) + f(s)$ implies a linear growth since it is uniquely given by the linear time function, $f(t+s) = (t+s)f(0)$.

This facet of 'linear growth of uncertainty' has been severely criticized as too simplistic, ignoring the long-term dependence of financial time series. Further, empirical evidence has shown that financial series are not always 'well-behaved' and thus, cannot be always predicted. For this reason, extensive research has been initiated seeking to explain, for example, the leptokurtic character of rates of returns distributions, the 'chaotic behaviour' of time series, underscoring the 'unpredictability of future asset prices'. These approaches characterize 'nonlinear science' approaches to finance. Practically, 'bursts' of activity, 'feedback volatility' and broadly varying behaviours by stock market agents, 'memory' etc., are contributing to processes which do not exhibit predictable price processes and therefore violate the presumptions of fundamental finance. The study of these series has motivated a number of approaches falling under a number of themes spanning: fat tails (or Pareto–Levy stable) distribution analysis characterized by infinite variance; long-term memory and dependence characterized by explosive growth of volatility; chaotic analysis; Lyapunov stability analysis; complexity analysis; fractional Brownian motion; multifractal time series analysis; R/S (range to scale) analysis etc. Extensive study has been devoted to these methods (see, for example, the review papers of Mandelbrot (1997a) and Lo (1997)).

Volatility modelling and estimation is often specialized to the second moment evolution of a price process, but it is much more. Generally, we say that a random variable, say the returns x is more volatile than a random variable y if for all $a > 0$, the cumulative density functions of the returns distributions $F_X(.)$, $F_Y(.)$ satisfies, $F_X(a) > F_Y(a)$. The mathematics of 'stochastic ordering' consisting in comparing and ordering distributions, as above, has focused financial managers' attention on such measurements using terms such as 'stochastic dominance' (or first, second and third degree), 'hazard rate dominance', convex dominance, etc. These techniques have the advantage of being utility-free, but they are not easy to apply, nor is it always possible to do so. A practical measurement of volatility is thus problematic. When the underlying distribution of a process is Normal, consisting of two parameters, the mean and the variance, it makes sense to accept the standard deviation as a measure of volatility. However, when the underlying distribution is not Normal (as with leptokurtic distributions, expressing asymmetry in the distribution), the definition of what constitutes volatility has to be dealt with carefully. An appropriate measure of volatility is thus far from being unique, albeit a process standard deviation is often used and will be used in this chapter. There are other indicators of volatility, such as the range R, the semi-variance, R/S statistics (see the last section in this chapter for a development and explanations of such statistics) etc. providing thereby more than one approach and more than one statistical measurement to express the volatility of a series.

Given the importance of volatility, a broad number of approaches and techniques have been applied to measure and model it. The simplest case is, of course, the constant (variance) volatility model implied in random walk models. When the variance changes over time (whether it is stochastic or not), models of volatility are needed that are both economically acceptable and statistically measurable. We shall provide a brief overview of these techniques in this chapter since they are currently a 'workhorse' of financial statistics.

9.2 MEMORY AND VOLATILITY

'Memory' represents quantitatively the effects of past states on the current one and how we use it to construct forecasts of future states. A temporal 'independence' is equivalent to a 'timeless' situation in which the events reached at one point in time are independent of past and future states. In this circumstance, there is no 'memory' and volatility tends to be smaller. A temporal dependence induces time correlations, however, and thereby a process variance (volatility) growth. Time and memory, in both psychological and quantitative senses, can also form the basis for distinguishing among past, present and future. Objectively, the present is now; subjectively, however, the present also consists of past and future. This idea has been stated clearly by St Augustine (*Confessions*, Book XI, xx):

> yet perchance it might be properly said, 'there be three times; a present of things past, a present of things present, and a present of things future.' For these three do exist in some sort, in the soul, but otherwise I do not see them; present of things past, memory; present of things present, sight; present of things future, expectation.

We are thus always in the present. But the present has three dimensions:

(1) The present of the past.
(2) The present of the present.
(3) The present of the future.

Technically, we construct the past with experiences and empirical observations of the (price) process as it unfolds over time; our construction of the future (prices), on the other hand, must be in terms of indeterminate and uncertain events which are our best assessment of the future (price) at a given (filtered) present time. To a large extent, 'technical analysis' in finance uses such an approach. We have different mechanisms for establishing things past and establishing things future. Our ability to relate the past and the future to each other – i.e. to make sense of temporal change – by means of a temporal 'sequentiality' is the prime reason for studying memory processes. For example, 'remembering that stock markets behave cyclically' might induce a cyclical behaviour of prices (which need not, of course, be the case). 'Remembering', i.e. recording the claims history of an insured over the last years, may be used to determine a premium payments schedule. The 'health' history of a patient might provide important clues to determining the probabilities of his survival over time as he approaches ages where a population has a tendency to be depleted. In finance, these issues are particularly relevant. Rational expectations and its risk-neutral pricing framework squarely states that 'there is no memory of the past' since all current price values are 'an estimate of future prices'. In this sense, in a rational expectations framework, the 'present is the anticipation of the future at the known risk-free rate'. By the same token, the SDF (stochastic discount factor) claims essentially the same but without specifying a deterministic kernel for discounting future states. By contrast, charting approaches in finance state that there is a memory of the past which is used through modelling based on past data to determine current prices. The financial dilemma regarding rational expectations and charting is thus reduced to a memory issue and how it affects the process of price formation. Potential approaches can be summarized by:

(1) No memory in which case the past and the future have no effect on current prices.
(2) Anticipative (rational expectations or SDF) memory in which current prices are defined in terms of a predictable 'expectation' of future prices.
(3) Long-run memory, expressing the inter-relationship of past events and current prices and therefore the omnipresent effects of the past in any present.

For example, if speculative prices exhibit dependency, then the existence of such dependency would be inconsistent with rational expectations and would thus make a strong case for technical forecasting on stock prices (contrary to the conventional assumption that prices fluctuate randomly and are thus unpredictable). In addition, the notion of market efficiency is dependent on 'market memory'. Fama (1970) defines explicitly an efficient market as one in which information is instantly reflected in the market price. This means that, provided all the past information $\mathbf{F}(t)$ at time t is used, a market is efficient if its expected price conditioned by this information equals the current price. Thus for a given time $t+T$ and price $p(t)$, we have, as seen previously: $p(t) = E[p(t+T)\,|\mathbf{F}(t)]$. As time goes by, additional information is obtained and $\mathbf{F}(t)$ grows to include more information (a new filtration) $\mathbf{F}(t+1)$ and thus $p(t+1) = E[p(t+1+T)\,|\mathbf{F}(t+1)]$. This property of markets efficiency (assuming that it exists) underlies the martingale approach to finance, as we saw earlier. Without it, markets have a measure of 'predictability' and can thus lead to some investor making arbitrage profits.

9.3 VOLATILITY, EQUILIBRIUM AND INCOMPLETE MARKETS

Volatility, and in particular stochastic volatility, is an increasingly important issue dealt with by financial managers. The *Financial Times* for example, reported in 1997 (although it could be any year):

> The New York Stock Exchange (NYSE) has been swinging far more wildly than in previous years. The events of 1997 that struck stock markets throughout Asia and subsequently in Europe and in the US are additional proof that volatility (stochastic) is becoming a determinant factor of stock values. This has an important effect on investments and investing behaviour. Some investors, for example, are 'tiptoeing' away from the stock market and sitting on cash rather than stocks. Others are lulled by the swings in stock values and as a result are becoming less sensitive to these variations (which may be a costly strategy to follow if the stock market were to decline significantly). This year for example, there were daily price drops of more than 3 %, a phenomenon which in years past would have attracted a great deal of attention and warnings. Past experience has also indicated that when the volatility increases, it may signal a downturn on the stock market (although, it has also signalled upturns on the stock market – but less often). In any case, a growth of volatility makes investors rethink their strategy and thereby, to change their portfolio holdings.

Stochastic volatility is often used as a proof that markets are incomplete (since the former implies the latter). In other words, it implies an underlying departure from conventional approaches to economics and finance that invalidates risk-neutral pricing. Incompleteness, thus, reflects our inability to explain uniquely prices' formation. Ever since the Second World War change has been plentiful, providing an opportunity to explain why volatility may have grown or changed. Some factors contributing to an appreciable change in economics and finance theories that seek to explain the behaviour of financial markets include among others:

- The demise of Bretton Woods.
- The liberalization of the financial sector worldwide.
- Globalization through the growth of multinational firms, cross-boundary capital flows etc.
- The growth of derivatives and related products that have enriched financial theories and financial markets but at the same time have allowed the use of financial products on an unprecedented scale.

Explicitly, economic theory has changed! Classical equilibrium precepts, coined by the Arrow–Debreu–Mackenzie studies have diverted attention to disequilibrium theories, information asymmetries, organization and the effects of contracts on economic behaviour. Economic and finance theories have recognized these changes that led to new approaches – both theoretical and practical and underlie to a large extent fundamental finance. The assumption of the rational expectations hypothesis that markets clear (i.e. decision and expectations are compatible both in current and derivatives markets in the present and the future), the assumption that decision makers are homogeneous, self-interested, rational and informed, with common knowledge of the market statistics came in some cases to be doubtful. As a result, the study of incompleteness and situations involving bounded rationality, information asymmetry, utility maximizing decision makers etc. have also become important elements to reckon with in devising a mechanism for the valuation and pricing of assets.

Although financial economics has greatly contributed to finance practice, both through its approach to valuation by risk-neutral pricing and in a better understanding of financial market mechanisms, there are some problems to be reckoned with. First, financial theory is based on assumptions that are not always right. In this case, we ought to develop other theories to compensate for theoretical imperfections. For these reasons, making sure that financial theory assumptions are validated is essential for making money using 'complete market models'.

9.3.1 Incomplete markets

Markets are incomplete when any random cash flow cannot be generated by some portfolio strategy. The market is then deemed 'not rich enough'. Technically, this means that the number of assets that make up a portfolio is smaller than the number of market risk sources plus one, or:

$$\textit{number of assets} \leq \textit{Number of risk sources} + 1$$

When this is not the case, we cannot replicate, for example, an option's implied cash flow and thus, are unable to value the option uniquely. For this, as well as other reasons, incompleteness, implying non-uniqueness in pricing, is particularly important. Non-uniqueness can arise for many reasons, however, including for example issues:

- Due to pricing, rationality and psychology.
- Due to information asymmetries and networking.
- Due to transaction costs.
- Due to stochastic volatility.

If markets are not complete or close to it, financial markets have problems to value assets and investments. Some cases are well studied, however (transaction costs for certain types of assets, some problems associated with stochastic volatility), where one uses additional sources of information to replicate a derived financial asset. Financial markets may be perceived as too risky, perhaps 'chaotic', and therefore profits may be too volatile; the risk premium would then be too high and investment horizons smaller, thereby reducing investments. Finally, contingent claims may have an infinite number of prices (or equivalently an infinite number of martingale measures). As a result, valuation becomes forcibly utility-based, which is 'subjective' rather than based on the market mechanism. In these circumstances, the SDF (stochastic discount factor) framework presented in Chapter 3 is particularly useful, providing an empirical approach to pricing (risk-discounting) financial assets.

Example: Sources of incompleteness

(1) Incompleteness can arise in many circumstances. Below, a few are summarized briefly:
   - Because of lack of liquidity (leading to market-makers' and bid/ask spreads – for which trading micro-models are constructed).
   - Because of excessive friction defined in terms of: taxes; indivisibility of assets; varying rates for lending and borrowing, such as no short sales and various portfolio constraints.
   - Because of transaction costs leading to 'friction' in market transactions.
   - Because of insiders trading introducing a risk originating in information asymmetries and leading thereby to assets mis-pricing.

(2) *Arbitrage*: The existence of arbitrage opportunities implies nonviable markets rendering the unique determination of contingent claim prices impossible. If there is arbitrage, there will be trade only out of equilibrium and thus the fundamental theory of finance will not be again applicable and risk-neutral pricing cannot be applied.

(3) *Network and information asymmetries*: Networks of hedge funds, communicating with each other and often coordinated explicitly and implicitly and herding into speculative activities can lead to market inefficiencies, thus contradicting a basic hypothesis in finance assuming that agents are price takers. In networks, the information exchange provides a potential for information asymmetries or at least delays in information. In this sense, the existence of networks in their broadest and weakest form may also cause market incompleteness.

(4) *Pricing and classical contract theory*: Transaction costs, informational asymmetries in the Arrow–Debreu paradigm, lead to significant amendments of classical analysis. For example, analysis of competition in the presence of moral hazard and adverse selection lead to stressing substantial differences between trade on 'contracts' and trade on contingent commodities. The profit associated with the sale of one unit of a (contingent) good depends then only on its price. Further, the profitability of the sale of one contract may also depend on the identity of the buyer. Identity matters, either because the buyer has bought other contracts (the exclusivity problem) or because profitability of the sales depends on the buyer's characteristics, also known as the screening problem. These issues relate to financial intermediation too, where special attention must be given to the effects of informational asymmetries to better understand prices and how they differ from the 'social values of commodities'.

(5) *Psychology and rationality*: The *Financial Times* has pointed out that some investment funds seek to capitalize on human frailties to make money. For example: Are financial managers human? Are they always rational, mimicking *Star Trek's* Mr Spock? Are they devoid of emotions and irrationality?

Psychological decision-making processes integrated in economic rationales have raised serious concerns regarding the rationality axioms of DM processes, as was discussed in Chapters 2 and 3. There are, of course, many challenges to reckon with in understanding human behaviour. Some of these include:

- Thought processes based on decision-making approaches focusing on the big picture versus compartmentalization.
- The effects of under- and over-confidence on decision making.
- The application of heuristics of various sorts applied to trading.

These psychological aspects underpin an important trend in finance called 'Behavioral Finance' and at the same they provide and presume important sources of incompleteness, stimulating research to bridge observed and normative economic behaviour.

9.4 PROCESS VARIANCE AND VOLATILITY

Say that a stock price has a time-variant mean and standard deviation given by (μ_t, σ_t). In other words, if we let z_t be a standard random variable then the record of the series can be written as follows : $x_t = \mu_t + \sigma_t z_t$. When the standard deviation is known, the time series can be used to estimate the mean parameter (even if it is time-variant). When the variance is not known, it is necessary to estimate it as well. Such estimation is usually difficult and requires that specific models describing the evolution of the variance be constructed. For example, if we standardize the time series, we obtain a standard normal probability random variable for the error as seen below,

$$z_t = \frac{x_t - \mu_t}{\sigma_t} \sim N(0, 1)$$

We can rewrite this model by setting $\varepsilon_t = \sigma_t z_t$ where the error has a zero mean (usually obtained by de-trending the time series). If the standard deviation is not known, then of course the error is no longer normal and therefore there are statistical problems associated with its estimation. Models of the type ARCH and GARCH seek to estimate this variance by using the residual squared deviations. There are many ways to proceed, however, from both a modelling and a statistical point of view, rendering volatility modelling a challenging task. Empirical finance research has sought to explain volatility in terms of the randomness of incoming information and trading processes. In the first instance, volatility is explained by the effects of external events which were not accounted for initially, while in the latter instance it is based on the behaviour of traders, buyers and sellers that induce greater volatility (such as herd or other systematic and unsystematic behaviours). The number of approaches and statistical techniques one may use for estimating volatility vary as well. For this reason, we shall consider some simple cases, although numerous studies, both methodological and empirical, abound. Many references related to these topics are included as well in the 'References and additional reading at the end of the chapter.

Example

Let R_{t+1}, the returns of a firm at time $t+1$, be unknown at time t and assume that mean returns forecast at time t are given by the next period expectation $\mu_t = E_t(R_{t+1})$.This means that the conditional expectation of the one-period returns 'forecast' can be calculated. Such a model assumes rational expectations since current returns are strictly an expectation of future ones. At present, hypothesize a model for the error, given by ε_t – also called the innovation. Thus, a one-period ahead return can be written by:

$$R_{t+1} = \mu_t + \varepsilon_{t+1}$$

The volatility (or the return variance) is by definition:

$$\sigma_t^2 = E_t\left(R_{t+1}^2\right) - \mu_t^2 = E_t\left(\varepsilon_{t+1}^2\right)$$

which is presumed either known or unknown, in which case it is a stochastic volatility model. A simple variance estimate can be based on statistical historical averages. That is to say, using closing daily financial prices P_t (spot on stocks for example) and in particular using daily proportional price change: $R_t = \ln P_t - \ln P_{t-1}$, we obtain (historical) estimates for the mean and the variance:

$$\mu = \frac{1}{T}\sum_{t=1}^{T} R_t;\ \sigma^2 = \frac{1}{T-1}\sum_{t=1}^{T}(R_t - \mu)^2$$

By the same token we can use the daily range (or a Hi, Lo statistic) for volatility estimation. This is justified by the fact that for identically and independently distributed (iid) large sample statistics, the range and the variance have, approximately, equivalent distributions. Then,

$$\sigma = \frac{0.627}{T-1}\sum_{t=1}^{T} \ln\left(H_t/L_t\right)$$

where (H_t, L_t) are the high and low prices of the trading day respectively.

Historical estimation can be developed further by building weighted estimation schemes, giving greater prominence to recent data compared to past data. In other words, say that a volatility estimate is given by a weighted sum of squares of past returns:

$$\hat{\sigma}_t^2 = E_t\left(R_{t+1}^2\right) = w_0 + \sum_{i=1}^{k} w_i(t) R_{t+i-1}^2$$

where $w_i(t)$ denotes the weight at time t associated to past returns. Variance models may be differentiated then by the weighting schemes we use. For the naïve historical model, we have:

$$\hat{\sigma}_t^2 = E_t\left(R_{t+1}^2\right) = \frac{1}{T}\sum_{i=1}^{T} R_{t+i-1}^2;\ w_i(t) = \frac{1}{T}$$

For an exponential smoothing of volatility forecasts, as done by Riskmetrics, we have:

$$\hat{\sigma}_t^2 = E_t\left(R_{t+1}^2\right) = \sum_{i=1}^{\infty} \theta^{i-1} R_{t+i-1}^2;\ w_i(t) = \theta^{i-1}, 0 \leq \theta^{i-1} \leq 1$$

since:

$$\hat{\sigma}_t^2 = \sum_{i=1}^{\infty} \theta^{i-1} R_{t+i-1}^2 = R_{t-1}^2 + \theta \sum_{i=1}^{\infty} \theta^{i-1} R_{t-1+i-1}^2$$

and we obtain the recursive scheme:

$$\hat{\sigma}_t^2 = R_{t-1}^2 + \theta \hat{\sigma}_{t-1}^2$$

Extensions were suggested by Engle (1987, 1995) (ARCH models) and Bollerslev (GARCH models). There are other estimation techniques such as nonparametric models that are harder to specify. In these cases, the weighting function $w(x_{t-i})$ expresses a memory based on a number of state variables. Such approaches are in general difficult to estimate. The importance of ARCH and GARCH modelling in financial statistics cannot be overestimated, however. Econometric software makes it possible to perform such statistical analyses with great ease, using general models of the variance. For further study we refer to Bollerslev (1986), Nelson and Foster (1994), Taylor (1986), and Engle and Bollerslev (1986).

Example: Stochastic volatility and process discretization

A stochastic volatility model can be obtained by discretization of a plain vanilla continuous-time model. This demonstrates that in handling theoretical models for practical ends and discretizing the model we may also introduce problems associated with stochastic volatility. Say that an asset price is given by the often-used lognormal model:

$$\frac{\mathrm{d}S}{S} = \mu\, \mathrm{d}t + \sigma\, \mathrm{d}W$$

where μ is asset rate of return and σ is its volatility. An application of Ito's differential rule to $Y = \ln S$, yields:

$$\mathrm{d}Y = \left(\mu - \frac{\sigma^2}{2}\right) \mathrm{d}t + \sigma\, \mathrm{d}W$$

A simple discretization with a time interval Δk, needed for estimation purposes, yields:

$$Y_k - Y_{k-1} = \left(\mu - \frac{\sigma^2}{2}\right) \Delta k + \left(\sigma\sqrt{\Delta k}\right) Z_k;\ Z_k \sim \mathrm{N}(0, 1);\ k = 1, 2, \ldots$$

A linear regression provides an estimate of $\left(\mu - \sigma^2/2\right)$, requiring that the volatility be presumed known and constant for the estimate to be meaningful. If the volatility is not known but it is also estimated by the data at hand, then another regression is needed, supplied potentially by the ARCH–GARCH apparatus and providing a simultaneous estimation of the model's parameters. Such estimation subsumes, however, a stochastic volatility (since the volatility is error-prone and estimated using historical values). As a result, discretization, even when it is properly done, can lead to estimation problems that imply stochastic volatility.

9.5 IMPLICIT VOLATILITY AND THE VOLATILITY SMILE

It is possible to estimate volatility for traded stocks, exchange rates and other financial instruments using the Black–Scholes (BS) equation. Note that the BS equation is given as a function of volatility and a number of other variables which are recorded easily or market-specified. As a result, we can use recorded option prices to calculate, other things being equal, the corresponding volatility. This is also called the implicit volatility. When there is no arbitrage (and the BS equation provides the option price), the implicit volatility corresponds to the actual volatility. Otherwise, there may be some opportunity for arbitrage profit. Ever since the stock market crash of 1987, it has been noted that options' implicit volatility with the same maturity are a function of the strike. This is known as the volatility smile, shown graphically in Figure 9.1. It is believed that this effect is due to some extent to agents' willingness to pay to hedge their position in case of sudden and unpredictable market reversal. Of course, such a 'smile' has a direct effect on return distributions which may no longer be normal but rather be defined by a skewed distribution.

Explicitly, say that we use the BS formula, specified in Chapter 6:

$$W = F(p, t; T, K, R_f, \sigma) = p(t)\Phi(d_1) - K\,\mathrm{e}^{-R_f(T-t)}\Phi(d_2)$$

where σ is the volatility, T is the exercise (maturity) date, K is the exercise price and R_f is the risk-free interest rate and, of course, W, is the option price of the underlying asset whose price a time t is equal to $p(t)$ with:

$$\Phi(y) = (2\pi)^{-1/2} \int_{-\infty}^{y} e^{-u^2/2}\,\mathrm{d}u,$$

$$d_1 = \left[\frac{\log(p(t)/K) + (T-t)(R_f + \sigma^2/2)}{\beta\sqrt{T-t}}\right], \quad d_2 = d_1 - \sigma\sqrt{T-t},$$

A solution for σ, leads by implicit numerical techniques to a function which is

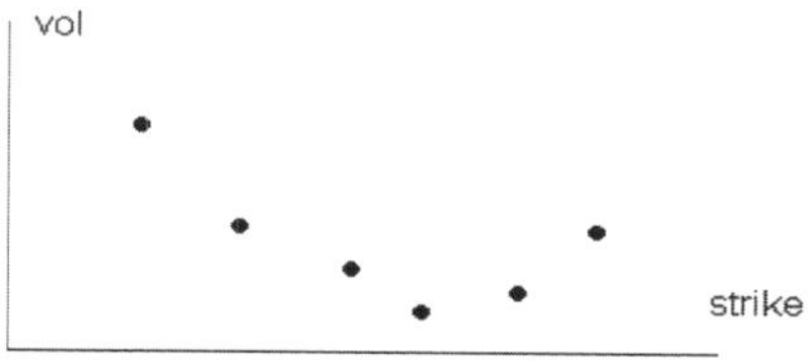

Figure 9.1

given by:

$$\sigma = \Psi(p, t; T, K, R_f, W)$$

In this manner, and using data for the option price, the volatility can be calculated. This analysis presumes, of course, that the BS option is the proper function for valuing an option on the stock exchange.

9.6 STOCHASTIC VOLATILITY MODELS

Stochastic volatility models presume that a process's volatility (variance) varies over time following some stochastic process, usually well specified. As a result, it is presumed that volatility growth increases market unpredictability, thereby rendering the application of the rational expectations hypothesis, at best, a tenuous one. Modelling volatility models might require then a broad number of approaches not falling under the 'random walk hypothesis'. Techniques such as ARCH and GARCH, we referred to, might be used to estimate empirically the volatility in such cases. Below, we consider a number of problems and issues associated with stochastic volatility in the valuation of financial assets.

Stochastic volatility introduces another 'source of risk', a volatility risk, when we model an asset's price (or returns). This leads to incompleteness and thus to non-unique asset prices. Risk-neutral pricing is no longer applicable since the probabilities calculated by the application of rational expectations (i.e. hedging to eliminate all sources of risk and using the risk-free rate as a mechanism to replicate assets) do not lead to risk-neutral valuation. For this reason, unless some other asset can be used to 'enrich' a hedging portfolio (for the volatility risk as well), we are limited to using approximations based on an economic rationale or on some other principles so that our process can be constructed (and on the basis of which risk-neutral pricing can be applied). A number of approaches can be applied including:

- time contraction,
- approximate replication,
- approximate risk--neutral pricing valuation,
- bounding.

These approaches and related ones are the subject of much ongoing research. Again, we shall consider some simple cases and, in some cases, define only a quantitative framework of the problem at hand.

9.6.1 Stochastic volatility binomial models*

Stochastic volatility has an important effect on the process underlying uncertainty, altering the basic assumption of 'normal' or 'binomial' driving disturbances. To see these effects we consider the simple binomial model we have used repeatedly

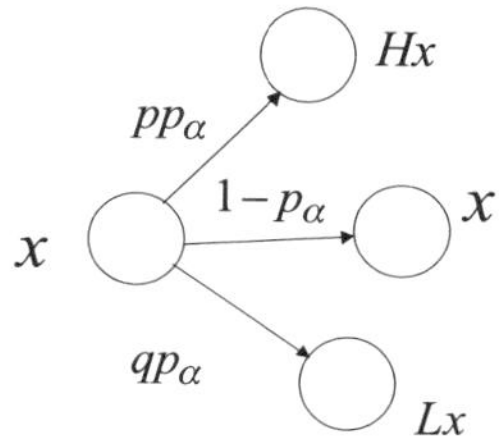

Figure 9.2

and given below:

$$\frac{x_{t+1}-x_t}{x_t}=\alpha\varepsilon_t;\ \varepsilon_t=\begin{cases}H & p\\ L & q\end{cases};$$

Here α denotes the process constant volatility. Now assume a mild stochastic volatility. Namely, we let the volatility α assume a value of 1 and zero with probabilities (p_α, q_α) leading to:

$$\frac{x_{t+1}-x_t}{x_t}=\tilde{\alpha}_t\varepsilon_t;\ \varepsilon_t=\begin{cases}H & p\\ L & q\end{cases};\ \tilde{\alpha}_t=\begin{cases}1 & p_\alpha\\ 0 & q_\alpha\end{cases}$$

In this case, the random (binomial) volatility is reduced to a trinomial model (see Figure 9.2) where p is the probability of the constant volatility model and p_α is the probability that volatility equals one and q_α is the probability that there is no volatility. For this simple case, already it is not possible to construct a perfect hedge for, say, an option as we have done earlier. This is because there are two sources of risk – one associated with the price and the other with volatility. Assuming one asset only, the number of risk sources is larger than the number of assets and therefore we have an incomplete market situation where prices need not be unique.

When volatility is constant, note that $\alpha_t\varepsilon_t$ is a random walk, but when volatility is a random variable, the process $\tilde{\alpha}_t\varepsilon_t$ is no longer a random walk. Let $z_t=\tilde{\alpha}_t\varepsilon_t$ have a density function $F_{z_t}(.)$ and assume that the random walk and the volatility are statistically independent (which is a strong assumption). Using elementary probability calculations, we have:

$$F_{z_t}(z)=\int_0^\infty F_\alpha(\alpha)F_\varepsilon\left(\frac{z}{\alpha}\right)\frac{d\alpha}{\alpha}-\int_{-\infty}^0 F_\alpha(\alpha)F_\varepsilon\left(\frac{z}{\alpha}\right)\frac{d\alpha}{\alpha}$$

For example, if the random walk $\varepsilon_t=(+1,-1)$ is biased and with probabilities $(p, \bar{p})$, while volatility assumes two values $\tilde{\alpha}=(a,b)$ with probabilities $(q,\bar{q})$, the following quadrinomial process results:

$$z_t=\begin{cases}+a & w.p. & pq\\ +b & w.p. & p\bar{q}\\ -b & w.p. & \overline{pq}\\ -a & w.p. & \bar{p}q\end{cases}$$

In other words, stochastic volatility has generated incompleteness in the form of a quadrinomial process. By enriching the potential states volatility may assume we are augmenting the 'volatility stochasticity'. Note, that it is not the size of volatility that induces incompleteness but its uncertainty. We shall see below, using a simple example, that the option of a 'mild volatility' process is larger than a larger (constant) volatility process – hence, emphasizing the effects of incompleteness (stochastic volatility) on option prices which in some cases can be more important to greater (but constant) volatility.

Time contraction

The underlying rationale of 'time contraction' is a reverse discretization. In other words, *assuming* that at the continuous-time limit, the underlying price process can be represented by a stochastic differential equation of the Ito type, it is then reasonable to assume that there is some binomial process that approximates the underlying process. Of course, there may be more than one way to do so (thus leading, potentially, to multiple prices) and therefore, this approach has to be applied carefully to secure that the limit makes economic sense as well. In this approach, a multinomial process is replaced by a binomial tree, consisting of as many stages (discretized time) as are needed to replicate the underlying model. For example, the trinomial process considered earlier can be reduced to a two-stage tree as shown in Figure 9.3, where (p_1, p_2, p_3) are assumed to be risk-neutral probabilities, appropriately selected by replication. Note that we have necessarily:

$$\begin{aligned}
&p_1 p_2 = p p_\alpha \\
&p_1 \bar{p}_2 + \bar{p}_1 p_3 = 1 - p_\alpha \\
&\bar{p}_1 \bar{p}_3 = q \bar{p}_\alpha
\end{aligned}$$

Since there are only two independent equations, we have in fact a system of three variables in two equations that can be solved in a large number of ways (for example as a function of $\bar{p}_3$). This means also that there is no unique price. In this case,

$$\begin{aligned}
&p_2 = [p p_\alpha / (p \bar{p}_\alpha / \bar{p}_3)] \\
&p_1 = 1 - (p \bar{p}_\alpha / \bar{p}_3)
\end{aligned}$$

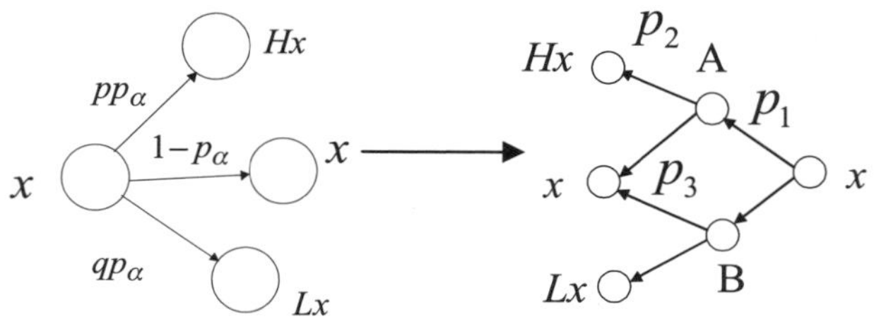

Figure 9.3

However, if we assume rational expectations, then:

$$A = \frac{1}{1+R_f}\,[p_2(1+H)x + \bar{p}_2 x] \text{ and } B = \frac{1}{1+R_f}\,[p_3 x + \bar{p}_3(1+L)x]$$

and

$$1 = \left(\frac{1}{1+R_f}\right)^2 [p_1 p_2(1+H) + p_1\bar{p}_2 + p_3\bar{p}_1 + \bar{p}_1\bar{p}_3(1+L)]$$

which provides a third equation in (p_1, p_2, p_3). Of course, for $p_\alpha = 1$ we have nonstochastic volatility and, therefore, we can calculate the approximate risk-neutral probabilities as a function of $0 < p_\alpha < 1$. For simplicity, set $p_1 = p_2 = p_3$, then we have a quadratic equation:

$$0 = \left[p^2 - 2\frac{L}{(H+L)}p - \frac{(1+R_f)^2 - 1 - L}{(H+L)}\right]$$

whose solution is given by:

$$p = \frac{L}{(H+L)} \pm \sqrt{\left(\frac{L}{H+L}\right)^2 + \frac{(1+R_f)^2 - 1 - L}{(H+L)}}$$

If we use the following parameters as an example, $1 + H = 1.4$, $1 + L = 0.8$, $R_f = 0.04$, the only feasible solution is:

$$p = \sqrt{1 + \frac{(1+0.04)^2 - 0.8}{(0.2)}} - 1 \quad \text{or} \quad p = 0.55177$$

Inserting in our equations:

$$A/x = 0.9615\,[(0.55177)(1.4) + 0.44823] = 1.1737$$
$$B/x = 0.9615\,[(0.55177) + 0.44823(0.8)] = 0.8753$$

In this particular case, the option price is given by:

$$C = \frac{1}{(1+r)^2}[p^{*2}(H-K)x] = 0.9245[0.3(0.55177)^2 x] = 0.084439x$$

We consider next the problem with mild volatility (see Figure 9.4) and set: $p' = p^{*2}/p_\alpha = 0.3044/p_\alpha$. If $p_\alpha = 0.6$, $p' = 0.50733$. If we assume no stochastic volatility but $\alpha = 1$ with probability 1 (that is a process more volatile than the previous one), then the value of an option is calculated by $p' = p^{*2}/p_\alpha = 0.3044/p_\alpha$. In addition, since $p_\alpha = 1$ we have $p' = 0.3044$. As a result, the option price is:

$$C = \frac{1}{1+R}[p'(H-K)x] = 0.9245\,[0.3044(1.4 - 1.1)x] = 0.08442x$$

compared with an option price with mild volatility given by $C = 0.084439x$.

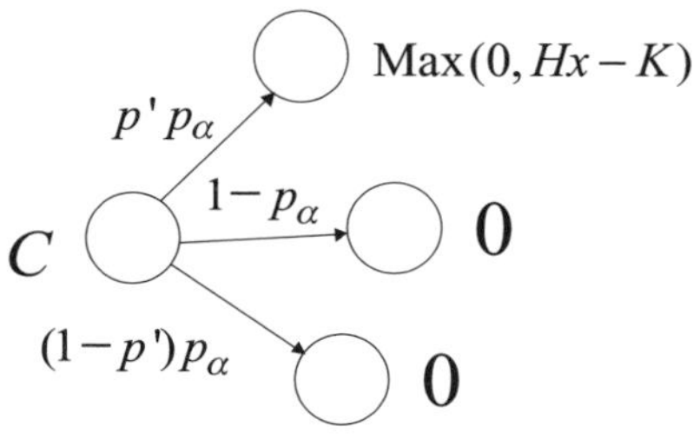

Figure 9.4

Thus, the difference due to the stochastic volatility growth is equal to $0.08442x - 0.084439x = 0.00003x$. In other words the value of an option increases both with volatility and with stochastic volatility.

We can generalize this approach further. For example, if the volatility can assume a number of potential values, say, $\tilde{\alpha} = (0, 1, 2, 3, 4, 5)$, then it is possible to reduce the ten-nomial process to a ten-stage binomial process as shown below. Mathematically, this is given by:

$$x_{t+1} = x_t + \tilde{\alpha}\varepsilon_t \quad \text{and} \quad \varepsilon_t = \begin{cases} +1 & 1/2 \\ -1 & 1/2 \end{cases}; \ \tilde{\alpha} = \begin{cases} 0 & w.p. \ p_0 \\ \cdots & \cdots \ \cdots \\ 5 & w.p. \ p_5 \end{cases}$$

An analysis similar to the previous one provides the mechanism to calculate the approximate binomial risk-neutral probability. Of course, in this model, the single stage ten-nomial process is transformed into a nine-stage binomial process (see Figure 9.5). The number of ways to do so might be very large, however. Additional information and assumptions might then be needed to reduce the number of possibilities and thereby constrain the set of prices the financial asset can assume.

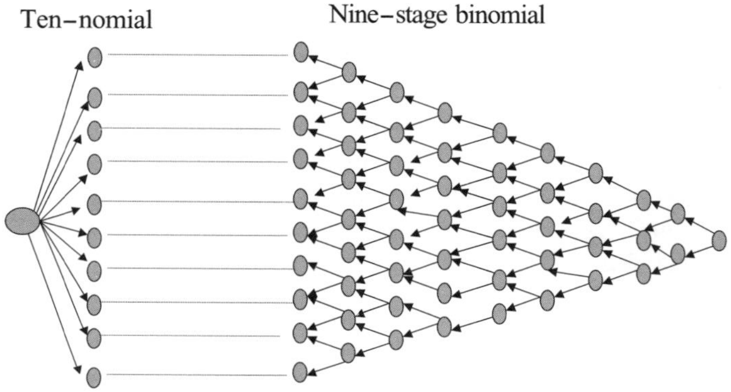

Figure 9.5

Problem

Consider the following stochastic volatility process:

$$x_{t+1} = x_t + \tilde{\alpha}\varepsilon_t; \qquad \varepsilon_t = \begin{cases} +1 & 1/2 \\ -1 & 1/2 \end{cases}; \ \tilde{\alpha} = \begin{cases} 2 & q \\ 3 & 1-q \end{cases}$$

Construct the equivalent binomial process and find the risk-neutral probabilities coherent with such a process.

$$x^{\alpha}_{t+1} = x^{\alpha}_t + z_t; \ z_t = \begin{cases} +3 & w.p.\ (0.5)(1-q) \\ +2 & w.p.\ (0.5)q \\ -2 & w.p.\ (0.5)q \\ -3 & w.p.\ (0.5)(1-q) \end{cases}$$

This approach can be extended to continuous-time models. For example, assume a mean reversion interest-rate model:

$$\mathrm{d}S = \mu(\alpha - S)\,\mathrm{d}t + V\,\mathrm{d}W \quad \text{or} \quad S(t) = \alpha + [S(0) - \alpha \pm V\varepsilon(t)]\ \mathrm{e}^{-\mu t}$$

where $\varepsilon(t)$ is a standard random walk and V is a stochastic volatility given by:

$$\mathrm{d}V = \theta(\beta - V)\,\mathrm{d}t + \kappa\,\mathrm{d}W \quad \text{or} \quad V(t) = \beta + [V(0) - \beta \pm \kappa\eta(t)]\ \mathrm{e}^{-\theta t}$$

We assume that $V \geq 0$ for simplicity. Note that the interest-rate process combines two sources of risk given by $(\eta(t), \varepsilon(t))$ and therefore:

$$\begin{aligned} S(t) = {} & S(0)\,\mathrm{e}^{-\mu t} + \alpha(1 - \mathrm{e}^{-\mu t}) \\ & \pm \left[V(0)\,\mathrm{e}^{-(\theta+\mu)t} + \beta(\mathrm{e}^{-\mu t} - \mathrm{e}^{-(\theta+\mu)t})\right]\varepsilon(t) \pm \kappa\,\mathrm{e}^{-(\theta+\mu)t}\eta(t)\varepsilon(t) \end{aligned}$$

The mean rate is therefore:

$$\hat{S}(t) = S(0)\,\mathrm{e}^{-\mu t} + \alpha(1 - \mathrm{e}^{-\mu t}) + \kappa\,\mathrm{e}^{-(\theta+\mu)t}E\left[\eta(t)\varepsilon(t)\right]$$

Since $[\eta(t), \varepsilon(t)]$ are standard random walks their covariation $E[\eta(t)\varepsilon(t)]$ is equal to 1/4, thus:

$$\hat{S}(t) = \alpha + \mathrm{e}^{-\mu t}[S(0) - \alpha + \kappa\,\mathrm{e}^{-\theta t}/4]$$

Since the resulting process is given explicitly by a quadrinomial process:

$$\begin{aligned} S(t) = {} & S(0)\,\mathrm{e}^{-\mu t} + \alpha(1 - \mathrm{e}^{-\mu t}) \pm \\ & + \mathrm{e}^{-(\mu+\theta)t}[V(0) + \beta(\mathrm{e}^{\theta t} - 1) + \kappa]\ w.p. \quad 1/4 \\ & + \mathrm{e}^{-(\mu+\theta)t}[V(0) + \beta(\mathrm{e}^{\theta t} - 1) - \kappa]\ w.p. \quad 1/4 \\ & - \mathrm{e}^{-(\mu+\theta)t}[V(0) + \beta(\mathrm{e}^{\theta t} - 1) - \kappa]\ w.p. \quad 1/4 \\ & - \mathrm{e}^{-(\mu+\theta)t}[V(0) + \beta(\mathrm{e}^{\theta t} - 1) + \kappa]\ w.p. \quad 1/4 \end{aligned}$$

The interest-rate variance can be calculated easily since:

$$\mathrm{Var}(S(t)) = \mathrm{e}^{-2(\theta+\mu)t}[V(0) + \beta(\mathrm{e}^{\theta t} - 1) - (\kappa)/4]^2 + \kappa^2\,\mathrm{e}^{-2(\mu+\theta)t}$$

$$\begin{array}{ccc} W(0) & \Leftarrow & \tilde{W}(1) \\ \Downarrow & & \Updownarrow \\ C(0) & & \tilde{C}(1) \end{array}$$

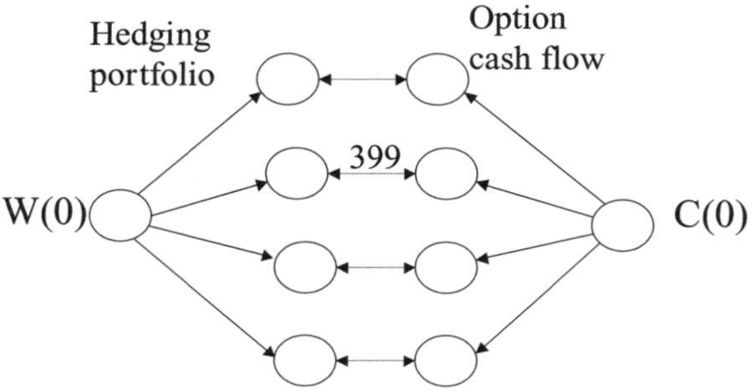

Figure 9.6

This process can be replaced by an approximate risk-neutral pricing process by time contraction, by mean–variance hedging or by applying the principle of least divergence as we shall see below.

(2) Mean variance replication hedging

This approach consists in the construction of a hedging portfolio in an incomplete (stochastic volatility) market by equating 'as much as possible' the cash flows resulting from the hedging portfolio and the option's contract. We seek to do so, while respecting the basic rules of rational expectations and risk-neutral pricing. In particular, say that we consider a two-stage model (Figure 9.6; see also Chapter 6). Thus, assuming that a portfolio and an option have the same cash flow, (i.e. $\tilde{W}(1) \equiv \tilde{C}(1)$), their current price are necessarily equal, implying that: $W(0) = C(0)$. Since, by risk-neutral pricing,

$$C(0) = \frac{1}{1+R_f} E\tilde{C}(1); \; W(0) = \frac{1}{1+R_f} E\tilde{W}(1) \text{ and } C(0) = W(0)$$

or equivalently:

$$E\tilde{W}(1) = \tilde{E}C(1) \quad \text{and further,} \quad E\tilde{W}^2(1) = \tilde{E}C^2(1)$$

These equations provide only four equations while the number of parameters might be large. However, since a hedging portfolio can involve a far greater number of parameters, it might be necessary to select an objective to minimize. A number of possibilities are available.

A simple quadratic optimization problem consisting in the minimization of the squared difference of probabilities associated to the binomial tree might be used. Alternatively, the minimization of a hedging portfolio and the option ex-post values of some option contract leads to the following:

$$\underset{p_1,\ldots,p_n}{\text{Min}} \; \Phi = E(\tilde{W}(1) - \tilde{C}(1))^2$$

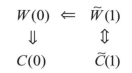

subject to: $W(0) = C(0)$ or

$$\frac{1}{1+R_f}[E\tilde{W}(1)] = \frac{1}{1+R_f}[E\tilde{C}(1)] \text{ and } E\tilde{W}^2(1) = E\tilde{C}^2(1)$$

Of course the minimization objective can be simplified further to:

$$\text{Min}\underset{p_1,\ldots,p_n}{\Phi} = E\tilde{C}^2(1) - E\tilde{W}(1)\tilde{C}(1) \quad \text{or} \quad Min\underset{p_1,\ldots,p_n}{\Phi} = \sum_{i=1}^{n} p_i(C_{1i}^2 - W_{1i}C_{1i})$$

where W_{1i}, C_{1i} are the hedging portfolio and option outcomes associated with each of the events i that occurs with probability p_i, $i = 1, 2, \ldots, n$. For example, consider the four-states model given above and assume a portfolio consisting of stocks and bonds $aS + B$. Let K be the strike price with $S_i - K > 0$, $i = 1, 2$ and $S_i - K < 0$, $i = 3, 4$ then the cash flows at time '1' for the portfolio and the option are respectively $aS_i + (1 + R_f)B$, $i = 1, 2, 3, 4$ and $(S_1 - K, S_2 - K, 0, 0)$ as shown in Figure 9.7. The following and explicit nonlinear optimization problem results:

$$\text{Min}\underset{p_1,\ldots,p_4}{\Phi} = \sum_{i=1}^{2} p_i(S_i - K)[(1-a)S_i - K - (1+R_f)B)]$$

Subject to:

$$a\sum_{i=1}^{4} p_i(S_i - S) + R_f B = \sum_{i=1}^{2} p_i(S_i - K) \quad \text{and}$$

$$S = \frac{1}{1+R_f}\sum_{i=1}^{4} p_i S_i, \sum_{i=1}^{4} p_i = 1; p_i \geq 0$$

Explicitly, say that we have the following parameters:

$$S_1 = 110,\ S_2 = 100,\ S_3 = 90,\ S_4 = 80,\ S = 90,\ K = 95,\ R_f = 0.12$$

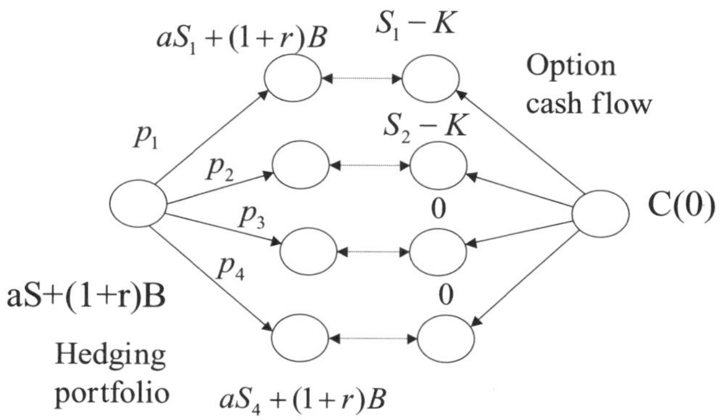

Figure 9.7

Then, our problem is reduced to a nonlinear optimization problem:

$$\underset{p_1,\ldots,p_4}{\text{Min}} \ \Phi = 3p_1(110a - 15 - 1.12B)) + p_2(100a - 5 - 1.12B))$$

Subject to:

$$4ap_1 + 2ap_2 - 2ap_4 + 0.024B = 3p_1 + p_2$$

$$1.09126p_1 + 0.99206p_2 + 0.8928p_3 + 0.79365p_4 = 1$$

$$p_1 + p_2 + p_3 + p_4 = 1; p_i \geq 0$$

The problem can of course be solved easily by standard nonlinear optimization software.

Example

Let $S_j,\ j = 1, 2, \ldots, n$ be the n states a stock can assume at the time an option can be exercised. We set, $S_0 < S_1 < S_2 < \cdots < S_n$ and define the buying and selling prices of the stock by: S^a, S^b respectively. By the same token, we define the corresponding observed call option prices C^a, C^b. Let p be the probability of a price increase. Of course, if the ex-post price is, S_n, this will correspond to the stock increasing each time period with probability given by: $P_n = \binom{n}{n} p^n (1-p)^{n-n} = p^n$. By the same token, the probability of the stock having a price S_j corresponding to the stock increasing j times and decreasing $n - j$ times is given by the binomial probability:

$$P_j = \binom{n}{j} p^j (1-p)^{n-j}$$

As a result, we have, under risk-neutral pricing:

$$S = \frac{1}{1+R_f} \sum_{j=0}^{n} P_j S_j = \frac{1}{1+R_f} \sum_{j=0}^{n} S_j \binom{n}{j} p^j (1-p)^{n-j}; S^a \leq S \leq S^b$$

and the call option price is:

$$C = \sum_{j=0}^{n} C_j = \frac{1}{(1+R_f)^n} \sum_{j=0}^{n} \binom{n}{j} p^j (1-p)^{n-j} \operatorname{Max}(S_j - K, 0)$$

subject to an appropriate constraint on the call options values, $C^a \leq C \leq C^b$. Note that S and C, as well as p, are the only unknown values so far. While the buy and sell values for stock and option, the strike time n and its price K, as well as the risk-free discount rate and future prices S_j, are given. Our problem at present is to select an objective to optimize and calculate the risk-neutral probabilities. We can do so by minimizing the quadratic distance between a portfolio of a unit of stock and a bond B. At time n, the portfolio is equal $aS_j + (1+r)^n B$ if the price is, S_j. Of course, initially, the portfolio equals:

$$S = \frac{1}{(1+R_f)^n}\left(a\sum_{j=0}^{n} P_j S_j + (1+R_f)^n B\right)$$

As a result, the least squares replicating portfolio is given by:

$$\Phi = \sum_{j=1}^{n} P_j(aS_j + (1+R_f)^n B - \mathrm{Max}(S_j - K, 0))^2$$

leading to the following optimization problem:

$$\underset{1\geq p\geq 0, C, S}{\mathrm{Min}} \ \Phi = \sum_{j=1}^{n} \binom{n}{j} p^j (1-p)^{n-j} [aS_j + (1+R_f)^n B - \mathrm{Max}(S_j - K, 0)]^2$$

Subject to:

$$S = \frac{1}{(1+R_f)^n}\sum_{j=0}^{n}\binom{n}{j} p^j(1-p)^{n-j} S_j;\ S^a \leq S \leq S^b$$

$$C = \frac{1}{(1+r)^n}\sum_{j=0}^{n}\binom{n}{j} p^j(1-p)^{n-j}\,\mathrm{Max}(S_j - K, 0);\ C^a \leq C \leq C^b$$

$$aS + B = C$$

This is again a tractable nonlinear optimization problem.

(3) Divergence and entropy

Divergence is defined in terms of directional discrimination information (Kullback, 1959) which can be measured in discrete states by two probability distributions say, (p_i, q_i), $i = 1, 2, \ldots, m$, as follows.

$$I(p,q) = \sum_{i=1}^{m} p_i \log\frac{p_i}{q_i};\ I(q,p) = \sum_{i=1}^{m} q_i \log\frac{q_i}{p_i}$$

while the divergence is:

$$J(p,q) = I(p,q) + I(q,p) = \sum_{i=1}^{m} p_i \log\frac{p_i}{q_i} + \sum_{i=1}^{m} q_i \log\frac{q_i}{p_i}$$

and finally,

$$J(p,q) = \sum_{i=1}^{m}(p_i - q_i)\log\frac{p_i}{q_i}$$

This expression measures the 'distance' between these two probability distributions. When they are the same their value is null and therefore, given a distribution 'p', a distribution 'q' can be selected by the minimization of the divergence, subject to a number of conditions imposed on both distributions 'p' and 'q' (such as expectations, second moments, risk-neutral pricing etc.). For example, one distribution may be an empirical distribution while the other may be theoretical, providing the parameters needed for the application of approximate risk-neutral pricing.

This approach can be generalized further to a bivariate state, involving time as well as states. In this case, we have for each time period:

$$I(p,q) = \sum_{t=1}^{T}\sum_{i=1}^{n} p_{it}\log\left(\frac{p_{it}}{q_{it}}\right); \quad J(p,q) = \sum_{t=1}^{T}\sum_{i=1}^{n} (p_{it} - q_{it})\log\left(\frac{p_{it}}{q_{it}}\right)$$

where the following constraints must be satisfied:

$$\sum_{i=1}^{n} p_{it} = 1; \sum_{i=1}^{n} q_{it} = 1; p_{it} \geq 0, q_{it} \geq 0$$

Moments conditions as well as other constraints may be imposed as well, providing a 'least divergence' risk-neutral pricing approximation to the empirical (incomplete) distribution.

Example

Consider the following random volatility process

$$x_{t+1}^{\alpha} = x_t^{\alpha} + z_t; \; z_t = \begin{cases} +3 & \text{w.p. } (0.5)(1-q) \\ +1 & \text{w.p. } (0.5)q \\ -1 & \text{w.p. } (0.5)q \\ -3 & \text{w.p. } (0.5)(1-q) \end{cases}$$

A three-stage standard binomial process leads to:

$$x_{t+1}^{\alpha} = x_t^{\alpha} + z_t; \; z_t = \begin{cases} +3 & \text{w.p.} & (0.5)(1-q) \leftrightarrow \pi^3 \\ +1 & \text{w.p.} & (0.5)q \leftrightarrow 3\pi^2(1-\pi) \\ -1 & \text{w.p.} & (0.5)q \leftrightarrow 3\pi(1-\pi)^2 \\ -3 & \text{w.p.} & (0.5)(1-q) \leftrightarrow (1-\pi)^3 \end{cases}$$

As a result, we can calculate the probability π by minimizing the divergence J which is given by an appropriate choice of π:

$$J = [\pi^3 - (0.5)(1-q)]\log\left(\frac{2\pi^3}{1-q}\right) + [3\pi^2(1-\pi) - (0.5)q]$$
$$\times \log\left(\frac{6\pi^2(1-\pi)}{q}\right) + [3\pi(1-\pi)^2 - (0.5)q]\log\left(\frac{6\pi(1-\pi)^2}{q}\right)$$
$$+ [(1-\pi)^3 - (0.5)(1-q)]\log\left(\frac{2(1-\pi)^3}{1-q}\right)$$

Of course other constraints may be imposed as well. Namely, if the current price is \$1, we have by risk-neutral pricing the constraint:

$$1 = \left(\frac{1}{1+R_f}\right)^2 [3\pi^3 + \pi^2(1-\pi) - \pi(1-\pi)^2 - 3(1-\pi)^3]; \; 0 \leq \pi \leq 1$$

As a result, the least divergence parameter π is found by solving the optimization problem above.

9.7 EQUILIBRIUM, SDF AND THE EULER EQUATIONS*

We have seen, throughout Chapters 5, 6 and 7, the importance of the rational expectations hypothesis as a concept of equilibrium for determining asset prices. In Chapter 3, we have also used the maximization of the expected utility of consumption to determine a rationality leading to a pricing mechanism we have called the SDF (stochastic discount factor). In other words, while in rational expectations we have an asset price determined by:

$$\text{Current price} = \frac{1}{1+R_f} E^* \{\textit{Future}\ \text{Prices}\}$$

where E^* denotes expectation with respect to a 'subjective' probability (in J. Muth, 1961 words) which we called the risk-neutral probability and R_f is the risk-free rate. In the SDF framework, we had:

$$\text{Current price} = E\left\{\textit{Kernel}^*\ \textit{Future}\ \text{Prices}\right\}$$

In this section, we extend the two-period framework used in Chapter 3 to multiple periods. To do so, we shall use Euler's equation, providing the condition for an equilibrium based on a rationality of expected utility of consumption. Let an investor maximizing the expected utility of consumption:

$$V_t = \text{Max} \sum_{j=0}^{T-1} \rho^j u(c_{t+j}) + \rho^T G(W_T)$$

where $u(c_{t+j})$ is the utility of consumption at time $t + j$, T is the final time and $G(W_T)$ is the terminal wealth state at time T. At time t, the change rate in the wealth is:

$$W_t - W_{t-1} = \Delta W_t = q_t c_t - R_t \quad \text{and therefore } c_{t+j} = \frac{\Delta W_{t+j} + R_{t+j}}{q_{t+j}}$$

We insert this last expression in the utility to be maximized:

$$V_t = \text{Max} \sum_{j=0}^{T-1} \rho^j u\left(\frac{\Delta W_{t+j} + R_{t+j}}{q_{t+j}}\right) + \rho^T G(W_T)$$

Application of Euler's equation, a necessary condition for value maximization, yields:

$$\frac{\partial V_t}{\partial W_{t+j}} - \Delta\left(\frac{\partial V_t}{\partial \Delta W_{t+j}}\right) = 0$$

Since

$$\frac{\partial V_t}{\partial W_{t+j}} = 0, \Delta\left(\frac{\partial V_t}{\partial \Delta W_{t+j}}\right) = 0$$

and therefore we have the following 'equilibrium':

$$\frac{\partial V_t}{\partial \Delta W_{t+j}} = \frac{\rho^j}{q_{t+j}}\frac{\partial u(c_{t+j})}{\partial \Delta W_{t+j}} = \text{constant or } \frac{\partial u(c_{t+j-1})}{\partial \Delta W_{t+j-1}} = E\left\{\rho \frac{q_{t+j-1}}{q_{t+j}}\frac{\partial u(c_{t+j})}{\partial \Delta W_{t+j}}\right\}$$

In other words, the marginal utility of wealth increments (savings) equals the discounted inflation-adjusted marginal utilities of consumption. If wealth is invested in a portfolio of assets such that:

$$\Delta W_t = (N_t - N_{t-1})p_t = p_t \Delta N_t \quad \text{and} \quad \frac{\partial V_t}{\partial \Delta W_{t+j}} = p_t \frac{\partial u(c_{t+j})}{\partial \Delta N_{t+j}}$$

and therefore,

$$\frac{\partial u(c_{t+j-1})}{\partial \Delta N_{t+j-1}} p_{t-1} = E\left\{\rho \frac{q_{t+j-1}}{q_{t+j}}\frac{\partial u(c_{t+j})}{\partial \Delta N_{t+j}} p_t\right\}$$

and

$$p_{t-1} = E\left\{\rho \frac{q_{t+j-1}}{q_{t+j}}\frac{u'(c_{t+j})}{u'(c_{t+j-1})} p_t\right\}; \; u'(c_{t+j}) = \frac{\partial u(c_{t+j})}{\partial \Delta N_{t+j}}$$

since at time $t-1$, the future price at time t is random, we have:

$$p_{t-1} = E\{M_t p_t\}; \; M_t = \rho \frac{q_{t+j-1}}{q_{t+j}}\frac{u'(c_{t+j})}{u'(c_{t+j-1})}$$

where M_t is the kernel, or the stochastic discount factor, expressing the 'consumption impatience'. This equation can also be written as follows:

$$1 + R_t = \frac{p_t}{p_{t-1}}; \; 1 = E\left\{M_t \frac{p_t}{p_{t-1}}\right\} \to 1 = E\{M_t(1+R_t)\,|\Phi_t\}$$

which is the standard form of the SDF equation.

Example: The risk-free rate

If p_t is a bond worth \$1 at time t, then for a risk-free discount rate:

$$\frac{1}{1+R_f} = E\{M_t\}(1) \quad \text{and therefore } E\{M_t\} = \frac{1}{1+R_f}$$

This leads to:

$$\frac{M_t}{E(M_t)} = \rho(1+R_f)\frac{q_{t+j-1}}{q_{t+j}}\frac{u'(c_{t+j})}{u'(c_{t+j-1})} \text{ and finally to } p_{t-1} = E_t^*\left(\frac{p_t}{1+R_f}\right)$$

where E_t^* is a modified (subjective) probability distribution.

Example: Risk premium and the CAPM beta
For a particular risky asset, the CAPM provides a linear discount mechanism which is:

$$M_{t+1} = a_t + b_t R_{M,t+1}$$

In other words, for a given stock, whose rate of return is $1 + R_{t+1} = p_{t+1}/p_t$, we have:

$$1 = E\{M_{t+1}(1 + R_{t+1})\} \rightarrow E(1 + R_{t+1}) = \frac{1}{E(M_{t+1})} - \frac{\text{cov}(M_{t+1}, 1 + R_{t+1})}{E(M_{t+1})}$$

and therefore, upon introducing the linear SDF, we have:

$$E(1 + R_{t+1}) = \frac{1}{E(M_{t+1})} - \frac{\text{cov}(M_{t+1}, 1 + R_{t+1})}{E(M_{t+1})}$$

After we insert the linear model for the kernel we have:

$$E(1 + R_{t+1}) = (1 + R_{f,t})[1 - \text{cov}(M_{t+1}, 1 + R_{t+1})] \quad \text{and}$$
$$E(1 + R_{t+1}) = (1 + R_{f,t+1})[1 - \text{cov}(a + bR_{M,t+1}, 1 + R_{t+1})]$$

which is reduced to:

$$E(R_{t+1} - R_{f,t+1}) = \frac{\text{cov}(R_{M,t+1} - R_{f,t+1}, R_{t+1} - R_{f,t+1})}{\text{var}(R_{M,t+1} - R_{f,t+1})} E_t(R_{M,t+1} - R_{f,t+1})$$

or to

$$E(R_{t+1} - R_{f,t+1}) = \beta E_t(R_{M,t+1} - R_{f,t+1})$$

However, the hypothesis that the kernel is linear may be limiting. Recent studies have suggested that we use a quadratic measurement of risk with a kernel given by:

$$M_{t+1} = a_t + b_t R_{M,t+1} + c_t R^2_{M,t+1}$$

In this case, the skewness of the distribution also enters into the determination of the value of the stock.

9.8 SELECTED TOPICS*

When a process has more sources of risk than assets, we are, as stated earlier, in an incomplete market situation. In such cases it is possible to proceed in two ways. Either find additional assets to use (for example, another option with different maturity and strike price) or approximate the stochastic volatility process by another risk-reduction process. There are two problems we shall consider in detail, including (stochastic) jumps and stochastic volatility continuous type models.

Problems are of three types: first, how to construct a process describing reliably the evolution of the variance; second, what are the sources of uncertainty of volatility; and, third, how to represent the stochastic relationship between the underlying process and its volatility. These equations are difficult to justify analytically and therefore we shall be satisfied with any model that practically can be used and can replicate historical statistical data. There are, however, a number of continuous-time models for stocks, returns, interest rates and other prices and their volatility that are often used. We shall consider such a model in the appendix to this chapter in detail to highlight some of the technical problems we must resolve in order to deal with such problems.

9.8.1 The Hull and White model and stochastic volatility

Hull and White (1987) have suggested a stochastic volatility model in which volatility is a geometric Brownian motion. This is written as follows:

$$\mathrm{d}S/S = \alpha\,\mathrm{d}t + \sqrt{V}\,\mathrm{d}w,\ \ S(0) = S_0;\ \mathrm{d}V/V = \mu\,\mathrm{d}t + \xi\,\mathrm{d}z,\ \ V(0) = v_0;$$
$$E(\mathrm{d}w\,\mathrm{d}z) = \rho \mathrm{d}t$$

where V is the volatility while dw and dz are two Wiener processes, with correlation ρ. A call option would in this case be a function of both S and V, or $C(t, S, V)$. Since there are two sources of risk, the hedging (replicating) portfolio must reflect this multiplicity of risks. Hull and White assume that the volatility risk is perfectly diversifiable, consequently the volatility risk premium is null. Using a Taylor series development of the option's price allows the calculation of the value of a call option as a function of small perturbations in volatility. The resulting solution turns out to be (see the Appendix for a mathematical development):

$$\frac{1}{\mathrm{d}t}E\left(\frac{\mathrm{d}C}{C}\right) = R_f + \left[(\alpha - R_f)\frac{S}{C}\frac{\partial C}{\partial S} + \frac{\lambda_V}{C}\frac{\partial C}{\partial V}\right]$$
$$= \frac{1}{\mathrm{d}t}E\left\{\frac{\partial C}{C\,\partial t}\mathrm{d}t + \frac{\partial C}{C\partial\, S}\mathrm{d}S + \frac{\partial C}{C\partial\, V}\mathrm{d}V + \frac{1}{2}\frac{\partial^2 C}{C\partial\, S^2}[\mathrm{d}S]^2 + \frac{1}{2}\frac{\partial^2 C}{C\partial\, V^2}[\mathrm{d}V]^2 + \frac{\partial^2 C}{C\,\partial\, S\partial V}[\mathrm{d}S\,\mathrm{d}V]\right\}$$

After some additional manipulations, we obtain a partial differential equation in two variables:

$$-R_f - \frac{\partial C}{C\partial\, t} = \frac{\partial C}{C\partial\, S}R_f S + \frac{\partial C}{C\partial\, V}(\mu V - \lambda_V) + \frac{1}{2}\frac{\partial^2 C}{C\partial\, S^2}S^2 V$$
$$+\frac{1}{2}\frac{\partial^2 C}{C\partial\, V^2}\xi^2 V^2 + \frac{\partial^2 C}{C\,\partial S\,\partial V}\rho\xi S V^{3/2} C(S, V, T) = \mathrm{Max}\,(S(T) - K, 0)$$

where K is the strike price, $\lambda_V = (\mu - R_f)V\beta_V$ and while β_V is the beta of the volatility. The analytical treatment of such problems is clearly difficult. In 1976, Cox introduced a model represented generally by:

$$\mathrm{d}S = \mu(S, t)\,\mathrm{d}t + V(S, t)\,\mathrm{d}W$$

Additionally a volatility state that $V(S,t) = \sigma S^{\delta}$ with δ a real number between 0 and 1. Application of Ito's Lemma, as seen in Chapter 4, leads to:

$$\mathrm{d}V = \frac{\partial V}{\partial S}\left[\mu(S,t) + \frac{1}{2}\frac{\partial^2 V}{\partial S^2}V^2(S,t)\right]\mathrm{d}t + V(S,t)\,\mathrm{d}W$$

and in this special case, we have:

$$V(S) = \sigma S^{\delta},\ \frac{\partial V}{\partial S} = \delta\sigma S^{\delta-1},\ \frac{\partial^2 V}{\partial S^2} = \delta(\delta-1)\sigma S^{\delta-2}$$

which we insert in the equation above to obtain a stochastic volatility model:

$$\mathrm{d}V/V = \delta S^{-1}\left[\mu(S,t) + \frac{1}{2}\delta(\delta-1)S^{-2}V(S,t)\right]\mathrm{d}t + \mathrm{d}W$$

A broad number of other models can be constructed. In particular, for interest rate models we saw in Chapter 8, mean reversion models. For example, Ornstein–Uhlenbeck models of stochastic volatility are used with both additive and geometric models for the volatility equation. The additive model is given by:

$$\frac{\mathrm{d}S}{S} = \mu\,\mathrm{d}t + \sqrt{V}\,\mathrm{d}W_t \quad \text{and} \quad \mathrm{d}V = \alpha(\theta - V)\,\mathrm{d}t + \xi\,\mathrm{d}W_t'$$

while the geometric model is:

$$\frac{\mathrm{d}S}{S} = \mu\,\mathrm{d}t + \sqrt{V}\,\mathrm{d}W_t \quad \text{and} \quad \frac{\mathrm{d}V}{V} = \alpha(\theta - V)\,\mathrm{d}t + \xi\,\mathrm{d}W'$$

In both cases the process is mean-reverting where θ corresponds to a volatility, a deviation from which induces a volatility movement. It can thus be interpreted as the long-run volatility. α is the mean reversion driving force while ξ is the stochastic effect on volatility. The study of these models is in general difficult, however.

9.8.2 Options and jump processes (Merton, 1976)

We shall consider next another 'incomplete' model with two sources of risk where one of the sources is a jump. We treat this model in detail to highlight as well the treatment of models with jumps. Merton considered such a problem for the following price process:

$$\frac{\mathrm{d}S}{S} = \alpha\,\mathrm{d}t + \sigma\,\mathrm{d}w + K\,\mathrm{d}Q$$

where $\mathrm{d}Q$ is an adapted Poisson process with parameter $q\,\Delta t$. In other words, $Q(t+\Delta t) - Q(t)$ has a Poisson distribution function with mean $q\,\Delta t$ or for infinitesimal time intervals:

$$\mathrm{d}Q = \begin{cases} 1 & \text{w.p. } q\,\mathrm{d}t \\ 0 & \text{w.p. } (1-q)\,\mathrm{d}t \end{cases}$$

Let $F = F(S, t)$ be the option price. When a jump occurs, the new option price is $F[S(1 + K)]$. As a result,

$$\mathrm{d}F = [F(S(1 + K)) - F]\,\mathrm{d}Q$$

When no jump occurs, we have a process evolving according to the diffusion process:

$$\mathrm{d}F = \frac{\partial F}{\partial t}\,\mathrm{d}t + \frac{\partial F}{\partial S}\,\mathrm{d}S + \frac{1}{2}\frac{\partial^2 F}{\partial S^2}\,(\mathrm{d}S)^2$$

Letting $\tau = T - t$ be the remaining time to the exercise date, we have:

$$\mathrm{d}F = \left[-\frac{\partial F}{\partial \tau} + \alpha S\frac{\partial F}{\partial p} + \frac{1}{2}S^2\sigma^2\frac{\partial^2 F}{\partial S^2}\right]\mathrm{d}t + S\sigma\frac{\partial F}{\partial S}\,\mathrm{d}w$$

Combining these two equations, we obtain:

$$\mathrm{d}F = a\,\mathrm{d}t + b\,\mathrm{d}w + c\,\mathrm{d}Q$$

$$a = \left[-\frac{\partial F}{\partial \tau} + \alpha S\frac{\partial F}{\partial S} + \frac{1}{2}S^2\sigma^2\frac{\partial^2 F}{\partial S^2}\right];\ b = S\sigma\frac{\partial F}{\partial S};\ c = F[S(1 + K)] - F$$

with

$$E(\mathrm{d}F) = [a + qc]\,\mathrm{d}t \quad \text{since} \quad E(\mathrm{d}Q) = q\,\mathrm{d}t$$

To eliminate the stochastic elements (and thereby the risks implied in the price process) in this equation, we construct a portfolio consisting of the option and a stock. To eliminate the 'Wiener risk', i.e. the effect of '$\mathrm{d}w$', we let the portfolio Z consist of a future contract whose price is S for which a proportion v of stock options is sold (which will be calculated such that this risk disappears). In this case, the value of the portfolio is:

$$\mathrm{d}Z = S\alpha\,\mathrm{d}t + S\sigma\,\mathrm{d}w + SK\,\mathrm{d}Q - [va\,\mathrm{d}t + vb\,\mathrm{d}w + vc\,\mathrm{d}Q]$$

If we set $v = S\sigma/b$ and insert in the equation above (as done by Black–Scholes), then we will eliminate the 'Wiener risk' since:

$$\mathrm{d}Z = S(\alpha - \sigma a/b)\,\mathrm{d}t + (S\sigma - vb)\,\mathrm{d}w + S(K - \sigma c/b)\,\mathrm{d}Q$$

or

$$\mathrm{d}Z = S(\alpha - \sigma a/b)\,\mathrm{d}t + S(K - \sigma c/b)\,\mathrm{d}Q$$

In this case, if there is no jump, the evolution of the portfolio follows the differential equation:

$$\mathrm{d}Z = S(\alpha - \sigma a/b)\,\mathrm{d}t$$

However, if there is a jump, then the portfolio evolution is:

$$\mathrm{d}Z = S(\alpha - \sigma a/b)\,\mathrm{d}t + S(K - \sigma c/b)\,\mathrm{d}Q$$

Since the jump probability equals, $q\,\mathrm{d}t$,we obviously have:

$$\frac{E(\mathrm{d}Z)}{\mathrm{d}t} = S(\alpha - \sigma a/b) + Sq(K - \sigma c/b)$$

There remains a risk in the portfolio due to the jump. To eliminate it we can construct another portfolio using an option F' (with exercise price E') and a future contract such that the terms in $\mathrm{d}Q$ are eliminated as well. Then, constructing a combination of the first (Z) portfolio and the second portfolio (Z'), both sources of uncertainty will be eliminated. Applying an arbitrage argument (stating that there cannot be a return to a riskless portfolio which is greater than the riskless rate of return) we obtain the proper proportions of the riskless portfolio.

Alternatively, finance theory (and in particular, application of the CAPM (capital asset pricing model) state that any risky portfolio has a rate of return in a small time interval $\mathrm{d}t$ which is equal the riskless rate plus a premium for the risk assumed. Thus, using the CAPM we can write:

$$E\frac{\mathrm{d}Z}{Z\,dt} = R_f + \lambda\frac{S(K - \sigma c/b)}{Z}$$

where λ is assumed to be a constant and expresses the 'market price' for the risk associated with a jump. This equation can be analysed further, leading to the following partial differential equation which remains to be solved (once the boundary conditions are specified):

$$-\frac{\partial F}{\partial \tau} + (\lambda - q)\left\{SK\frac{\partial F}{\partial S} - (F[S(1+K) - F)\right\} + \frac{1}{2}\frac{\partial^2 F}{\partial S^2}S^2\sigma^2 - R_f F = 0$$

with boundary condition:

$$F(T) = \mathrm{Max}\,[0, S(T) - E]$$

Of course, for an American option, it is necessary to specify the right to exercise the option prior to its final exercise date, or

$$F(t) = \mathrm{Max}[F^*(t), S(t) - E]$$

where $F^*(t)$ is the value of the option which is not exercised at time t and given by the solution of the equation above. The solution of this equation is of course much more difficult than the Black–Scholes partial differential equation. Specific cases have been solved analytically, while numerical techniques can be applied to obtain numerical solutions.

9.9 THE RANGE PROCESS AND VOLATILITY

The range process of a time series is measured by the difference between the largest and the lowest values the time series assumes within a given time interval. It provides another indication for a process volatility with some noteworthy differences between the range and the process standard deviation (or variance). Explicitly, when a series becomes more volatile, the series standard deviation estimate varies more slowly than that of the range. Thus, a growth surge in volatility might be detected more quickly using the range. By the same token, when the volatility declines, the range process will be stabilized. These properties have been used, for example, in the *R/S* (range to standard deviation statistic) applied in financial analysis to detect volatility shifts. Both the variance and the range processes are therefore two sources of information which are important. The Bloomberg, for example, provides such a statistic for financial time series, also named the Hurst exponent (Hurst, 1951) or the *R/S* index. This index is essentially a parameter that seeks to quantify the statistical bias arising from self-similarity power laws in time series. In other words, it expresses the degree of power nonlinearity in the variance growth of the series. It is defined through rescaling the range into a dimensionless factor.

Calculations for the range and the *R/S* statistic are made as follows. Samples are of fixed length N are constructed, and thus the sample range is given by:

$$R_{t,N} = \text{Max}\{y_{t,N}\} - \text{Min}\{y_{t,N}\}$$

while the sample standard deviation is calculated by:

$$S_{t,N} = \sqrt{\frac{\sum_{i=1}^{N} (y_{t,i} - \widehat{y}_{t,N})^2}{N-1}}$$

where $\widehat{y}_{t,N}$ is the sample average. A regression, $(R/S) = (\text{Const}^* N)^H$ provides an estimate of H, the Hurst exponent, or using a logarithmic transformation:

$$\text{In}\left(\frac{R_N}{S_N}\right) = a + bH;\ b = \log(\alpha N);$$

With the notation: H = Hurst exponent, R = sample's range, S = sample's standard deviation and finally α = a constant. For random (Normal) processes, the Hurst index turns out to equal 0.5. While for any values larger than 0.5 obtained in a regression, it may indicate 'long-term dependence'. Use of the Hurst index should be made carefully and critically, however. The origins of the Hurst exponent are due to Hurst who began working on the Nile River Dam project and studied the random behaviour of the dam and the influx of water from rainfall over the thousand years data have been recorded. The observation was made that if the series were random, the range would increase with the square root of time – A result confirmed by many time series as well as theoretically for normal processes. Hurst noted explicitly that most natural phenomena follow a biased random walk and thus characterized it by the parameter H expressing as well a series' dependence called by Mandelbrot the 'Joseph effect' (Joseph interpreted Pharaolc's dream as seven years of plenty followed by seven years of famine). Explicitly, a correlation C between disjoint increments of the series is given by $C = 2^{2H-1} - 1$. Thus, if $H = 0.5$, the disjoint intervals are uncorrelated. For $H > 0.5$, the series are correlated, exhibiting a memory effect as stated above (which tends to amplify patterns in time series). For $H < 0.5$, these are called 'anti-persistent' time series. Such analyses require large samples N, however, which might not be always available. For this reason, such analyses are used when series are long, such as sunspots, water levels of rivers, intra-day trading stock market ticker data etc.

An attempt to represent these series, expressing a persistent behaviour (or alternatively a nonlinear variance growth) was reached by Mandelbrot who introduced a fractional Brownian motion, denoted by $B_H(t)$ (see also Greene and Fielitz (1977, 1980) for an application in finance). A particular relationship for fractional Brownian motion which is pointed out by Mandelbrot and Van Ness (1968) is based on the self-similarity of the power law for such processes which means that the increment for a time interval s are in distribution proportional to s^H, or:

$$B_H(t+s) - B_H(t) \underset{\text{i.d.}}{\rightarrow} s^H \left[B_H(t+1) - B_H(t)\right]$$

where i.d. means in distribution. Furthermore, the increments variance is:

$$E\left[B_H(t+s) - B_H(t)\right]^2 = s^{2H} E\left[B_H(t+1) - B_H(t)\right]^2$$

which means that the variance for any time interval s is equal to s^{2H} times the variance for the unit interval. Of course, it is now obvious that for $H = 0.5$, the variance is linear (as is the case for random walks and for Brownian motion) and it is nonlinear otherwise. In this sense, assuming a relationship between the Hurst exponent (which is also a power law for the series) and the notion of long-run dependence of series (modelled by fractional Brownian motion), an estimate of the one is indicative of the other. From the finance point of view, such observations are extremely important. First and foremost, long-run dependence violates the basic assumptions made regarding price processes that are valued under the assumption of complete markets. As such, they can be conceived as statistical tests for 'fundamental' assumptions regarding the underlying process. Second, the Hurst index can be used as a 'herd effect' index applied to stocks or other time series, meaning that series volatility that have a tendency to grow, will grow faster over time if the index is greater than 0.5 and vice versa if the index is smaller than 0.5. For these reasons, the R/S index has also been associated to 'chaos', revealing series that are increasingly unpredictable.

REFERENCES AND ADDITIONAL READING

Adelman, I. (1965) Long cycles – fact or artifact ? *American Economic Review*, **60**, 440–463.

Amin, K. (1993) Jump diffusion option valuation in discrete time, *Journal of Finance*, **48**, 1833–1863.

Amin, K.I., and V.K. Ng (1993) Option valuation with systematic stochastic volatility, *Journal of Finance*, **48**, 881–909.

Andersen, T.G. (1994) Stochastic autoregressive volatility: A framework for volatility modeling, *Mathematical Finance*, **4**, 75–102.

Anis, A.A., and E.H. Lloyd (1976) The expected values of the adjusted rescaled Hurst range of independent normal summands. *Biometrika*, **63**, 111–116.

Baillie, R.T., and T. Bollerslev (1990) A multivariate generalized ARCH approach to modeling risk premia in forward foreign rate markets, *Journal of International Money and Finance*, **97**, 309–324.

Ball, C., and W. Torous (1985) On jumps in common stock prices and their impact on call option prices, *Journal of Finance*, **40**, 155–173.

Bera, A.K., and M.L. Higgens (1993) ARCH models: Properties, estimation and testing, *Journal of Economic Surveys*, **7**, 305–366.

Beran, Jan (1994) *Statistics for Long-Memory Processes*, Chapman & Hall, London.

Bhattacharya, R.N., V.K. Gupta and E. Waymire (1983) The Hurst effect under trends, *Journal of Applied Probability*, **20**, 649–662.

Biagini, F., P. Guasoni and M. Pratelli (2000) Mean–variance hedging for stochastic volatility models, *Mathematical Finance*, **10**(2), 109–123.

Blank, S.C. (1991) 'Chaos' in futures markets? A nonlinear dynamical analysis, *The Journal of Futures Markets*, **11**, 711–728.

Bollerslev, T. (1986) Generalized autoregressive conditional heteroskedasticity, *Journal of Econometrics*, **31**, 307–328.

Bollerslev, T. (1990) Modeling the coherence in short run nominal exchange rates: A multivariate generalized ARCH model, *The Review of Economics and Statistics*, **72**, 498–505.

Bollerslev, T., and R.F. Engle (1993) Common persistence in conditional variances, *Econometrica*, **61**, 167–186.

Bollerslev, T., R.Y. Chou and K. F. Kroner (1992) ARCH modeling in finance: A review of the theory and empirical evidence, *Journal of Econometrics*, **52**, 5–59.

Bollerslev, T., R.F. Engle and D. B. Nelson (1994) ARCH models, in *Handbook of Econometrics*, Vol. 4, R. F. Engle and D.McFadden (Eds), North Holland, Amsterdam.

Booth, G., F. Kaen and P. Koveos (1982) R/S analysis of foreign exchange rates under two international monetary regimes, *Journal of Monetary Economics*, **10**, 407–415.

Breeden, Douglas T. (1979) An intertemporal asset pricing model with stochastic consumption and investment opportunities, *Journal of Financial Economics*, **7**(3), 265–296.

Breeden, Douglas T., and Litzenberger, Robert H. (1978) Prices of state-contingent claims implicit in option prices, *Journal of Business*, **51**, 621–651.

Brock, W.A., and P.J. de Lima (1996) Nonlinear time series, complexity theory and finance, in G. Maddala and C. Rao (Eds), *Handbook of Statistics,* Vol. 14, *Statistical Methods in Finance*, North Holland, Amsterdam.

Brock, W.A., and M.J. P. Magill (1979) Dynamics under uncertainty, *Econometrica*, **47** 843–868.

Brock, W.A., D.A. Hsieh and D. LeBaron (1991) *Nonlinear Dynamics, Chaos and Instability: Statistical Theory and Economic Evidence*, MIT Press, Cambridge, MA.

Cao, M., and J. Wei (1999) Pricing weather derivatives: An equilibrium approach, Working Paper, Queens University, Ontario.

Campbell, J., and J. Cochrane (1999) By force of habit: A consumption-based explanation of aggregate stock market behavior, *Journal of Political Economy*, **107**, 205–251.

Cecchetti, S., P. Lam and N. Mark (1990) Mean reversion in equilibrium asset prices, *American Economic Review*, 80, 398–418.

Cecchetti, S., P. Lam and N. Mark (1990) Evaluating empirical tests of asset pricing models, *American Economic Review*, **80**(2), 48–51.

Cheung, Y.W. (1993) Long memory in foreign exchange rates, *Journal of Business Economics and Statistics*, **11**, 93–101.

Cotton, P., J.P. Fouque, G. Papanicolaou and K.R. Sircar (2000) *Stochastic Volatility Correction for Interest Rates Derivatives*, May, Stanford University.

Cox, D.R. (1991) Long range dependence, nonlinearity and time irreversibility, *Journal of Time Series Analysis*, **12**(4), 329–335.

Cvitanic, J., W. Schachermayer and H. Wang (2001) Utility maximization in incomplete markets with random endowment, *Finance Stochastics*, **5**(2), 259–272.

Davis, M. H. A. (1998) Option pricing in incomplete markets, in *Mathematics of Derivatives Securities*, M.A.H. Dempster and S. R. Pliska (Eds), Cambridge University Press, Cambridge.

Diebold, F., and G. Rudebusch (1989) Long memory and persistence in aggregate output, *Journal of Monetary Economics*, **24**, 189–209.

Diebold, F., and G. Rudebusch (1991) On the power of the Dickey–Fuller test against fractional alternatives, *Economic Letters*, **35**, 155–160.

El Karoui, N., and M. C. Quenez (1995) Dynamic programming and pricing contingent claims in incomplete markets, *SIAM Journal of Control and Optimization*, **33**, 29–66.

El Karoui, N., C. Lepage, R. Myneni, N. Roseau and R. Viswanathan (1991) The valuation of hedging and contingent claims Markovian interest rates, Working Paper, Université de Paris 6, Jussieu, France.

Engle, R.F. (1995) *ARCH Selected Reading,* Oxford University Press, Oxford.

Engle, R.F., and T. Bollerslev (1986) Modeling the persistence of conditional variances, *Econometric Reviews*, **5**, 1–50.

Feller, W. (1951) The asymptotic distribution of the range of sums of independent random variables, *Annals of Mathematical Statistics*, **22**, 427–432.

Feller, W. (1957, 1966) *An Introduction to Probability Theory and its Applications*, Vols I and II, John Wiley & Sons, Inc., New York.

Fouque, J. P., G. Papanicolaou and K. R. Sircar (2000) *Stochastic Volatility*, Cambridge University Press, Cambridge.

Frank, M., and T. Stengos (1988) Chaotic dynamics in economic time series, *Journal of Economic Surveys*, **2**, 103–133.

Fung, H.G. and W.C. Lo (1993) Memory in interest rate futures, *The Journal of Futures Markets*, **13**, 865–873.

Fung, Hung-Gay, Wai-Chung Lo, John E. Peterson (1994) Examining the dependency in intra-day stock index futures, *The Journal of Futures Markets*, **14**, 405–419.

Geman, H. (Ed.) (1998) *Insurance and Weather Derivatives: From Exotic Options to Exotic Underlying*, Risk Books, London.

Geske, R., and K. Shastri (1985) Valuation by approximation: A comparison of alternative option valuation techniques, *Journal of Financial and Quantitative Analysis*, **20**, 45–71.

Ghysels, E., A.C. Harvey and E. Renault (1996) Stochastic volatility, in C. R. Rao and G. S. Maddala (Eds), *Statistical Methods in Finance*, North-Holland, Amsterdam.

Gourieroux, C. (1997) *ARCH Models and Financial Applications*, Springer Verlag, New York.

Granger, C.W., and T. Trasvirta (1993) *Modeling Nonlinear Economic Relationships*, Oxford University Press, Oxford.

Green, M.T., and B. Fielitz (1977) Long term dependence in common stock returns, *Journal of Financial Economics*, **4**, 339–349.

Green, M.T., and B. Fielitz (1980) Long term dependence and least squares regression in investment analysis, *Management Science*, **26**(10), 1031–1038.

Harvey, A.C., E. Ruiz and N. Shephard (1994) Multivariate stochastic variance models, *Review of Economic Studies*, **61**, 247–264.

Helms, B., F. Kaen and R. Rosenman (1984) Memory in commodity futures contracts, *Journal of Futures Markets*, **4**, 559–567.

Hobson, D., and L. Rogers (1998) Complete models with stochastic volatility, *Mathematical Finance*, **8**, 27–48.

Hsieh, D.A. (1989) Testing for nonlinear dependence in daily foreign exchange rates, *Journal of Business*, **62**, 339–368.

Hsieh, D.A. (1991) Chaos and nonlinear dynamics application to financial markets, *Journal of Finance*, **46**, 1839–77.

Hull, J., and A. White (1987) The pricing of options on assets with stochastic volatilities, *Journal of Finance*, **42**, 281–300.

Hurst, H.E. (1951) Long-term storage capacity of reservoirs, *Transactions of the American Society of Civil Engineers*, 770–808.

Imhoff, J. P. (1985) On the range of Brownian motion and its inverse process, *Annals of Probability*, **13**(3), 1011–1017.

Imhoff, J. P. (1992) A construction of the Brownian motion path from BES (3) pieces, *Stochastic Processes and Applications*, **43**, 345–353.

Kullback, S. (1959) Information Theory and Statistics, Wiley, New York.

LeBaron, B. (1994) Chaos and nonlinear forecastability in economics and finance, *Philosophical Transactions of the Royal Society of London, A*, **348**, 397–404.

Liu, T., C.W.J. Granger and W. P. Heller (1992) Using the correlation exponent to decide whether an economic series is chaotic, *Journal of Applied Econometrics*, **7**, Supplement, S23–S39.

Lo, Andrew W. (1992) Long term memory in stock market prices, *Econometrica*, **59**(5), 1279–1313.

Lo, Andrew W. (1997) Fat tails, long memory and the stock market since 1960's, *Economic Notes*, **26**, 213–245.

Mandelbrot, B. (1971) When can price be arbitraged efficiently? A limit to the the validity of the random walk and martingale models, *Review of Economics and Statistics*, **53**, 225–236.

Mandelbrot, B. (1972) Statistical methodology for non-periodic cycles: From the covariance to R/S analysis, *Annals of Economic and Social Measurement*, **1**, 259–290.

Mandelbrot, B.B. (1971) Analysis of long run dependence in economics: The R/S technique, *Econometrica*, **39**, 68–69.

Mandelbrot, B. (1997a) Three fractal models in finance: Discontinuity, concentration, risk, *Economic Notes*, **26**, 171–212.

Mandelbrot, B.B. (1997b) *Fractals and Scaling in Finance: Discontinuity, Concentration, Risk*, Springer Verlag, New York.

Mandelbrot, B., and J. W. Van Ness (1968) Fractional Brownian motions, fractional noises and applications, *SIAM Review*, **10** 422–437.

Mandelbrot, B., and M. Taqqu (1979) Robust R/S analysis of long run serial correlation, *Bulletin of the International Statistical Institute*, **48**, Book 2, 59–104.

Merton, R. (1976) Option pricing when underlying stock returns are discontinuous, *Journal of Financial Economics*, **3**, 125–144.

Naik, V., and M. Lee (1990) General equilibrium pricing of options on the market portfolio with discontinuous returns, *Review of Financial Studies*, **3**, 493–521.

Nelson, Charles, and Charles Plosser (1982) Trends and random walks in macroeconomic time series: Some evidence and implications, *Journal of Monetary Economics*, **10**, 139–162.

Nelson, Daniel B., and D. P. Foster (1994) Asymptotic filtering theory for univariate ARCH model, *Econometrica*, **62**, 1–41.

Otway, T.H. (1995) Records of the Florentine proveditori degli cambiatori: An example of an antipersistent time series in economics, *Chaos, Solitons and Fractals*, **5**, 103–107.

Peter, Edgar E. (1995) *Chaos and Order in Capital Markets*, John Wiley & Sons, Inc., New York.

Renault, E. (1996) Econometric models of option pricing errors, in D.M. Kreps and K.F. Wallis (Eds), *Advances in Economics and Econometrics: Theory and Applications*, Cambridge University Press, Cambridge.

Sandmann, K., and D. Sondermann (1993) A term structure model and the pricing of interest rates derivatives, *The Review of Futures Markets*, **12**(2), 391–423.

Scheinkman, J.A. (1994) Nonlinear dynamics in economics and finance, *Philosophical Transactions of the Royal Society of London*, **346**, 235–250.

Scheinkman, J.A., and B. LeBaron (1989) Nonlinear dynamics and stock returns, *Journal of Business*, **62**, 311–337.

Schlogl, Erik, and D. Sommer (1994) On short rate processes and their implications for term structure movements, Discussion Paper B-293, University of Bonn, Department of Statistics, Bonn.

Siebenaler, Yves (1997) Etude de l'amplitude pour certains processus de Markov, Thesis, Department of Mathematics, University of Nancy I, France (under supervision of Professor Pierre Vallois).

Tapiero, C.S., and P. Vallois (1997) Range reliability in random walks, *Mathematical Methods of Operations Research*, **45**, 325–345.

Taqqu, M. S. (1986) A bibliographical guide to self similar processes and long range dependence, in *Dependence in Probability and Statistics*, E. Eberlein and M.S. Taqqu (Eds), Birkhäuser, Boston, pp. 137–165.

Vallois, P. (1995) On the range process of a Bernoulli random walk, in *Proceedings of the Sixth International Symposium on Applied Stochastic Models and Data Analysis*, Vol. II., J. Janssen and C.H. Skiadas (Eds), World Scientific, Singapore, pp. 1020–1031.

Vallois, P. (1996) The range of a simple random walk on Z, *Advances in Appllied Probability*, **28**, 1014–1033.

Vallois, P., and C. S. Tapiero (1996) The range process in random walks: Theoretical results and applications, in *Advances in Computational Economics*, H. Ammans, B. Rustem and A. Whinston (Eds), Kluwer, Dordrecht.

Vallois, P., and C.S. Tapiero (1996) Run length statistics and the Hurst exponent in random and birth–death random walks, *Chaos, Solutions and Fractals*, September. **7**(9), 1333–1341.

Vallois, P., and C.S. Tapiero (2001) The inter-event range process in birth–death random walks, *Applied Stochastic Models in Business and Industry*.

Wiggins, J. (1987) Option values under stochastic volatility: theory and empirical estimates, *Journal of Financial Economics*, **5**, 351–372.

APPENDIX: DEVELOPMENT FOR THE HULL AND WHITE MODEL (1987)*

Consider the stochastic volatility model in which volatility is a geometric Brownian motion. This is written as follows:

$$
\begin{aligned}
\mathrm{d}S/S &= \mu\,\mathrm{d}t + \sqrt{V}\,\mathrm{d}w,\ S(0) = S_0 \\
\mathrm{d}V/V &= \alpha\,\mathrm{d}t + \beta\sqrt{V}\,\mathrm{d}z,\ V(0) = V_0
\end{aligned}
$$

where (w, z) are two Brownian motions with correlation ρ. A call option price $C(t, S, V)$ would in this case be a function of time, S and V. Since there are two sources of risk, the hedging (replicating) portfolio must reflect this multiplicity of risks. Use Ito's Lemma and obtain:

$$
\begin{aligned}
\mathrm{d}C = {} & \frac{\partial C}{\partial t}\,\mathrm{d}t + \frac{\partial C}{\partial S}\,\mathrm{d}S + \frac{\partial C}{\partial V}\,\mathrm{d}V + \frac{1}{2}\frac{\partial^2 C}{\partial S^2}\,[\mathrm{d}S]^2 + \frac{1}{2}\frac{\partial^2 C}{\partial V^2}\,[\mathrm{d}V]^2 \\
& + \frac{\partial^2 C}{\partial S\,\partial V}\,[\mathrm{d}S\,\mathrm{d}V]
\end{aligned}
$$

It is a simple exercise to show that we have:

$$\mathrm{d}C = \begin{bmatrix} \dfrac{\partial C}{\partial t} + \dfrac{\partial C}{\partial S}\alpha S + \dfrac{\partial C}{\partial V}\mu V \\ + \dfrac{1}{2}\dfrac{\partial^2 C}{\partial S^2}[\sigma S]^2 + \dfrac{1}{2}\dfrac{\partial^2 C}{\partial V^2}[\xi V]^2 \\ + \dfrac{\partial^2 C}{\partial S\,\partial V}(\mu V\,\mathrm{d}t + \rho\xi\sigma SV) \end{bmatrix} \mathrm{d}t + \frac{\partial C}{\partial S}[\sigma S\,\mathrm{d}W] + \frac{\partial C}{\partial V}[\xi V\,\mathrm{d}Z]$$

The first term in the brackets is a deterministic component while the remaining ones are stochastic terms that we seek to reduce by hedging. For example, if we construct a replicating portfolio consisting a riskless asset and a risky one, we have a portfolio whose value is $X = nS + (X - nS)$ where the investment in a bond is $B = (X - nS)$. Further, we equate the replicating portfolio X and its differential $\mathrm{d}X$ with the option value and its differential $(C, \mathrm{d}C)$, leading to:

$$\mathrm{d}X = n\,\mathrm{d}S + \mathrm{d}B = \mathrm{d}C$$

Since for the riskless bond $\mathrm{d}B = R_f B\,\mathrm{d}t$, we have also $\mathrm{d}C - n\,\mathrm{d}S = R_f B\,\mathrm{d}t$ which leads to:

$$\begin{aligned} 0 &= \mathrm{d}C - n\,\mathrm{d}S - R_f B\,\mathrm{d}t \\ &= \left[-R_f B + \left(\frac{\partial C}{\partial S} - n\right)\alpha S + \frac{\partial C}{\partial V}\mu V\,\mathrm{d}t + \frac{\partial C}{\partial t} + \frac{1}{2}\frac{\partial^2 C}{\partial S^2}[S\sigma]^2 \right. \\ &\quad \left. + \frac{1}{2}\frac{\partial^2 C}{\partial V^2}[S\xi]^2 + \frac{\partial^2 C}{\partial S\,\partial V}[\rho S V\sigma\xi]\right]\mathrm{d}t + \left(\frac{\partial C}{\partial S} - n\right)\sigma S\,\mathrm{d}W + \frac{\partial C}{\partial V}\xi V\,\mathrm{d}Z \end{aligned}$$

For a hedging portfolio we require:

$$\left(\frac{\partial C}{\partial S} - n\right)\sigma S = 0 \to n = \frac{\partial C}{\partial S} \quad \text{as well as} \quad \frac{\partial C}{\partial V}\xi V \neq 0$$

which is clearly a nonreplicating portfolio. Of course, if the volatility is constant then, $\partial C/\partial V = 0$. In this case we have a replicating portfolio, however. For this reason it is essential to seek another asset. There are a number of possibilities to consider but for our current purpose we shall select another option. The replication portfolio is then:

$$X = n_1 S + n_2 C_2 + (X - n_1 S - n_2 C_2), \; B = (X - n_1 S - n_2 C_2)$$

where n_1, n_2 are the number of stock shares and an option of different maturity. In this case, proceeding as before and replicating the option cash process by the portfolio, we have $\mathrm{d}C_1 = \mathrm{d}X$ which implies:

$$\mathrm{d}C_1 - n_1\,\mathrm{d}S - n_2\,\mathrm{d}C_2 = R_f B\,\mathrm{d}t = R_f(C_1 - n_1 S - n_2 C_2)\,\mathrm{d}t$$

and therefore,

$$(\mathrm{d}C_1 - R_f C_1\,\mathrm{d}t) - n_1(\mathrm{d}S - R_f S\,\mathrm{d}t) - n_2(\mathrm{d}C_2 - R_f C_2\,\mathrm{d}t) = 0$$

This provides the equations needed to determine a hedging portfolio. Set,

$$\begin{aligned}
\mathrm{d}\Phi_1 &= (\mathrm{d}C_1 - R_f C_1\,\mathrm{d}t) \rightarrow \Phi_1 = \mathrm{e}^{-R_f t} C_1\\
\mathrm{d}\Phi_2 &= (\mathrm{d}p - R_f S\,\mathrm{d}t) \rightarrow \Phi_2 = \mathrm{e}^{-R_f t} S\\
\mathrm{d}\Phi_3 &= (\mathrm{d}C_2 - R_f C_2\,\mathrm{d}t) \rightarrow \Phi_2 = \mathrm{e}^{-R_f t} C_2
\end{aligned}$$

and

$$\mathrm{d}\Phi_1 = n_1\,\mathrm{d}\Phi_2 + n_2\,\mathrm{d}\Phi_3 \quad \text{or} \quad \mathrm{d}\left(\frac{C_1}{e^{R_f t}}\right) = n_1\,\mathrm{d}\left(\frac{S}{e^{R_f t}}\right) + n_2\,\mathrm{d}\left(\frac{C_2}{e^{R_f t}}\right)$$

Further, if we write:

$$\begin{aligned}
\frac{\mathrm{d}C_1}{C_1} &= \mu_1\,\mathrm{d}t + \sigma_1\,\mathrm{d}W_1 \text{ and set } \lambda_1 = \frac{(\mu_1 - R_f)}{\sigma_1} \text{ or } \frac{\mathrm{d}C_1}{C_1} - R_f\,\mathrm{d}t\\
&= \sigma_1\,(\lambda_1\,\mathrm{d}t + \mathrm{d}W_1) = \sigma_1\,\mathrm{d}W_1^*
\end{aligned}$$

With $(\lambda_1\,\mathrm{d}t + \mathrm{d}W_1) = \mathrm{d}W_1^*$, the risk-neutral measure. Applying a CAPM risk valuation, we have:

$$\frac{1}{\mathrm{d}t} E\left(\frac{\mathrm{d}C_1}{C_1} - R_f\right) = \sigma_1\lambda_1 = [(R_p - R_f)\beta_{cp} + (R_V - R_f)\beta_{cV}]$$

where (R_p) is the stock mean return,

$$\beta_{cp} = \frac{p}{C_1}\frac{\partial C_1}{\partial p}\beta_p$$

is the stock beta, (R_V) is the volatility drift, while

$$\beta_{cV} = \frac{V}{C_1}\frac{\partial C_1}{\partial V}\beta_V$$

is the beta due to volatility. We therefore obtain the following equations:

$$\begin{aligned}
&\frac{1}{\mathrm{d}t} E\left(\frac{\mathrm{d}C_1}{C_1}\right) = R_f + \left[(\alpha - R_f)\,\frac{S}{C_1}\frac{\partial C_1}{\partial S}\beta_p + (\mu - S)\,\beta_V \frac{V}{C_1}\frac{\partial C_1}{\partial V}\right]\\
&R_p = \alpha;\ R_V = \mu,\ \lambda_V = (\mu - R_f)V\beta_V
\end{aligned}$$

where λ_V is the risk premium associated to the volatility. Thus,

$$\frac{1}{\mathrm{d}t} E\left(\frac{\mathrm{d}C_1}{C_1}\right) = R_f + \left[(\alpha - R_f)\frac{S}{C_1}\frac{\partial C_1}{\partial S} + \frac{\lambda_V}{C_1}\frac{\partial C_1}{\partial V}\right]$$

which we equate to the option to value. Since

$$\frac{1}{\mathrm{d}t} E\left(\frac{dC}{C}\right) = R_f + \left[(\alpha - R_f)\frac{S}{C}\frac{\partial C}{\partial S} + \frac{\lambda_V}{C}\frac{\partial C}{\partial V}\right]$$

we obtain at last:

$$\frac{1}{\mathrm{d}t}E\left(\frac{\mathrm{d}C}{C}\right) = \frac{1}{\mathrm{d}t}E\left\{\frac{\partial C}{C\partial t}\,\mathrm{d}t + \frac{\partial C}{C\partial S}\,\mathrm{d}S + \frac{\partial C}{C\partial V}\,\mathrm{d}V + \frac{1}{2}\frac{\partial^2 C}{C\partial S^2}\,[\mathrm{d}S]^2 \right.$$
$$\left. + \frac{1}{2}\frac{\partial^2 C}{C\partial V^2}\,[\mathrm{d}V]^2 + \frac{\partial^2 C}{C\,\partial S\,\partial V}\,[\mathrm{d}S\,\mathrm{d}V]\right\}$$

We equate these last two expressions and replace all terms for $(\mathrm{d}S,\ \mathrm{d}V)$ leading to a partial differential equation in (t, S, V) we might be able to solve numerically. This equation is given in the text.

CHAPTER 10

Value at Risk and Risk Management

10.1 INTRODUCTION

The definition of risk and its 'translation' into a viable set of indices is of practical importance for risk management. Risk is generally defined in terms of the returns variance. This is not the only approach, however. The range, semi-variance (also called the downside risk), duration (measuring, rather, an exposure to risk as we saw in Chapter 8) and the value at risk (VaR or the quantile risk) are some additional examples. Stone (1973) presented a class of risk measures which include most of the empirical and theoretical class of risk measures. It is specified by the following:

$$R_t(W_0, k, A; f) = \left(\int_{-\infty}^{A} |W - W_0|^k \, f_t(W)\,\mathrm{d}W \right)^{1/k}$$

Note that $R_t(\hat{W}, 2, \infty; f)$ denotes the standard deviation. The parameter A can be used, however, to define semi-variance and other potential measures. This class of risk measures is closely related to Fishburn's (1977) measure, widely used in economics and finance, or

$$R_{\alpha,t}(f) = \int_{-\infty}^{t} (t - W)^{\alpha} \, f(W)\,\mathrm{d}W$$

It includes the variance and the semi-variance as special cases and α and t are two parameters used to specify the attitude to risk. Practical measurements of risk are extremely important for financial risk management. Artzner *et al.* (1997, 1999) and Cvitanic and Karatzas (1999), for example, have sought to provide a systematic and coherent definition of risk measurement, consistent with applications to financial products, and related to issues of:

Risk and Financial Management: Mathematical and Computational Methods. C. Tapiero
 ISBN: 0-470-84908-8

- diversification (through portfolio design),
- the use of derivatives for risk management,
- real options valuation,
- risk transfer and sharing (in insurance for example).

Artzner *et al.* (1997, 1999) have proposed a definition in terms of the required reserve for the wealth state W of an insurance firm (and thus appropriate for a regulator), denoted by $R(W)$. The required properties from such a 'reserve function' are as follows:

(1) Monotonicity, meaning that the riskier the wealth, the larger the required reserve.
(2) Invariance by drift, meaning that if wealth increases by a fixed quantity, then the required reserve remains the same.
(3) Homogeneity, meaning that reserves are proportional to wealth, or in mathematical terms $R_t(aW) = aR_t(W)$.
(4) Sub-additivity, and therefore $R_t(W + W^*) \leq R_t(W) + R_t(W^*)$.

In particular, properties (3) and (4) imply convexity of the reserve function while (4) implies the economic usefulness of merging portfolio holdings (i.e. diversification). These axioms can be problematic, however (Shiu, 1999). For example, homogeneity is not always reasonable. An insurance company may charge X dollars to insure a person for a million dollars. It will not charge 1000X to insure the same person for a billion dollars. Another way to put it is that an insurance company would rather insure one-tenth of each of ten ships rather than insure all of one ship.

Although VaR or 'value at risk', is a widely applied measure of risk, it does not satisfy all the properties specified by Artzner *et al.* (1997). It is essentially a quantile measure of risk expressing the expected loss resulting from potential adverse market movements with a specified probability over a period of time. The advantage of VaR is that it provides a single number which encapsulates the portfolio risk and which can be applied easily by non-technically minded financial risk managers. Its origin can be traced to the '4:15 report' of Dennis Weatherstone, chairman of JP Morgan who demanded that a one-page report be delivered to him every day summarizing the company's market exposure and providing an estimate of the potential loss over the next trading day. By the same token, CAR, or capital adequacy ratio (Jorion, 1997) is also used to compensate for market risk exposure. In this case, financial institutions set aside a certain amount of capital so that the probability that the institution will not survive adverse market conditions remains very small.

Although the concept of VaR has its origin in the investment banking sector, the recent generalization of its use by financial institutions is largely due to regulatory authorities. In April 1995, the Basle Committee announced that commercial banks could use the results given by their internal model to compute the level of regulatory capital corresponding to their market risk. The Basle Committee officially recognized VaR as sound risk-management practice as it adopted the

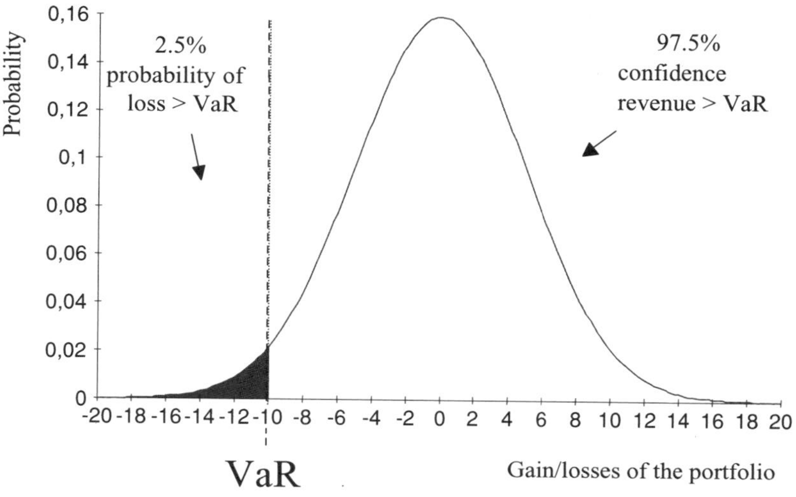

Figure 10.1 The VaR.

formula for the level of capital C given by:

$$C_{t+1} = \text{Max}\left[VaR_t, (M+m)\frac{1}{60}\sum_{j=t-60}^{t-1} VaR_j \right]$$

where M is a factor whose value is arbitrarily set to 3 to provide near-absolute insurance for bankruptcy, m is an additional factor whose value is between 0 and 1 and depends on the quality of the prediction of the internal model developed by the institution. Finally, VaR_j is the VaR calculated in the jth day. Recently, Basle capital 2 has suggested a uniform specification (.99, 10), meaning a safety factor of 99 % over 10 days.

A graphical representation of the VaR can be shown by the probability distribution of Figure 10.1, expressing the potential (and probabilistic) variations in assets holdings over a specified period of time. Practically, it is not simple to assess such a distribution, since many adverse and unpredictable events can be rare and, therefore, mostly ignored. In addition, risks may arise from several independent sources that may be difficult to specify in practice and the interaction effects of which may be even more difficult to predict. Extensive research is pursued that seeks to provide better probability estimates of the risks firms are subject to.

10.2 VaR DEFINITIONS AND APPLICATIONS

VaR defines the loss in market value of say, a portfolio, over the time horizon T that is exceeded with probability $1 - P_{VaR}$. In other words, it is the probability that returns (losses), say ξ, are smaller than $-$VaR over a period of time (horizon) T, or:

$$P_{VaR} = P(\xi < -VaR) = \int_{-\infty}^{-VaR} P_T(\xi)\,\mathrm{d}\xi$$

where $P_T(.)$ is the probability distribution of returns over the time period $(0, T)$. When the returns are normal, VaR is equivalent to using the variance as a risk measure. When risk is sensitive to rare events and extreme losses, we can fit extreme distributions (such as the Weibull and Frechet distributions) or build models based on simulation of VaR. When risks are recurrent, VaR can be estimated by using historical time series while for new situations, scenarios simulation or the construction of theoretical models are needed.

The application of VaR in practice is not without pitfalls, however. In the financial crises of 1998, a bias had made it possible for banks to lend more than they ought to, and still seem to meet VaR regulations. If market volatility increases, capital has to be put up to meet CAR requirements or shrink the volume of business to remain within regulatory requirements. As a result, reduction led to a balance sheet leverage that led to a vicious twist in credit squeeze. In addition, it can be shown, using theoretical arguments, that in nonlinear models (such long-run memory and leptokurtic distributions), the exclusive use of the variance–covariance of return processes (assuming the Normal distribution) can lead to VaR understating risk. This may be evident since protection is sought from unexpected events, while the use of the Normal distribution presumes that risk sources are stable. This has led to the use of scenario and simulation techniques in calculating the VaR. But here again, there are some problems. It is not always possible to compare the VaR calculated using simulation and historical data. The two can differ drastically and therefore, they can, in some cases, be hardly comparable. Nonstationarities of the underlying return processes are also a source of risk since they imply that return distributions can change over time. Thus, using stationary parameters in calculating the VaR can be misleading again. Finally, one should be aware that VaR is only one aspect of financial risk management and therefore overfocusing on it, may lead to other aspects of risk management being neglected. Nonetheless, VaR practice and diffusion and the convenient properties it has, justifies that we devote particular attention to it. For example, the following VaR properties are useful in financial risk management:

- Risks can be aggregated over various instruments and assets.
- VaR integrates diversification effects of portfolios, integrating a portfolio risk properties. This allows the design of VaR-efficient portfolios as we shall see later on.
- VaR provides a common language for risk management, applicable to portfolio management, trading, investment and internal risk management.
- VaR is a simple tool for the selection of strategic risk preferences by top management that can be decomposed in basic components and applied at

Table 10.1 VaR methods and parameters in a number of banks.

Bank	Technique applied	Confidence interval	Holding period
Abbey National	Variance/covariance	95 %	1 day
Bank Paribas	Monte Carlo simulations	99 %	1 day
BNP	Variance/covariance	99 %	1 day
Deutsche Bank Paris	Variance/covariance	99 %	1 day
Société Générale	Historical simulation	99 %	10 days
Basle Committee	Variance/covariance, historical simulation, Monte Carlo	99 %	10 days or 1 (days$*\sqrt{10}$)

various levels of a portfolio, a firm or organization. Thus, risk constraints at all levels of a hierarchical organization can be used coherently.

- Finally, VaR can be used as a mechanism to control the over-eagerness of some traders who may assume unwarranted risks as this was the case with Barings (see also Chapter 7).

Example: VaR in banking practice
Generally, the VaR used by banks converges to the Basle Committee specifications as shown in Table 10.1. Most banks are only beginning to apply VaR methodologies, however. At the Société Générale, for example, market risk measurement consists in measuring the potential loss due to an accident that may occur once in 10 years (with a confidence interval of 99.96 %). Correlations are ignored, leading thereby to an underestimation of this risk. This too, led to the introduction of the VaR approach including correlations, with a probability of 99 %. The first technique was maintained, however, in order to perform stress tests required by regulators. In most banks the time period covered was one day, while the Basle Committee requires 10 days. This means that VaR measures must be aggregated. If risks are not correlated over time, then aggregation is simple, summarized by their sum (and thus leading to a linear growth of variance). In this case, to move from a one day to a 10-day VaR we calculate:

$$VaR_{10\,days} = \text{VaR}_{1\text{ day}} * \sqrt{10}$$

When there is an inter-temporal correlation, or long-run memory, the calculation of the VaR is far more difficult (since the variance is essentially extremely large and the measurement of VaR not realistic). Regulatory institutions, operating at international, EU and national levels have pressured banks to standardize their measurements of risks and the application of controls, however. In particular, the CRB operating by a regulation law of 21 July 1995 defines four market risks that require VaR specification:

(1) Interest-rate risk: relating to obligations, titles, negotiable debts and related instruments.

(2) Price risk of titles and related instruments.
(3) Regulated credit risk
(4) Exchange risk.

Financial institutions have to determine market positions commensurate with Basle (VaR) regulation for these risks.

Numerical examples
(1) Say that a firm buys in a position in DM that produces a loss if the dollar appreciates (and a profit if it depreciates). We seek to determine the VaR corresponding to the maximum loss that can be sustained in 24 hours with a 5 % probability. Assume that the returns distributions are stable over time. Based on data covering the period of 2000 for the DM/USD exchange rate, a histogram can be constructed and used to calculate the probability of an appreciation larger than 5 %. The critical value for such dollar appreciation turns out to be 1.44 %; which is applied to a market value of \$1 million, and thereby leading to a VaR of \$14 400.

(2) A US investor assumes a long position of DM 140 millions. Volatility in DM/USD FX is: 0.932 while the exchange rate is 1.40 DM/USD. As a result, the VaR for the US investor is calculated by:

$$VaR_{USD} = 140 \text{ millions DM} * 0.932\,\%/1.40\text{DM/USD} = 932\,000 \text{ USD}$$

(3) We define next two positions, each with it own VaR calculation, VaR_1, VaR_2 respectively. The VaR of both positions is thus

$$VaR = \sqrt{VaR_1^2 + VaR_1^2 + 2\rho_{12} VaR_1 VaR_2}$$

where $\rho_{1,2}$ is the correlation of the positions. The US investor assumes then a long position of DM 140 million in a German bond for 10 years (thereby maintaining a long position in DM). The bond volatility is calculated to be: 0.999 % while the DM/USD volatility is: 0.932 % with correlation −0.27. The interest-rate and exchange-rate risks are thus:

$$\text{Interest-rate risk: 140 million DM} * 0.999\,\% \;/\; 1.40 = 999\,000 \text{ USD}$$
$$\text{Exchange-rate risk: 140 million DM} * 0.932\,\%/1.40 = 932\,000 \text{ USD}$$

As a result, the VaR in both positions is:

$$\begin{aligned} VaR_{USD} &= \sqrt{(999\,000)^2 + (932\,000)^2 + 2 * (-0.27) * 999\,000 * 932\,000} \\ &= 1.17 \text{ million USD} \end{aligned}$$

Non-diversifiable risks that have perfect correlation have a risk which is simply their sum. A correlation 1 would thus lead to a risk of 999 000 USD + 932 000 USD = 1.93 million USD. When risk is diversifiable the risk can be reduced with the difference benefiting the investor. In our case, this is equal to: (\$1.93 million −\$1.17 million = \$760 000).

10.3 VaR STATISTICS

VaR statistics can be calculated in a number of ways. These include the analytic variance–covariance approach, the Monte Carlo simulation and other approaches based on extreme distributions and statistical models of various sorts. These approaches are outlined next.

10.3.1 The historical VaR approach

The historical VaR generates scenarios directly from historical data. For each day, returns and associated factors are assessed and characterized probabilistically. This approach generates a number of errors, however. First, markets may be nonstationary and therefore the scenarios difficult to characterize. This has the effect of overestimating the leptokurtic effect (due to heteroscedasticity) while application of the VaR using simulation tends to underestimate these effects because it uses theoretical distributions (such as the normal, the lognormal etc.)

10.3.2 The analytic variance–covariance approach

This approach assumes the normality of returns given by a mean and a standard deviation. Thus, if X has a standard Normal distribution, $N(0, 1)$, then $P(X > 1.65) = 5\,\%$. If we consider again the preceding example, the standard deviation relative to the DM/USD exchange can be easily calculated. For 1995, it has a value of 0.7495 %. Assuming a zero mean we can calculate for a risk of 5 % the value 1.645 as seen above. As a result, we obtain a critical value of $0.7495 * 1.645 = 1.233\,\%$; and therefore a VaR of 12 330 USD.

In trading rooms, VaRs are the outcomes of multiple interacting risk factors (exchange-rate risk, interest-rate risk, stock price risks etc. that may be correlated as well). Their joint effects are, thus, assessed by the variance–covariance matrix. For example, a portfolio consisting of a long position in DM and a short in yen with a bivariate normal distribution for risk factors has a VaR that can be calculated with great ease. Let σ_1 and σ_2 be the standard deviations of two risk factors and let $\rho_{1,2}$ be their correlation, the resultant standard deviation of the position is thus:

$$\sigma_{\rm p} = 0.5\left(\sigma_1^2 + \sigma_2^2 + 2 * \rho_{1,2}\sigma_1\sigma_2\right)^{1/2}$$

This is generalized to more than two factors as well. Problems arise regarding the definition of multivariate risk distributions, however. Mostly, only marginal distributions are given while the joint multivariate distribution is not easily observable – theoretically and statistically. Recent studies have supported the use of Copula distributions. These distributions are based on the development of a multivariate distribution of risks based on the specification of their marginal distributions. Such an approach is attracting currently much interest due to the interaction effects of multiple risk sources where only marginal distributions are directly observable. There are other techniques for deriving such distributions, however. For example, the principle of maximum entropy (which we shall elaborate below) also provides a mechanism for generating distributions based on the partial information regarding these distributions.

10.3.3 VaR and extreme statistics

The usefulness of extreme statistics arises when the normal probability distribution is no longer applicable, underestimating the real risk to be dealt with, or when investors may be sensitive to extreme losses. These distributions are defined over the Max or the Min of a sample of random returns and have 'fatter' tails than say the Normal distribution. For example, say that a portfolio has a loss density function $F_R(x)$ while we seek the distribution of the largest loss of n periods:

$$Y_n = \text{Max}\{R_1, R_2, \ldots, R_n\}$$

Assuming all losses independent (which is usually a strong assumption), let $R_{(j)}$ be the jth order statistic, with $R_{(1)} < R_{(2)} < R_{(3)} < \cdots < R_{(n)}$ over n periods. Let $F_j(R)$, $j = 1, 2, \ldots, n$ denote the density function of $R_{(j)}$. Then, the distribution function of the maximum $Y_n = R_{(n)}$ is given by:

$$Y_n = R_{(n)} = \text{Max}\{R_1, R_2, \ldots, R_n\}$$

If we set a target maximum loss VaR with risk specification P_{VaR}, then we ought to construct a portfolio whose returns distribution yields:

$$P_{VaR} = P(Y_n < -\text{VaR}) = \int_{-\infty}^{-VaR} \mathrm{d}F_n(\xi)$$

In probability terms:

$$F_n(R) = P\{R_{(n)} < R\} = P\{\text{all } R_i < R\} = [F(R)]^n$$

Similarly for the minimum statistic: $Y_1 = R_{(1)} = \text{Min}\,[R_1, R_2, \ldots, R_n]$, we have,

$$\begin{aligned} F_1(R) &= P\{R_{(1)} < R\} = 1 - \Pr\{R_{(1)} > R\} = 1 - \Pr\{\text{ all } R_i > R\} \\ &= 1 - [1 - F(R)]^n \end{aligned}$$

and generally, for the jth statistic,

$$F_{(j)}(R) = \sum_{j=i}^{n} \binom{n}{i} [F(R)]^i \, [1 - F(R)]^{n-i}$$

By deriving with respect to R, we have the distribution:

$$f_j(R) = \frac{n!}{(j-1)!(n-j)!} f(R)\, [F(R)]^{j-1}\, [1 - F(R)]^{n-j}$$

If we concentrate our attention on the largest statistic, then in a sample of size n, the cumulative distribution of the Max statistic is: $F_{Y_n}(x) = [F_R(x)]^n$. At the limit, when n is large, this distribution converges to Gumbel, Weibull or a Frechet distributions, depending on the parameters of the distribution. This fact is an essential motivation for their use in VaR analyses. These distributions are given by:

Gumbel: $F_Y(y) = \exp(-\mathrm{e}^{-y}),\ y \in IR$

Weibull: $F_Y(y) = \begin{cases} \exp\left[-(-y)^{-k}\right] & \text{for } y < 0 (k < 0) \\ 0 & \text{for } y \geq 0 \end{cases}$

Frechet: $F_Y(y) = \begin{cases} 0 & \text{for } y \leq 0 \\ \exp\left(-y^{-k}\right) & \text{for } y > 0\ (k > 0) \end{cases}$

The VaR is then calculated simply since these distributions have analytical cumulative probability distributions as specified above and as the example will demonstrate.

Example: Weibull and Burr VaR

Assume a Weibull distribution for the Max loss, given by:

$$f(x/\beta) = a\beta x^{a-1}\,\mathrm{e}^{-\beta x^a},\ x \geq 0, \zeta, \mathrm{a} > 0;\ \ F(x) = 1 - \mathrm{e}^{-\beta \mathrm{x}^{\mathrm{a}}}$$

where the mean and the variance are:

$$E(x) = \beta^a \Gamma\left(\frac{1}{a} + 1\right),\ \mathrm{var}(x) = \beta^{2a}\left[\Gamma\left(\frac{2}{a} + 1\right) - \Gamma^2\left(\frac{1}{a} + 1\right)\right]$$

Say that the parameter β has a Gamma probability distribution given by: $g(\beta) = \delta\beta^{\nu-1}\,\mathrm{e}^{-\delta\beta}$, a mixture (Burr) distribution results, given by:

$$f(x) = \frac{\alpha\nu x^{\alpha-1}}{(x^\alpha + \delta)^{\delta+1}},\quad F(x \geq \Lambda) = \left[1 + \frac{\Lambda^\alpha}{\delta}\right]^{-\nu}$$

Thus, if the probability of a loss over the $\Lambda =$ VaR is P_{VaR}, we have:

$$P_{VaR} = F(x \geq VaR) = \left[1 + \frac{VaR^\alpha}{\delta}\right]^{-\nu} \text{ and therefore,}$$

$$VaR = [\delta\{[P_{VaR}]^{-1/\nu} - 1\}]^{1/\alpha}$$

The essential problem with the use of the extreme VaR approach, however, is that for cumulative risks or stochastic risk processes, these distributions are not applicable since the individual risk events are statistically dependent. In this sense, extreme distributions when they are applied to general time-varying stochastic portfolios have limited usefulness.

Example: The option VaR by Delta–Gamma

Consider an option value and assume that its underlying asset has a Normal distribution. For simplicity, say that the option value is a function of an asset price and time to maturity. A Taylor series approximation for the option price is then:

$$V_{t+h} - V_t \approx \delta(P_{t+h} - P_t) + \tfrac{1}{2}\Gamma(P_{t+h} - P_t)^2 + \theta h$$

$$\delta = \frac{\partial V_t}{\partial P_t}; \Gamma = \frac{\partial^2 V_t}{\partial P_t^2}, \theta = \frac{\partial V_t}{\partial \tau}, \tau = T - t$$

The option's rate of return is therefore:

$$R_V = \frac{V_{t+h} - V_t}{V_t}, R_p = \frac{P_{t+h} - P_t}{P_t}, \quad \text{or}$$

$$R_V \approx \delta R_p \eta + \frac{1}{2}\Gamma R_p^2 \eta P^2 + \theta h / V_t; \eta = P_t / V_t$$

Explicitly, we have then:

$$R_V = \eta\,\delta R_P + 0.5\eta\Gamma P_t R_P^2 + n\left(\frac{\theta}{V_t}\right)$$

which can be used to establish the first four moments of the rates of return R_V of the option:

	Option ROR	Underlying
	R_V	R_P
Mean:	$0.5\tilde{\Gamma}\sigma_P^2 + \tilde{\theta}n$	0
Variance:	$\tilde{\delta}^2\sigma_P^2 + 0.5\tilde{\Gamma}^2\sigma_P^4$	σ_P^2
Skewness:	$3\tilde{\delta}^2\tilde{\Gamma}\sigma_P^4 + \tilde{\Gamma}^3\sigma_P^6$	0
Kurtosis:	$12\tilde{\delta}^2\tilde{\Gamma}^2\sigma_P^6 + 3\tilde{\Gamma}^4\sigma_P^8 + 3\sigma_P^4$	$3\sigma_P^4$

In practice, after we calculate the Greek coefficients delta, gamma, theta and the variance of the underlying furnished by Risk Metrics, the four moments allow an estimation of the distribution. Such an approach can be generalized further to a portfolio of derivatives and stocks, albeit calculations might be difficult.

10.3.4 Copulae and portfolio *VaR* Measurement

In practice, a portfolio consisting of many assets may involve 'many sources of risks', which are in general difficult to assess singly and collectively. Empirical research is then carried out to specify the risk properties of the underlying portfolio. These topics are motivating extensive research efforts spanning both traditional statistical approaches and other techniques. Assume that the wealth level of a portfolio is defined as a nonlinear and differentiable function of a number of assets written as follows:

$$W_t = F(P_{1t}, P_{2t}, P_{3t}, \ldots, P_{nt}, B)$$

with $P_{1t}, P_{2t}, P_{3t}, \ldots, P_{nt}$, the current prices of these assets, while B is a riskless asset. The function F is assumed for the moment to be differentiable. Then, in a small interval of time, a first approximation yields:

$$\Delta W_t = \frac{\partial F}{\partial P_{1t}} \Delta P_{1t} + \frac{\partial F}{\partial P_{2t}} \Delta P_{2t} + \cdots + \frac{\partial F}{\partial P_{nt}} \Delta P_{nt} + \frac{\partial F}{\partial B} \Delta B$$

If we set $N_{it} = \partial F/\partial P_{it}$ to be the number of shares bought at t, we have a portfolio whose process is:

$$\Delta W_t = N_{1t}\,\Delta P_{1t} + N_{2t}\,\Delta P_{2t} + \cdots + N_{nt}\,\Delta P_{nt} + C\,\Delta B$$

where C is the investment in a riskless asset. Assume for simplification that the rate of return for each risky asset is $\alpha_{it}\,\Delta t = E(\Delta P_{it}/P_{it})$ while volatility is $\sigma_{it}^2\,\Delta t = \text{var}(\Delta P_{it}/P_{it})$. A normal approximation, $\Delta W_t \sim \mathbb{N}(\mu_t\,\Delta t, \Lambda_t\,\Delta t)$ of the portfolio with mean μ_t and variance–covariance matrix Λ_t is:

$$\mu_t = C\,\Delta B + \sum_{i=1}^{n} N_{it}\alpha_{it}P_{it};\ \Lambda_t = \frac{1}{2}\sum_{i=1}^{n}\frac{\partial^2 F}{\partial P_{it}^2}\sigma_{it}^2 + \sum_{i=j}^{n}\sum_{i=1}^{n}\frac{\partial^2 F}{\partial P_{it}\,\partial P_{jt}}[\rho_{ij}\sigma_{it}\sigma_{jt}]$$

and, the VaR can be measured simply, as we saw earlier. This might not always be possible, however. If the marginal distributions only are given (namely the mean and the variances of each return distribution) it would be difficult to specify the appropriate distribution of the multiple risk returns to adopt.

Copulae are multivariate distributions constructed by assuming that their marginal distributions are known. For example (see Embrecht *et al.*, 2001), given two sources of risks (X_1, X_2), each with their own VaR appropriately calculated, what is the VaR of the sum of the two, potentially interacting sources of risks $(X_1 + X_2)$? Intuitively, the worst case VaR for a portfolio $(X_1 + X_2)$ occurs when the linear correlation is maximal. However, it is wrong in general. The problem at hand is thus how to construct bounds for the VaR of the joint position and determine how such bounds change when some of the assets may be statistically dependent and when information is revealed, as the process unfolds over time. The Copula provides such an approach which we briefly refer to. For example, this approach can be summarized simply by a theorem of Sklar (see Embrechts *et al.*, 2001), stating that a joint distribution F has marginals (F_1, F_2) only if there is a Copula such that:

$$F(x_1, x_2) = C\,[F_1(x_1), F_2(x_2)]$$

Inversely, given the marginals (F_1, F_2), the joint distribution can be constructed by choosing an appropriate Copula C. This procedure is not simple to apply, but recent studies have contributed to the theoretical foundations of this evolving area of study. For example, if the marginal distributions are continuous, then there is a unique Copula. In addition, Copulae structures can contribute to the estimation of multiple risks VaR. Embrecht, Hoing and Juri (2001) and Embrecht, Kluppelberg and Mikosch (1997) have studied these problems in great detail and the motivated reader ought to consult these references first. Current research is ongoing in specifying such functions and their properties.

10.3.5 Multivariate risk functions and the principle of maximum entropy

When some characteristics, data or other information regarding the risk distribution are available, it is possible to define its underlying distribution by selecting that distribution which assumes the 'least', that is the distribution with the greatest variability, given the available information. One approach that allows the definition of such distributions is defined by the 'maximum entropy principle'.

Entropy (or its negative, the negentropy) can be simply defined as a measure for 'departure from randomness'. Explicitly, the larger the entropy the more the distribution departs from randomness and vice versa, the larger the negentropy, the greater a distribution's randomness. The origins of entropy arose in statistical physics. Boltzmann observed that entropy relates to 'missing information' inasmuch as it pertains to the number of alternatives which remain possible to a physical system after all the macrospically observable information concerning it has been recorded. In this sense, information can be interpreted as that which changes a system's state of randomness (or, equivalently, as that quantity which reduces entropy). For example, for a word, which has k letters, assuming zeros and ones, and one two, define a sequence of k letters, $(a_0, a_1, a_2, \ldots, a_k)$,

$$a_i = \begin{cases} 1 \\ 0 \end{cases}$$

for all $i \neq j, a_j \neq a_i = 2$ and one j. The total number of configurations (or strings of $k + 1$ letters) that can be created is N where, $N = 2^k(k + 1)$. The logarithm to the base 2 of this number of configurations is the information I, or $I = \log_2 N$ and, in our case, $I = k + \log_2(k + 1)$. The larger this number I, the larger the number of possible configurations and therefore the larger the 'randomness' of the word. As a further example, assume an alphabet of G symbols and consider messages consisting of N symbols. Say that the frequency of occurrence of a letter is f_i, i.e., in N symbols the letter G occurs on the average $N_i = f_i N$ times. There may then be W different possible messages, where:

$$W = \frac{N!}{\prod_{i=1}^{N} N_i!}$$

The uncertainty of an N symbols message is simply the ability to discern which message is about to be received. Thus, if,

$$W = e^{HN} \quad \text{and} \quad H = \lim_{N \to \infty} \frac{1}{N} \log(W) = \sum p_i \log(1/p_i)$$

which is also known as Shannon's entropy. If the number of configurations (i.e. W) is reduced, then the 'information' increases. To see how the mathematical and statistical properties of entropy may be used, we outline a number of problems.

(1) Discrimination and divergence

Consider, for example, two probability distributions given by $[F, G]$, one theoretical and the other empirical. We want to construct a 'measure' that makes it possible to discriminate between these distributions. An attempt may be reached by using the following function we call the 'discrimination information' (Kullback, 1959):

$$I(F, G) = \int F(x) \log \frac{F(x)}{G(x)} \, dx$$

In this case, $I(F, G)$ is a measure of 'distance' between the distributions F and G. The larger the measure, the more we can discriminate between these distributions. For example, if $G(.)$ is a uniform distribution, then divergence is Shannon's measure of information. In this case, it also provides a measure of departure from the random distribution. Selecting a distribution which has a maximum entropy (given a set of assumptions which are made explicit) is thus equivalent to the 'principle of insufficient reason' proposed by Laplace (see Chapter 3). In this sense, selecting a distribution with the largest entropy will imply a most conservative (riskwise) distribution. By the same token, we have:

$$I(G, F) = \int G(x) \log \frac{G(x)}{F(x)} \, dx$$

and the *divergence* between these two distributions is defined by:

$$J(F, G) = I(F, G) + I(G, F) = \int [F(x) - G(x)] \log \frac{F(x)}{G(x)} \, dx$$

which provides a 'unidirectional and symmetric measure' of distribution 'distance', since $J(F, G) = J(G, F)$. For a discrete time distribution (p, q), we have similarly:

$$I(p, q) = \sum_{i=1}^{n} p_i \log \frac{p_i}{q_i}; \; I(q, p) = \sum_{i=1}^{n} q_i \log \frac{q_i}{p_i}$$

$$J(p, q) = \sum_{i=1}^{n} p_i \log \frac{p_i}{q_i} + \sum_{i=1}^{n} q_i \log \frac{q_i}{p_i} = \sum_{i=1}^{n} (p_i - q_i) \log \frac{p_i}{q_i}$$

For example, say that $q_i, i = 1, 2, 3, \ldots, n$ is an empirical distribution and say that $p_i, i = 1, 2, 3, \ldots, n$ is a theoretical distribution given by the distribution:

$$p_i = \binom{n}{i} p^i (1 - p)^{n-i}$$

then:

$$J(p, q) = \sum_{i=1}^{n} \left[\binom{n}{i} p^i (1 - p)^{n-i} - q_i \right] \log \frac{\binom{n}{i} p^i (1 - p)^{n-i}}{q_i}$$

which may be minimized with respect to parameter p. This will be, therefore, a binomial distribution with a parameter which is 'least distant' from the empirical distribution. For a bivariate state distribution, we have similarly:

$$I(p,q)=\sum_{j=1}^{m}\sum_{i=1}^{n}p_{ij}\log\left(\frac{p_{ij}}{q_{ij}}\right);\ J(p,q)=\sum_{j=1}^{m}\sum_{i=1}^{n}(p_{ij}-q_{ij})\log\left(\frac{p_{ij}}{q_{ij}}\right)$$

while, for continuous distributions, we have also:

$$I(F,G)=\iint F(x,y)\log\frac{F(x,y)}{G(x,y)}\,\mathrm{d}x\,\mathrm{d}y$$

as well as:

$$J(F,G)=\iint [F(x,y)-G(x,y)]\log\frac{F(x,y)}{G(x,y)}\,\mathrm{d}x\,\mathrm{d}y$$

This distribution may then be used to provide 'divergence-distance' measures between empirically observed and theoretical distributions.

The maximization of divergence or entropy can be applied to constructing a multivariate risk model. For example, assume that a nonnegative random variable $\{\theta\}$ has a known mean given by $\hat{\theta}$, the maximum entropy distribution for a continuous-state distribution is given by solution of the following optimization problem:

$$\text{Max } H=-\int_0^\infty f(\theta)\log[f(\theta)]\,\mathrm{d}\theta$$

subject to:

$$\int_0^\infty f(\theta)\,d\theta=1,\hat{\theta}=\int_0^\infty \theta f(\theta)\,\mathrm{d}\theta$$

The solution of this problem, based on the calculus of variations, yields an exponential distribution. In other words,

$$f(\theta)=\frac{1}{\hat{\theta}}\mathrm{e}^{-\theta/\hat{\theta}}\ ,\theta\geq 0$$

When the variance of a distribution is specified as well, it can be shown that the resulting distribution is the Normal distribution with specified mean and specified variance. The resulting optimization problem is:

$$\text{Max } H=\int_{-\infty}^{\infty} f(\theta)\log\frac{1}{[f(\theta)]}\,\mathrm{d}\theta$$

subject to:

$$\int_{-\infty}^{\infty} f(\theta)\,\mathrm{d}\theta=1,\hat{\theta}=\int_{-\infty}^{\infty}\theta f(\theta)\,\mathrm{d}\theta,\sigma^2=\int_{-\infty}^{\infty}(\theta-\hat{\theta})^2 f(\theta)\,\mathrm{d}\theta$$

which yields (as stated above) the normal probability distribution with mean and variance $(\hat{\theta}, \sigma^2)$. If an empirical distribution is available, then, of course, we can use the divergence to minimize the distance between these distributions by the appropriate selection of the theoretical parameters. This approach can be applied equally when the probability distribution is discrete, bounded, multivariate distributions with specified marginal distributions etc. In particular, it is interesting to point out that the maximum entropy of a multivariate distribution with specified mean and known variance–covariance matrix turns out also to be a multivariate normal, implying that the Normal is the most random distribution that has a specified mean and a specified variance. Evidently, if we also specify leptokurtic parameters the distribution will not be Normal.

Example: A maximum entropy price process:
Consider a bivariate probability distribution (or a price stochastic process) $h(x, t), x \in [0, \infty), t \in [a, b]$. The maximum entropy criterion can be written as an optimization problem, maximizing:

$$\text{Max } H = \int_0^\infty \int_a^b h(x, t) \log \frac{1}{[h(x, t)]} \, dt \, dx$$

Subject to partial information regarding the distribution $h(x, t), x \in [0, \infty), t \in [a, b]$. Say that at the final time b, the price of a stock is for sure X_b while initially it is given by X_a. Further, let the average price over the relevant time interval be known and given by $\bar{X}_{(a,b)}$, this may be translated into the following constraints:

$$h(X_a, a) = 1, h(X_b, b) = 1 \quad \text{and} \quad \frac{1}{b-a} \int_a^b \int_0^\infty x h(x, t) \, dx \, dt = \bar{X}_{(a,b)}$$

to be accounted for in the entropy optimization problem. Of course, we can include additional constraints when more information is available. Thus, the maximum entropy approach can be used as an 'alternative rationality' to constructing probability risk models (and thereby modelling the uncertainty we face and computing its VaR measure) when the burden of explicit hypothesis formulation or the justification of the model at hand is too heavy.

Problem: The maximum entropy distribution with specified marginals
Define the maximum entropy of the joint distribution by specifying the marginal distributions as constraints. Or

$$\text{Max } H = -\int_0^\infty \int_a^b F(x, y) \log [F(x, y)] \, dy \, dx$$

subject to:

$$F_1(x) = \int_a^b F(x, y)\,dy;\ F_2(y) = \int_0^\infty F(x, y)\,dx$$

and apply the multivariate calculus of variations to determine the corresponding distribution.

10.3.6 Monte Carlo simulation and VaR

Monte Carlo simulation techniques are both widely practised and easy to apply. However, applications of simulation should be made critically and carefully. In simulation, one generates a large number of market scenarios that follow the same underlying distribution. For each scenario the value of the position is calculated and recorded. The simulated values form a probability distribution for the value of a portfolio which is used in deriving the VaR figures. It is clear that by using Monte Carlo techniques one can overcome approaches based solely on a Normal underlying distribution. While such distributions are easy to implement, there is an emerging need for better models, more precise, replicating closely market moves. But where can one find these models? Underlying distributions are essentially derived from historical data and the stochastic model used to interpret data. The same historical data can lead to very different values-at-risk, under different statistical models, however. This leads to the question: what model and what distribution to select? Monte Carlo simulation does not resolve these problems but provides a broader set of models and distributions we can select from. In this sense, in simulation the GIGO (garbage in, garbage out) principle must also be carefully apprehended. The construction of a good risk model is by no means a simple task, but one that requires careful thinking, theoretical knowledge and preferably an extensive and reliable body of statistical information.

10.4 VaR EFFICIENCY

10.4.1 VaR and portfolio risk efficiency with normal returns

VaR, either as an objective or as a risk constraint, has become another tool for the design of portfolios. Gourieroux *et al.* (2000), Basak and Shapiro (2001) have attracted attention to this potential usefulness of the VaR. To see how VaR-efficient portfolios can be formulated we proceed as follows: let $W_t(.)$ be the value of a portfolio at time t and let, at a time $t + h$ later, wealth be $W_{t+h}(.)$. The probability of the value at risk in the time interval $(t, t + h)$ is then given by:

$$\Pr[W_{t+h} - W_t < -Var_W(h)] = P_{VaR}$$

Thus, if the return is $R_W(h) = W_{t+h} - W_t$ we can write instead:

$$\Pr[R_W(h) < -Var_W(h)] = P_{VaR}$$

Models will thus vary according to the return model used in value at risk estimation. For our purposes, let N be the number of financial assets, $P(i, t)$ be the price of asset i at time t and $B(t)$ be the budget at time t. We also set $a_1, a_2, a_3, \ldots, a_N$ be the number of shares held of each asset at time t. As a result, the wealth state at time t is:

$$W_t(a) = \sum_{i=1}^{n} a_i p_{i,t} = a^T p_t$$

while the return in a time interval is:

$$W_{t+1}(a) - W_t(a) = a^T (p_{t+1} - p_t)$$

The loss probability $\alpha = P_{VaR}$ implies a VaR calculated in terms of the portfolio holdings and the risk constraint, or $VaR_t(a, \alpha)$ is defined by:

$$P_t \left[W_{t+1}(a) - W_t(a) < -\mathrm{VaR}_t(a, \alpha) \right] = \alpha$$

where P_t is the conditional distribution of future asset prices given the information available at time t. Set the price change $y_{t+1} = p_{t+1} - p_t$. Then we have equivalently,

$$P_t[-a^T y_{t+1} > VaR_t(a, \alpha)] = \alpha$$

In other words, the portfolio VaR is determined in terms of information regarding past prices, the portfolio composition and the specified loss probability α. When the price changes are normally distributed:

$$y_{t+1} \sim \mathbb{N}(\mu_t, \Omega_t) \text{ then } VaR_t(a, \alpha) = -a^T \mu_t + \left(a^T \Omega_t a\right)^{1/2} Z_{1-\alpha}$$

where $Z_{1-\alpha}$ is the $1 - \alpha$ quantile of the standard normal distribution. We might consider then a number of portfolio design approaches.

Problem 1
Maximize expected returns subject to a VaR constraint.

Problem 2
Minimize a VaR risk subject to returns constraints.
The first problem is stated as follows:

$$\underset{a}{\text{Max}}\, E\left(a^T y_{t+1}\right) \text{ subject to } VaR_t(a, \alpha) \leq VaR_0$$

where VaR_0 is a VaR constraint specified by management. The solution of this problem will turn out to be a portfolio allocation which is necessarily a function of the risk parameters $\{\alpha, VaR_0\}$. By the same token, we can state the second problem as follows:

$$\underset{a}{\text{Min}}\, VaR_t(a, \alpha) \text{ subject to } E\left(a^T y_{t+1}\right) \geq \bar{R}$$

where $\bar{R}$ is the required mean return. A solution to these problems by analytical or numerical means is straightforward.

10.4.2 VaR and regret

The Savage (1954) regret criterion has inspired a number of approaches called 'regret–disappointments models' (Loomes and Sugden, 1982, 1987). According to Bell (1982), disappointment is a psychological reaction to an outcome that does not meet a decision maker's expectation. In particular, Bell assumed that the measurement of disappointment is assumed to be proportional to the difference between expectation and the outcome below the expectation. Elation, may occur when the outcome obtained is better than its expectation. Usually (for risk-averse decision makers), the 'cost of reaching the wrong decision' (disappointment) may be, proportionately, greater than the payoff of having made the right decision (elation). In other words, managers abhor losses, valuing them more than they value gains for having made the right decision. The inverse may also be true. A trader whose income is derived from trade-ins only may be tempted to assume risks which may be larger than the investment firm may be willing to assume. For example, a stock performance below that expected by analysts can have disproportionate effects on stock values while an uncontrollable trader may lead to disaster. To see how to proceed, consider the utility of an investment given by $u(x)$ and let CE be its certainty equivalent. That is to say, as we saw in Chapter 3:

$$u(CE) = Eu(x) \quad \text{and} \quad CE = u^{-1}\left[Eu(x)\right]$$

Say that 'disappointment' occurs when the outcome is below CE, resulting in an expected utility that is depreciated due to disappointment. As a result, an adjusted expected utility, written by $V(b)$ with a factor b can be specified by:

$$V(b) = Eu(x) - bE\left[u(CE) - u(x)\,|CE \geq x\right]$$

Explicitly, it equals the expected utility less the expected utility loss when the ex-post event is below the certainty equivalent,

$$V(b) = Eu(x) - b\left[Eu(x) - u(x)\,|CE \geq x\right]$$

Since $Eu(x) = u(CE)$, we also have:

$$\begin{aligned} V(b) &= Eu(x) - b\left[u(CE) - u(x)\,|CE \geq x\right] \\ &= u(CE)(1 - bF(CE)) + bEu\left(x\,|x \leq CE\right) \end{aligned}$$

This procedure allows us to resolve the problem of a reference point endogenously. For example, consider a two-event process:

$$x = (x_1, x_2) \text{ with prob } (\alpha, 1 - \alpha) \text{ with } x_1 > x_2$$

which corresponds to a binomial model. Then, the expected utility is:

$$Eu(x) = u(CE) = \alpha u(x_1) + (1 - \alpha)u(x_2)$$

while

$$V(b) = [\alpha u(x_1) + (1 - \alpha)u(x_2)]\,[1 - b(1 - \alpha)] + bu(x_2)(1 - \alpha)$$

and therefore:

$$V(b) = \alpha\left[1 - b(1 - \alpha)\right]u(x_1) + (1 - \alpha)\,(1 + b\alpha)\,u(x_2)$$

which we can write as follows:

$$V(b) = \alpha\beta u(x_1) + (1-\alpha)(\beta + b)\, u(x_2); \beta = [1 - b(1-\alpha)]$$

Since $b > 0$, it provides a greater weight to the lower outcome. Further,

$$\Delta = V(b) - Eu(x) = (1-\beta)\,[u(x_1) - \alpha u(x_2)]$$

By applying the same argument to the VaR by assuming a linear truncated utility, we have:

$$V(b, VaR) = E(x) + bE\,[-VaR + x\,|-VaR > x\,]$$

As a result, the expected value is:

$$V(b, VaR) = E(x) + b \int_{-\infty}^{-VaR} (x - VaR)\, \mathrm{d}F(x)$$

For example, say that an option price is $C(0)$ while the strike price over one period is K and assume two states $x_1, x_2, x_1 \le K \le x_2$. Thus, the expected value of the option price is:

$$C(0) = \frac{1}{1+r}\left\{\frac{\alpha}{1+b(1-\alpha)}u(0) + \frac{(1+b)(1-\alpha)}{1+b(1-\alpha)}u(x_2 - K)\right\}$$

If the option price can be observed together with the utility of $u(0)$, then the equivalent parameter b is found to be equal to:

$$0 < b = \frac{\alpha u(0) + (1-\alpha)u(x_2 - K) - C(0)(1+r)}{(1-\alpha)\,[C(0)(1+r) - u(x_2 - K)]}$$

Note that $\alpha u(0) + (1-\alpha)u(x_2 - K) < C(0)(1+r)$, meaning that we require for a positive disappointment parameter (since we have necessarily, $C(0)(1+r) < u(x_2 - K)$) that the option price be larger than the expected value of the utility of the option lottery. In other words, the option price also includes the investor's desire for the potential to reduce deception by allowing him to regret and reverse a decision badly taken.

REFERENCES AND ADDITIONAL READING

Alexander, C.O. (1998) *The Handbook of Risk Management and Analysis*, John Wiley & Sons, Inc., New York.

Arrow, K.J. (1971) *Essays in the Theory of Risk Bearing*, Markham, Chicago.

Artzner, P., F. Delbaen, J.M. Eber and D. Heath (1997) Thinking coherently, *Risk*, **10**, 67–71.

Artzner, P., F. Delbaen, J. Elber and D. Heath (1999) coherent measures of risk, *Mathematical Finance*, **9**, 203–228.

Basak, S., and A. Shapiro (2001) Value-at-risk-based risk management: Optimal policies and asset prices, *The Review of Financial Studies*, **14**, 371–405.

Basle Committee (1996) Amendment to the capital accord to incorporate market risks, on banking supervision, January.

Bauer, C. (2000) Value at risk using hyperbolic distributions, *Journal of Economics and Business*, **52**, 455–467.

Bell, D.E. (1982) Regret in decision making under uncertainty, *Operations Research*, **30**, 961–981.

Bell, D.E. (1983) Risk premiums for decision regrets, *Management Science*, **29**, 1156–1166.

Bell, D.E. (1985) Disappointment in decision making under uncertainty, *Operation Research*, **33**, 1–27.

Cvitanic, J., and I. Karatsas (1999) On dynamic measures of risk, *Finance and Stochastics*, **3**, 451–482.

Dowd, K. (1998) *Beyond Value at Risk, The New Science of Risk Management*, John Wiley & Sons, Ltd, Chichester.

Duffie, D., and J. Pan (1997) An overview of value at risk, *Journal of Derivatives*, **4**, 7–49.

Embrecht, P., A. Hoing and A. Juri (2001) Recent advances in the application of copulae to non-linear Value at Risk, ETH Department of Mathematics, CH-8092 Zurich, April.

Embrecht, P., C. Kluppelberg and T. Mikosch (1997) *Modelling Extremal Events*, Springer-Verlag, Berlin.

Feller, W. (1957, 1966) *An Introduction to Probability Theory and its Applications*, Vols 1 and 2, John Wiley & Sons, Inc., New York.

Fishburn, P. (1977) Mean risk analysis with risk associated with below target returns, *American Economic Review*, **67**(2), 116–126.

Gourieroux, C., J.P. Laurent and O. Scaillet (2000) Sensitivity analysis of values at risk, *Journal of Empirical Finance*, **7**, 225–245.

Gul, Faruk (1991) A theory of disappointment aversion, *Econometrica*, **59**, 667–686.

Jia, J., and J.S. Dyer (1994) Risk-value theory, Department of Management Science and Information Systems, The Graduate School of Business, University of Texas, Austin, WP, 94/95-3-4.

Jorion, P. (1997) *Value at Risk: The New Benchmark for Controlling Market Risk*, McGraw-Hill, Chicago.

Jorion, P. (1999) Risk management lessons from long term capital management, Working Paper, University of California, Irvine.

Kullback S. (1959) Information Theory and Statistics, Wiley, New York.

Loomes, G., and R. Sugden (1982) Regret theory: An alternative to rational choice under uncertainty, *Economic Journal*, **92**, 805–824.

Loomes, G., and R. Sugden (1987) Some implications of a more general form of regret theory, *Journal of Economic Theory*, **41**, 270–287.

Morgan, J.P. (1995) *Introduction to RiskMetrics*^{TM}, 4th Edition, November.

Robinson, Gary (1995) Directeur, R&D Division, Global Market Risk Management, BZW, *Value-at-Risk Analysis: Its Strengths and Limitations*, October.

Savage, L.J. (1954) *The Foundations of Statistics*, John Wiley & Sons, Inc., New York.

Schachter, B. (2002) All about value at risk, www.GloriaMundi.org/var.

Shiu, E.W. (1999) Discussion of Philippe Artzner's, 'Application of coherent risk measures to capital requirements in insurance', *North American Actuarial Journal*, **3**(2).

Shore, H. (1986) Simple general approximations for a random variable and its inverse distribution function based on linear transformations of nonskewed variate, *SIAM Journal on Scientific and Statistical Computing*, **7**, 1–23.

Shore, H. (1995) Identifying a two parameter distribution by the first two sample moments (partial and complete), *Journal of Statistical and Computational Simulation*, **52**, 17–32.

Stone, B. (1973) A general class of three parameter risk measures, *Journal of Finance*, **28**, 675–685.

Tapiero, C.S. (2003) VaR and inventory control, *European Journal of Operations Research*, forthcoming.

Telser, L.G. (1956) Safety first and hedging, *Review of Economic Studies*, **23**, 1–16.

Author Index

Akerlof, G. 68, 226
Allais' Paradox 52–3
Amikam, Meir 199
Aquila 197
Arnold, L. 92
Arrow, K.J. 44, 212
Artzner, P. 309, 310
Augustine, St 273

Bachelier, L. 16, 84
Barrois, T. 7
Basak, S. 324
Bell, D.E. 29, 326
Bensoussan, A. 92, 139, 154
Bismut, J.M. 92
Black, F. 226, 252
Bloomberg 300
Bolder, David 253
Bollerslev, T. 280
Borch, K.H. 8
Born, M. 79
Broadie, M. 154
Brock, W.A. 92
Brown, Robert 88

Capocelli, R.M. 209
Cardano, Girolamo 81–2
Carr, P. 154, 157
Cochrane, John H. 66, 67
Connolly, K.B. 184
Cotton, Peter 258
Cox, D.R. 92, 192, 296
Cox, J.C. 226
Cox–Ingersoll–Ross (CIR) model 257
Cramer, Gabriel 51–2
Cramer, H. 8
Cvitanic, J. 309

Darling, D.A. 209
Delbaen, F. 253
Detemple, J. 154
Dixon, Hugo 68
Donsker Theorem 99–100
Doob, J.L. 82
Duff & Phelps 12
Duffie, G.R. 230

Eeckoudt, L. 44
Einstein, Albert 84
Elliot, R.J. 92
Embrecht, P. 319
Engle, R.F. 280

Fama, E.F. 88, 112, 274
Feller, W. 51–2, 194
Fielitz, B. 301
Filipovic, D. 253
Fishburn, P. 309
Fisher, Irving 212
Fisher, R.A. 6
Fokker–Planck equation 88
Foster, D.P. 280
Fouque, Jean-Pierre 258
Friedman, Milton 112

Ghashghaie, S. 80
Goldman, M.B. 184
Gourieroux, C. 324
Green, M.T. 301

Hogan, M. and Weintraub, K. 252
Hoing, A. 319
Huang, J. 154
Hull, J. 258
Hurst, H.E. 300

Ito, K. 92

Jarrow, R. 227, 242
Jorion, P. 310
Julien, H. 139
Juri, A. 319

Kahnemann, D. 53
Kalman, R.E. 79
Kappa 183
Karasinski, P. 252
Karatzas, I. 309
Kay, John 9
Kim, I.J. 156–7
Kimball, M. 44, 45
Kluppenberg, C. 319
Koch 197
Kolmogorov 79, 84
Kortanek, K.O. 253
Kullback, S. 291, 321

Leeson, Nick 202
Leland, H.E. 226
Levy, P. 82, 84
Lintner, J. 63
Lo, Andrew W. 272
Longstaff, F. 226
Loomes, G. 29, 326
Lorenz, Edward 79, 80
Lorimier, S. 253
Lucas, R.E. 112, 113
Luce, R.D. 7
Lundberg, F. 8

Machina, M.J. 53
Magill, M. 111, 212
Malkiel, Burton 13
Malliaris, A.G. 92
Mandelbrot, B. 88, 272, 300, 301
Markowitz, Harry 62–3
McKean, H.P. 92
Medvedev, V.G. 253
Merrill Lynch 68, 200
Merton, R.C. 153, 226
Mikosch, T. 319
Miller, H.D. 92, 192
Miller, M.H. 112
Milshtein, G.N. 100
Morgan, J.P. 197, 200
Morgenstern Oskar, 41
Muth, John 111, 112, 293

Nelson, C.R. and Siegel, A.F., approach to interest rates 253
Nelson, Daniel 96, 280
Nelson, Erick 199

Papanicolaou, George 258
Pearson, Karl 6
Pliska, S.R. 139, 146
Pratt, J.W. 44

Quiggin, J. 53
Quinzii, M. 111, 212

Raiffa, H. 7
Ramaswamy, K. 96
Reyniers, D. 73, 75
Ricciardi, L.M. 209
Roll, Richard 112
Rubinstein, Ariel 20

Samuelson Paul 112
Sandmann, K. 252
Sargent, T.J. 112
Savage, L.J. 29
Schlaiffer, R. 7
Schwartz, E. 226
Shapiro, A. 324
Sharpe, William 63
Shiu, E.W. 310
Shreve S.E. 139
Siegel, A.F. 253
Siegert, A.J.F. 209
Singleton, K.J. 230
Sircar, K. Ronnie 258
Sklar 319
Sonderman, D. 252
Soros, George 84
Spence, M. 72
Stone, B. 309
Sugden, R. 29, 326
Svensson 253
Szekely, Gabor 82

Tapiero, Charles 92, 257
Taylor H.M. 280

Tetens, J.N. 7
Tobin, James 63
Turnbull, S. 227
Tverski, A. 53

Van Ness, J.W. 301
Vasicek 254–8, 258
Ville, J. 82
Von Neumann John 41

Walras, L. 212
Weatherstone, Dennis 310
Wets, Roger J.B. 253
White, A. 258
Wiener, Norbert 84
Willasen, Y. 45
Wilmott, P. 182, 183, 262

Yu, G. 156–7

Index compiled by Annette Musker

Subject Index

ABN AMRO Asset Management 21, 202
absolute risk aversion, index of 44
absorbing state 207
accounting rate of return (ARR) 54–5
actuarial science, insurance and 7–10
adverse selection 68–9, 71
affine structure 254, 255
affine term structure (ATS) model 261
Allais' Paradox 52–3
Allied Irish Bank (AIB) 24
AMBAC 231
American call option 116
American put option 149, 154–7
AMEX 118
Aquila 197
arbitrage 6, 119, 277
  arbitrage-based strategies 121–2
  arbitrage pricing theory (APT) 64
ARCH 7, 80, 278, 280, 282
Arrow–Debreu paradigm 277
Arrow–Pratt index of risk aversion 45, 47
Artificial Intelligence 15
Asian financial crisis 70
autocovariance function 91

Baltic Exchange (London) 116
Bank of Canada 253
Bankers Trust 199–200
barrier option 164–5
basis risk 115
Basle Committee 310–11, 313
Bayes criterion 26–7
Bayes decision making 22–6
Bayes theory 7
Bayesian rationality 22
bear spread 162, 179
Bermudan option 118
Bernouilli, Daniel 51–2
beta factor 60, 61, 63
binomial random walk 96
binomial tree, multi-period 140–1
black sheep syndrome 21
Black–Scholes option formula 6, 147–57, 168, 196
  volatility and 281
bond(s)
  arbitrageurs 214
  convertible 114, 261–2
  corporate 217
  coupon-bearing 215–17
  default 216, 224–30
  definition 211
  fixed rate 211
  floating rate 211
  forward rate and 222–4
  junk 216
  markets 212
  mathematics 218
  options on 260–4, 268–9
  ratings 231
  risk 161
  strips 211
  valuation, interest-rate processes, yields and 251–9
  see also rated bonds
bonus-malus 8
bounded rationality 4, 19, 20–2
bounds, option 152–3
Brownian motion 16, 80, 88, 92, 99–100
bull spread 162, 178–9
butterfly effect 80
butterfly spread 162, 181

calendar spread 162
call–call spread strategy 178
call option 116, 117, 133–4, 149, 169, 172–4

call option price 135
calls on calls 164
calls on forwards 164
calls on puts 164
cap 117, 163, 262
capital adequacy ratio (CAR) 310
capital adequacy requirements (CAR) 12, 13, 73
Capital Asset Pricing Model (CAPM) 54, 56, 187
 capital markets and 63–4
 investment and 59–61
 risk-premium and 295
capital market equilibrium 56
caplet 262
caption 163, 264
cash or nothing 163
CBOE 117
Central Limit Theorem 16
certain cash equivalent 57
certainty equivalence 43–4, 45
certification 271
CFTC 126, 127
chaos 79–80
chaotic analysis 272
cherry-picking 71
Chicago Board of Trade (CBOT) 114, 117
Chicago Mercantile Exchange (CME) 13, 114, 117, 126, 197
chooser option 164
claims 8
classical discounting 65
classical rationality 21
climatic option 163
climatic risks 196–7
collar 262
collateral call 124
Commodity Exchange Act 125
Commodity Futures Trading Commission 170
commodity option 163, 169
commodity pool 125
commodity pool operators (CPOs) 125
competition risk 11
complete market 119, 146
complexity analysis 272
compound option 162, 165–70
Computational Tool 15
condor 162
condor spread strategies 81
consumption, individual investment and 57–9
contingent claim assets 114
contract 114
control 24
convertible bond 114, 261–2
convertible preferred stock 114
convex dominance 272
convexity 218–22
copula distributions 315, 318–19
corporate bond 217
correlation analysis 122
correlation coefficient 91, 92
correlation function 91
correlation risk 197
corridor (range note) 163, 262
counter strategies of hedge funds 124
coupon-bearing bond 215–17
coverage 225
covered call 162, 176–7
Cox–Ingersoll–Ross (CIR) model 257
credit ratings, downgrading 231–2
credit risk 12–13
crowd psychology 21
cubic utility function 47
currency option 163
currency risk 12

Daiwa Bank 13
data mining 3, 15
dead value 141
death spiral 200
decision criteria 26–31
Decision Making Under Uncertainty (DMUU) 4, 7
decision table 31–3
default bond 216
 risky debt and 224–30
default, definition 227–8
default rate 225
degree of absolute prudence 45
delivery price of stock 141
Delta 151, 174, 182, 199
dependencies 8, 9
derivative asset 114
derivative contract 114
derivative security 6
derivatives losses 198
destabilizing strategies of hedge funds 124
deterministic analysis 34

Deutsche Bank 202
Deutsche Terminbrose (DTB) 118
diffusion–stochastic differential equation models 96
diffusion–volatility 89
disappointment 326–7
discrete time approximations 100–3
discrimination information 321
divergence-distance measures 322
diversification, principle of 62
Donsker Theorem 99–100
downside risk 309
duration 218–22, 309
dynamic hedging 124
Dynegy Inc. 231

efficient markets hypothesis 89
endowment effect 21
Enron 69, 197, 232
entropy, maximum, principle of 320–4
  discrimination and divergence 320–3
  distribution with specified marginals 323–4
  price process 323
environmental litigation 67
environmental protection 25
environmental risk 9–10, 196–7
EOE (European Options Exchange, Amsterdam) 118
equally likely (Laplace) criteria 34
equilibrium 293–5
error normality 88
Euler equation 65–7
  equilibrium and 293–5
Eurodollar deposits 162
European call option 116, 148, 149
European Monetary System 120
event-related funds 123
ex-ante risk management 23
excess discount rate 56
exchange option 163, 164
exercise date 116
exercise value 141
expectation criterion 43
expected cost criterion 32
expected monetary cost 32
Expected Monetary Value (EMV) 26–7, 33–7, 39, 40
expected opportunity loss (EOL) 33–7
expected payoff criterion 32
Expected Profit under Perfect Information (EPPI) 33–7
expected utility 40, 41, 45
  bounds 45–6
  critiques of 51–3
  finance and 53–67
  insurance, risk management and 48–50
expected value (Bayes) criterion 26–7
expected value of sample information (EVSI) 33–7
exponential utility function 46–7
ex-post risk management 23

fair competition 67
fair game 81–2
'fallen angels' 225
fat tails 81, 89, 272
FGIC 231
filter rule 191
financial instruments 113–19
financial physics 15–16
finite difference model 103
firm liquidity 8
First Boston 200
first passage time 207–9
Fitch Investor Service 12, 217
fixed equivalence rate of return (FIE) 55
flexibility 29
flipping 70
floating rate bond 211
floor 117, 163, 262
floorlet 262
floortion 264
Fokker–Planck equation 88
forcing contract 75
forecasts, optimal, rational expectations and 146–7
foreign exchange (FX) trading 195
  price of a European call option on 196–7
foreign exchange risk 10–11
forward contract 114–16, 141–5
forward rate agreement (FRA) 115–16
forward rate, bonds and 222–4
forward starts 162
fractional Brownian motion 272, 301
franchises 75
Frechet distribution 312, 317
Freddie Mac 203
front-running 70
FSA 231

fundamental factor model 64
fundamental theorem of asset pricing 83
fundamental theory, definition 3
funding risk 201
futures
  contract 114–16, 141–5
  market 115
  occurrences, example 145
  price 143–4

gambler's ruin problem 191
gambling 81–2
Gambling Act (1774) 8
Gamma 151, 182
GARCH 7, 80, 278, 280, 282
generalized method of moments (stochastic discount factor; SDF) 54, 65–7, 113, 274
genetic testing, insurance and 71
Gibson Greeting 199–200
Girsanov Theorem 104–8
Granite Partners 203
'Greeks' 151–2, 174
  dynamic strategies and 181–4
Gumbel distribution 317

HARA (hyperbolic absolute risk aversion) 47
hazard rate dominance 272
hedge fund fees 127
hedge funds 120–5
hedging 6, 119, 120–3
hedging portfolio 6
herd effect index 301
herding 21, 124–5
holding cost 139
homogeneity of reserves 310
Hong Kong Futures Exchange 118
Hurst exponent 300, 301

IBM 198
immediacy, financial instruments and 5
immunization 221–2
in the money (ITM) 151
incentive compatibility constraint 75
incompleteness in pricing 276, 277–8
independent increments 92
individual rationality constraint 75
inflation 14–15
inflation risk 14, 15
information asymmetry 67–75, 113
insider trading 113
insufficient reason (Laplace), principle of 27, 321
insurance 24
  actuarial science and 7–10
  contract 8
interest rate modelling 251–3
interest rate swap 262–4
interest rates and term structure 259, 267–8
internal rate of return (IRR) 55
International Monetary Fund (IMF) 70
invariance by drift 310
investor protection rules 125–7
irrational behaviour 39
Ito's Lemma 93–4, 147, 148, 170, 191, 251, 252
  applications of 94–108
  application to wealth process of portfolio of stocks 95–6

Jensen's inequality 45, 46
Joseph effect 300
junk bonds 216
Junk–Treasury Spread (JTS) 224, 225

knock-out option 162
Kolmogorov 79, 84
Kolmogorov (Fokker–Plank) equation 208

Laplace (equally likely) criteria 34
Laplace principle of insufficient reason 27, 321
Laplace transform 208
laps and flex 162
law of the single price 24, 83, 84, 134, 146
least divergence, principle of 288
lemon phenomenon 68–9
leptokurtic distributions 81, 88, 89
leverage, hedge fund and 127
linear growth of uncertainty 272
linear regression 7
linear risk-sharing rule 74
Lipschitz conditions 90, 91
liquidity 5
liquidity risk 14
logarithmic utility function 47
London International Financial Futures Exchange (LIFFE) 118, 197
London's Baltic Exchange 116

long call option 172
long put option 174
Long-Term Capital Management (LTCM) 120
lookback option 163
loss prevention 24–5
loss rate 225
lotteries, utility functions and 40–2
lower bounds of option 152
Lyapunov stability analysis 272

macro funds 121
macro hedged funds 121
macro specific fund strategies 122
macroeconomic approach 64
man-made risks 8, 9
margin calls 124
Marine Stewardship Council (MSC) 271
market efficiency 112–13, 184
  market memory and 274
market fluctuations risk 201
market gladiators 161, 162
market hedge funds 123
market inefficiencies 25, 184–5
market integrity protection rules of hedge funds 126
market rationality 123
market risk 14
Markov chain 96, 224, 230
Markov stochastic process 89
martingale 82–3
  Girsanov Theorem and 104–8
  market efficiency and 112
  Wiener process as 106–7
MATHEMATICA 152
MATIF (Marché à Terme International de France) 118
MATLAB 152
maturity 141
maximax (minimin) criterion 28, 34
maximin (minimax) criterion 28, 34
MBIA 231
mean variance replication hedging 288–91
mean reversion model 252
memory, volatility and 273–4
Merrill Lynch 68, 200
Merton model 170, 251, 297–9
Metalgesellschaft (MG) 201–2
  MG Refining & Marketing (MGRM) 201–2
minimax (maximin) criterion 28
minimax regret 28–31, 334
minimin (maximax) criterion 28
min-max objectives 26
monotonicity 310
Monte Carlo simulation 163
  VaR and 324
Moody's 12, 212, 217, 224, 225, 227
  bond ratings 231, 232
moral hazard 25, 69–71
multifractal time series analysis 272
multiple-factor model 64
mutual funds regulation 127

net present value (NPV) 55–6, 217–18
network and information asymmetries 277
neural networks 3
New York Federal Reserve Bank 200
New York Mercantile Exchange (NYMEX) 201
New York Stock Exchange (NYSE) 70, 118
Nikkei Index 202–3
Nippon Investor Service 12
no arbitrage assumption 134, 135, 241, 242
nonlinear (non-Gaussian) distributions 81
non-Gaussian distributions 81
non-uniqueness in pricing 276
Normal distribution 88, 89–90, 312, 322–3
Normal–Wiener processes 89
numeraire 137

obligation to sell 116
offshore funds 162
one-factor models (market premium in the CAPM) 63–4
opportunity loss 3
opportunity loss table 32–3, 35
option misses 197–204
option writers 161
option(s) 13, 114, 116–19
  asset or nothing 163
  as you like it 165
  Asian 118, 162, 164
  barrier 164–5
  Bermudan 118
  on bonds 260–4, 268–9
  bounds 152–3
  call 116, 117, 133–4, 148, 149, 169, 172–4
  chooser 164
  climatic 163

option(s) (*Cont.*)
commodity 163, 169
on corporate bonds 162
currency 163
on dividend-paying stocks 169
exchange 163, 164
on foreign currencies 169
on futures 163, 170
on a futures price 170
jump processes and 297–9
knockout 162
long call 172
lookback 163
on options 164
over-the-counter 118
packaged 162, 163–5
passport 165
path-dependent 162
practice and 171–84
rational expectations 135–7
stock 162, 165–70
upper bounds of 153
valuation 131–41
value at risk,, by Delta–Gamma 317–18
see also put options
Orange County (California) case 199, 200–1
Ornstein–Uhlenbeck process 97, 254, 297
application of Ito's Lemma to 94–5
Osaka Stock Exchange (OSE) 203
out of the money (OTM) 21, 151
over-insurance 25
over-the-counter (OTC) derivatives 114
over-the-counter option 118

packaged option 162, 163–5
Pareto–Levy distributions 88, 89
Pareto optimal risk sharing 74
participation constraint 75
passport option 165
path-dependent options 162
payback period 54
payments based in effort 75
payoff table 29, 31
Philadelphia Exchange 195, 196
PHLX 118
plain vanilla strategies 172–5
planning experiments 6–7
platokurtic distributions 88
Poisson Jump 92
positive feedback trading 124
power utility function 47
predictability of markets 274
preferred probability distribution 53
premium 116
premium payments 8, 48–50
premium principles 8
prepayment risk 122
present value 12
price risk 197
PriceWaterhouseCoopers 203
pricing and classical contract theory 277
pricing insurance 8
principal–agent problem 73–5
private placements 125
probabilistic analysis 34–7
Procter & Gamble 199–200
production-efficiency function 25
proper hedge risk 201
protective put 124, 162, 177
prudence 44–5
psychology and rationality 277–8
put option 116, 117, 169, 172–4, 175
American 149, 154–7
buying 177
calls on 164
long 174
protective 124, 162, 177
put–call parity 153, 165
puts on calls 164

quadratic utility function 47
quantile risk *see* value at risk (VaR)
quantity risk 197
quantos 163

R/S index 300, 301
random cash flows 55–6
random walk 84–91, 272, 282
randomness, definition 79
range 309
range note 163, 262
range to scale (R/S) analysis 272, 273
rare events 8–9
rated bond 212, 216
default and 230–51
duration 235–9
Markov chain and 233–5
pricing for risk-free rates 239–44
no-arbitrage condition 241–2
valuation of two-rates rated bond 242–4

valuation of default-prone rated bonds 244–51
cash value of rated firm 250–1
two-rated default bond 248–9
rational expectations 21–2, 83–4, 111–13, 131
optimal forecasts and 146–7
rationality, principles of 4, 20–2
recovery ratio 226
reduced-form models of default 226, 227–8
cf structural models 228–30
reflexivity 84
regret (Savage regret) 28–31
ex-post optimization and 26
valuation of firms and 30
VaR and 326–7
regret–disappointments models 326
reinsurance 49
remembering 274
repos 211
reverse mortgage 215
reward 40
rho 151
right to exercise 116
risk aversion 43–5
risk behaviour, utility and 42–8
risk, definition 5–6, 23, 309
risk diversification 24
risk loving 43
risk management 22, 23–6
Risk Metrics 318
risk-neutral pricing 22, 53, 112, 131, 137–9, 166, 184, 185, 186, 188
risk-neutral probabilities 6, 119, 136, 137, 145–7, 167
risk-neutral world 142–3
risk premium 24, 43–4, 45
risk sharing 8, 47–8, 49
risk tolerance 42
risk transfer 49
robustness 23, 25–6, 44–5

Savage regret criterion *see* regret
scenario analysis 31–3, 122
scenario optimization 26, 62
screening 72–3, 277
SEC 126
sector funds 123
sector risk 14
Security Exchange Act (SEA) 126
self-selection constraints 72–3
semi-variance 309
sensitivity equations 151–2
September 11th, 2001 9
sequentiality 274
Shannon's entropy 320, 321
shape risk 14
short selling 123, 174
signalling –3
SIMEX (Singapore International) 118, 203
smile 6, 151
smoothing splines 253
Société Générale 313
Southern Energy 197
spatial diversification 5
speculation 119
Spence, M. 72
spillover events 9
spread strategies 178–9
square root process 257
St Petersburg Paradox 51–2
stabilizing strategies of hedge funds 124
stacked rollover hedge 201
Standard & Poors 12, 212, 217, 224, 227, 231
state discretization 96–9
stationarity 91–2
statistical approach, APT and 64
statistical data analysis 6
stochastic calculus 92–4
stochastic differential equations (SDEs) 88, 89, 90, 96
stochastic discount factor (SDF) (generalized method of moments) 54, 65–7, 113, 274, 276, 293
stochastic dominance 272
stochastic ordering 272
stochastic processes, properties of 91–2
stochastic volatility binomial models 282–93
stochastic volatility interest-rate models 258–9
stochastic volatility models 89, 282–93
stochastic volatility, process discretization 280–1
stock, buying 177
stock option 162, 165–70

stopping time sell and buy strategies 184–95, 226
buying/selling on a random walk 189–90
pricing a buy/sell strategy on a random walk 193–5
straddle 179, 180
strangle 179, 180
strap 180–1
stress test 122
strict stationarity 91
strike price 116
strip 180, 181
strips bonds (Separate Trading of Registered Interest and Principal of Securities) 211
structural models of default 226–7
cf reduced form models 228–30
sub-additivity of reserves 310
subjective probability distribution 111
sub-martingale 82
sunk costs 20
super-martingale 82
supershare 163
sure-thing principle 52
Svensson procedures 253
swaps 117, 163
contracts 196
interest rate 262–4
swaption 163, 264
switch 163
systematic risk 199
systemic risk 125
reduction rules of hedge funds 126–7

Taylor series approximation 100, 101
Taylor series expansion 93, 151
Tchebycheff bounds on expected utility 45
technical analysis 273
technological innovation 25
term structure 56
interest rates and 267–8
theta 151, 182
Thomson Bank Watch 12
time contraction 284–6
time (process) discretization of continuous-time finance models 96–103
time phasing of events 5
Tokyo International Financial Futures Exchange (TIFFE) 118
too big to fail syndrome 70
Toronto Stock Exchange (TSE) 118
trade on volatility 151
trading strategy 131
traditional valuation 54–6
tragedy of the commons 21
transaction risk 11
transition bond rating matrix 244
transivity axiom 42
transparency 70
Treasury bills 212
trinomial walk 96, 191–3
truth-in-lending 67
two-fund separation theorem 63

upper bounds of option 153
utility
concept of 19, 39–42
lotteries and 40–2
maximization 61–3
risk behaviour and 42–8
see also expected utility

value at risk (VaR) (quantile risk) 23, 26, 46, 185, 199, 309, 310–11
analytic variance–covariance approach 315–16
in banking practice 313–14
Burr 317
copulae and portfolio measurement 318–19
definition and applications 311–15
efficiency 324–7
extreme statistics and 316–17
Frechet distribution and 317
Gumbel distribution and 317
historical approach 315
Monte Carlo simulation and 324
negentropy 320
numerical examples 314–15
option, by Delta–Gamma 317–18
portfolio risk efficiency with normal returns and 324–5
regret and 326–7
statistics 315–24
Weibull distribution and 317
value investing funds 123
Vasicek interest-rate model 254–8
nonstationary 258

Vega 151, 174
Vivendi Universal SA 231
volatility 6, 88
  definition 271–3
  equilibrium and 275–8
  implicit 281–2
  incomplete markets and 276–8
  memory and 273–4
  modelling and estimation 272–3
  process variance and 278–81
  range process and 299–301
  risk 14
  smile 281–2

Wald's identity 190
Walras, L. 212
warrants 114, 162, 163, 168–9
wealth, definition 8, 131
wealth process of portfolio of stocks
  application of Ito's Lemma to 95–6
Weibull distribution 312, 317
Wiener (Levy) process 88, 90, 92, 95–6
  as a martingale 106–7
Wiener risk 298
'wild money' 13
Willasen inequalities 46
World Bank 162
WorldCom Inc. 231

yield curve estimation 24
zero-coupon bonds 211
zero-coupon, default-free bond 213–14

Index compiled by Annette Musker

Transit, Land Use & Urban Form

Transit, Land Use & Urban Form

Wayne Attoe, Editor

Patricia Henderson,
Assistant to the Editor

Center for the Study of American Architecture
School of Architecture, The University of Texas at Austin
Austin, Texas 78712

Library of Congress Catalog Card Number: 88-070662
ISBN: 0-934951-01-2

Printed in the United States of America

Funding provided by the Mike Hogg Endowment for Urban Governance
and Capital Metropolitan Transportation Authority, Austin

Design by William Benedict

Contents

Editor's Preface vii

Introduction 1
Simon Atkinson

Part I

Transit's Role in Shaping Urban Character . . 7

Exploring the Land Development and Transit Connection 9
King Cushman

Traffic and Transit Futures 37
C. Kenneth Orski

Transit and Changing Urban Character 47
Ken Greenberg

Part II

Case Studies in Transit Development 55

Toronto: Thirty Years of Transit Development 57
Juri Pill

The Portland Light Rail Experience 63
John R. Post

Portland and Its Unique Traditions 73
Greg Baldwin

The San Diego Trolley 81
Robert Robenhymer

Transit and Planning Techniques in San Diego 89
Paul D. Curcio

Sacramento Light Rail: Lessons and Advice 101
Wendy Hoyt

Adelaide's Automated Busways 107
Alan Wayte

Bus Technology as Rapid Transit in Ottawa-Carleton . . 119
Ian Stacey

Denver's 16th Street Mall 127
Gary Zehnpfennig

The Vancouver Skytrain 135
L. E. Miller

Part III

Issues in Implementation 153

Community Involvement and Planning for Transit 155
Thomas C. Parker, Jr.

Infrastructure Financing and Joint Development 161
Jon W. Martz

Value Capture and Benefit Sharing for
Public Transit Systems 171
Jane Howard

Joint Development at Transit Stations 179
King Cushman

Transit, Urban Life and Development in the Sunbelt . . 189
Anthony James Catanese

Reflections 197
Simon Atkinson

Editor's Preface

The essays and reports that follow are drawn from a conference conceived by the Capital Metropolitan Transportation Authority, which serves the Austin, Texas, area. Like other fast-growing Sunbelt cities, Austin is on the verge of reintroducing mass transit to the region. The central question facing the conference was, "How can the Austin area use new transit development as a positive force in shaping the city and the region?"

Recognizing the importance and timeliness of a discussion of transit and urban design, the Center for the Study of American Architecture joined Capital Metro in programming and staging the conference, with support from the Public/Private Transit Network (a branch of the Urban Mass Transit Administration) and the City of Austin. Experts from other cities in the United States, Canada and Australia that have built new transit systems were invited to share their experiences, and other speakers were called on to address land use, development, and community involvement themes. The two-day conference was held May 15-16, 1987, on the University of Texas at Austin campus.

Many people helped plan and implement the conference. John Calkin, then Director of Systems Development at Capital Metro, and Simon Atkinson, Associate Dean of the School of Architecture at the University of Texas at Austin, took the lead in conceiving the conference as an opportunity to see how transit development, land use planning and urban design goals can be coordinated to shape life-enhancing urban form and rational transportation systems. Deborah Glover, Herb Hudson and Ed Taylor at Capital Metro and Patricia Henderson at the Center carried the burden of detailed conference planning. Their extra efforts ensured a smooth-running event. Staff from the School of Architecture skillfully provided for the technical requirements of the conference.

While some conference speakers tailored their remarks to address Austin-area issues, much of what they said is pertinent to many American cities. In fact, this gathering of experience and wisdom from across North America and Australia seemed historic and worthy of broad dissemination in book form. Grants from the Mike Hogg Endowment for Urban Governance and Capital Metro were obtained, making publication of the revised and expanded papers possible. Their assistance is gratefully acknowledged.

Lawrence W. Speck, Director of the Center for the Study of American Architecture, and Hal Box, Dean of the School of Architecture, gave full support throughout the conference planning and publication processes. Simon Atkinson, in addition to his written contributions to the book, played a key role at every step of the process, helping shape the conference program, introducing the impressive line-up of speakers, and skillfully leading end-of-conference discussions through which attendees could voice their concerns and enthusiasms. Perhaps most important, he continually emphasized the significance of the entire endeavor for Austin and other cities faced with what amounts to a transportation/quality of life crisis.

The versatility and dependability of Patricia Henderson made conference planning and book editing a virtual breeze. The hats she wore throughout the process were numerous, and the justice she did them was considerable. In producing this book we have relied on the knowledge and skill of William Benedict for design and Ann Cuthbertson for final proofreading.

Wayne Attoe

Austin, Texas
March, 1988

Introduction

Simon Atkinson
Mike Hogg Professor,
School of Architecture,
The University of Texas at
Austin; Principal, Black
Atkinson Vernooy

In recent decades a number of cities in the United States, and many more in Europe, have made bold commitments to rapid transit. Most of these in the United States are in large, dense metropolitan areas, yet a noteworthy trend is that mid-sized cities are making or seriously considering an equivalent commitment to public transit. These actions fly in the face of both current trends in cities and their decision-making behavior. Residential density has decreased in most American cities. In Austin, for example, it ranges from approximately ten houses per acre in inner neighborhoods to approximately four in new subdivision development. Consequently, suburban areas and outlying business/retail centers are, apparently, more impractical to serve with public transit.

It is also true that many people in the United States are automobile users. Since the price of gasoline in the U.S. is considerably less than in most other countries, the illusion exists that automobile use is less expensive than public transit. Add to this the apparent convenience and flexibility of the automobile and it is difficult to build a persuasive case for transit system development. A related argument is that the poor of America do not stop driving, they just drive older and cheaper vehicles.

A further obstacle to effective transit is the evolution of major peripheral commercial centers. At one time restricted to retailing, these developments now include extensive office uses often rivaling downtown. Capitalizing on lower land values and an ability to intercept commuter traffic well before the downtown, such developments have captured up to 60% of urban commercial growth in recent years. One could rightly question the sense of focusing expensive transit systems on downtown in a radial fashion when growth points could be 10 to 20 miles away and patterned concentrically.

Commitments to public transit would also appear to be contrary to established patterns of bureaucratic expediency exhibited by urban and regional policymakers. The pattern of decision making in most American cities could be described as a form of disjointed incrementalism. In other words, problems are "solved" in an isolated and short term manner to bring maximum political benefits in minimal time.

Transit development also faces a set of entrenched public attitudes that cannot be ignored. Recently, I asked an elderly gentleman what he thought about public transit in Austin. "Well," he said, addressing me as "Sonny" (which I took as a great compliment), "us Texans, we have wheels welded to our boots. Don't expect us to change our habits. Didn't you know this was the frontier state?" This type of sentiment is evident in many cities. Americans enjoy the liberation and freedom cars seem to offer. And we are incredibly selfish when it comes to cars. Stand by any highway going into downtown at 8 o'clock any morning and you will see freedom-loving commuters, one person per car, bumper to bumper.

Other views reflect prejudices about public transit. "Buses are for poor people" means that public transport is for somebody else. "It's okay if it doesn't cost too much, but it's really for them," couples with, "Well, look at all those big, empty buses." Misconceptions about transit are as significant as other factors that argue against transit development.

Why, then, is it that some cities have decided to take an alternative path, to consolidate and connect important parts of their regions through long term transit plans and coordinated land use? This is one of the basic questions that stimulated the writing of this book. Here, case studies drawn from a cross-section of cities provide a basis for evaluating the prospects for public transit in view of trends that seem contrary to transit development. The value of these case studies extends beyond the decision to build transit systems to the question, "How can transit development help accomplish additional goals in cities and regions?" Of particular interest is the mid-sized city experiencing considerable growth—some of which is positive, but much an embarrassment and dreadful waste of natural environment—and facing a restructuring of its infrastructure, land mass and built form.

What are some of the technical options in transit now? One of the simplest and most economical systems is to create bus priority lanes that other vehicles cannot use during rush hour. Another approach dedicates certain downtown streets to exclusive bus and pedestrian use. A related strategy is to dedicate corridors along freeways to express-bus use, with inflow in the morning and outflow in the afternoon, providing considerably faster travel times for buses during peak commuter hours than for automobiles. Each of these options means that other traffic is slightly more congested. In political terms, mass transit steals road space from individual cars.

A very different system is the use of microbuses, small, "neighborhood-friendly" vehicles that circulate through residential areas and then travel nonstop to the downtown. They have almost the convenience of a

A dedicated lane within a freeway gives priority to buses. Houston, Texas.

Microbuses are an appropriately scaled means of collecting people in residential areas for express service to downtown or regional nodes.

Light rail vehicles are compatible with traffic on city streets and yet can use separate rights-of-way for speedier movement outside downtown. Tri-Met system, Portland, Oregon.

High-speed rail and bus systems and automated vehicles require a separate right-of-way which often is elevated. Walnut Creek, California, station, Bay Area Rapid Transit.

taxi and can operate with great frequency over a longer time span than conventional buses.

Historically, many cities have employed a variety of rail systems for public transit. Some use a fixed track, often within city streets. A main feature of these systems is that they are compatible with the city fabric in inner areas and can move onto faster, separated rights-of-way as they serve suburban areas. With many of these systems, new, outlying stations have become generators of real estate development, as have mixed transit/traffic streets in the downtown area. These systems are, however, costly to implement and take time before showing a return.

Finally, there are heavy rail and automated rail systems which, for safety, require completely separated rights-of-way. Typically, they are technically sophisticated and are highly visible. The disadvantage of these systems is the cost of separating them from the rest of the fabric of the city, and the costs associated with automation.

An examination of cities where a commitment to public transit has been made raises a number of important questions: Do the transit systems work effectively and efficiently? Do they serve the city well? What is the public perception of the transit system? Does it work for "yuppies" as well as blue collar workers? Can systems be integrated comfortably into the city? In short, how will transit affect the quality of the city and its life?

As the papers here demonstrate, partnerships of city and development interests can produce quality environments while at the same time serving transit needs. Questions of quality of life and a sense of security regarding the future development of the city are thus of paramount importance. Also addressed is the matter of paying for transit and transit-related development.

Our aim is to focus on the urban aspects associated with public transit proposals, the problems and benefits associated with particular systems, and the consequences for urban economies and the quality of life in cities. These cases do not endorse a particular solution, but offer a range of experience. It will become obvious, however, that the case studies evidence a conviction that public transit is more than a transportation issue; it is a land use, economic development, urban design, and quality of life issue as well. It is hoped that these essays offer a basic set of considerations for those wondering whether a major public transit commitment is appropriate to a city's and region's future.

Part I

Transit's Role in Shaping Urban Character

These essays characterize the transportation crisis in American cities, how the crisis developed, and the history of our efforts to rationalize transportation services with both public and private mechanisms. While the traditional challenge in transit was to move large numbers of people to and from a city center, the problem now is that of moving people between several high-density subcenters. The history of a single intersection in Toronto demonstrates the potential of coordinating public, private and neighborhood interests and of making transit policy a part of a broader planning and urban design effort.

Exploring the Land Development and Transit Connection

A Walk in the Past, A Challenge for the Future

King Cushman
Director of Development
and Community Affairs,
Pierce Transit,
Tacoma, Washington

Although managing growth and consciously designing the urban transport system to help shape an area's future have been strong Canadian and European traditions, they have been little understood and even less practiced in the United States. Public transport systems were fairly successful in the United States before World War II, but their use was beginning to decline due to increased competition from automobiles. Suddenly, during World War II, "private" public transit became quite successful because the government required the auto industry to stop making cars, and then it rationed gasoline. Those were the golden days of transit, supply-side business economics with a vengeance.

Being a nation that believes in fair play, we did not let the transit industry maintain this "unfair" competition for long. However, the swing of the pendulum towards use of the auto for personal transportation seems to have gone well past center. To avoid more exaggerated swings in the future, we need to know why and how we developed as we have and decide whether we are fully satisfied with the results before we can make informed, rational choices for future directions. To this end, let us take a walk in the past to review the evolution and development of our transit systems and their role in and relationship to the shape and form of cities in the United States.

Then let us consider the "transit marketplace," a little understood reality that encompasses the character of land use in urban areas, and which largely determines whether anyone will want to use transit services. This holds true regardless of whether those transit services are buses, rail cars, vans, commuter trains, etc. If the transit market has the "right" form and ingredients, almost any transit option will work. Conversely, if the urban form or transit "market" is not right, transit can do very little that will be effective. When that is the case, and a decision is made to run transit services in the wrong kind of urban transit market, then incredibly high tax subsidies for public transportation must be implemented. Many systems around the country do this.

Finally, I will conclude with a few thoughts on the critical need to establish policies and directions early, before continuing development activity creates the degree of "traffic gridlock" suffered by most major metropolitan areas. Many of these areas have found it too late to "plan ahead" and have been forced to develop effective transportation and land use coordination after the fact.

Transit and highways should receive conscious consideration in planning an updated land use plan, but the mobility emphasis should be on people rather than vehicles. We tend to forget that people can be pedestrians as well as drivers; especially forgetful are the designers of most new suburban commercial and residential developments. The effort required to coordinate transportation and land use planning is well worth the trouble, not only in terms of greatly increasing mobility, commitments and investment, but also in terms of encouraging the development of better and more exciting places in which to live and work.

A Walk in the Past: A Historical Perspective on Transit and Development Relationships

Transportation systems and land use have always been interdependent and related in the development patterns of cities, but consciously trying to plan for and control that relationship is far from a science. The natural marketplace has usually dominated the situation, for better or worse. Historians and economists tend to cite three distinguishable but overlapping phases in the evolution of the United States into an urban society:

(1) the agrarian period in which cities were essentially commercial ports or centers for trade of agricultural products;
(2) the industrial period which brought cities into their own as economic centers of industry; and finally
(3) the period in which our cities became progressively dispersed, blurred into broad urban areas with economic bases that are decreasingly industrial and increasingly oriented toward service and information.

The largely pastoral agrarian period of society in America lasted well into the early 1800s. Cities were small centers of commercial activity, usually located on a body of water to aid in shipping agricultural products. (Notable is the fact that, even today, of the 100 largest North American cities, only three are not situated on navigable water.) Our early cities were typically not much bigger than 20 square miles, only about four to five miles from edge to edge, and walking was the dominant mode of transportation. This placed most people at well under an hour's walk from the city's residential fringe to its commercial center for most trade and employment, similar to today's typical commuting time, 30-60 minutes.

The settlers of this country were attracted primarily by the opportunity to acquire land. The freehold tenure concept of private ownership emerged and had a strong and long term influence on future land use policies. Like current residents, the "new" Americans did not like or want a lot of controls on the use of their lands—most still resist. Opportunities to own and control land were much in contrast to the land-scarce European countries they had left; private land ownership and control were not part of the feudal system for the common man. At the end of the eighteenth century, about 95% of our country's 3.9 million people lived outside cities (defined as any community with over 2,500 residents). Most people were farmers living in independent, single family cottages and cabins. The early cities, were in fact, villages consisting of clusters of wooden houses, sometimes with tight wall-to-wall row houses for the common folk in the larger ones. Transportation involved feet, two if walking and four if riding.

The next phase of city formation coincided with the Industrial Revolution, fueled also by massive immigration to the United States in the first decades of the 1800s—"land of plenty and plenty of land." As many as 60 acres of land existed per capita for the nearly 32 million people in the U.S. by the time of the Civil War. (For comparison, we have slightly less than 10 acres per capita today, but that still generously beats the less than one acre per person in India or two in China.) During this industrial period, cities took on new shapes and real economic purposes. The industrialization of America channeled the population growth into the cities for jobs. By 1820, nearly 20% of the population was living in cities; this rose to about 50% by 1860. New technological developments with steel structures and mechanical elevators supported amazing new skyscrapers and unique building forms for compact, higher density cities. Until the late 1800s, people still needed to live close to their jobs.

The beginning of the public transit industry in the United States is usually fixed around 1827 when Abraham Brower established horse-drawn omnibuses in New York. They were not much faster than walking, but they provided a new and welcome alternative for many. The horsecar transit systems spread slowly at first, but shortly after the Civil War they were found in almost every city.

Steam locomotion also thrived at this time and became the principal mode of intercity travel, but it was considered too dirty and noisy for use in the heart of cities. Systems expanded, knitting the country together and opening up land for development all across the country. The fast trading

Horse-powered, "fixed guideway" omnibus, Tacoma, Washington. Courtesy of Washington State Historical Society, Tacoma.

deals, mergers and takeovers in the rail industry of the 1800s resembled the recent fallout of airline deregulation. In 1870 the "Pullman Hotel Express," a grand and elaborately outfitted old train, left Boston and became the first train to travel coast-to-coast.

Cable cars were the next form of urban mass transportation. This technology had shortcomings, particularly safety problems when the cables broke or the "gripper" brakes failed. But cable systems provided a means of using the power of the steam engine in a stationary location out of the way of busy streets. These systems could run a cable out five miles from their center in each direction, but they were not appropriate for every city. Hills, allowing counterbalancing, and a mild climate played big roles in the limited success of the systems. Cable cars made their mark on San

There was a time when automobiles mingled with cable cars and streetcars in cities like Tacoma. Courtesy of Washington State Historical Society, Tacoma.

Francisco in the 1870s, and they are still rolling as a unique historic landmark and an operating part of the transit system.

The advent of widely available municipal electric power generating capabilities towards the end of the nineteenth century to support the new industrial economy in cities also caused the rapid expansion of urban transportation systems. The first development and expansion of urban transportation with electric streetcars occurred in Cleveland and Richmond in 1884. Electric trolleys using "clean" power were cheaper to operate (five-cent fares instead of a dime for the horse cars), and they quickly sprang up in almost every city of over 25,000 people with a flurry of private franchises. Between 1890 and 1920, the "golden age" for electric street railways, electric trolley ridership grew from two billion annual riders to 15.5 billion. These figures are impressive, for at the end of 1986 the American Public Transit Association (APTA) proudly reported that approximately eight billion total annual passenger trips were made, half of the 1920 figure. And that 1986 figure includes all of our public transportation systems, bus and rail. The all-time "peak" for transit ridership in the United States occurred in 1945 when, with few automobiles and little gasoline, there were almost 19 billion passengers carried by mass transit systems.

By 1900, street railway systems reached out like spokes from central cities into the fringes of urban areas. These were the early "radially-oriented" cities with electric trolleys carrying people out to the expanding residential areas, the early suburbs. What made people want to leave central cities at the turn of the century? One hears much reminiscing about the "good old days" at the turn of the century, but in actuality they were anything but good for most working people stuck in sweatshop factories and tenements. Filth, poverty and crowding were common and worth working hard to escape. A housing reformer in New York, Jacob Riis, complained in the late 1800s that New York's Lower East Side contained 290,000 persons per square mile! New York's density has diminished now to only about 23,500 persons per square mile (compared to Dallas, Houston and Seattle with approximately 2,700, 2,900 and 3,400 persons per square mile, respectively).

The sad state of affairs in New York around the turn of the century becomes vivid when quantified. In the late 1800s, horses were still the primary mode of personal transportation and goods movement in the cities. (The automobile did not make its marketable debut until about 1920.) These were the days before zoning and before the need for public health standards was recognized. It was reported that 2.5 million pounds of manure and 60,000 gallons of urine were dumped onto the streets of New York each day; over 15,000 dead horses were hauled away from the streets each year. In terms of sheer numbers of people, New York may have been one of the worst examples, but these same pathetic conditions existed in most industrial cities at that time. This bleak picture helps us understand why so many people were motivated to find a better life in the progressively expanding residential developments away from central cities. Transit systems at the turn of the century helped fulfill the desire to escape deplorable industrial city conditions and exploited the transit and land development connection that quickly followed.

Transit's Contribution to Early "Suburbia"

By World War I, the decentralization of American cities was well under way. Development of new suburbs reached 10-15 miles from downtowns. The proliferation of electric trolley systems around the country made commuting these longer distances into the city centers a viable and common daily affair. Riders' tolerances of up to a one-hour commute endured, since these trolleys averaged a speed of about 8-15 miles per hour. The beginnings of congestion were evident, but by 1910 congestion was still largely due to the myriad of streetcars competing for space on downtown streets.

Interurban electric railway promoters and land developers had mutual economic interests and began to join forces, hustling a competitive market for riders and real estate sales. In Chicago, by the turn of the century, there were 15 separate electric street railway companies competing for space in the downtown area. These railway and land partnerships were typically more interested in profits from land sales than they were in coordinating their transit services. Growing congestion and competition for space on teeming city streets caused city engineers, planners and utility regulators to look for ways to separate trolleys from pedestrians, horses and cars. In Chicago they put trolleys up in the air and the "L" (elevated electric railway) took its place in history and literature. New York put its trolleys underground in the 1890s with the initial construction of its now extensive subway system.

The combination of city regulation of street railway franchises and the beginning of mass-produced automobiles by the 1920s began to slow the pace of electric railway system development. It also slowed some of the scandals that went with the politics of the times as street railway operators regularly bribed officials to get franchises. The merger of transit and land use interests in electric railway syndicates provided profits from land sales that supported many years of hidden operating subsidies for the private street railway companies. In Los Angeles, one famous entrepreneur, Henry E. Huntington, took advantage of growing street railway operations. His syndicate bought 50,000 acres of land in the San Fernando Valley, ran out the rails, and became one of the big players in the Los Angeles land boom. Huntington's Pacific Electric Railway Company became the most extensive interurban railway system under one owner in the world. The Los Angeles area was said to have had over 1,400 miles of street railway lines in operation by the 1940s. The infamous Los Angeles "sprawl" was initially a product of public transit in the first four decades of this century. Electric railway systems reached out like a web from the heart of Los Angeles to the San Fernando and San Gabriel valleys, to coastal towns and on southeast into Orange County.

Automobility and Roadbuilding Begins

By 1920, Henry Ford and his famous Model "T" had made a splash in the commercial mass-produced automobile market that was felt around the world. He sold his first cars for the exorbitantly high price in 1913 of $600, but lowered it to $393 in 1923. Chevrolet followed suit and lowered its prices, along with other ambitious and colorful competitors such as Overland Motors. Automobiles began competing for surface street space with electric trolleys, unleashing complaints about the congestion and slowness of trolleys; the American love affair with personal mobility and speed had begun. With the availability of cars affordable to the working man (women were still generally relegated to the home), and with more land going into housing in the suburbs, pressure built for more and better roads.

Eventually, roads played a much greater role than transit in shaping our development patterns and influencing the style of our commerce. Sometimes we think of federal involvement in U.S. highway programs and policies only in the last half of the twentieth century. Surprisingly, it was as early as 1797 that the newly formed Congress authorized construction of a "National Pike," as it was called, to connect major population centers in the east. By 1830, although crude by today's engineering standards (mostly dirt, with some cobblestone or brick surfaces) and built for "primitive" modes of transport (horse and wagon or walking), approximately 27,000 miles of highways had been developed. Due to some not unusual disputes over states' rights, federal involvement stopped until 1916. The Federal Aid Road Acts of 1916 and 1934 became the forerunners of the extensive highway program in effect today. That 1916 Act was a landmark since it defined an approach to funding and strategies for national transportation programs that is still in use; it required development of a State program to allocate and administer formula funding for construction of road projects in each state.

The first major commitments to national highways came with those Federal Aid Road Acts. The New Deal programs during the Depression of the 1930s resulted in the extension of county roads and urban streets to support growing automobile traffic. The coming of World War II may have temporarily slowed down road building efforts, but not the desire for them.

With the luxury of historical hindsight, we now have a better understanding of the significant impacts that major road building had on the form of our urban areas. However, at the turn of the century and through the 1950s, the contemporary wisdom of most city planners and engineers placed great faith in and promoted the automobile for the "health" of our cities. Greater use of automobiles would relieve congestion in central cities, ameliorate social disorders and decentralize housing. The Burnham Plan for Chicago was one of the more ambitious in its anticipation of the needs of the automobile, advocating in the early 1900s an extensive radial and circumferential network of roadways. Although he was ahead of his time and surely did not realize the problems the automobile would soon generate, Burnham's even more far-sighted contribution was his attempt to use a road plan to direct land use rather than to react only to existing conditions. In Seattle, a man named Virgil Bogue proposed a major roadway design in 1911 that assumed true accommodation for many automobiles—a waterfront boulevard 300-500 feet wide and other city boulevards 160 feet wide, clearly much more ambitious than the modest 60-100 foot arterial streets that Seattle and most other cities eventually built.

A European visitor at a 1915 U.S. planning conference espoused a contrary and perhaps prophetic view of the automobile and transit issues. Dr. Werner Hegemann from Berlin charged that the automobile was dividing American city dwellers into two classes, the "barons," riding automobiles instead of horses, and the "common people," who depended upon public transportation. He feared that once the wealthier, more influential classes adopted the automobile, they would have little incentive to secure improved public transportation. Hegemann felt that promoting the auto would hurt mass transit, and that living spaces for the masses would be restricted to the central cities which could best be served by streetcars. This was in 1915.

Transition to a Service- and Information-Based Economy: Autos, Buses, Autos, Some Transitways, More Autos

In the 40 or so years since World War II, the United States has experienced a great shift in the character and location of economic and population growth. These major changes have reshaped our urban areas and cities. Economically, we have moved from a predominantly industrial base in the first half of the century to a broader service- and information-based economy. This has had as profound an influence on American life as did the transition from an agriculturally-based society to an industrial society in the early nineteenth century. The move away from industry, when combined with a massive national road building program begun in the 1950s, allowed much greater latitude for the growth of widely dispersed residential and commercial developments throughout expanding urbanized areas. A federal tax deduction even subsidized the housing sprawl.

In the 1950s and early 1960s, there was an understandably supportive attitude among most local governments that growth, almost any growth any place, was good for the economy and the tax base. And since it was going to go some place, why not in our respective city, township, county or parish? The dispersed development and consumption of suburban and rural land in the last 40 years has been impressive, to say the least. Look at the Standard Metropolitan Statistical Areas (SMSAs). These are metropolitan areas defined as a county or group of counties containing at least one city with a population of 50,000 or more. The United States had 215 SMSAs in 1960, increasing to 273 by 1970. The defined amount of urban land area in the same 273 SMSAs increased over 38% between 1970 and 1980, from 35,090 square miles to 48,465 square miles. The corresponding population densities of our cities and urban areas has experienced a progressive decline while population itself has grown. In 1920, urban areas reported having 6,580 people per square mile. This declined to 4,230 people per square mile by 1960 and to less than 2,000 by 1980.

With very few exceptions, accompanying this expansive atmosphere over the past decades across the United States has been an absence of carefully conceived land use plans or growth management policies that might channel or direct the spatial or geographic location of this mostly non-industrial growth. This absence of growth management or direction has probably been the single greatest cause of the subsequent decline and disappearance of most privately owned public transportation systems, street railways and city bus systems. And with so many of our urban areas still not cognizant of the direct relationship between the productivity and performance of transit systems and urban land use density and form, it is also the major cause for our naive surprise at the declining performance and productivity of our now publicly owned urban mass transportation systems.

Please note that I am not at all suggesting that we need to stop or even slow growth. Growth can do wonderful things for our economies and public transit systems if we plan wisely to accommodate, support and manage development. Economic growth and development are essential for our local and national economy and tax base. But growth has been much more dynamic and effective in those urban areas that have recognized commerce as a critical ingredient for the health of our society and still had the wisdom to manage and foster such growth for the greater public interest of all concerned. Atlanta and Portland are particularly good examples of American cities that have a long tradition of coordinated regional planning and have attempted to marry their land use and transit plans to achieve innovative developments in a "people oriented" environment. Although such planning has long been a Canadian and European tradition, our commitment to a largely laissez faire, "pioneer" spirit has not been sympathetic to the idea of controls on land.

The Changing Character of American Public Transit

It is amazing how many academicians, politicians and even transportation professionals still believe that the demise of "private" mass transit (especially the old electric street railways) was due to one of three causes. First, they say, transit's decline was partly attributable to escalating wage rates for labor in the 1950s and 1960s, causing a cost-push price spiral that lost ridership when transit companies needed to raise prices to cover increasing costs, thus losing more riders and eroding system productivity. This camp of "demise" theorists also says that poorly coordinated or ineffective regional planning failed to bring many diverse transit systems together in the larger urban areas.

Second, and quite out of kilter with chronological reality, cities were accused of starting up their own "public" mass transit systems, initiating "unfair" competition with the private sector by publicly subsidizing transit operations and taking away private business. Third, and by far the most colorful of the three myths, is the old "conspiracy" theory. It is held by many that the electric railways in America were systematically "done in" by a cabal of General Motors and Standard Oil. General Motors wanted to sell buses and cars and Standard Oil wanted to fuel them.

These interesting theories ignored the radically changing transit marketplace itself, a changing urban form that resulted in a very unlevel playing field. Since World War II, transit's competitive position as an option for personal travel *vis a vis* the automobile in the United States could be likened to sending an athletic team into competition wearing straightjackets and blindfolds, and then complaining that the team's performance was inadequate and ineffective. It is really the conditions of competition, the character of the urban environment, that is to say, transit's marketplace, that sealed the U.S. transit team's fate and effectively guaranteed its resulting comparative inadequacy.

I noted above that our government ungraciously, although in the national interest for the war effort, removed transit's competition from the automobile during World War II. After the war, our government proceeded to inadvertently but overwhelmingly tip the scales to radically unbalance the playing field in the opposite direction for the next 40 years. Governmental programs and policies that reflected contemporary values and the will of the people at those times had an incredible long term impact on our country's development pattern in urban and suburban areas. The absence of a growth management mentality, along with the presence of a few other policies and programs, has created a string of negative effects that have hurt the health and productivity of most of our nation's public transportation systems.

Another radical change in the marketplace was (and still is) a national policy for subsidized housing and sprawl. The first major dent in the urban transit market came when our government passed a national policy after World War II providing tax breaks on the interest homeowners pay on mortgages through the federal housing and GI loan programs. It created pressure to develop more land and build more roads. As this preceded the era of shared ownership of multi-unit dwellings, today's condominiums, we essentially provided an economic incentive for buyers of "the American dream," the single family, detached home. This homeowners' tax deduction on the interest payments for their mortgages was, and still is, a federal tax subsidy of over $40 billion per year, available only to homeowners. It was almost removed by the Reagan Administration in an early draft of the 1987 Tax Reform Act; we came close to accidental land use planning at the national level.

Tax subsidized homeowners' financing came with the first wave of the post-World War II "baby boom" and spurred an unprecedented era of building that has yet to subside for large scale housing developments on cheaper suburban and often rural land. This national "homeowners only" tax deduction policy, unique to the United States, had the effect of an economic penalty on those who remained renters in cities. After a few decades, who was left in central cities to ride the trolleys and buses, pay the taxes to support public schools and services, and patronize the downtown commercial and retail businesses? Except for a small percentage of affluent families who went on living in the beautiful, grand old homes in the neighborhood fringes of many old downtown areas, it was largely the poor who were left behind in central cities. They still could not afford "the American dream" in the suburbs nor the car to get there. They did not have the money to keep up their neighborhoods, which they often did not own anyway, nor could they adequately support downtown businesses that typically, although apologetically, started to relocate and follow the money to the suburbs. There were more than 5,000 new shopping centers built in suburban communities in the 25 years between the end of World War II and 1970.

A third radical change in the transit marketplace came about with interstate highways. To further aggravate the situation for our cities, we created the Federal Highway Trust Fund in 1956. This highway program became the transportation corollary to the subsidized housing loan program. It designated the creation of 42,500 miles of interstate highways, making it much more convenient for those who could afford it to flee central urban areas, buy subsidized homes in the suburbs, and commute back into cities to work. This 1956 road program legislation became one of the largest and most ambitious public works projects ever undertaken in the world, an incredible public investment in a network of concrete and asphalt, coast-to-coast and border-to-border.

The federal interstate highway program was initially conceived in 1944, although not funded until the Eisenhower years in 1956. It was proposed as a key component of our national defense system, assuring logistical intercity access and connections all across the country and it was originally intended as an intercity, not intracity, highway system. After a few years when the ribbons of highway began to play out and consume large swaths of our cities in many states, often removing and breaking up low income communities and cutting through parks and old historic buildings, the well financed highway program became a political catalyst, precipitating the 1968 Environmental Protection Act of Congress, an effort to minimize or avoid further negative impacts on our urban areas. The concept and the political will to be on top of such problems through integrated and coordinated growth management for urban land and transportation system development was not yet acceptable nor practiced much in the United States. Funded at roughly $5-6 billion per year, this national highway program will easily exceed $100 billion in cost and tax revenues before the system is completed. It will take even more taxes for our states to adequately maintain, repair and rehabilitate this now aging but essential transportation network in order to assure its continued economic viability and reliability as part of our urban infrastructure.

The cumulative effect of the lack of land use planning to effectively channel directions for growth, the subsidized housing loan program with its decades of uncontrolled urban sprawl, and the massive federal highway building program which created unprecedented mobility for independent travel by automobile has been dramatic and staggering for much of our urban infrastructure. For transit, the urban marketplace for productive service has become very different from the simple "golden days" of single centered urban areas with radially-oriented transit systems serving smaller and more compact cities with little or no real competition.

Another change in the marketplace has been the development of "transit unfriendly" environments. Urban areas were undergoing great changes during the 1950s and 1960s in their forms and types of growth. The spatial layouts of new, often even "walled," residential housing tracts with their cul-de-sac streets were difficult, if not impossible, for conventional transit buses to serve. Many downtown merchants abandoned central cities to find new markets in the suburbs. The seemingly random dispersal of commercial development throughout urban areas in commercial strips along arterial roads and highways—often without any sidewalks—and in multiple new centers proved a real challenge and a losing battle for productive transit. These new forms of development were totally oriented to the automobile and quite hostile for pedestrians, especially those who were disabled. Changing growth and travel patterns eventually spelled the downfall of most conventional radial transit systems in terms of productivity; they did not and could not respond to the new urban forms. Most continued to focus on the downtown hub and made a few token but ineffective efforts to serve the rest.

The cumulative effect of all these changes in the urban marketplace was the demise of "private" public transportation. In 1940, only 2% of 1,000 public transportation systems in the United States were "publicly" owned. By 1980, over 55% of 1,047 public transit systems were publicly owned. The decline of private city transit systems and the corresponding public acquisition and subsidization of those systems in hundreds of cities across the country in the 1950s and 1960s was all too common. Increasingly dispersed populations and ever-sprawling suburban areas led to a vicious downward spiral in the transit market and began the end of "private" city transit. And so it went: reduced ridership and fare revenue, increased fares to cover revenue losses, more reduced levels of ridership with increased fares, more people moving to the suburbs, and further reductions in service to offset declining productivity and fare revenues. Ultimately, private transit in cities either went under or was saved by new local tax subsidies provided to prop up or "buy out" troubled private transit systems. People unfairly criticized and attacked their cities' private and public transit systems for not adequately "doing their jobs" or "making a profit" (even when they were private, transit fares were typically not the source of the transit system's profits). The urban transit marketplace had changed. It would never be the same. Transit would also have to begin to change.

Cities Introduce the U.S. Congress to Transit

In the 1950s and 1960s, most cities lost their "private" city transit systems. Cities either began subsidizing them with public resources or lost transit services altogether. In the same period, cities became focal points for major social upheavals in terms of protests, civil rights movements and even riots. The federal government was called upon to "do something" about cities' social and economic problems and growing environmental deterioration. The multi-billion dollar annual investment that was going steadily forward in the national highway program contrasted sharply with a record of "no support" for urban public transportation systems. Blame for the decline, decay and disarray of urban public transportation systems was laid at the feet of Congress by the nation's cities, and the federal response was the Urban Mass Transportation Act of 1964. This act, ironic in view of subsidized urban sprawl with a housing loan program, began the transit program in the U.S. Department of Housing and Urban Development (HUD). There was no formal comprehension of the possible land use connection at that time; HUD was simply the federal government's "urban" department.

By 1968 Congress had authorized a federal reorganization and created the U.S. Department of Transportation with its Urban Mass Transportation Administration (UMTA). Prior to 1975 all funding for urban transportation assistance was restricted to capital projects. In the last half of the 1960s and early 1970s, a significant infusion of federal funds started flowing out to cities to begin buying out some private transit systems that had all but gone under and replacing and upgrading buses, rail cars and other capital equipment. But transit systems continued to lose ridership. It was during this same period that a few visionary (or, as some then said, foolhardy) cities—San Francisco, Atlanta and Baltimore—passed local tax initiatives to start building new major high speed rail transit systems to offer a competitive alternative to the automobile and to seek to maintain and revitalize their core areas. Washington, D.C., also got into the act with a major new high speed rail system, not with the same local tax initiative, but rather a singular "national" commitment to the nation's capital. Whatever one thinks about these systems, they are doing a major job now in their cities, carrying as many as 500,000 passengers daily in Washington, D.C., and, with the help of local zoning support, attracting new higher density developments around their stations.

Incidentally, around 1950 most large U.S. cities hit their peak population growth with the post-World War II baby boom. During the 1950s a lot of returning GIs were completing trade school or college educations, expanding their families, and entering the job market to buy automobiles to get out to the new suburbs and buy those "no down" GI Loan homes with tax subsidized mortgages. After the 1950s most large U.S. cities began to see a steady population decline, aided by the growing interstate highway network. Between 1960 and 1970, 15 of the nation's largest cities experienced a net loss of population. Since the urban areas still did not make the direct connection between population losses, suburban growth and development and transit ridership declines, the U.S. transit industry simply held to one of the three "myths" outlined above to rationalize their problems.

After much discussion and debate during the first few years of the 1970s, the national diagnosis of the American transit problem was to make a full scale commitment to treating the symptoms of mass transit's continuing dilemma instead of its causes. We began pouring federal money into the problem. It was believed that we needed to hold down or even reduce fares to attract passengers back to transit, but this approach could not cover operating costs. The solution was an infusion of operating subsidies from federal sources, since local tax subsidies were either nonexistent or inadequate to cover costs. There were some interesting and expensive demonstrations tried in the 1970s to test the "lower fares attract riders" theory, but these demonstrations did not prove much and were usually abandoned. Lower fares might be economically more attractive, but they were not the needed incentive to disengage people from their automobiles. They simply further subsidized those who were already riding transit because they had little choice.

The start of the federal subsidization of local transit operations, while doing little to bring back transit riders, contributed significantly to the rapid labor cost escalation within the U.S. transit industry that took off at rates far in excess of the Consumer Price Index or any other index of inflation rates. Unfortunately, this federal transit subsidy championed by cities and local transit officials regrettably introduced a national "deep pockets" dilemma, a situation that essentially and effectively stripped those same local officials and transit managers of the "bottom line" clout they needed to credibly hold the line politically on inflating labor costs, the largest component of a transit system's operating costs.

The more recent threats and efforts by the Reagan Administration to eliminate this operating subsidy, although not successful, have begun to bring costs more into line with reality. Although certainly not a popular point of view among my colleagues in the transit industry, one of the best ideas that would ultimately give local policy officials and managers more real control and incentives to deliver more effective and productive transit services is the total elimination of federal operating assistance for transit. I take this position because it would then require all of us to begin looking at and running the transit industry more like a proper, productive public enterprise. We would have to quit looking for external crutches to prop up some of the inherent inefficiences in serving a dispersed and expanding urban marketplace. Perhaps then we would start paying attention to the effects and configurations of land use and begin working together to shape policies for managing land development and transit systems and services that will offer the public true mobility options, now and into the next century.

The Transit Marketplace

What does it take for public transportation to be well used, efficient and effective? The answer is indicated in a term frequently used to describe transit—mass transportation. In essence, transit should use the same principles of market economics that are practiced in any business or enterprise. The enterprise of transit is, quite simply, productively moving people. The more people transit can move per vehicle, per hour or per mile of service, the more effective and productive it is. The more densely concentrated those people are when you pick them up, whether at home, at work or out shopping, the more people you are going to carry per unit of service offered, the more revenue you are going to bring in and the more productive your transit system is going to be.

Productive "Market" Factors*

The following are the three most critical land use/development factors that make transit productive and effective. As we look at these factors, recall the historical description of transit, when it was and was not effective.

Residential Density: This is usually stated and measured as net residential density per square mile or per acre. To support local neighborhood transit service, a minimum threshold of seven dwellings (household units) per acre or 2,400-3,700 persons per square mile should be present. (Smaller area measurements such as "per acre" are recommended because analysis and operational service planning will be more realistic and meaningful and more readily understood.) When densities rise as high as 30 dwelling units per acre, transit usage has been found to triple; transit trips can outnumber auto trips at 50 dwelling units or more per acre. But this does not mean that less dense areas should not be served. Since most suburbs have only about two to four dwellings per acre, it is a waste of time and money to go into such neighborhoods with big buses. Instead, the people in these low density suburban developments can use their cars to go to the transit system. Convenient park-and-ride lots can be developed

* I would like to give credit to and recommend an excellent reference document on this subject called "Encouraging Public Transporation Through Effective Land Use Actions" by Metro Seattle, May 1987.

near good access points to highways, high occupancy vehicle lanes or transitways (bus or rail). To make the transit system productive when homes are not intensely concentrated, facilities like park-and-ride lots should be provided to encourage people to get themselves concentrated, but on their time, not the transit system's. In this way, a transit system can start out well loaded, offering reasonably fast, high quality service that will be productive, even in low density areas.

Better designs and layouts of housing developments that allow all residents to conveniently walk a short distance to the transit bus stop or station are possible and would vastly improve transit effectiveness. Local governments should require developers and designers to seriously plan for pedestrians and prepare designs that recognize alternative mode access. Even uniformly requiring sidewalks would be great progress for many areas.

***Employment Density*:** Whether the public will want to ride transit depends more on the density at the other end of the trip away from the home. The real payoff in urban form for a productive transit marketplace is activity centers—concentrations of employment or college students, whether they are in a downtown metropolitan core or in fringe areas or suburban activity centers. The minimum desirable employment/student concentrations for these centers to attract respectable transit ridership would be at least 50 employees per net employment acre *and* greater than 10,000 jobs in those centers. When dealing with college areas, "students" and "jobs" tend to be interchangeable for this measure of density and transit effectiveness. The area around the University of Washington in Seattle has about 65 jobs per acre for a total number of 42,000 jobs in the area. Downtown Seattle has about 710 jobs per net acre, with a total of 137,000 employees. (Seattle Metro carried about 40% of all work trips on transit in its core downtown area by the end of 1980. It is an all-bus system with a strong park-and-ride lot orientation from the lower density outlying suburban areas—about 41 lots with just under 12,000 parking stalls.)

In my own transit district in Tacoma, 35 miles south of Seattle, we have only one true activity center that meets this density definition, downtown Tacoma. In this center, we have developed an effective "timed transfer" system of bus routes, serving multiple suburban centers outside of the downtown area. This is one way to make transit more productive and to serve the truly dispersed travel patterns of a sprawling suburban county. We manage the system with a heavy emphasis on productivity and performance. With 41 routes and about 110 buses in the peak hour, we get about 10.8 million annual riders in our 275 square mile service area with a population of approximately 435,000 people. This results in carrying about 29 passengers per revenue bus hour operated as an annual system-wide average.

Much of our service area is in low density areas like farmlands or in the woods. Our system achieves its respectable productivity by using a multi-centered network of routes serving five subsidiary transit centers located in concentrated activity centers: regional and community shopping centers, a community college, etc. This tailors service appropriate to the dispersed suburban densities, at least for now. The first three centers are small in scale and are examples of "joint development." They were developed on sites of from one to three acres obtained from a national shopping mall developer, the State (on a community college site) and another local public agency; each one has a 20- or 30-year lease at $1 per year. These other property owners saw the benefit of a transit presence on their land. This orientation of service at subregional centers also provides visible focal points around which a land use plan can be developed to try to create the preferred higher residential and employment densities necessary to attract more riders and a greater farebox return on operating costs.

***Parking Costs and Parking Management*:** This factor varies with higher densities of employment. Transit has increasingly absorbed the employment growth travel demands in cities like Portland, San Francisco and Seattle in the United States and Calgary, Toronto and Vancouver in Canada. These cities are now putting the concept of "maximum" allowable parking into their central business district zoning ordinances, which has positive results for the cities and the transit systems. While still allowing economic and employment growth, parking limits minimize growth in automobile traffic and congestion by shifting greater demand for travel onto the urban transit system. With public zoning policy, cities are managing parking demand and improving the quality and desirability of the central business district environment.

Some cities, like Miami, have failed in this area. They defeat their own citizens' investment in transit by requiring more asphalt and concrete for moving and parking more automobiles, the foolish result of divorcing parking policy and parking management from transportation and land development policies. This shortsightedness is not limited to cities. In my own state, Washington, we get a normal degree of scrutiny from the state legislature from time to time about the effectiveness and productivity of some of our smaller transit systems. The irony, of course, is that the legislature and all State employees have traditionally had free or low cost parking in and around the State Capitol complex. At a cost of $10,000-$15,000 per stall in a parking structure, that amounts to a large public tax subsidy for automobile parking and it diminishes the effectiveness of the local transit system that serves the State Capitol. We have been working for over two years to try to sensitize state government to the importance of parking policies and the need to create pricing incentives for carpooling and vanpooling, and to offer transit passes to employees inside the Capitol to reduce the public costs for parking space development and to encourage riding transit.

I do not suggest eliminating parking or treating all parking as a problem for transit. Rather, I would advocate minimizing long term parking for employees in the downtown area and any other activity centers that have existing transit service or the potential for ride sharing. It is fine to encourage and recognize the economic benefit for commercial or retail businesses of maintaining and developing needed short term parking using meters or well enforced time limitations to make sure it is not abused. Short term parking is very much in the interest of private sector employers. But they get no return or benefit from building or developing parking spaces for their employees at a cost of from $1,000 to $15,000 per parking stall.

Trends in the Contemporary Urban Marketplace

The symptoms, realities and consequences of contemporary development trends are well known. We experience, read or hear about aspects of them every day, for example, "traffic congestion is getting worse—to avoid traffic jams this morning stay home!" This seems to be the theme reported everywhere on the car radio for commuters (from airborne reporters, of course, as they are the only ones who can move around freely over so much land). Development projects are being built in great numbers one year, yet many still stand vacant the next. Our economies are becoming more and more international, more interdependent and less tied to the old certainties of simpler domestic markets.

Even the *Wall Street Journal* and *Atlantic Monthly* are reporting on the growing state of suburban traffic congestion and the creative and controversial approaches being taken by public agencies across the country to mitigate these congestion problems. Some of these ideas involve such means as development fees, traffic management ordinances, transportation "fees" paid to public agencies for related traffic or transit improvements to offset a development's traffic impacts, or elaborate ride sharing requirements for new developments. One developer in Los Angeles has paid for the State Department of Transportation to build new freeway ramps to mitigate the estimated local traffic impacts from proposed new high-rise buildings.

The *Los Angeles Times* recently reported that the morning and afternoon traffic crunch is beginning to blend together with no breaks on some freeways throughout the day. The congestion seems to be intensifying faster than the rate of population growth. Some "transportation officials" are saying that a current rush hour trip on the freeway that now takes 1-1/2 hours will take three hours by the year 2000. A three-hour commute, six hours of a person's day spent getting to and from work? Transportation planners come up with this conclusion with the help of computer models that simulate travel. But we should not take them too seriously when they project "extrapolated extremes." Computers do not program and compute the well known will of the public not to tolerate intolerable situations for very long. When commuters finally have had enough, they will foil the computer models and change their jobs, relocate to new homes, stop new developments, demand that elected officials "do something!", or even elect new officials. Imagine the impact on any given dispersed urban area if toll roads were reintroduced. The California State Legislature has been toying with such ideas.

The new economic growth trends that John Naisbitt and others have pointed out so well in books like *Megatrends* no longer need industrial-type factory space. "Clean" service- and information-based employment growth associated with high technology offices and plants in commercial campus-like "parks" can be located on less expensive land in suburbs. These locations also get jobs closer to newer housing markets. This new growth does not need to be in traditional downtown centers; these developments are creating their own centers which include regional shopping centers surrounded by mid- and high-rise commercial buildings and hotels and entertainment complexes.

The degree and speed with which this growth has taken place is itself impressive. Robert Cervero, a professor in the Department of City and Regional Planning at the University of California at Berkeley, has done a lot of research in the area of suburban studies. He has noted that in 1970 only 25% of all office floor space nationwide was located outside of city centers. By 1980, 43% of this office space was in suburbs, and by

1986 about 60% of the office space was in suburbs—"from bedroom to boardroom."

A couple of years ago the jargon for these new forms of metropolitan development was "mega centers" since they are typically developing with more than 10 million square feet of space among the varied uses. Now the term "urban villages" is used to describe this new urban cluster development form. For a sense of the magnitude and subtleties of this development phenomenon, I recommend a concise article by Christopher B. Leinberger and Charles Lockwood called "How Business is Reshaping America," in the October 1986 issue of *Atlantic* magazine. They note the astounding rate of growth that has resulted in the blending of the Costa Mesa/Irvine/Newport Beach commercial areas into a new metropolitan subcenter (urban village) of 21.1 million square feet. While less than downtown Los Angeles's 36.6 million square feet, it is approaching San Francisco's 26.8 million square feet.

I have difficulty with the term "urban village" since it appears that these development complexes have a nearly universal lack of any pedestrian character or linkages that were a natural part of true villages. These new centers are organized for automobiles. For me, the term "village" conjures images of something less than mirrored, high tech, glass, mid- or high-rise buildings, and more of a cluster of several blocks with tight rows of shops, stores and offices on relatively slow and narrow streets (narrow by today's standards for arterials). In the villages I knew growing up outside New York, streets allowed parking, but people were always walking around. A pedestrian character was the most dominant feature.

How does transit respond to all this? Professor Peter Muller of the Geography Department at the University of Miami says that we are "building 1920s-style mass transit systems for our 1990s metropolitan areas" when speaking about trying to "fit" major fixed rail projects into areas with multiple urban centers with widely dispersed origins for the trips to those multiple centers. This was the Miami Metrorail situation, a classic example of a significant lack of comprehensive transit and land use planning. They had no dedicated financing for the operation of the system and proceeded to strip the bus system to help operate the rail line. There are fewer buses operating today in Miami than there were 15 years ago, only 395, and their plan called for 1,000 to provide critical feeder service. There was also no coordination with a key land use variable for the transit market, namely parking supply. When the Metrorail opened, over 8,000 new subsidized parking spaces were available in downtown Miami, built by the independent public parking authority which was charging as little as two dollars per day. The system never had a fair chance.

The Rediscovery of Metropolitan Centers

Just when we think we have begun to grasp what is going on in urban development, that is, to realize that it looks like all the action is going out to the suburbs in urban villages, we begin to see another development pattern emerging. There is a parallel rediscovery and renaissance in many of our older metropolitan centers. In the 1970s we thought we would have to "write them off." But there have been significant new developments and even large scale rehabilitations of older buildings in and around a number of the older U.S. city centers—New York, Boston, Pittsburgh, Atlanta, Washington, D.C., Los Angeles, San Francisco, Seattle and Portland, to name a few. Part of this rebirth relates to a renewed cosmopolitan interest in cities themselves, something of a gentrification cycle. Then, too, part of the renaissance may well be a simple and rational human reaction to the economic success of the new suburban centers which have managed to clog up all the roads in suburbia. Now, a little late, suburban cities are looking into transportation and land use/growth management planning.

Many of these large new centers are inadequately designed, sterile places to work or shop, much less live. The new urban villages and suburban megacenters often appear as unconnected commercial palaces lacking the interest of a more natural and eclectic human scale pedestrian environment. People tend to be drawn to and enjoy the kind of urban environment found along active and vibrant streets, the pedestrian friendly streets that transit used to call "home" in the real urban villages of old and downtown centers.

An interesting twist can be seen in some of these new downtown revitalization projects and developments. Developers such as the Rouse Company of Columbia, Maryland, and Ernest Hahn in California have been sharp enough to lead the way in capitalizing on the suburbanization of downtowns. Ironically, as the "centers" have sprouted in the suburbs, the new retail and commercial development complexes in the old downtown centers have taken on the essence of what worked well in the suburbs, covered and integrated mixed use shopping malls, although now tied in better with other adjacent blocks and developments. Some are even finally consciously planning to accommodate pedestrians since there is not much room left for many automobiles.

The apparent contradiction of these trends in development is the product of different analytical "camps." One sees the emergence of new economic centers rising as urban villages in suburbia and the progressive clogging of suburban streets with new traffic from people in cars who cannot benefit from or be served by the existing radially oriented transit systems that go downtown. The new centers cannot be served well by any transit services, at least not as they are currently being designed with total auto orientation. This is, of course, correctable with a little careful thought about how one parcel relates to the next parcel, and so on, allowing for a pedestrian layout among the site plans that begins to work well for transit accessibility, just like the old villages and cities used to—places for people. In contrast to this urban village phenomenon, other trend spotters see the rediscovery of the traditional downtowns. They are being revived and have been even more successful than suburban centers in attracting residential cluster developments. Perhaps the new urban village designers should take another look at the "old stuff" and how it works.

Both of these trends are occurring, which means that creative development design, and land use, growth and transportation management and plans, must emerge to deal effectively and quickly with the urban villages/new centers trend. Otherwise, a greater price will be paid later for "band aid" retrofit fixes to improve lousy infrastructure systems for circulation and access. For the old downtown centers, new growth and modest suburbanization simply require a reconsideration of the conventional radial bus and rail corridor types of transit system plans to find modifications or new slants that might be appropriate for these already proven approaches and technologies.

Transit and Land Use Directions: A Challenge for the Future

As with many other things on life's carousel, transit and development seem to be evolving in somewhat repetitive cycles; we rediscover old forms, systems and concepts as "new," and make new ones look "old." Ignoring the great benefits and mobility options that a real first class bus or rail transit system can offer is truly a missed opportunity. We see from our Canadian colleagues that they did not miss it, not with an average of three to eight times our average transit utilization in terms of transit trips per capita, and with over 60% of their operating costs derived from operating revenues. The transit and land development relationship offers one of those significant opportunities that, in the United States, has most often been overlooked, unacknowledged or, when it was exploited by the street railway and real estate barons, kept relatively invisible for the sake of profits. Sometimes narrow or linear thinking looks only at the transit system plan or the land use or development plan and thus overlooks the most important and critical relationship and opportunity. How is it that U.S. cities have such a difficult time getting this act together when the Canadians seem to do it with apparent ease?

In most states, it is our own institutional structures that limit how we look at, think about and approach urban problems. This lack of transit (or transportation) and land use relationship has been largely due to the earlier creation of independent and vertically related cities, towns and counties that are indeed broadly affected by metropolitan area forces such as transportation and land development but are totally unequipped to deal with them from an institutional perspective. Decision making bodies in the United States are quite fragmented in their realm of authority. Overlaid on the geography of various governmental jurisdictions in our metropolitan areas have also been a multitude of special purpose governmental institutions dealing with highways, transit, sewers, water, power, ports, etc., that have no mandate, authority or responsibility to deal with or care about cross-jurisdictional, multi-dimensional issues like transportation and land use.

In spite of this seemingly impossible morass of bureaucratic entities with which to deal in our metropolitan areas, some regions have shown significant fortitude, managing to do wonderful and effective things for their people with transit system development and economic development by taking action to establish moderate but enlightened policies for directions in growth management. The State of Oregon, the City of Portland and its three county region, and the City of Atlanta are noteworthy exceptions and positive U.S. models to emulate.

One's expectations and judgments of performance in any business or governmental venture or program should have a sound understanding of marketplace realities. We should not be surprised at changes in transit's performance, good or bad, when we radically alter the environment and conditions in which it functions. If we committed ourselves to the time and effort of a more rational and comprehensive business-like approach, evaluating and understanding the environment in which transit must function, would we not want to shape that environment so that our transit investments and system designs will be much better and system performance more productive and predictable? We would do so much better if we could agree about the value of managing the form and shape of the urban environment in which transit works. We would have everything to gain and nothing to lose but our rather high rates of transit operating subsidies, especially high in comparison to Canadian and European transport systems. Canadians and Europeans have long accepted and practiced the concept of metropolitan area land use and growth management, and have coordinated the design and development of their urban transport systems to benefit transit *and* highway travel.

References

Cervero, Robert. "Curbing Traffic in Fast-Growing Suburbs." *ITS Review 9* (May 1986): 4-8.

Cervero, Robert. "Urban Transit in Canada: Integration and Innovation at Its Best." *Transportation Quarterly* 40 (July 1986): 293-316.

Charles River Associates. *Characteristics of Urban Transportation Demand: Second Edition, A Handbook for Transportation Planners.* Prepared for the Urban Mass Transportation Administration, December 1985.

Farris, Martin T., and Forrest E. Harding. *Passenger Transportation.* Englewood Cliffs, New Jersey: Prentice Hall, 1976.

Foster, Mark S. *From Streetcar to Superhighway: American City Planners and Urban Transportation 1900-1940.* Philadelphia: Temple University Press, 1981.

Jackson, Richard H. *Land Use in America.* New York: John Wiley and Sons, 1981.

Lave, Charles A., ed. *Urban Transit: The Private Challenge to Public Transportation.* San Francisco: Pacific Institute for Public Policy Research, 1985.

Leinberger, Christopher B., and Charles Lockwood. "How Business is Reshaping America." *Atlantic Monthly* 258 (October 1986): 43-52.

Lieb, Robert C. *Transportation: The Domestic System, Second Edition.* Reston, Virginia: Reston Publishing, 1981.

McDonnell, James J. *National Trends: Population, Employment and Vehicle Ownership 1970-1985.* Office of Highway Information Management,

Federal Highway Administration. Presented before the 1987 Transportation Research Board Meeting, January 15, 1987.

METRO Transit. *Encouraging Public Transportation Through Effective Land Use Actions*. Seattle, Washington: May 1987.

Parody, Thomas E. *Metropolitan Area Changes in the Characteristics of Travel Demands*. Boston, Massachusetts: Charles River Associates, 1986.

Parsons Brinckerhoff Quade and Douglas. *Perspectives on Transit and Land Use Relationships: How Transit Systems and Services Appear Most Compatible with Various Long-Range Land Use Policies*. Prepared for METRO Transit, Seattle, Washington, March 8, 1978.

Smerk, George M. *Urban Mass Transportation: A Dozen Years of Federal Policy*. Bloomington, Indiana: Indiana University Press, 1974.

Traffic and Transit Futures

C. Kenneth Orski
President, Urban Mobility Corporation,
Washington, D.C.

We find ourselves today in the midst of what has rightfully been called the "second suburban migration," a demographic change whose implications promise to be equally profound as those of the original suburban migration immediately following World War II. The present migration is not one of people but of jobs and, more specifically, of office and high technology jobs. The scale of this relocation is nothing short of dramatic. Increases in exurban and suburban employment have been outpacing increases in central business district employment in virtually every metropolitan area in the nation. In many markets, as much as 60-90% of all office space expansion during the last four years has occurred outside the downtown area, and this has radically modified the structure of metropolitan areas. Just five years ago, 57% of all office space in the United States was located in urban centers, and 43% in the suburbs. Today the situation is virtually reversed: 58% of total office space is in the suburbs and 42% is in the core cities.

Behind this massive locational shift lie some powerful economic and technological forces. Typically, office space in suburban locations can be obtained for $20-22 per square foot. This compares to $30-45 per square foot in choice downtown locations. What does this differential in office space mean for an average employer? According to an analysis performed by the Office Network, an employer moving to the suburbs in 1985 saved almost $4000 per worker per year in San Francisco, $3400 in Miami, $2900 in Washington, D.C., $2300 in Los Angeles. Thus, a large bank or insurance company employing a clerical staff of several thousand people can save millions of dollars annually in office rent by moving to the suburbs. This provides a tremendous, almost irresistible, incentive for businesses to move to the suburbs.

Even so, many businesses would not be leaving central business districts, at least not in such massive numbers, were it not for recent advances in telecommunications. Modern suburban office buildings and office parks are wired for an array of sophisticated communication systems that have enormously simplified operations in remote locations. Thanks to a full array of devices such as digital PBX and private fiber optic cable systems, suburban office workers can communicate with the outside world as easily as their downtown-based counterparts.

The suburban office boom is characterized not just by an increase in the number of jobs but also by a significant increase in the density of employment. Modern suburban office complexes are no longer scattered across the metropolitan landscape; they tend to cluster in giant enclaves, many of which exceed in size and density the central business districts of medium sized cities. What is more, market pressures seem to be driving the densities of suburban development upwards. Ten years ago it was rare for a suburban office complex to be developed at densities higher than a floor area ratio (FAR) of 0.2 or 0.3. Today, suburban FARs of 0.8 or even 1.0 are not uncommon. This creates a tremendous increase in vehicle trip generation.

What are these new suburban complexes like? They give every appearance of a traditional central business district. They offer the same high quality construction, architectural distinction and building amenities that are available in downtown office buildings. They have sophisticated hotels, upscale shopping facilities, diverse restaurants. Some suburban centers, such as South Coast Metro in Orange County, California, even boast art galleries and performing arts centers. However, unlike traditional downtowns that evolved slowly over many decades, the new suburban downtowns have been growing at a dizzying speed. For example, the North Park area in Dallas has more than quadrupled in size in just six years, from 1980 to 1986. Central Contra Costa County in northern California has more than tripled its office space in four years, and in just ten years of explosive growth it has reached one-third the size of San

Francisco's financial district. Tyson's Corner in northern Virginia, my favorite example, took only 15 years to grow from a semi-rural crossroads—a gas station and a tavern—to a bustling suburban downtown that already has more office space than downtown Miami.

In contrast, traditional downtowns took many generations to reach their present size and density. This allowed them to accommodate rising traffic volumes in a much more graceful, orderly manner. The modern suburban center does not enjoy the luxury of gradual adaptation, so the instant downtowns of today's suburbia give rise to equally instant congestion. To compound the problem, the new suburban megacenters are being superimposed on what is essentially a mature highway system whose expansion since the late 1970s has slowed to a mere trickle. So it seems that suburban America must learn to live with a highway system that is largely already in place. This, of course, means increasingly crowded, increasingly congestion-prone roadways.

Suburban Traffic

All this has fundamentally altered the nature of the "traffic problem." In the past, traffic congestion was associated with downtown commuting and the radial routes leading into the city. Today, some of the worst traffic snarls occur far from the urban core, on circumferential highways, on roads leading to suburban office parks, and in suburban centers. In Los Angeles, for example, the busiest stretch of the Ventura Freeway, the busiest freeway in the world, is in Encino, and Encino is far from anyplace even remotely considered to be a "downtown."

Congestion has also lost its directional bias. People commuting from one suburb to another or driving from their suburban homes to a shopping center are just as likely to encounter heavy traffic as are commuters on the way to the central business district. The days of the leisurely reverse commute that we knew in the 1970s are fast drawing to an end. Indeed, on many urban freeways, inbound and outbound traffic during the rush hours is virtually identical. Driving along Maryland's I-270, a major commuter road in Washington, D.C., it would be difficult to determine which way was the downtown and which way was the suburbs because the flows are virtually identical in both directions.

Suburban traffic congestion also seems to be rapidly spreading in space. In many metropolitan areas, endemic traffic congestion is no longer confined to main radial corridors but instead pervades the entire highway network. Nor is traffic any longer confined to the densely populated portions of metropolitan areas. Country roads on the outer edges of suburbia—for example, western DuPage County, Illinois; Princeton, New Jersey; or Lowdon County, Virginia, roads which only a few years ago had very little traffic at all—now seem hopelessly clogged as thousands of commuters try to find shortcuts from their suburban homes to their suburban office destinations. In the past, commuters were able to avoid traffic by taking the back roads, but in today's suburbia in most of the country there are no "back roads." A two-lane country road is just as likely to be clogged with traffic as a freeway. Many commuters find themselves faced with the disorienting experience of being locked in bumper-to-bumper traffic while surrounded by bucolic vistas of corn fields, farmhouses and grazing cattle.

The new American city is multinucleated and increasingly dependent on a less and less adequate infrastructure of highways. Photo by Landis Aerial, Dallas.

As traffic invades formerly tranquil suburbs, it intrudes on the lives of an ever-expanding universe of people. Not surprisingly, public opinion surveys indicate that traffic congestion has become a top ranking concern of suburban voters, superceding such traditional concerns as unemployment, air pollution, housing and crime. In the San Francisco Bay Area, voters have singled out traffic as their highest priority problem for four consecutive years. In Washington, D.C., Atlanta and Long Island, public opinion surveys echo these results. Indeed, one can think of few issues in recent years that have aroused such concern and stirred so much passion among the suburban electorate as has the rising tide of traffic.

One tangible manifestation of this discontent has been an increasingly vocal grassroots opposition to commercial development. Anti-growth sentiments are not new. But unlike the growth control movement of the early 1960s and 1970s, which was distinctly a "liberal" cause fueled by a desire to preserve the bucolic character of suburban communities, the current traffic-induced anti-development fever seems to cut across ideological lines. Indeed, as one California politician put it, "Growth control has become a banner held up by everyone, from San Diego conservatives to San Francisco liberals." The problem of escalating traffic dominates the agendas of neighborhood associations, city councils and county commissions across the nation and has become the subject of intense political debate in such widely dispersed locations as Contra Costa County, California; Fairfax County in northern Virginia; DuPage County in Illinois; Princeton, New Jersey; and Atlanta, Georgia.

Let me cite a few specific examples of how this citizen discontent translates itself into political action. In Walnut Creek, California, a bustling suburban center in Contra Costa County, voters have recently approved a proposition that prohibits any commercial development of over 10,000 square feet until peak hour traffic volumes at 75 critical intersections have been rolled back to level of service D, a virtually impossible goal.

In San Francisco, voters recently passed Proposition M, a sweeping growth control initiative that effectively limits office development to no more than one medium sized and perhaps two smaller office buildings per year for the next 10-15 years. In Los Angeles, Proposition U cut the allowable FAR ratios by half, from 3.0 to 1.5. The ordinance applies to the entire Los Angeles area, with the exception of the Wilshire corridor and the central business district.

In Fairfax County, Virginia, which already boasts 40 million square feet of office space, an anti-development backlash is sweeping the county. Traffic and growth control dominate local election campaigns and development approvals have come to a grinding halt. In Bethesda, a suburb of Washington, D.C., there is an effective moratorium on any development that generates more than five vehicle trips in the peak hour. No further construction will be authorized until development is brought more into balance with the supporting infrastructure. All this suggests that public discontent with traffic congestion can indeed turn to political action, and not necessarily of the most productive kind.

But the traffic crisis has also produced a positive backlash. Aroused public opinion has finally jolted local officials, developers and employers into action and has stimulated a set of innovative responses. These responses are now coalescing into a new strategy that offers promise to control our biggest suburban problem. This strategy has been aptly named "congestion management." The word "management" truly reflects the thrust of the concept. We cannot roll back congestion; we probably cannot even stop increases in congestion. What we *can* do is keep traffic congestion under control.

The congestion management strategy has four distinct components. The first is the principle of cost sharing, requiring private developers to share in the cost of transportation improvements. Impact fees are now being routinely imposed in at least 15 or 20 jurisdictions. The intent of cost sharing is to shift more of the cost of transportation infrastructure from the general public to those whose actions have made the improvements necessary in the first place, and to those who are most able to pay and most likely to benefit from the transportation improvements.

While some observers have likened cost sharing requirements to extortion (public officials prefer to call them "enlightened extortion"), no one denies that impact fees and negotiated development contributions have become an accepted feature of the land development process. Just how important these contributions are can be seen in Fairfax County, Virginia, where over $100 million has been collected from private developers during the last five years. This compares to approximately $130 million that has been gained through a bond referendum which will be paid off over a period of 40 years.

Developer involvement is not limited to the payment of impact fees. In Colorado, a group of private investors, frustrated by the slow pace of public highway improvements, has proposed a privately funded toll road from Fort Collins to Pueblo, roughly 300 to 350 miles in length, at a cost of more than half a billion dollars. Whether the project will ever come to fruition is uncertain, but the very fact that a group of businessmen has publicly announced a project of this kind is significant. In Washington, D.C., a group of entrepreneurs announced a plan to build, finance and operate a ten-mile private toll road that will serve as an extension to the Dulles toll road and, incidentally, open up some attractive opportunities for new development. So a lot of activity is occurring in the area of private financing.

The second element of the congestion management strategy is growing private sector involvement in traffic mitigation through transportation management associations (TMAs). In a matter of a few short years, TMAs have evolved from mere curiosities to established institutions. Today, more than 30 TMAs are operating nationwide, and their number is growing. They are a new breed of private organization designed to facilitate private sector involvement in local traffic and transportation matters. TMAs give the business community a voice in local transportation decision making and provide an organizational framework for private sector participation in traffic mitigation programs. They also serve as a forum for consultations between local public officials and the private sector on common transportation concerns.

TMAs attempt to fill an institutional vacuum that often pervades the new suburbs. Indeed, in many contemporary "megacounties" the power structure is a kind of modern feudalism, a collection of small self-centered units of governance, often no larger than the homeowners association, that lacks the authority and vision to address public needs on a regionwide basis. TMAs offer a forum in which fragmented suburban interests can rally around a common concern and act as a voice and a conscience for a larger grouping of interests. Indeed, some TMAs have been used to successfully bridge, without complex political negotiations, the interests of several local political jurisdictions.

Most of the TMAs launched so far have been spurred by a mounting concern over suburban traffic and congestion. Not surprisingly, TMAs are particularly prevalent in high growth metropolitan areas where traffic conditions have reached critical proportions, such as Dallas, Los Angeles, northern New Jersey, Orange County, San Francisco's East Bay, and Washington, D.C. The initiative to form a TMA may be sparked by a variety of circumstances. In some cases, such as Tyson's Corner, one of the largest and most traffic clogged suburban centers in the nation, the catalysts were local employers and property owners who feared that traffic congestion could adversely affect the productivity of their operations and stifle future economic prospects of the area.

In other cases, such as Pleasanton and Irvine, California, and North Bethesda, Maryland, the need for a TMA has risen out of local ordinances that set a legal cap on trip generation and obliged local developers to come up with transportation management strategies that would keep traffic within the prescribed limits. In yet other cases, such as Princeton, New Jersey; the Parkway Center in North Dallas; Bellevue, Washington; and the Baltimore-Washington International Airport, the TMAs have been the product of a joint decision by developers, employers and local governments to establish a means for addressing local transportation problems on a cooperative, area-wide basis.

In short, each association is individually crafted to respond to the special needs and circumstances of the area. Some focus primarily on policy leadership and advocacy and serve as a voice of the business community in local decisions about highway and transit improvements. Other TMAs give priority to traffic mitigation. They encourage and facilitate ride sharing and transit usage among employees, coordinate alternative work hours programs, administer parking management programs, and conduct "travel audits" for local employers. Still other TMAs concentrate on service provision, such as internal circulators, park-and-shuttle systems, car rental services and subscription buses. Many TMAs also try to nurture a positive "mobility image," an important but often unstated objective, especially in places where intense traffic and congestion could adversely affect the leasing potential. Most associations engage in all four types of activities, with differing degrees of emphasis.

Whatever their primary mission, all TMAs are grounded in a common principle, that the public and private sectors must share the responsibility for dealing with traffic congestion. Unhampered by bureaucratic constraints, they can be freewheeling and entrepreneurial, devising solutions that might be difficult to implement in the more conservative environment of local government. Borne of the traffic congestion crisis, transportation management associations could mature into a form of local shadow government well suited to the realities of contemporary suburbia.

The third element of the congestion management strategy is regulatory innovations, new mechanisms whose purpose is to induce, persuade or influence greater private participation in traffic mitigation. Trip reduction ordinances, originated in California, have now spread to the

East. Also included are transportation management districts, transportation corporations and other legal and regulatory mechanisms whose intent is to involve the private sector more intensively in the control of congestion.

Can Transit Help?

What does all this mean for transit? Back in the 1970s, when I was in the Urban Mass Transit Administration (UMTA), we all had great faith that, given enough money, transit would "decongest" metropolitan areas by providing attractive, convenient, less costly alternatives to the automobile. We were convinced that transit also held great promise for revitalizing our cities and reducing pollution and congestion. Today, many of us tend to be more skeptical. We are faced with growing evidence that mass transit can play only a modest role in relieving suburban congestion.

I am not arguing that there is no reason for mass transit. I am simply saying that those old arguments that mass transit will solve congestion problems should be put to rest. The fact of the matter is that the migration of jobs to the suburbs has enormously complicated the job of mass transportation. As long as the majority of jobs was in the downtown area, public transportation could, and indeed did, function quite effectively. Buses and trains collected people at staying points such as park-and-ride lots or rail stations and whisked them directly to their places of employment in the central business district. But today, with the growing proportion of commuter trips ending as well as beginning in dispersed suburban locations, there simply is not enough "mass" to make mass transit work effectively. Indeed, some planning professionals argue that densities of FAR 0.5 to FAR 2.0 are intrinsically dysfunctional from a transportation standpoint. They produce too many trips to be accommodated by a highway network, but they do not produce enough trips to warrant efficient, effective mass transit service. Unfortunately, these are precisely the densities favored by the market forces in the suburbs.

A pessimist would say that there is no solution, no highway solution and no mass transit solution, and that we are growing in a way that is intrinsically unmanageable. I am more optimistic. I think there will always be a need for buses and trains to serve the downtown. Indeed, I am tremendously impressed by the experience of San Diego and Portland, which have generated a consensus for their light rail systems and built them on time and under cost. But Census Bureau data indicate the kind of situation mass transit is up against. In a typical metropolitan area, suburb-to-suburb commuters outnumber suburb-to-downtown commuters by 6 to 1. Even in the Washington, D.C., metropolitan area, with its heavy concentration of federal employment in the District of Columbia, only 22% of all commuter trips in the region have a downtown destination. Recent data show that the proportion of commuters who live and work in the suburbs continues to rise. In a typical Sunbelt city probably fewer than 20% of commuter trips have a central business district destination. Again, I am not arguing that there is no role for mass transit; but mass transit cannot solve the congestion problem.

Transit and Changing Urban Character

The Evolution of an Intersection in Toronto

Ken Greenberg
Partner, Berridge Lewinberg Greenberg,
Toronto, Canada

Traditionally, streets have functioned as social space, a public realm. The automobile in concert with other factors changed that.
Courtesy Barker Texas History Center, The University of Texas at Austin.

A traditional street offers a well-balanced set of conditions. It is not just a channel for movement but a social space, where people live over shops that front right onto streets and have the opportunity to meet each other frequently on street corners. It is very easy to imagine the series of daily routines that occur in that kind of setting. In this century, as a result of numerous psychological, political and social changes, there has been a fundamental challenge to that kind of built form.

The Dream of Mobility

In turn-of-the-century "transit fantasies," one can see every conceivable means of moving around at all levels. Such images are symptomatic of a fascination with the possibilities for personal mobility during most of this century, a fascination that has contributed to the distintegration of urban form. We are finally coming out of that period.

Dreams of mobility have challenged traditional urban form and the character of streets. From "King's Views of New York," 1915, by Moses King.

When I was a kid I used to read *Popular Mechanics* and *Popular Science* and, like everyone else, I fantasized about what the future would be. I imagined that my house would look like a combination airport-transportation terminal, and every member of my family would be flying off every which way in one or another type of exotic vehicle. Popular advertising reinforced the fantasy. The traditional city seemed to evaporate along with that mobility. But all of that fantasizing did not deliver what it promised, and instead we had fractured, fragmented and destroyed urban form. By the late 1960s and early 1970s, downtowns took on less evidence of human inhabitation (as a result of dispersal) and went into a tremendous tailspin.

Evolution of an Intersection

The current phase of reconsolidation of the fragmented North American city is pointedly evident in the life of a particular intersection, Yonge Street and Eglington Avenue in Toronto. The role of transit in affecting this change has been crucial. In the 1890s when the original concessions were laid out, this intersection reflected a rural pattern. By 1910 that pattern was already beginning to take an urban character and horse-drawn streetcars operated on Yonge Street. Then, between 1949 and 1953, the subway was built. Through a combination of planning policy and economic forces, a new, dynamic pattern was laid over the original, a pattern generated by subway lines that carried up to 30,000 people per hour in one direction.

The new pattern had very large scale buildings popping up at the subway stations, punctuating the traditional fabric of the city. The first generation of new buildings was extremely crude and problematic. The intersection of Yonge and Eglington is about seven kilometers from the city center along the subway line, with a periphery of single family

Large-scale new buildings began to occur at subway nodes.

houses, modest lots, tree-lined streets. The introduction of entirely new pressures into these areas was difficult for people to cope with. There was an intense struggle between the people who wanted to stay in their homes and the planning policies and economic pressures that were driving towards higher densities. Surprisingly, many of the people who lived around these areas did not abandon them to speculators or allow the areas to be cleared completely.

Sometimes these pressures resulted in large, surface parking lots that were real intrusions in the fabric of the streets and high-rise buildings drawn mostly from the repertory of second-rate Modernist design. Occasionally the buildings were a little smaller and fit better into the fabric of the streets, but for the most part, the radical juxtaposition of scales between the two- and three-story buildings and new large structures was all too evident. In short, the raw ingredients for a new urban phenomenon were in place, albeit in very strange and awkward containers. About 15,000 people now worked within walking distance of the subway station, about 60% of whom used the subway to get to and from work. Within the catchment area of that subway station lived perhaps 25,000 people.

Radical and unfortunate juxtapositions of scale occurred near the intersection.

The next stage of the process was the more graceful blending of those raw, sometimes crude, ingredients into what started to become a real urban place. In the 1970s there was a standoff between the people remaining in the neighborhood and the continuing pressures to develop the subcenter. The formulation of what is called a secondary plan was begun to try to make sense out of the confusion.

In addition to the outward expanding tendencies of large blockbuster buildings was the problem that individual projects were based on suburban models rather than urban models. These buildings had features like internalized shopping areas or an underground shopping mall with austere and lifeless plazas. In some cases, such buildings were good from a land use standpoint, concentrating people above the subway, but the way they related to streets made for a hostile environment. Also, with the elimination of previously existing streets, the blocks were too large to be workable in urban terms.

Superblock developments that emphasized interiors instead of street frontage produced austere and lifeless spaces.

As the secondary plan process began to unfold, the urban design strategy was based on an examination of how the intersection had worked originally. In searching for a more carefully balanced relationship between uses, one of the critical ingredients urban designers and planners became aware of was the immediate relationship of the ground floors of buildings to the street. It was evident that the portions of the street remaining from the streetcar period are very lively and interesting places, even at lower densities. There is fine grain and individuality in retailing as opposed to a shopping mall full of franchise outlets.

In the period since the 1970s, a process has occurred that could be equated to that of a painter who throws great gobs of pigment on the canvas in the first instance and then goes back with a more sensitive hand at a later period and reworks those gobs of paint. One aspect of this reworking is the retrofitting of the first generation of large scale buildings to give them more urbane qualities. A second is a new generation of buildings that are filling in vacant spaces. The key factor is that the overall building fabric through both retrofits and new construction is re-establishing the essential relationship of street fronts to public spaces rather than trying, in suburban fashion, to develop internal places within each project.

Another concern in dealing with density at the subcenter was the need for an intermediate scale of building, between the scale of houses and that of later towers. The intersection has started to feel more comfortable as buildings of three to eight stories containing a mix of uses have begun to mediate between these two, creating the range and variety of scales we associate with "city."

The "Mature" Transit Node

The old streetcar-generated fabric is coming back to life due to the presence of so many people on the sidewalks. The tall towers and large scale developments had tended to follow fairly rapidly on the heels of the subway development, and then slowly the fabric has filled in with intermediate scale buildings that contain an almost equal ratio of people living

The process of retrofitting the first generation of buildings is well under way. As the transit node matures, new park spaces are being introduced.

and people working. At this point in the process, a kind of fusion or synthesis occurs. What had been a series of isolated objects and discrete groups of people became something greater than the sum of its parts. From an economic standpoint, interesting relationships have emerged among the activities in the subcenter. Uses are not just randomly located because the office space is cheap; rather, as in the traditional city, the subcenter begins taking on particular economic vocations. People want to be at a certain subcenter because that is where all the other people in a certain business or profession are located.

Then night life appears. Enough people work and live there to support nightclubs, cinemas and restaurants. Similarly, enough people live there to support a network of community institutions and retail facilities that are characteristic of a real city. The cycle continues to build with development of the remaining parcels of land. High quality residential projects crop up in places where no one imagined they would go.

The Yonge Street/Eglinton Avenue intersection in 1987. Courtesy Toronto Transit Commission.

A corollary concern relates to what happens when large numbers of people are concentrated in areas that historically did not have public parks and public spaces. The suburban fabric with its low density inevitably changes in character with the imposition of intensive, rapid development. Clearly defined public spaces, both simply for visual relief as well as for active use, must be created.

In describing this history of development at one intersection in Toronto, it becomes clear that the suburban fabric in North American or certainly in Canadian cities, which was largely a post-war phenomenon, is experiencing a process not unlike that of European cities in the nineteenth century, with the emergence of *faubourgs*, urban modes outside the town center. These are "urbanizing" places, and one of the key roles that we can play—those of us who are in planning and design professions or involved in public transit—is to act as midwives for that process. Through a combination of land use policies, design initiatives aimed at strengthening the role of streets, filling in vacant parcels and an emphasis on transit, we are trying to shape the urban places emerging in suburbia, having learned from the intersections described. In fact, the greater concept of "city" to which this leads is that of a series of centers, some of which are larger than others, all of which possess the full range of rewarding urban characteristics.

At the same time this "urbanization" is occurring, the center city is becoming less and less purely "central business district." In fact, we are encouraging people to live downtown. Symptomatic of this shift is the fact that in the last few years the residential population in the most central political ward of Toronto is now as great as that in the purely "residential neighborhoods" that surround it. So on the one hand city centers are becoming neighborhoods and on the other hand more and more people are working in traditional neighborhoods.

Part II

Case Studies In Transit Development

In the last decade and a half, a number of cities have responded to the need for mass transit in a variety of ways. These case studies describe strategies and technologies for re-introducing mass transit into cities which, in most cases, had workable transit systems prior to the widespread use of automobiles. Comparisons can be drawn between technologies selected (varieties of rail and bus) and the rationales for their selection. The ways the vital economic potential of transit development is being harnessed also can be compared. In many cases, an insistent citizenry ensured that planning and implementation be responsive to a broad set of issues—urban design, land use, environmental quality—and not to transit alone.

Toronto: Thirty Years of Transit Development

Juri Pill
General Manager, Planning,
Toronto Transit Commission

In transit and urban planning, Toronto has often taken a unique approach. In the late 1940s when everybody was building expressways, we built our first subway, and when everybody was abandoning streetcars, we kept ours. We still have over 200 streetcars in service and most of our major downtown surface routes are operated by these vehicles. Some people claim that transit cannot work in the suburbs, yet true to Toronto tradition we are putting transit into the suburbs.

The first subway line in Metropolitan Toronto was a happy accident. During World War II transit had little competition from automobiles. The Toronto Transit Commission (TTC) developed a large surplus in revenues because the price controls that existed in Canada did not apply to transit fares, and after the war the transit commission went to the people of Toronto and said, "We have this money and we would like to build a subway with it. What do you think?" With the money and financing to build that subway, the busiest streetcar line in Toronto was replaced by a subway that opened in 1954. That first north-south line was about 4-1/2 miles long, and based on its success a second, east-west, subway, was built in the mid-1960s. As in most North American cities, the per-capita ridership in Metropolitan Toronto had been declining. We cannot prove it was the cause, but when the east-west line opened and transit travel speed became competitive with the car in a second major corridor, the decline in per-capita ridership stopped.

Currently, the TTC has the highest per-capita ridership in North America; residents of metropolitan Toronto take an average of 200 trips per year on the transit system. The operating revenue/cost ratio is about 70%. The employee turnover at the TTC is about 3%, half the industry average in Ontario. The industrial accident rate is also about half the average for industry in Ontario, and the TTC has won the top American Public Transit Association traffic safety awards 15 times in the last 18 years as the top transit system above one million population.

But in 1946, the TTC was no different from any other transit system in North America. How we got where we are now is a matter of strategy: Provide the best service possible and people will use it. With a first-class product, quality rather than price is the issue, and the TTC is a transportation business, not a social service agency. We fulfill a social function but we try to operate in as businesslike a way as possible, and as a result about 65% of metropolitan Toronto residents over the age of 15 use transit at least once a week. As of 1984, 22% of our riders had family incomes over $40,000, compared to 27% of the entire population. The latest figures indicate that, in terms of family income, about two-thirds of the top 20% of the population in metropolitan Toronto use transit at least once a week.

At the lower income levels, transit use is higher, but we have consciously chosen to cater to those who have a choice as well as those who use transit by necessity. Instead of a downward spiral, transit can have an upward spiral, in that the more riders you have, the closer the headways are and the more riders you attract. There is a feedback loop that works in both directions, such that if you get everybody to use transit then those who *must* use transit end up with better service. We have special subsidies for low income people instead of having a low fare for everyone, since it would not make sense to subsidize fares for those with high incomes.

As for financing, we get about 70% from the farebox. Three-quarters of our capital financing is provided by the provincial government. We have no federal government involvement in transit in Canada, and 25% of the capital cost is paid by the metropolitan government through property tax assessment and business taxes. We work very closely with our financial partners and since 1977 we have had what we call the "Users' Fair Share Formula." That is, we set a revenue/cost ratio target and then impose regular small fare increases to keep up with inflation.

Transit Shapes Development

The subways in metropolitan Toronto have shaped development, because the system was built incrementally over a period of about 35 years at the same time that the population of metropolitan Toronto doubled. When we started, the population was about one million; it is now approximately 2.2 million, so the two worked together during that time. However, major transit-related development in many cases took 10 to 20 years to materialize. For instance, at the intersection of the two major subways—the Yonge Line, the first north-south line, and the Bloor-Danforth Line, which opened in the mid-1960s—development took about 20 years to materialize, and even now only two of the four corners have been developed to a high density.

Metro has had strong planning incentives for transit-oriented development at subway nodes, such as zoning bonuses for developing

Keele Station on the Bloor-Danforth subway line, 1966, prior to opening. Courtesy Toronto Transit Commission.

Keele Station and subsequent adjacent development, 1987. Courtesy Toronto Transit Commission.

around subway stations. Most of that development took place during a 10 to 15-year period starting in the 1960s under a zoning policy that encouraged high density apartment redevelopment around subway stations. But there was a reaction against that scale of redevelopment, so it generally has stopped, as Ken Greenberg recounts in Part I.

The Official Plan for metropolitan Toronto is also transit-oriented, and since the major political reaction of the early 1970s was against rapid growth in the downtown area, limits were set and expressway construction in metropolitan Toronto ended. Part of the Metropolitan Official Plan now includes a provision that there will be no more major transportation links into the downtown.

Mixed use development at Yonge Street and St. Clair Avenue, 1987. Courtesy Toronto Transit Commission.

The Scarborough Rapid Transit Line to the eastern suburbs was built to guide line use to an eastern subcenter under a subcenters policy adopted about 10 years ago. We are hoping to guide subnode development to achieve a higher transit modal split. Our goal is about 50% at three subscenters, but these subcenters will not take all the suburban development riders by any means. There is considerable office park and auto-oriented development as well, but by placing limits on downtown development and structuring growth in the suburbs, we hope suburban transit use will also increase.

Plans are under way for two metro subcenters. The North York Subcenter, which is in the northern part of Metro, grew from about 750,000 square feet of commercial space in 1976 to 2.6 million square feet of commercial space in 1985, with approved proposals bringing the total to approximately 3.5 million square feet of gross office floor area. Employment was 15,000 jobs in 1985 and was projected to increase by the turn of the century to 47,000 jobs. Scarborough City Center to the east grew from 380,000 square feet in 1976 to about 2 million square feet, with approvals for another 500,000 square feet. Employment in the Scarborough City Center will change from 7,000 in 1985 to 13,000 in 1990 and 40,000 by the turn of century.

Long Term Land Use and Transit Planning

Has the TTC been successful in orienting a city to transit use through metropolitan land use policies, or are the transit-related developments successful because we have good transit service? The answer is "both." Transit and land use have to work together, and Toronto is a case study of how it has worked, where over a period of about 40 or 50 years both policies have headed in the same direction at a time when the population doubled.

Every community has to make conscious decisions about urban form, about whether it is going to have a single center, multiple centers, or sprawl. For rapid transit to work, development has to be high density, preferably clustered. The single-center, "downtown" option is inefficient because only half the capacity is used. If there is only a strong downtown center with radial lines then the off-peak direction is not used; therefore, only half of the investment is used. A balanced mix seems best and that is the basis of MetroPlan, the Metropolitan Toronto official plan.

To manage development associated with transit planning, we put calls out for proposals and maintain explicit criteria for evaluating these proposals. We use present value and equivalent annual return in evaluating the financial analysis. We tightly define the net return so as to share in the return from development. One of our more complicated agreements took about two years to negotiate but in the end gave us a nearly 50% net return. And we never sell land that relinquishes the right-of-way. Instead, we lease air rights and look for return on the profit from related development. Refinancing, as detailed in a development agreement, has to be very tightly defined or there will be no profits to share.

Joint development is not a financial panacea in terms of financing the transit system. Its main benefits come as much in system ridership to the transit system as in direct financial return, and in our case the financial return comes through the overall tax base rather than through any earmarked development levies.

The integration of transit planning with land use planning must be part of an overall, long term strategy. This overall strategy has to be adopted by the community as a whole, not by the transit system alone. The transit system cannot do it by itself. A city must consider its intentions: Why do we have or why do we want a transit system? Is the transit system intended mainly as a social service, that is, as a charity operation in the crassest sense, or as a business operation? If it is intended purely as a social service, then joint development and land development must be based on this role, for example, with low income housing related to the transit system. In that case, the overall joint development options are going to be very limited. With a transit captive-only system, there is not much point in integrating overall transit planning and land use because that type of land use planning affects only a small portion of the entire community. The transit system becomes peripheral and the community will be almost totally auto-oriented. The integration of expressways and development will be the major issue until everything stops in one gigantic traffic jam.

The option that Toronto chose 40 or 50 years ago was to make transit part of the fabric of the community, to aim at the "by choice" riders, that is, people who have a choice, with quality, cleanliness, reliablity, safety, speed—all those boring sorts of things that make transit work. They may not be flashy, but they are critical. Details count; quality requires investment. It does not matter precisely what it is—LRT, busway, subway, commuter rail; that depends on the community, on the population distribution density, etc. But it does require investment in the real sense of that word, that is, putting money in for a payback later. The commitment has to be long term.

The Portland Light Rail Experience

John R. Post
Assistant General Manager,
Tri-Met, Portland, Oregon

We at Tri-Met like to speak of the success of the Portland light rail system in terms of ridership, improved productivity, service quality and reliability. It is easy for us to overlook the fact that such a project significantly affects the urban landscape of the community it serves. Portland has been proud to say that through the efforts of local elected officials, jurisdictional professional staffs, local business and citizen representatives and the committed professional design firms brought into the project, the project design has itself contributed to the success and acceptance of the project. A story reported in the local press illustrates the point with a former opponent of the light rail project, a local architect, declaring, "It fits; it looks as if it was always there." The point is, the project is a complement to the community and has been comfortably absorbed into the fabric of the city.

Project Background

The Portland project is a combined highway and light rail project. For the highway portion we upgraded a substandard inner-city freeway section five miles in length. The light rail project, totaling 15.4 miles from the center of downtown Portland to the heart of the suburban community of Gresham, runs adjacent to the upgraded freeway for a portion of its

Heading east on Interstate-84, Tri-Met's MAX travels up to 55 miles per hour. Courtesy Tri-Met.

alignment. We chose the combined project because it best fit the emerging transportation and land use policies of the region.

Features of Portland's Light Rail System

- 27 stations
- five stations that serve as bus centers providing connecting service for multiple bus lines (all but three stations outside the downtown have some level of bus service)
- five park-and-ride lots with a total of 1700 spaces
- a fare structure fully integrated with the bus fare system
- 26 bi-directional light rail vehicles with four double doors per side; the cars are single articulated, 88 feet in length, 8 feet 8 inches wide, and have a seating capacity of 76 (the cars were furnished by Bombardier of Canada and were assembled in Barre, Vermont)
- maximum operating speed of 55 miles per hour (speed varies within each segment depending upon the operating environment)
- travel time end-to-end of 43 minutes

In the downtown, the design was particularly sensitive to the need to accommodate pedestrians and automobiles. Sidewalks were upgraded and extensive use was made of different materials to delineate the areas for the rail vehicles, automobiles and pedestrians. Operating speed is 15-25 miles per hour in this section of the project. In the two downtown historic districts, care was taken to fit the light rail system into the character of the areas, including use of such materials as cobblestone within the trackway.

Outside the downtown, we used a more functional design in terms of rail operations, featuring a more positive segregation from adjacent automobile traffic. Along the freeway section, the rail line is totally segregated from the adjacent traffic and pedestrian movement. The top operating speed in this section is 55 miles per hour.

Along Burnside Street, a rebuilt suburban arterial street, the rail line is separated from traffic by a standard curb except at key intersections. Operating speed is 35 mph on this section, with intersections controlled by standard traffic signals that can be preempted by the rail vehicles.

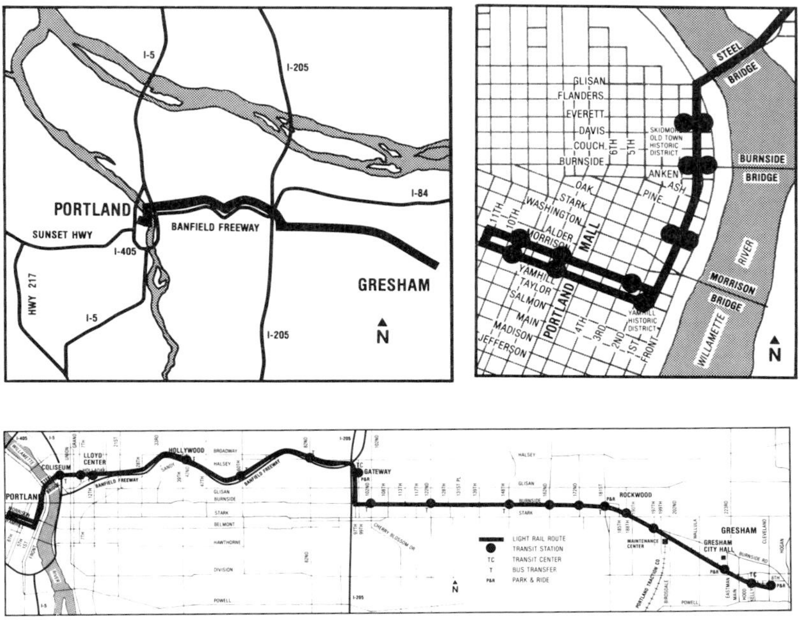

The Portland area light rail system connects downtown Portland with neighborhoods and towns to the east.
Courtesy Tri-Met.

The outer three miles of the project is on a separate alignment, an abandoned freight line. Operating speed on this section is between 35 and 55 miles per hour, with crossings handled by standard railroad gates actuated by the rail vehicles.

Handicapped access is provided by wayside lifts located at each station platform. The lifts represent one of only a very few project elements which were developmental, that is, not standard, off-the-shelf design. The lifts are located on the platforms and are operated by the vehicle operator.

Stations were designed to be simple, straightforward and low in maintenance requirements. They include features such as protective covers, information displays, trash receptacles, benches and fare vending machines.

The project was completed within the budget established in 1981, which specified $328.6 million for the light rail and highway improvements, 83% of which was available from Federal Interstate Transfer Funds and UMTA Section 3 and Section 9 sources. Local funds were provided from a variety of sources, including $11.6 million in state gas tax funds, $25.8 million in state general revenue funds, $15.1 million in Tri-Met funds, and $3 million from various private and local jurisdiction sources. Approximately $105 million of the budget was used for freeway construction items. It is anticipated that settlement of some construction claims will increase the final budget, but will not approach the available funding amount.

Project History

The planning basis for the project was established in the mid-1970s. During this period, the Portland area was rethinking its regional transportation plan which called for the construction of an extensive urban freeway system to handle projected travel demand to the year 2000 and beyond. The transportation issue was brought to a head when a major new freeway segment designed to serve the east side of the Portland region ran into stiff community and local political opposition, which eventually led to the withdrawal of the freeway segment from the interstate highway system. A revised transportation plan calling for a "balanced" transportation system was adopted, and the region set out to define a combination of highway and transit solutions that would best serve the needs of the region. Highest priority was assigned to finding a solution to the transportation problems of the east side of the region from which the previously mentioned freeway segment was withdrawn. Numerous highway and transit options were explored.

On the transit side, detailed exploration was undertaken on transportation system management (minor improvements on existing arterials), high occupancy vehicle lanes, busway and light rail options. These planning efforts, involving Tri-Met, the Oregon Department of Transportation, the regional Metropolitan Planning Organization and all affected local jurisdictions, took place during the period 1976-78. The results of these efforts were assembled in the federally required Draft Environmental Impact Statement in the spring of 1978. The remainder of 1978 was committed to the development of a Preferred Alternative Report and Staff Recommendation reports which recommended selection of the light rail alternative. The combined light rail/freeway option was adopted by the six local and state jurisdictions required to formally act on the recommendations.

The period 1979-81 was primarily spent pulling together funding commitments at the local, state and federal levels, although the preliminary design efforts were also advanced during this period. The keystone of local funding was a one-time general fund appropriation passed by the State Legislature in June 1979. Final federal funding commitments in the form of a Full Funding Agreement with the Urban Mass Transportation Administration was not received until February 1982, the day of groundbreaking for the light rail maintenance building.

Construction was completed during the summer of 1986, and the system was opened for revenue passengers on September 6, 1986. Start to finish, approximately 10 years were required, with four years devoted to the construction phase. As always, there were several unique circumstances that affected our schedule which might not be pertinent to other communities. One particular element was the fact that the project was the first combined transit/highway project utilizing a combination of Interstate Transfer and UMTA Section 3 funding to be jointly administered at the federal level by both UMTA and the Federal Highway Administration. Decisions at the federal level were slow at best and often had to be coordinated with opportune moments in the federal funding cycles. Local decisions also took considerable time, but it was regarded as time well spent in making a milestone community decision.

Why Light Rail Was Selected

Six separate jurisdictions were involved in selecting the light rail option from the list of available alternatives. With the possible exception of the county government, the light rail option did not start out as the favored alternative of the involved jurisdictions. In fact, it was twice dropped from consideration. From the perspective of Tri-Met, the transit operator, the most significant factor was the projected reduction in operating costs when compared to the costs to carry a comparable number of passengers with any of the bus alternatives.

Happily, we are able to report that our experience to date confirms that rail operating costs per both boarding and originating ride are well

Gateway Transit Center, about midway along the MAX line, affords a convenient transfer between Tri-Met's light rail service and a dozen bus lines. Courtesy Tri-Met.

below our experience on the bus side of the operation. Such results are driven by a strong ridership response and by below-budget operating expenses. Other factors of importance to Tri-Met in the selection of the rail option were the higher projected ridership; operating characteristics such as safety, speed and reliability; the fact that the rail option resulted in the least impact in terms of the housing and business relocations; and the fact that of the options under consideration light rail received the highest level of public support.

Other jurisdictions cited other factors in their decisions. The City of Portland felt the rail option was most consistent with its plans to reduce air and noise pollution and traffic congestion in the downtown area. In particular, the City had concerns about the impacts on Portland's confined downtown area if all the projected transit ridership was to be handled by additional diesel buses. The City also cited the rail option as offering the greatest opportunity to focus and enhance adopted development and redevelopment plans. The combined rail/highway alternative was also judged to be most consistent with the City's transportation plan, which called for diverting regional auto and transit trips off inner-city arterials and onto regional facilities.

The city and county governments both cited the environmental advantages of the rail option over the bus options. The primary factor in Multnomah County's favorable decision was the consistency of the rail alignment with emerging land use policies and plans. All jurisdictions required to take a final position on the project cited strong citizen support as a key factor in their choice of the light rail option.

Citizen Participation and Awareness

Oregon has a tradition of extensive citizen participation in all aspects of public and private development. It was assumed that the largest public works project in the state's history would involve an extensive citizen involvement program. The process was most intense during the project planning period leading up to the jurisdictions' decisions in the fall of 1978. As an example, the Oregon Department of Transportation (ODOT) and Tri-Met jointly sponsored 120 public meetings and workshops during 1976 and 1977. In addition, each local government jurisdiction involved in the decision process held its own series of public hearings prior to taking their final action.

A Citizen Advisory Committee (CAC) was active throughout this period with an appointed membership of 133. To effectively interact with the project technical staffs, the committee formed into eight subcommittees representing both geographical and topical areas of interest. The CAC was particularly active in designing a program to involve the citizenry and made sure comments received were considered. The CAC was also heavily involved in developing the information program for formal public hearings. The media and direct contact techniques were used extensively and resulted in 300 groups and individuals submitting comments during the hearings. During this phase, the CAC process proved itself successful in defining local concerns and identifying special problems and sensitive areas.

From 1979 through early 1982, when the project was in the design phase, the citizen involvement process was less intense but nonetheless extensive. Participation was focused in smaller groups, businesses and neighborhood representatives as design details were worked out. Central topics were land use planning in station areas, final alignment and station locations, and such diverse subjects such as the aesthetics and comfort of stations and light rail vehicles.

From mid-1982 to completion, Tri-Met sought to involve and consult citizens in the construction phase. Meetings with neighborhood groups, merchant associations and local service clubs were an ongoing process. The most impressive citizen outreach, however, was the Banfield Transitway Community Relations Program. This Tri-Met program employed a special staff team that worked with neighborhood groups and one-on-one with individual property owners to reach agreement on property changes made by the construction process. Property owners were able to review detailed design plans with Tri-Met before construction regarding issues of grading, sidewalk and driveway reconstruction, and the removal and replacement of trees and shrubs in the right-of-way. Complete agreement was reached with property owners before any construction took place. Unique to the Tri-Met program was the "Early Warning System," a 48-hour pre-notification procedure for notifying residents of utility shutoffs and new traffic routings. Special notice was given to businesses for any planned disruption of access. A hotline and 24-hour answering service were installed at the project office.

Since no streetcars had operated in Portland since 1958, an extensive public education program was also undertaken just prior to inaugurating the new light rail service. Particular attention was paid to safety, that is, how automobile drivers and pedestrians were to relate to this new vehicle running down their streets. All in all, the program used to involve citizens in the planning, design and construction of the project was well accepted and provided a basis for a smooth transition to the operating phase.

Land Use Issues

The light rail project was and is part of a conscious strategy to shape regional growth by coordinating transportation investments with land use policies. For more than a decade, land use issues have consistently been included in the light rail decision-making process. Land use considerations influenced the decision to substitute transit for a freeway, the choice of the light rail mode and the selection of route and station locations. Concern for the land use impacts of light rail resulted in a $1.2 million station area planning program. The program laid the foundation for development along the line by determining market potential, planning for the urban fit of the project, and rezoning station areas. That work is in place and although we would be the first to state that it is too early to provide a definitive report, based on the development community's response to the project, everything we have seen to date looks very good. Most important, the development and business communities seem satisfied with the results and have provided an upbeat attitude concerning the ability of light rail to be a positive factor in the economic development plans of the region.

System Advantages and Disadvantages

Given our experiences to date, it is fair to say that we continue to feel that the "advantages" side of the ledger is much longer than the disadvantages. Advantages include:

Public/Rider Acceptance: The general public and transit riders have enthusiastically embraced the project. Persons who will not give the bus system a try will use light rail.

Community Pride: While no one gets excited about the initiation of new bus service, light rail aroused interest. Beyond the 200,000 riders who showed up for the first 2-1/2 days of free service, there has been a continuing positive response from the community, business interests and the media. The light rail system is prominently featured in numerous publications attempting to interest businesses and visitors in the Portland area.

Operating Costs: On a per passenger basis, the light rail system's operating cost is approximately half that experienced with the bus system. To date, our first year operating costs are running below our projections. With increased ridership we anticipate the results will be even better.

Farebox Recovery: Because the fare structure for the rail operation is fully integrated with bus operations, it was not anticipated that the rail system would generate a signficantly higher farebox recovery ratio. But this has not been the case; with a higher-than-anticipated level of originating rides, the rail system is realizing a 51% recovery ratio compared to 27% for buses.

Service Quality: The rail service provides a ride that is smoother, quieter and faster than the previous bus service, and because of its attractiveness it is an easier service to market.

Permanence/Development Response: No matter what the volume of service and passengers, bus service is difficult to sell to the development community which is much more willing to respond to a rail project and the permanent commitment to high level service it represents.

Disadvantages include:

Capital Cost: This is an obvious item, but one that cannot be ignored. The up-front investment required to obtain the advantages outlined previously can be considerable.

Maintenance: Introduction of light rail into a previously all-diesel bus operation has created some difficulties. First, a whole new vehicle maintenance crew trained in electronics rather than diesel engines was required, which cuts down the flexibility in use of our maintenance forces. As opposed to the bus operation which requires very little on-street maintenance, the rail operation introduced a 15-mile corridor that must be continuously monitored and maintained. With the facility totally accessible to the public, the susceptibility to vandalism is significant.

Traffic Interface: Although segregated or semi-segregated from traffic for a good portion of the alignment, the rail operation's success, in terms of reliability, is heavily dependent upon smooth operation of those segments affected by automobile traffic and traffic control devices. A careless automobile operator, a malfunctioning signal and similar events do occur and do interrupt the rail operations. The success of a system with design features such as Portland's is heavily dependent upon the cooperation of the local jurisdictions and requires complete cooperation with local traffic engineers.

In summary, the entire process of reintroducing rail operations into Portland has been an effort that has paid off handsomely in terms of a community learning to work together and taking pride in the product of its efforts.

Portland and Its Unique Traditions

Greg Baldwin

Planning and Downtown Development Specialist, Zimmer Gunsul Frasca Partnership, Portland, Oregon

When we first contemplate a public capital improvement program for a city, it is important that we understand the personality of the city and its stage of development. For example, remember that American cities, regardless of how vital or how urbane, are very different from their European counterparts. We are not far removed from nature and therefore maintain an agrarian philosophy that espouses qualities of independence and self sufficiency. When you look at American settlements you will find that this philosophy is fundamental to their preservation, a persuasion very different from one promoting the ideal city of interdependence, the rich and sometimes chaotic character of communities on the other side of the Atlantic.

Remember that this is the country in which Ralph Waldo Emerson said that his happiest hours in Boston were spent in the Back Bay Station waiting to catch the train for Concord. Recognize also that most of our western cities are very young. As with our children, we have tried to nurture them, give them the kind of advice over time that will allow them to make intelligent decisions on their own (with an occasional exception). We have done this with the expectation that when they reach adolescence they will be able to establish, based on this advice and experience, a vision of what their own future might be.

Portland is that kind of city. We are still very much attached to our roots and have enjoyed only a brief period of adolescence. Consequently, we have just begun to chart a vision for our future. Portland is a port city with a very large hinterland that for the most part has been supportive and only occasionally hostile, a hinterland to which we are well connected by rail, water and air. Its downtown has a ubiquitous grid plan that has allowed it to grow as required—a fine architectural heritage, but one that, until recently, has made the future difficult to define.

In the late 1960s we learned four very important lessons affecting that future. One was that in a 120-acre area just outside of downtown, we found that with $14 million worth of public investment in streets and open space, one public building, and a few amenities, we could leverage more than $400 million worth of private investment which in turn complemented the original public investment. The second lesson was the potential threat of a competing retail center close to downtown that could effectively undermine downtown's potential. The third was that any new development downtown, if not carefully directed, could provide a mixed blessing for the future of downtown. Finally, whereas in the 1960s we were the U.S. city most committed to freeways on a per-capita basis, in the 1970s we produced one of the first two interstate transfer projects in the United States, providing funding and an incentive to do many more creative things with transportation projects.

Strategic Public Improvements

It was in this context that Portland began to plan for its future, not with prescriptive plans supported by restrictive development regulations, but with a plan based on some general goals and policies. The plan provided a framework for a series of key public capital investments such as the transformation of two streets in the heart of downtown into a transit mall, which today carries more buses than any comparable facility in the country. In less than a decade the transit mall has encouraged the development of almost five million square feet of public and private projects along its length, which otherwise would have located in suburban communities. There was a commitment to provide the focus and support for a revitalized downtown center and the short term parking it required, with the result that closing department stores were replaced. Others threatening to leave stayed, and downtown's share of the regional

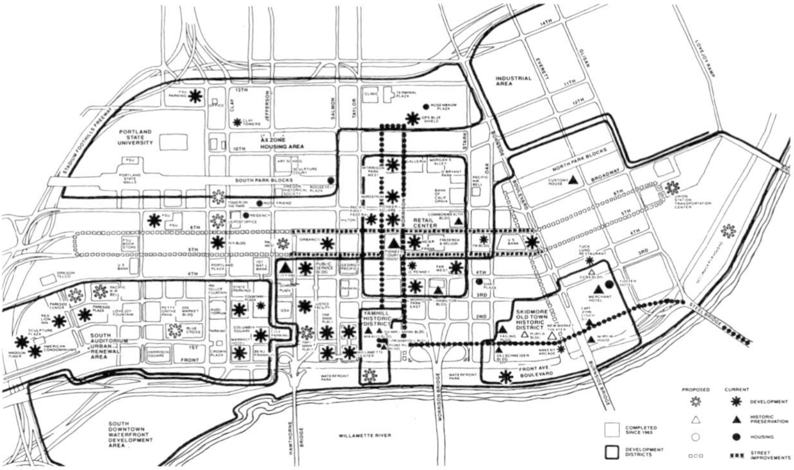

Transit development in Portland was part of a comprehensive effort to restructure and revive the central city.
Courtesy Portland Development Commission.

retail market grew from 7% to nearly 30% in a dozen years. A freeway and a parallel arterial road next to the Willamette River were removed and replaced by a park that had been programmed for regional use in hopes that adjacent historic districts would gradually be restored and new development would eventually reconnect the river and the downtown area.

Recently a new public project not contemplated in the early plans for Portland, but guided by their presence, has opened. As the first phase of a light rail system for Portland, the Banfield Transitway extends 15 miles through downtown along a freeway, through a major transfer station, and then down the middle of a street through a dozen suburban communities. In fact, it is much more than a transit project. Already it has transformed areas it serves. The heart of downtown has become more urbane and dynamic, both during the week and on weekends. The neighboring Yamhill Historic District has been altered architecturally and economically while benefitting from major new development stimulated on its periphery. The long-vacant Skidmore Historic District undertook concurrent improvements and is now active seven days a week and evenings.

The light rail system and its station areas avoid imposing a uniform "transit" image by assuming varied characteristics in particular neighborhoods, in downtown historic districts, in the retail core, at the downtown

Care was taken to reinforce the character of historic areas, like the Yamhill Historic District. Courtesy Tri-Met.

turnaround and maintenance area, in a quiet park, in the hostile environment of a freeway where a reasonable pedestrian environment must be created, in the wide open spaces of a major suburban transit center or in the middle of a suburban street. If more citizens are to be attracted and served by transit, the system must be inviting, conceived with care, and accommodating, while also contributing the unexpected. Even its construction should be an attraction. It should be conceptually simple, functionally prudent, and at the same time catholic in concept. In the process it should provide a very special place in the city.

Principles Learned in Portland

From these few public projects we have learned a number of principles that are significant, but not particularly surprising. The first and perhaps most important is that with a few well-placed dollars in public investment you can attract a significant and complementary private investment. In the Portland Transit Mall, for every public dollar expended we found that within a decade 30 to 50 private dollars were invested in response. Waterfront Park is drawing millions of square feet of new development towards the Willamette River. Retailers along the light rail line report sales up 30-100% over previous highs. More people ride the system on Saturday than ride during commuting hours or during weekdays.

Even in the continuing recession, construction cranes were visible, mostly along the light rail line. In one case, a developer built 40,000 square feet of retail space at the base of an office tower so that he could establish a rapport with light rail patronage and capitalize on the proximity of his project to the transitway. The federal government has recently completed a $65 million office structure next to a light rail station outside of downtown. A convention center now being designed has been located across the river from downtown and rationalizes its location by its proximity to the light rail line.

Second, we have found that, as in the past, the very best streets are those that accommodate the full range of activities that normally occupy an urban area. Single purpose streets, pedestrian malls, and exclusive transit malls have a number of deficiencies, especially a conspicuous lack of activity. The most vital streets are those that try to accommodate everything and everyone—transit vehicles and private vehicles in the same rights-of-way and intense pedestrian use. The design problem is to allocate space carefully and ensure enough for each activity, places for people

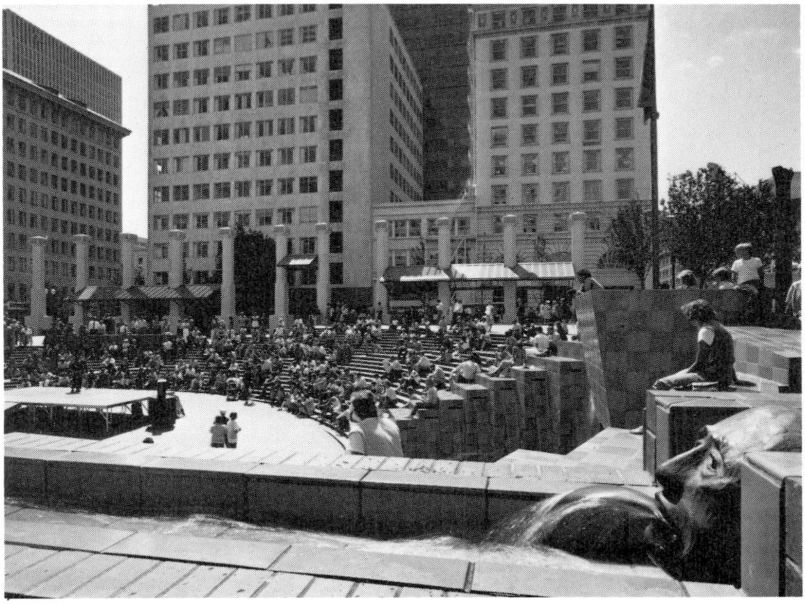

The hub of the light rail system is Pioneer Courthouse Square. Photo by Strode-Eckert Photographic.

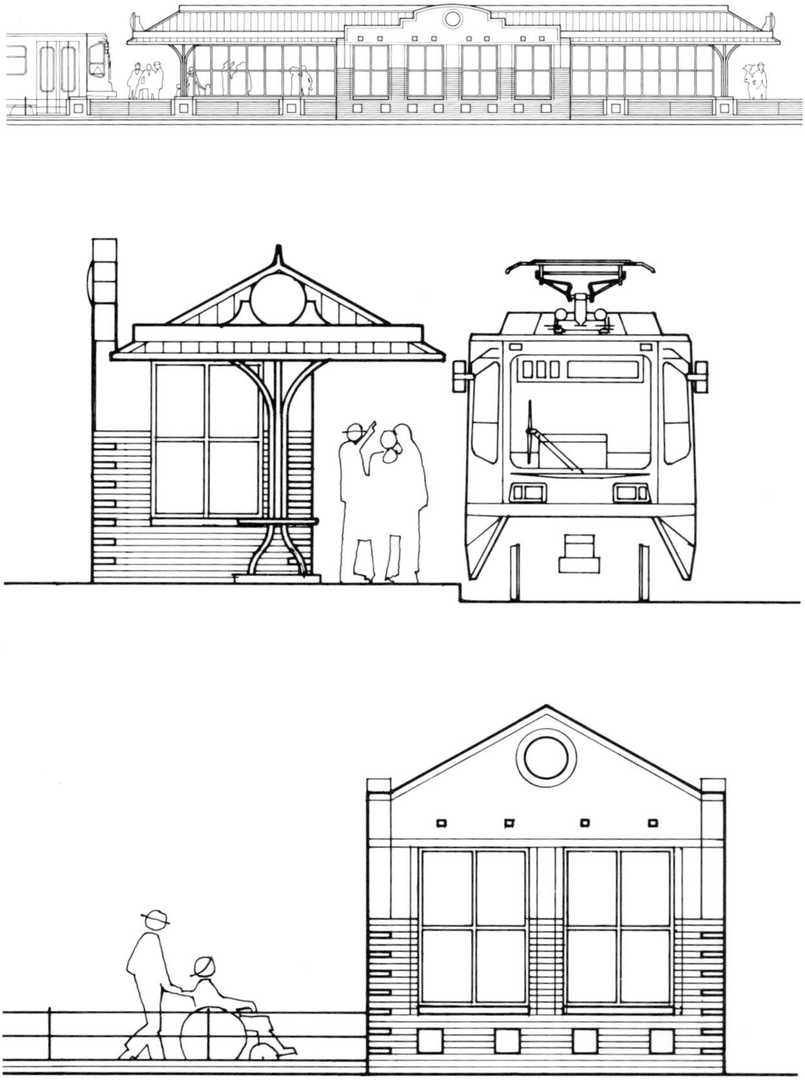

Suburban transit shelters are brick and expandable to accommodate changing patronage requirements. The masonry structure, laid on site, may be provided in different colors and executed in several shapes to permit each station its own architectural identity. Courtesy Zimmer Gunsul Frasca Partnership.

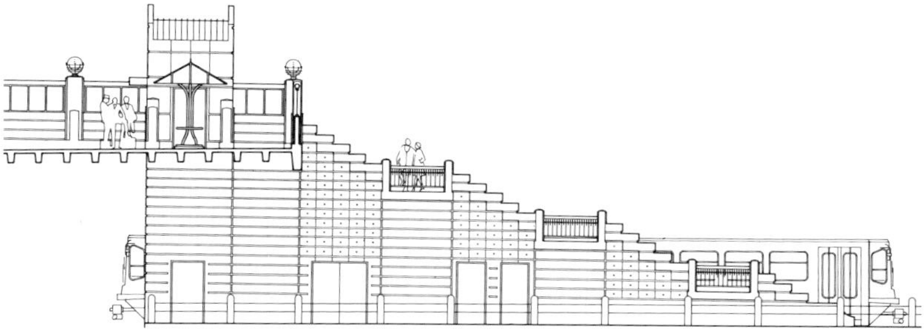

To reintroduce the bridge as a structure intended to accommodate pedestrians as well as vehicles, an early twentieth century State Highway bridge design common to the Portland area has been modified. The scale of the pedestrian has been emphasized in the design.
Courtesy Zimmer Gunsul Frasca Partnership.

to walk, for ground floor retailers to move their activities out onto the street, for fixed elements like light standards, trees and benches.

The third principle is that the construction process needs to be made more inviting and less repulsive. It is always a disruptive activity. However, depending on use of materials and scheduling, it can be an inviting process and its negative impacts on downtown business, particularly on retailing, can be minimized. A key factor is the involvement of the downtown business community in the implementation process.

Finally, we recognize that we are always creating a civic architecture, something that needs to be special and reflect the very best aspirations of the citizens we hope to serve. When we use special materials, it must be with conviction, conveying permanence. Whether employing a very simple array of materials or an elaborate palette, we must recognize that we are creating something permanent for the public and a precedent for future projects. Even when the components are modest, like a substation under a bridge, we must make sure that the design receives extra attention.

The development of projects in Portland is a very public process. What is important is that we have learned to apply that process without compromising the independence and integrity of the individual project, whether public or private. In Portland, we have learned that our very best projects are those that "talk" to each other, those that begin to communicate without compromising their own independence. Buildings in many other cities often stand alone at the expense of any kind of communication.

Although we are very serious about our goals, we do tend to approach them with anticipation and humor. Remember that Portland is the city where a local tavern owner discovered that with the proper public exposure he could learn the route to the mayor's office. And it was also in our city that our *magna mater*, Portlandia, was introduced with pomp and fanfare, as tens of thousands of citizens accompanied her down the Willamette River when she was brought to her final resting place on the public services building. It is a city where public projects are designed to attract and serve, where both public and private buildings are unusually gregarious, and yet a frontier spirit of independence and self-sufficiency is still uncompromised.

The San Diego Trolley

Robert Robenhymer
Senior Transportation Planner,
San Diego Metropolitan Transit
Development Board

The San Diego Metropolitan Transit Development Board (MTDB) is the umbrella agency over the Metropolitan Transit System established in 1975 by an act of the State Legislature with an emphasis on guideway development. MTDB is not an operator but owns the San Diego Trolley and San Diego Transit, the area's major bus transit supplier. It also owns the San Diego and Arizona Eastern Railway and contracts with the San Diego and Imperial Valley Railroad to provide freight service on our facilities. MTDB also has programming and administrative responsibilities over several smaller transit agencies. The objective of the Metropolitan Transit System is to present a single transit system to the riding public through the cooperative coordination of fares, transfers and services.

LRT Development in San Diego

Planning for San Diego's light rail transit (LRT) system, called the San Diego Trolley, was initiated in 1976. Because of limited capital resources, construction cost had to be low. In order to compete successfully with the automobile, the initial line had to be long and operate at a high speed, and a high speed could not be achieved without exclusive right-of-way. Acceptable environmental impacts were a necessity in order to gain community acceptance and operating costs had to be low to stay within known financial resources.

The initial LRT project, the South Line, was adopted by the MTDB Directors in June 1978, and the first construction contract was awarded in January 1979. Thirty months later, in July 1981, revenue service began on the South Line. It took only 54 months to go from inception of project planning to initiation of revenue service and the project was completed on schedule and under budget. A second line, the East Line, was opened in 1986, also on schedule and under budget.

Project Description

The South Line extends 16 miles from the Santa Fe Depot in downtown San Diego to the International Border with Mexico at San Ysidro. The South Line has 12 suburban stations and six downtown San Diego stations. Travel time is 42 minutes, with a top speed of 50 mph. Trains operate every 15 minutes from 5 a.m. to 7 p.m., then every 30 minutes until 1 a.m. The East Line added 4.5 miles and four new stations to the light rail system.

Light rail in San Diego shares track facilities with an operating freight railroad owned by the MTDB. Both the South Line and the East Line are

San Diego Trolley terminus at Santa Fe Depot, downtown. Courtesy MTDB.

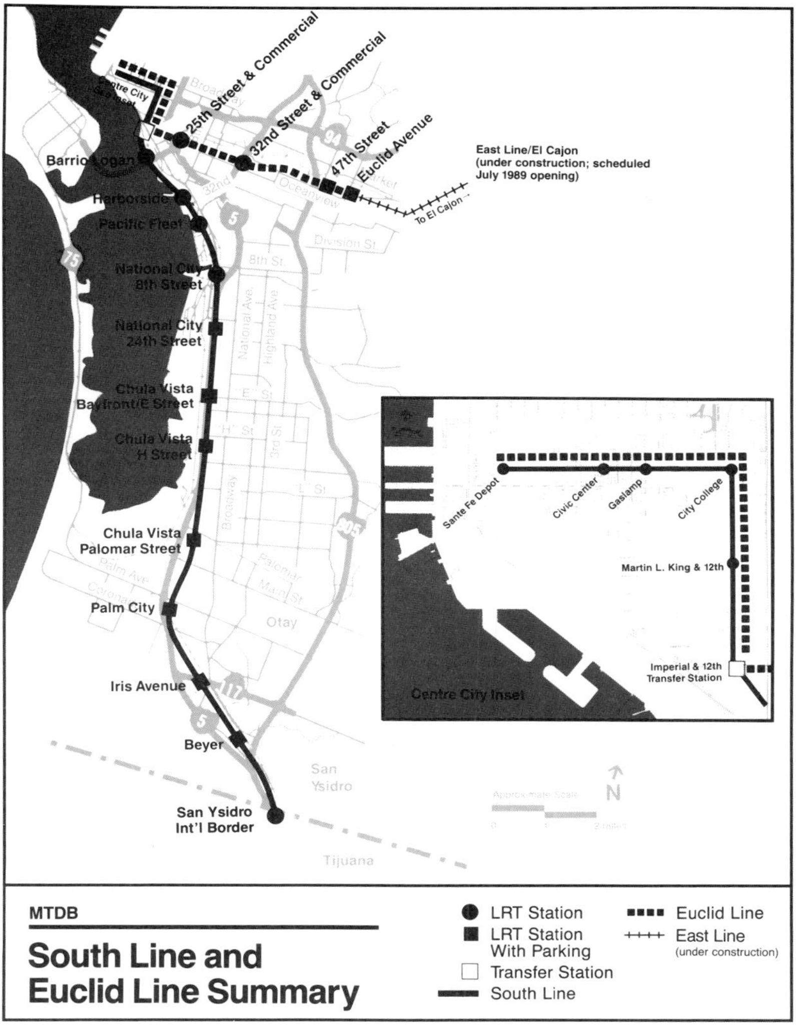

Route map, San Diego Trolley.
Courtesy MTDB.

built along a railroad right-of-way purchased in November 1979 for $18.1 million. The railroad right-of-way was completely rehabilitated, including replacing all the rail and most of the ties and reballasting the trackbeds. However, in keeping with our low-cost approach, some curves were left unchanged. Another cost savings was to use existing timber trestles if testing confirmed that they were structurally sound. On the East Line, rather than build a second bridge adjacent to an existing one-track bridge spanning an eight-lane freeway, a gauntlet track was built, providing double tracking at each end of the existing bridge and a single track over the bridge.

Downtown pedestrian/transit mall, San Diego. Courtesy MTDB.

In downtown San Diego, the Trolley operates within the street, with an exclusive path for LRT and auto traffic maintained alongside. However, four blocks along C Street have been closed to automobiles to create an LRT/pedestrian mall. Along the C Street transit/pedestrian mall, stations are simply an extension of the sidewalks. Within the downtown, cross-street traffic is controlled by signals and stop signs. Along the suburban portions, at-grade crossings are protected by gates and flashers.

LRT is characterized by its ability to use all kinds of rights-of-way. Also, LRT vehicles can operate singly or, in San Diego's case, in trains up to four cars. Operating costs in San Diego are kept low by adding or

Suburban station on the San Diego Trolley system.
Courtesy MTDB.

deleting cars in accordance with passenger demands. During peak periods, four-car trains are operated. However, because there are short blocks in downtown San Diego, four-car trains would block cross streets. Therefore, these are split into two-car trains as they approach downtown and are recoupled again after they leave downtown. This is another example of LRT's versatility of application.

Stations range from those with parking in suburban areas to simple platforms in downtown San Diego. The suburban stations offer modest shelters that provide overhead protection and minimum amenities including benches, transit information, telephones and fare vending equipment. Station design is consistent with our practical, low-cost approach. Bus service has been coordinated to provide timely transfer to and from the Trolley.

The Trolley uses a self-service fare collection system. One or more ticket vending machines are located at each station. Fares range from 50 cents to $1.50. The new fare machines accept $1 bills and change machines are located in some stations. Fare inspectors check passengers for proof of payment. About 25% are checked, with an evasion rate of 1.5%. Violators are issued a ticket similar to a parking citation, and these are handled through the court system just like a parking citation. The Trolley recovers over 85% of its operating cost from farebox revenue.

The cost of the initial South Line project was $86 million. This included construction of a mostly single-track line, purchase of 14 light-rail

Ticketing machines, San Diego.
Courtesy MTDB.

vehicles, and the acquisition of the railroad. Subsequently, the line was completely double-tracked, and six additional light rail vehicles were purchased, resulting in a total South Line cost of $116 million. The 4.5-mile East Line cost was $33.6 million, including the procurement of four more vehicles. The total cost of the South and East lines was $150 million, or $7.3 million per mile.

Construction recently began on an 11.1-mile extension of the East Line, which is scheduled for completion in July 1989. It will serve the eastern suburban cities of Lemon Grove, La Mesa and El Cajon. This extension will also utilize the railroad right-of-way and will cost $103.4 million, including the purchase of 15 light rail vehicles. For the first time, there will be federal funding involved, including $20 million in federal discretionary money and $38 million in formula money; 56% of the project will be funded with federal dollars.

Key Factors in Trolley Program Development*

We made a clear decision early in the process regarding our objectives. These were not stereotyped objectives such as improving air quality, trying to serve everybody within one-quarter mile, reducing traffic congestion, etc., but were more philosophical objectives such as providing an affordable system that will move people efficiently. Very specific objectives must be defined at the outset of a study and then abided by through project implementation and operation. Fortunately for us, a strong, clear legislative directive was part of the enabling legislation that created MTDB. This legislation stated that any fixed guideway that would be implemented must use off-the-shelf technology, lend itself to incremental development, and use existing rights-of-way to the greatest extent possible.

This legislative directive, plus the realization that transit ridership demands in San Diego could not justify a high capital-intensive system, resulted in a pragmatic approach laid down by our Board of Directors through their adoption of elements of feasibility, such as low construction cost, high operating speed, and cost-efficient operations. These have been translated into guiding principles that we have not abandoned.

Another key factor is that we had, and still have, a very limited budget. It was impossible for us to get into trouble by "gold-plating" our system, since the Board was very strict in seeing that the original project budget and schedule were adhered to. For this reason, our projects have come in under budget and on schedule. Also, our ability to purchase a 108-mile freight railroad for $18.1 million was fortuitous. This provided us with two corridors, south and east, that lent themselves to the implementation of joint light rail transit/freight operation at a very low cost.

MTDB had total local control over the implementation of the South and East lines. Both projects were funded entirely with state and local money; the Board realized early in the process that it was very unlikely for San Diego to successfully compete for federal discretionary dollars. However, for the new line to El Cajon we do have federal participation and will see if there are any budget or schedule impacts. Already there

*Tom Larwin, MTDB General Manager, authored a portion of this section.

have been inflationary impacts because the project startup had to be delayed while we sought federal approval.

It is clear that events shaping implementation vary from area to area due to differences in institutional arrangements, personalities, and levels of commitment of the people involved. We were fortunate in San Diego to have supportive leadership at the local, state and, more recently, federal levels at times when all three were necessary. This leadership translated into key actions being taken by various political bodies as critical decision points were reached. The personalities involved permitted a smooth coordination of bus and LRT planning from the start of initial planning studies through to operations. Because MTDB was the "new kid on the block," there could have been some major problems. However, all issues were worked out harmoniously to the mutual satisfaction of all involved, including the riding public. The staff and Board levels of commitment were very high. Supporting the political leaderhsip were a comprehensive technical effort and a substantive community outreach program. We were willing to meet any group of two or more people on their own turf.

Finally, from start to finish, the MTD Board, management and staff have had a united, pragmatic view of LRT. This view included low cost and quality functioning—speed, safety and reliability—as chief objectives. Budget and schedule objectives were firmly maintained by the Board. The Board principles adopted in 1976 were translated into design criteria in 1978 and have not been significantly altered over time. This consistency has permitted a clearly-defined focus for project implementation.

Transit and Planning Techniques in San Diego

Paul D. Curcio
Assistant to the Planning Director, Urban Design, City of San Diego

Mass transit is a system of transportation that is in complete opposition to the freewheeling personal mobility associated with Sunbelt cities. Like it or not, the car symbolizes America, personal wealth, freedom and power. If seen as a threat to these things people value, mass transit will not succeed. Instead, transit must be characterized as the only means of *maintaining* personal mobility. An educational process must clearly illustrate that mass transit and supportive urban development patterns are the only way of preserving mobility-oriented lifestyles.

From a policy perspective, force-feeding mass transit to an unwilling public is not a substitute for providing a transportation network that satisfies personal mobility demands. Sunbelt cities will continue to grow as automobile-dependent, polycentric areas. The planning challenge is threefold: to promote the highest quality automobile-oriented development, both environmentally and aesthetically, for areas of lesser intensity; to preserve the opportunity for future intensification by requiring development phasing plans that allow for diminished automobile dependency, and guaranteeing right-of-way reservations for future transit; and to establish development incentives to support transit by encouraging higher intensity development within potential transit corridors and transit nodes.

At a more physical level, the life-giving importance of circulation must be more fully realized. It determines the growth patterns of all cities. The architectural maxim "form follows function" has an urban corollary. In city design today, form will follow parking, street design and transit. Recognizing that this framework will generate the form of a city, it is imperative that land use and circulation be conceptualized simultaneously. From a technical point of view, the professions need to reorient their efforts; engineers need to accept the challenge of doing more within less space—less right-of-way, more traffic volume. Environmental designers need to advocate quality-oriented street standards, for it is essential to reestablish the role of the public right-of-way in offering open space and in serving pedestrians. Streets should be designed as pleasant and memorable places to walk and not just to move cars. Planners must institute a wide array of programs to improve personal mobility. Transit lines and transfer stations should be fixed early in the planning process to encourage mobility-enhancing land development patterns.

With regard to implementation strategies, power brokers are much more likely to endorse a transportation management program as a whole than separately presented, specific mass transit solutions. If $100 million is needed for mass transit, it is far better to request $300 million for "transportation improvements," allocating one-third to freeways, one-third to roads, and one-third to mass transit. For some reason, the fixed menu is much more palatable than the same items a la carte. In addition, comprehensive programs must be supported by quantified analysis to clearly illustrate the long term value created by balanced transportation systems and the marginal short term cost that is incurred. Meaningful and understandable statistics go a long way toward building consensus and devising a workable fiscal strategy.

Transportation Problems and Solutions in San Diego

The history of San Diego transit is literally a "Back to the Future" storyline. In 1886, the City of San Diego had several horsecar lines operated by the San Diego Streetcar Company. By 1890, the San Diego Cable Railway Company was running 12 cars on two lines. The San Diego Electric Railway Company, backed by John D. Spreckels, was formed in 1891. This new company purchased the San Diego Streetcar Company along with the trackage of several small firms and began work on an electric rail system. By the end of 1897, electrification was complete and the system continued to grow. It reached its peak size in 1925 with 107 miles of single track.

As prosperity and mass production of automobiles in the late 1920s quickly expanded ownership and use of cars, transit use began to decline. San Diego's post-World War II growth was rapid and largely suburban. By 1970, 80% of households owned one car, and population densities were relatively low. Core areas actually lost population. Communities became dispersed; federal funds subsidized freeways, and parking was inexpensive and often free. Economic forces including federal freeway funding and the merchandising of Ford Motor Company supported, if not induced, sprawl.

The 1960s and 1970s saw a shift back toward localized bus transit. Fragmented service emerged to serve the needs of each community. Had people lived, worked and played within a single jurisdiction, this system would have worked very well. But the reality was that despite balanced community planning, various recreational and employment opportunities tended to blend our region's population, so people do not live and work in single communities. Increased interregional trips created by dispersed communities and tremendous population growth gave birth to our current transportation problems and transit programs.

The problems in San Diego and other major metropolitan areas are not surprising. Failure of the 1972 Clean Air Act is imminent, accompanied by a potentially devastating loss of federal funds. The amount of land used to move and store automobiles is a more visible problem, affecting environmental quality, land values and the city's tax base. Additionally, the human resource lost to mindless commuting is immeasurable. Most importantly, the dependency of the entire system on the price of oil offers a frightening prospect. Currently, 10-15% of San Diego's urban freeway mileage is at an F level of service—bumper to bumper. This could increase to 39% by the year 2000 with current growth rates and proposals.

These problems and inspired political leadership gave birth to a series of plans and programs in the 1970s and 1980s. First and foremost is our public information effort, Ride Choice. The Ride Choice program promotes ride sharing through the commuter-computer program, park-and-ride interceptor lots, vanpools, dial- a-ride and general transit usage. The program has gone a long way in raising awareness of transit options and traffic impacts. San Diegans are very protective of their quality of life and the quality of their environment. Most are startled to learn that one bus will eliminate 60 unsightly parking spaces, that 80% of all trips are single occupancy, and that San Diegans average four trips a day for every man, woman and child in the city. In addition to publicizing information like this, the Metropolitan Transit Development Board (MTDB) provides programming coordination for all transit services. This allows for transit passes and compatible scheduling while preserving separate ownership of each line by its respective municipality.

In 1987, the citizens of San Diego endorsed a 1/2% sales tax, which is projected to generate $2.5 billion over the next 20 years. This ballot measure also earmarked the improvements for which the funds would be used, with approximately one-third going to mass transit. Of this $700 million, 80% will be used for the extension of the trolley line to the north and east. Additional bus service and diesel train operations to Oceanside along the Amtrak right-of-way will also be funded. The remaining $1.7 billion will be equally divided between state highways and local streets and roads. This tax is an indication not only of mass transit's growing popularity but of the politically significant fact that voters dislike traffic even more than taxes.

Existing and proposed facilities depend upon trolley and headway frequency to help make the transit option visible and viable. The synergy of linked systems provided for by MTDB's intermodal coordination efforts is substantiated by increased ridership on downtown buses. Plans in Centre City are proceeding for a major downtown transportation station including a trolley stop, bus exchanges and a mixed use office, retail and parking complex.

The introduction of the trolley gave birth not only to MTDB, but to its headquarters. The project started out as a modest six-story office building program. Not only has it almost doubled in size since its inception but the program has been expanded to include retail uses and interceptor parking.

The eastern communities of San Diego will soon be serviced by the Euclid line. In total there will be over 90 miles of track in place by the year 2000, ironically 17 miles less than what was operating in 1925.

Development and Incentives

Current policies reinforce transit stops with moderate density and mixed use zoning designations. In Centre City East, transit overlays have been instituted, limiting auto intensities within walking distance of trolley stops to encourage pedestrian-oriented development. The recently enacted Mid-City Plan District has density gradients directly related to transit corridors with reduced residential parking requirements.

Density bonuses, relaxed parking requirements and transit availability are slowly transforming miles of underutilized Mid-City strip commercial into the dynamic core of an urban community. For example, if someone desires to build a convenience store in a commercial strip, they are permitted an FAR of 0.25, basically all that is necessary for that type of development. However, should they wish to, they may build more commercial square footage as part of accompanying medium-density residential projects. This bonus, combined with automatic reductions in required parking for mixed use, makes the economics of intense development very attractive.

The city has also introduced citywide landscape requirements that assure quality automobile-oriented development. These requirements produce one tree for every four parked cars, a street tree for every 30 feet of frontage, landscaped parkways, a prohibition of front-yard parking in residential zones, and the provision of buffer strips between parking lots and the public right-of-way. These landscaping provisions, coupled with increased parking requirements for lower-density, single-use developments, are changing development pro formas. The economics of sprawl change radically when mandated to take the form of well-designed street development and enhance the attractiveness of more urban transit-oriented development patterns.

Finally, we have succeeded in exacting developer contributions to aid in the provision of the Mission Valley Line. This is a critical link in the system, tying together the north, south and east lines in a functional loop. This system will go a long way toward relieving congestion along one of the most heavily traveled freeways, Interstate 8. In anticipation of the 6% reduction in trip generation, the City Council has offered to developers bonuses of up to 10% in development intensity for "full participation." Full participation has come to mean dedication of 35 feet of right-of-way, funding station and track installations, and agreeing to participate in future assessment districts. Also, as part of the Mission Valley financing plan, an intervalley jitney will be provided, linking mandatory mixed use developments, transit stations and additional parking facilities. The basic strategy here is to allow development to proceed only at the rate that infrastructure is provided. Everyone is therefore motivated to have the system move forward as quickly as possible. We are currently projecting completion of the entire system between 1995 and 2005.

Mobility Planning

The development of our fixed rail system is only part of an extensive mobility planning program. This program is a comprehensive citywide planning effort. It will prescribe, on a community-specific basis, parking management and demand management programs to reinforce transit

Mobility Planning Process

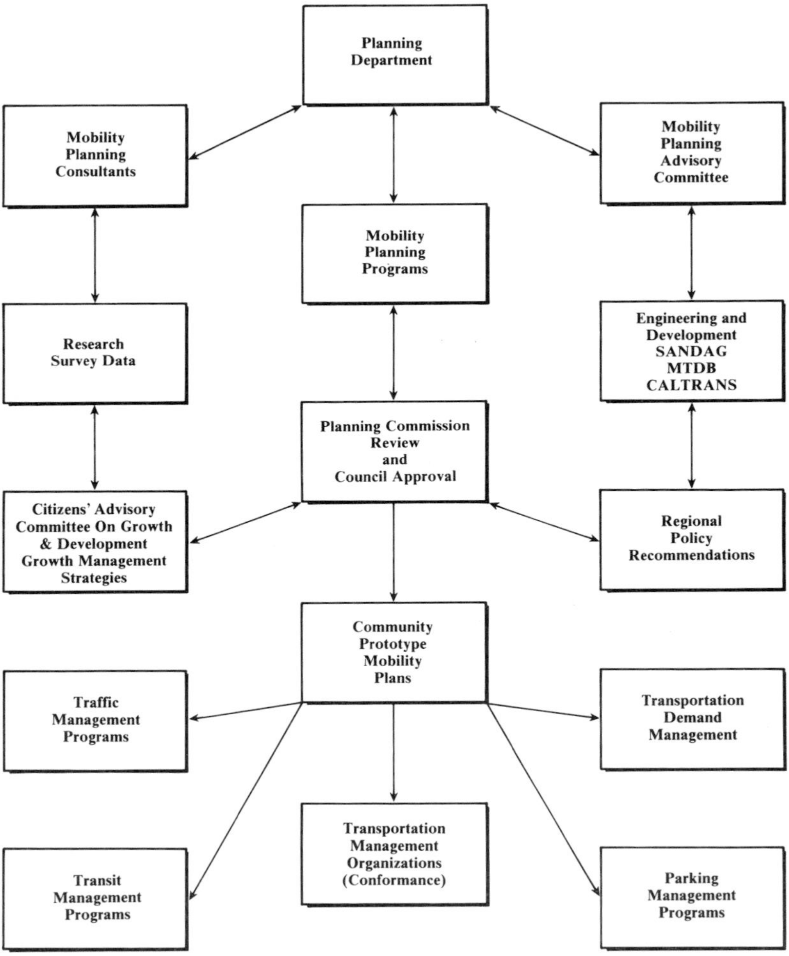

Figure 1

Mobility Planning Components

Components	*Implementation*	*Products*
Traffic Management	Traffic Operations	Intersection and roadway widening One-way streets Turn lane installation Turning movement and lane use restriction New freeway lane using shoulders
	Traffic Control	Local intersection signal improvement Arterial signal system Area signal system Freeway diversion and advisory signing Freeway surveillance and control
	Roadway Assignments	Exclusive bus lane — arterial Bus-only street Contraflow bus lane Reversible lane systems Freeway HOV bypass Exclusive HOV land — freeway On-street loading zones Off-street loading zones Peak hour on-street loading prohibition Truck route system
	Pedestrian and Bicycle	Widen sidewalks Pedestrian grade separation Bikeways Bike storage Pedestrian control barriers
Transit Management	Transit Operations	Bus route and schedule modifications Express bus service Bus traffic signal preemption Bus terminals Simplified fare collection
	Transit Administration	Marketing program Maintenance improvements Vehicle fleet improvements Operations monitoring program
	Inter-Modal Coordination	Park-n-ride facilities Transfer improvements
Demand Management	Paratransit	Carpool matching programs Vanpool programs Taxi/group riding programs Dial-a-ride Jitney service Elderly and handicapped service HOV preferential parking
	Work Schedule	Staggered work hours and flex-time Four-day week
	Pricing	Peak hour tolls Low-occupancy vehicle tolls Gasoline tax Peak/off-peak transit fares Elderly and handicapped fares Reduce transit fares
Parking Management	Parking Regulations	Parking requirements for new development Curb parking restrictions Residential parking control Off-street parking restrictions Parking rate changes
	Restricted Areas	Area licensing Auto restricted zones Pedestrian malls Residential traffic control
	Marketing Program	Maps Signage Subsidy system

Figure 2

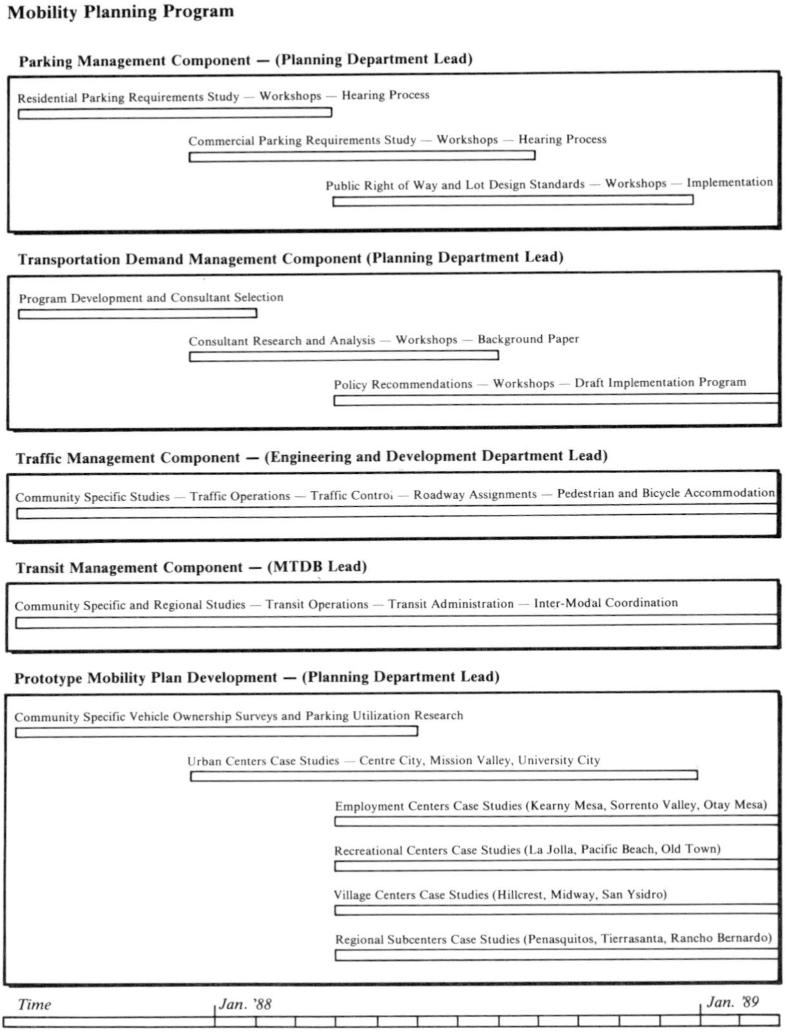

Figure 3

utilization and master plan traffic improvement projects (see Figures 1, 2 and 3). Current Council policy requires that all community plans consider transit an integral component and address how other plan elements specifically further the city's transit objectives. The hierarchy of planning policies and programs is illustrated in Figure 4.

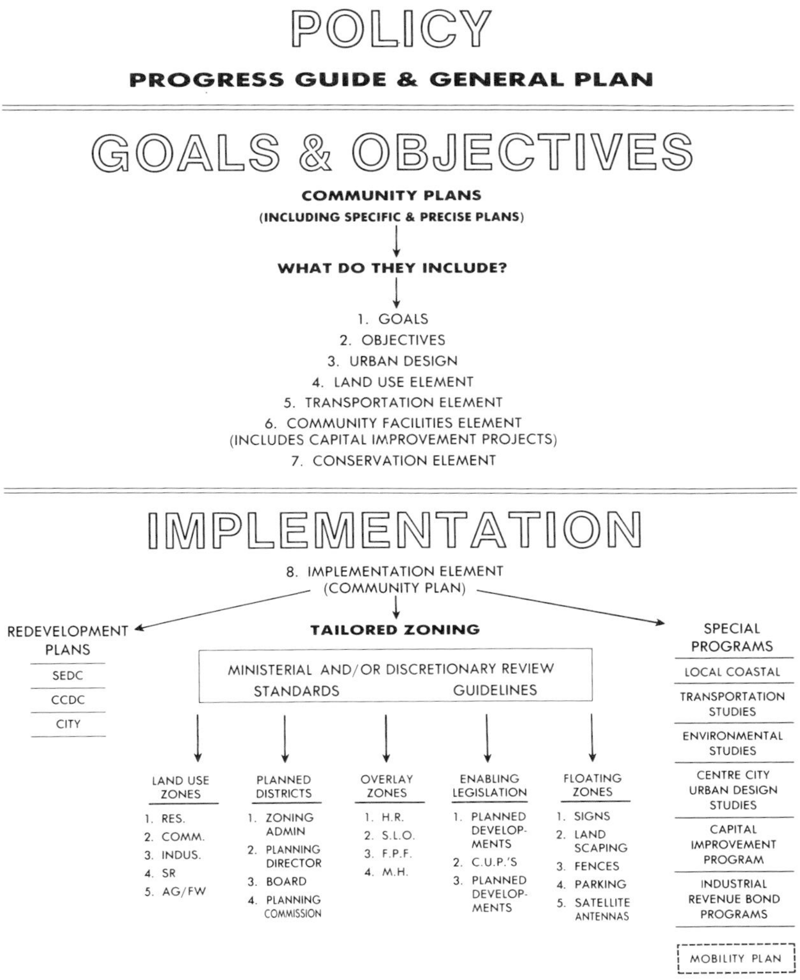

Figure 4

The General Plan establishes subregional policies to guide when and where the city should grow. The Growth Management Strategy designates urban reserve, urbanizing and urbanized areas. This strategy coupled with the Urban Design Element directs how quality development may occur. The next level within the hierarchy of planning programs is constituted by the 52 separate community plans. These community plans recognize the unique character and opportunities with regard to open space, housing, transportation and urban design found in each neighborhood. The Implementation Element of each plan specifies tools available to accomplish the planning concepts. Mechanisms vary from redevelopment and capital improvement programs to land use regulations and assessment districts. Current policy is to use ministerial or "as of right" zoning tailored to each specific neighborhood instead of indiscriminately applied citywide controls or mandatory discretionary review at the mercy of avenging "taste police."

No single "urban wizard" can be credited for the many successful programs already in place; they are the product of participatory planning. This critical element in San Diego planning, as illustrated in Figure 5, provides for the involvement of volunteer architects, planners, developers, and community interests in formulating visions and action plans.

The Participatory Process for Community Plan Development and Implementation

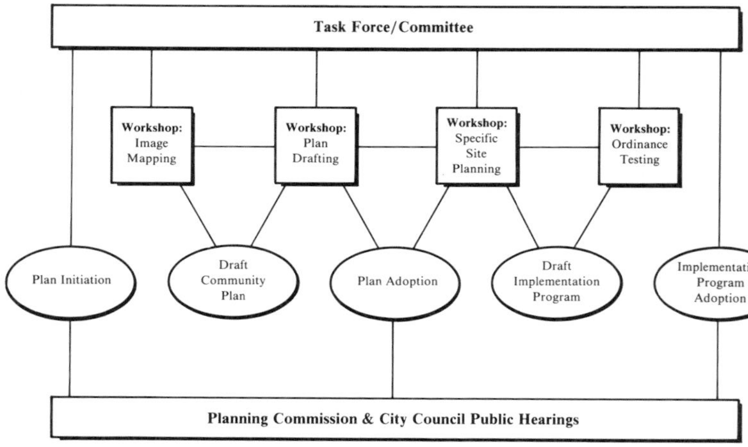

Figure 5

Broader Planning Context

In addition to the programs already discussed, the City of San Diego has initiated several other pro-active planning measures. These planning efforts deal with how, when and where the City should grow. Our basic growth management strategy is currently being reworked. Faced with the specter of building moratoria and ballot-box planning, the development industry is gladly accepting a radical shift from a "pay-as-you-go" philosophy to a "pay-for-where-you-went-before-you-go-any-further" mandate. The previous strategy of encouraging infill in other neighborhoods has worked all too well. What mass transit advocates might label as "a critical mass capable of supporting public transit" has been perceived by the residents as urban overfill, taxing schools, parks and infrastructure to the breaking point.

In response to these forces, several approaches to public facility financing are being considered, such as property transfer taxes, special assessment districts, tax increment instruments, increased sales taxes and general obligation bonds. Beyond these rather traditional direct methods are also some innovative approaches. Incentives for contributing either rights-of-way or capital for public improvements in exchange for density bonuses are being considered. Allowable development intensity is increasingly being tied to transit availability.

Some communities are actually advocating mandatory minimum development intensities along transit corridors to guarantee a shift of development pressures away from existing sensitive areas. All community plans are being formulated with a strong concept of balanced development. Application of this concept directs growth of commercial, residential and office space such that that their planned proximity to one another results in lower trip generation. It is important to note that these programs may result in fewer trips but probably will not drastically reduce average vehicle ownership.

An examination of citywide parking requirements is currently under way and we do not anticipate lowering resident requirements. Current proposals do, however, provide for substantial reductions in guest parking and commercial parking in recognition of mixed-use, high-intensity transit accessible development. In the downtown core area of San Diego, which is heavily served by transit, parking maximums have been advocated with no success. In 1985 a concept was presented to establish a maximum office parking ratio of one space per 1,000 square feet. This approach met with certain death due to claims that market demands are clearly four spaces per 1,000, even though developers were currently providing only 1-1/2 spaces per 1,000. In any case, four parking spaces per 1,000 square feet does not make a city. An alternative proposal is currently being developed to require the equivalent of the projected demand of four spaces per 1,000 square feet. The proposed concept would allow one of these spaces to be provided on-site when completely enclosed, one in the impact area in the form of public parking, and the other two as interceptor parking along existing or proposed transit lines. A system of "in-lieu-of" contributions will also be available, along with bonuses for reduced on-site parking.

In addition to these complex methods of making transit more appealing, the City will also be exploring increases in downtown parking rates for meters and lots, aggressive ticket enforcement programs such as booting the wheels of frequent offenders, and designation of high-occupancy vehicle lanes, in addition to citywide required demand management techniques such as employee transit subsidies, telecommuting, flex hours, shuttle buses and preferential carpool parking. It should be noted that incentive programs work only if base requirements are high enough to make them attractive.

In considering the full range of plans and programs, it is important to continue to ask the question, "If we get what we ask for, why is it we do not get what we want?" As more and more sophisticated land-use controls are created, sometimes basic objectives become obscured by or gradually lost within the complexity of the regulations. To quote the former Planning Director of San Francisco, Allan Jacobs, "Ultimately, there is no incentive like a requirement."

Sacramento Light Rail: Lessons and Advice

Wendy Hoyt
Assistant General Manager,
Planning and Development,
Sacramento Regional
Transit District

The Sacramento Transit Property serves both the City and County of Sacramento, a service area of 340 miles. The city of Sacramento, the state capital, has a population of approximately 400,000, and another 400,000 residents live in the unincorporated area of the county. Sacramento is not on the coast, but is inland, east of San Francisco in California's Central Valley, and was until recently primarily an agricultural land.

The Sacramento Transit Property is a small bus company with only 186 buses at the peak. Regional Transit carries about 16 million riders per year, 60,000 riders per day. We have 80 bus routes in the county traveling 30,000 miles daily. We carry commuters, shoppers, students and tourists. Previously, the I-80 and U.S. 50 arterials brought in 37 different bus routes per day, every day. They picked up passengers in all the suburban areas and then brought them in on those major arterials. We decided it was an inefficient use of operating dollars to have all those buses with all those drivers running in the same two corridors. So we built a trunkline rail system and feed it with a bus system. We also developed concentrated bus access points. We have only six major transit centers that allow passengers to transfer from bus to bus as well as from bus to rail.

The cost of this system was about $9.6 million per mile. The interstate highway project from which we diverted the funds would have cost $25 million a mile. We have the capacity to carry 10 times as many people as we did with our previous transit system, with the same number of drivers. Our rail operators all transferred from buses and earn the same hourly rate. Therefore, for that hourly rate, instead of carrying up to 70 people on the bus we are carrying 700 people on four-car trains. We are saving a lot of operating money, and operating dollars are very precious now.

In the last five or ten years, the community has changed tremendously, and from the planning perspective it will change tenfold more in the next decade because people have discovered that our labor pool is extensive and our land costs are low.

Anticipating Growth

Since Sacramento is presently one of fastest growing metropolitan areas in the United States, we are starting to think more in regional terms. The decisions made in Sacramento and other "Megatrend" cities are going to have a far greater impact than similar decisions made by planners in slow or non-growth areas. Implementation goes so quickly that there is no time to regroup and change course as there is in other communities.

It is very important in such a situation to have a focused direction that everybody can share—community, elected officials and staff alike. Agreement is needed about where you want to be in 15 or 20 years. Related to this is a need for coordination with local governmental agencies and the community. A transit agency cannot plan in isolation; it must involve city, county and state governments and regional councils of goverment. Staff people must be involved in the technical working groups that plan systems and consider alternative analysis and long range planning. Agreements about assumptions and methodology are necessary, although sometimes difficult to achieve. The downtown employment base must be considered. I have never been in a community without debate about how many people work downtown!

A transit plan needs to be coordinated and registered in everybody else's plan. In California, we have State mandated general and community plans at both the City and County levels. Also, there is the Parks and Recreation Department master plan and the streets and roadways long range plan. Each of these plans should include major transit corridors, whether rail or busway. They should include satellite transit facilities like major park-and-ride lots. Developers need to know about transit corridors and anticipated transit centers. These cannot appear on transit agency plans only. Early in the process it is wise to preserve precious rights-of-way, even if the technology to be used is uncertain. It will be too late in 5 or 10 or 15 years to go back and refine those corridors; they are going to be encroached upon as the community reaches build-out.

Regardless of the combination of financing, whether sales tax, assessment districts or land dedications from developers, a reasonable level of financing must be in place. My strong opinion is that, to the extent possible, the financial base be non-federal. San Diego was able to do much of what it did with its rail lines because local money was used. That gave them local control of what they wanted to do, where they wanted to do it, and most especially when they wanted to do it. Federal approval takes a long time.

The long range plan and the financing have to be in place early because if action is delayed until traffic problems become severe, before the general populace starts to complain, it is too late. That is the situation Los Angeles and Detroit are in now. They are so built out that there is no right-of-way, and widening is impossible because setbacks are not sufficient. It is better to take advantage of the opportunities to acquire land early and, when building or widening any major arterials or highways, to allow for at least a high-occupancy vehicle or light rail lane.

I spent seven years in Detroit working for a seven-county transit property. There are over five million people in that service area, and the infrastructure was built several decades earlier, when Detroit was going through the auto industry boom. The area grew very quickly, just as Sacramento is growing quickly today. A lot of my time in Detroit was spent correcting mistakes that had been made earlier. We had an existing infrastructure in an established community and needed to retrofit transit into it, contending with tremendous traffic and development impacts after the fact. That is a frustrating environnment to be in. From Detroit I learned that we cannot go back. Los Angeles's attempt to build a heavy rail system for $400 million a mile is a warning, as is Detroit's attempt to build a light rail system for almost $80 million per mile. It is much better and certainly less costly to build early in a community's growth cycle.

Working Cooperatively with Others

In our planning department we do not have control over land use decisions, which is typical for a transit agency. We talk to the development and planning communities about coordinating land use and transit. We have a process through which every development application that comes into the City or County is reviewed by us for transit impacts. We might suggest a modification in the design of a bus shelter contribution or, in a large project, a light rail station or right-of-way dedication. The important thing is that we ask for them right then, at the outset of the process.

We review between 100 and 150 development applications per month, a time consuming effort done with a very limited staff. But it has paid off well. Many of our requests have been granted. In particular, we try to determine the future necessity for transit in an area and whether a plan or design allows for it. In Detroit, a large portion of the community was built out in a way that precluded transit. The private sector came to us and said, "We really need transit here because we have traffic problems." But we could not serve them because they were too dispersed or too remote, with long access roads, or because turning radii were too tight or overhangs were too low for a bus to maneuver. Basic considerations like these determine whether transit can be incorporated later, even if funding is available.

It is wise to discourage applications for low density, scattered industrial complexes and office buildings. It is too much to ask a developer to completely change his plan, but alternatives can be encouraged, for example, clustering developments so there are small servicing nodes. Higher density development can be encouraged along the major arterials. Infill development along transit corridors is wise, not only for transit but because other infrastructure is in place—schools, library systems, parks, etc. We encourage high density development to locate at rail stations or transit centers or at least along major arterials.

We also stress joint use of parking facilities because numerous surface parking lots are not good uses of land and are costly for us to maintain. In shopping development proposals, we ask that a certain percentage of parking spaces be shared during the daytime, usually from 6 a.m. to 6 p.m. We have some shared parking facility commitments as large as 250 spaces. Our up-front capital costs are low, and we pay only for any additional cost to the developer.

We discourage excessive free parking by private employers. Instead, we encourage employers to subsidize monthly transit passes costing $40 per month rather than provide employees with free parking. The City and County have cooperated with us, but it is taking some time to convince developers that it costs more to build a parking space ($10-15,000) and maintain it over the next 20 years than to provide employees with free transit.

We now have a trip reduction ordinance for the city and county. This discourages auto use into the downtown and major activity centers. We also have a long range plan for right-of-way preservation that is in line with our rail corridors. When major development proposals were submitted, we got as much as six miles of right-of-way dedicated for transit use, although we may not know when transit will be provided in that corridor. One of the things I tell developers is that if they ever want to see transit in the north end of the service area, they are going to have to give us an incentive to go out there (that is, right-of-way dedication and contributions for stations) rather than someplace else and that it is to their advantage to preserve that right-of-way for us now.

In reviewing development applications and dealing with the City and County planning staffs, it is necessary to comment early and consistently in the process, at the application and environmental impact stages and with the community advisory council, planning commission, City Council or Board of Supervisors. This requires a lot of staff time, but for us it has paid off. We have received many capital commitments from the County, whereas in the past we had not been successful. They recently required dedication of four acres of land to us for a park-and-ride lot, in an area where acreage sells for $350,000 an acre. We got it free and clear because we started working for it a year earlier, because we were there every time the development passed a milestone, and because we consistently asked for the same things.

At the same time, we did not ask for anything we did not need. We plan our requests before making them: "Are we being idealistic?", we ask ourselves. "Do we really need six acres, or can we manage with 3-1/2?". We ask only for what we need so we can justify our requests to the developer and elected officials. And we say specifically what we are going to do with it and why we need it. That establishes credibility with community and elected boards, and consistency pays off.

Timing the Transit Decision

Many regions in America claim to be uniquely car-oriented. Automobiles represent our freedom, our individuality as Californians, as Texans or Detroiters. I have never been any place that people have not told me that their town is different because their people are auto-oriented;

in Detroit, the "Motor City:" "Transit will never work here because we are the auto capital of the world;" in Sacramento: "California is so auto-oriented. We all have our BMW's; no one is going to ride transit." But transit has nothing to do with the value system or where a city is geographically. It has much more to do with economics, with convenience, and with where the community is in its growth cycle.

The stage in a growth cycle affects the palatability of transit in the community and can change dramatically in a short time. In Sacramento, people who would have refused to use transit a few years ago are now riding the rail, and in five years we are going to see many more people using transit than we have now. People will use transit when it can offer something over the automobile. If traffic congestion increases, then transit is a hassle-free alternative, a time management tool. When travel time savings are greater by transit than by auto, then people will use public transit. When there are financial savings, when parking rates are so high in downtown or in the major suburban activity centers that it costs too much money to park a car, then people will use public transit.

During my tenure in Detroit we tried to build a rail project in a transit corridor that already carried 80,000 people a day just on buses. Detroit still does not have that rail corridor because they had a system that was going to cost about $1 billion. But in Sacramento, where we carry only 60,000 people a day in the entire region, we now have an 18-mile-long rail system that was built for $170 million. We were able do that because we built it early in our growth cycle. I do not know that any of us will ever have the luxury of building another Atlanta, Miami or Washington system. I doubt very much that funding would be available. It is better to preserve existing right-of-way now rather than wait to locate transit in ideal locations at some later time. The price may be too high, which is what Detroit discovered.

My predecessors in Sacramento decided very early to transfer monies from an interstate project and put it into a light rail system in hopes of eventually carrying 20,000 people a day. They had no intention of being a major heavy rail system. They did it when the land was still available. The cost was so low that they built the entire system for $170 million—$9.6 milion per mile including a maintenance facility and all the light rail vehicles. That is the least expensive rail system ever built in the United States with federal dollars. The San Diego system was less expensive, but they built it with local funding and therefore bypassed some federal requirements.

Adelaide's Automated Busways

Alan Wayte
Director, Project Team,
Adelaide Automated Busway

Australia's Adelaide is a city with a metropolitan population of about one million. Personal transport in the metropolitan area is heavily dominated by the automobile, which accounts for about 83% of all journeys. The remaining trips are made on a relatively well developed public transport system with the emphasis on buses. Four heavy rail lines serve the more remote residential areas. An old light rail line operates between the city center and one of the beachside suburbs on its own right-of-way except in the city center. Rail travel carries about 20% of all public transport journeys; the remainder use conventional buses on streets and highways. The whole transit system operates with a large deficit; cost recovery is approximately 40%. Cost recovery is considerably better on the bus portion of the system than on the rail elements; the total deficit is nearly equally shared by the two modes despite the 80:20 patronage split.

In the 1960s various transportation studies resulted in proposals for major freeway development to handle the predicted increases in traffic, and land reservation and purchase commenced. By the 1970s the concerns for environmental degradation and energy conservation resulted in a public and political backlash against unlimited provision for automobiles, and none of the freeways were built except on the interstate exit from the city through the hills.

The same concerns also resulted in major efforts to improve the quality of service in the public transport system, initially by expansion and renewal of the bus fleet (and, to a lesser extent, the rail car fleet) and by the extension of bus routes and the southern rail line to the developing outer suburbs. Experiments in the type of services offered were undertaken and various marketing studies were conducted. Fare increases were deliberately held below the general inflation rate, converting a break-even operation into one with a poor rate of cost recovery.

But the overall effect was small. Patronage increases generally correlated with population growth and most people continued to drive automobiles. Fuel consumption fell, but principally as a result of increases in the use of smaller vehicles.

Freeway Corridors Become Transit Corridors

One of the products of this period was a review of the previously reserved freeway corridors to assess their potential as transit corridors. The northeast corridor clearly had the highest priority because it served an area of major population growth with a high demand for travel to the central business district and was not served by any of the existing rail lines. It relied solely on street buses, with long and uncomfortable journeys. This conclusion led to a major public transport review of this segment of the metropolitan area between 1977 and 1979. It was aimed at establishing public transport needs and included research and analysis of social, economic, environmental and land use issues in addition to the testing of various transit options. Public involvement was sought through the publication of working papers for community distribution, public meetings and by consultation with the elected local government bodies, local residents associations and public transport groups.

The options included were: heavy rail within the corridor; heavy rail branching from the existing Northern rail line; light rail within the corridor; busway within the corridor; upgrading of existing street bus services by road modification; and various forms of more advanced guided and automated systems. It became clear that the only viable options were either to upgrade the existing bus services or to build a new, exclusive route in the corridor, with a preference for the corridor options. It was less clear which technology, light rail or busway, was more appropriate.

Land use issues were important in reaching these conclusions. Development in the metropolitan area is controlled by a fairly rigid development plan that designates the type of development that can occur in any particular area. By the time of the studies, the Regional Centre of the Northeast was already well established on its designated site with major retail, local government and medical facilities in existence. This effectively dictated that the terminus of the new system should be at that regional center, fed by the bus system that served it. Some opportunities for development at intermediate stations were studied but eventually not actively pursued.

Environmental issues were also significant. The principal concerns were aesthetics, noise impacts, air and other pollution and the effect on established residential areas. A particular issue arose from the alignment of the corridor along a two-mile length of Adelaide's only river, the Torrens. This stream is small and flows in summer only as a result of controlled releases of water from reservoirs in its upper reaches. Nevertheless, it is of importance in a low rainfall city and generated considerable emotion when threatened. All issues were addressed in a Draft Environmental Impact Statement that comprised a comprehensive summary of the findings in all matters studied, together with the measures that would be undertaken to mitigate impact on the urban environment as a whole.

The Choice Between Light Rail and Busway

By 1978 a decision was made to proceed with preliminary design of a light rail (LRT) system in the corridor, a decision based primarily on environmental concerns. Light rail would occupy less right-of-way than a conventional busway, would be less intrusive, was expected to be quieter in operation, would result in less air pollution along the route, and would offer a more comfortable ride. Economic evaluation favored light rail, although none of the options rose higher than being marginally justified by the conventional economic analysis used. A public preference for light rail also emerged from surveys.

During the preliminary design work for the favored LRT option, the cost estimate rose significantly, largely due to the need to include additional grade separations at critical road crossings, upgrading of environmental protection measures, and the addition of a central city tunnel rather than on-street operation. This, coupled with a change in the governing party in the 1979 election, resulted in a decision to suspend the work and to again review the options.

By this time the O-Bahn guided bus system had been developed in Stuttgart, Germany, by Daimler-Benz and Ed Zublin to the point that it was considered suitable for public transit application. This form of busway was added to the options. After an initial review of the technical aspects of the new technology, work proceeded to recompare the corridor options against the base case and one of the heavy rail options. By late 1980, a decision was made to build a busway in the corridor that would incorporate a section of O-Bahn in the river valley area. Preliminary design in the period up to mid-1981 resulted in this decision being modified to use "O-Bahn" track for the entire seven-mile length of the system, from the central city edge to the Northeast Regional Center at Tea Tree Plaza. Factors leading to this decision included:

Busway versus Light Rail:

(1) Initial capital cost was approximately 50% of the competing light rail option.

(2) The system's potential capacity was equivalent to that of light rail and in excess of likely future demand.

(3) Its economic merit was marginally inferior to light rail, but this was considered to be outweighed by the financial considerations.

(4) A busway eliminated the need for passenger transfer at the stations, as buses perform both the feeder and the line haul role as one through operation.

(5) A busway eliminated the need for new construction in the central business district since buses can operate on street in a conventional manner (this was one reason for the major reduction in cost).

(6) Overall travel time (with the elimination of transfer time) was superior.

Guided Busway (O-Bahn) versus Busway/Light Rail:

(1) O-Bahn eliminates most environmental objections to conventional busway.

(2) Its width requirement is the same as light rail.

(3) Noise levels are lower than conventional busway and comparable to light rail.

(4) The quality of ride is equal to rail and can be maintained at that standard over its life. (The Adelaide system uses diesel-powered buses but is designed for conversion to electric traction when justified by such factors as relative power costs or liquid fuel supply limitations. In the meantime, it has been argued that the air pollution of the bus fleet is insignificant compared with the 500,000 automobiles in use in Adelaide.)

(5) The track can be operated at high speed (62 miles per hour) with greater safety than manually steered buses.
(6) The track design ensures minimum maintenance expenditure throughout its life.
(7) The track offers potential for development of alternative traction systems (electrification).
(8) The system can be readily adapted for a range of operational practices and vehicles (for example, high capacity, double articulated or coupled vehicles) to suit future needs.

The Adelaide System

The guideway track links the city center to the Northeastern Regional Center, a distance of 7.5 miles. It is totally grade-separated throughout, including pedestrian crossings. Structures include 10 river bridges, 14 busway/road separations, 4 pedestrian overpasses, and 4 pedestrian underpasses. The system is designed for operation at 62 miles per hour cruising speed.

Busway corridor and access routes, Adelaide, Australia.

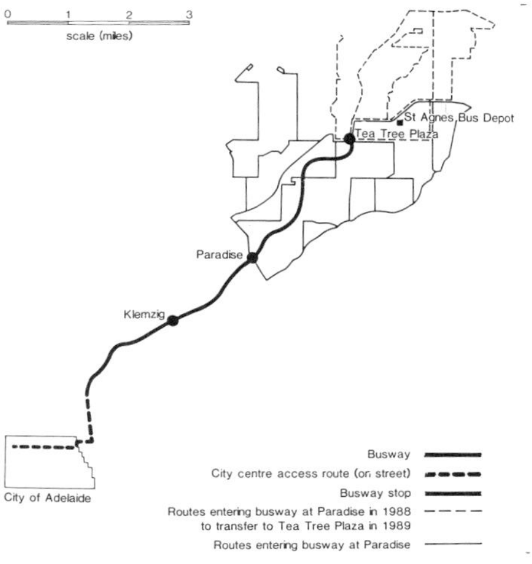

Twelve bus routes operating on-street in the northeast suburbs use the busway for the high speed run to the central business district, joining the busway at its outer terminal or at a midpoint station. Only three stations are provided, including the terminal interchange. Areas between stations are served by on-street buses which join the track at the first available station. Stations provide for bus access and exit from the track, for non-busway bus services to connect with the system, for walk-in passengers, and for park-and-ride and kiss-and-ride passengers. A fleet of 41 conventional buses and 51 articulated buses is in use.

Operating schedules provide for individual bus route headways averaging 10-15 minutes at peak hours and 30-45 minutes off-peak. This results in intervals of about one minute (peak) and five minutes (off-peak) on the busway itself where the routes coalesce. Most routes operate as through-services from their outer suburban terminal to the city center. Some routes in the off-peak period revert to feeders to the stations.

Adelaide's Busway is independent of existing streets and highways. Objections to its presence along the river were overcome by improving the river course visually and providing for recreational uses.

The major environmental objections to the construction of a transport system in the river section of the corridor have been overcome by developing the river valley into a linear park for recreational use. Bank regrading to improve water access, major tree planting and land shaping, and the construction of foot and bicycle paths were the main elements in this work. In addition, the remainder of the busway route remote from the river has been similarly treated, taking advantage of some of the excess land available as a result of the wide freeway reservation.

The overall cost of the system is about US$70 million, broken down as follows (in millions of US dollars):

Structures	$12.3
Other civil works	$11.2
O-Bahn track	$9.5
Stations	$4.0
Vehicles	$16.8
Land purchase	$4.0
Landscape works (busway)	$3.1
Landscape works (river)	$4.2
Administration and supervision	$4.2

The total cost per mile is therefore about US$9.5 million, of which the O-Bahn track cost is $1.4 million per mile of double track. (This represents cost since 1982. Cost at 1987 prices is $1.7 million per mile. All costs are based on conversion to US dollars at the exchange rate of 0.7).

Construction was initially intended to occur between 1982 and 1986, a period achievable in terms of construction requirements. However, adverse economic circumstances in this decade reduced the overall ability of the government to fund major capital works, so the system is not likely to be complete before mid-1989. The first stage of four miles of busway to the mid-station was opened for traffic in early 1986. However, the whole of the planned service is in operation. In the second stage, section buses currently operate on the road system until the track is complete.

The Technology

O-Bahn is a track-guided busway system developed by Daimler-Benz and Ed Zublin of Stuttgart, West Germany. The buses are standard production, city service vehicles modified only by the addition of horizontal, solid-rubber-tired guide rollers connected to the front steering knuckles. In addition, to meet the designed speed performance for Adelaide, uprated engine power (280 horsepower for articulated buses) and antilock brakes were specified. The rollers are not retractable and project beyond the bus body about two inches when the wheels are in the straight ahead position. The bus is steered manually on street.

The track consists of precast concrete elements assembled in a similar manner to railroad track. Concrete cross beams (ties) at 13-foot centers are supported on bored piles to ensure long term stability and carry the concrete L-shaped running slabs 40 feet in length. The vertical part of these slabs forms a guide surface to engage the guide roller. The gauge between the opposite guide faces in slightly smaller than the roller gauge so that the rollers remain in contact with the surface at all times. It is necessary to construct the track to tolerances of +/-2mm to achieve high comfort levels.

The bus operator steers the bus into a funnel-shaped area for track entry and thereafter is not required to steer. No action is required at track exit other than to take control of the steering wheel again. Entry speeds

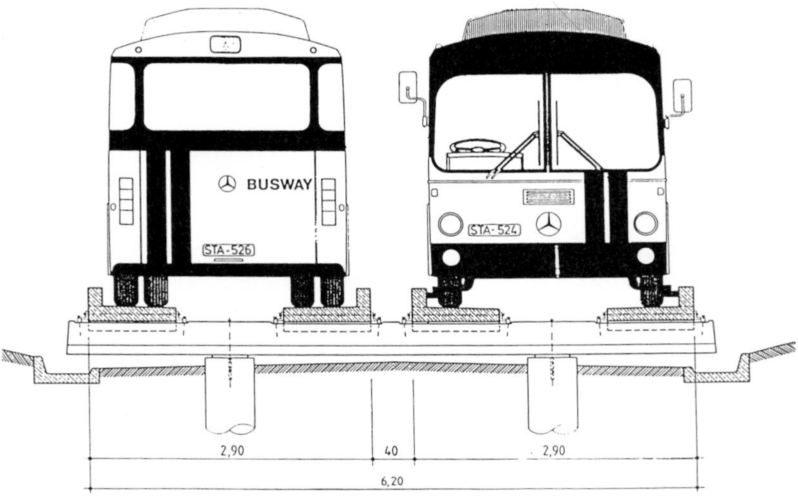

Structural section, Adelaide Busway.

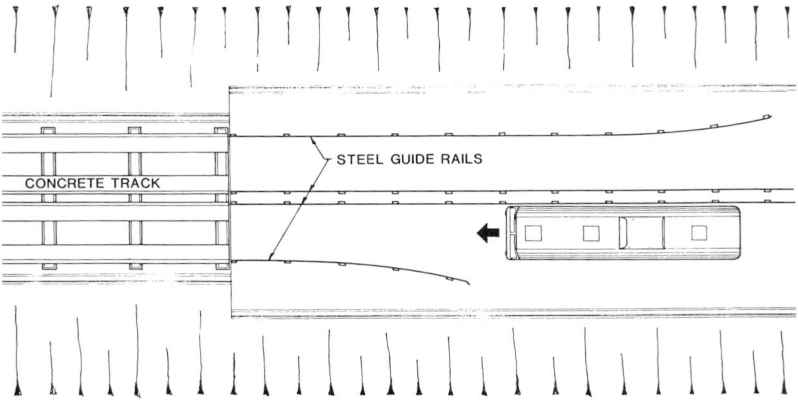

Steel guide beams direct buses onto the busway.

are designed for 28 miles per hour and track cruising speeds are 62 miles per hour in the Adelaide system.

Other general features of this system include:

- A narrow right-of-way is required since track gauge is four inches wider than the bus.
- The high quality ride surface results in good passenger comfort.
- Rigid track ensures long term stability and low maintenance.
- Precise location at platforms assists passenger entry and exit. In some applications high level platforms can be used to avoid steps.
- A high level of safety is achieved since manual steering errors are eliminated.
- There is the possibility of high capacity vehicles or coupled vehicles if restricted to track.
- Standard vehicles can operate equally well on streets, avoiding the need for passenger transfers at stations.
- Full flexibility of bus routing off track is retained.
- Adaptability of the vehicle to either road or track use allows full route operation prior to completion of track. This enables staged construction, beginning with critical sections and continual extension up to full route coverage as circumstances and finances permit.
- This adaptability also permits the vehicle to use a combination of right-of-way conditions ranging from local roads to bus/high-occupancy vehicle lanes on freeways to guided track and to central city streets.
- The cost of new infrastructure can be minimized and funds applied only where most effective.

Standard buses and articulated buses are currently in public use on O-Bahn systems. The developers have also built a double articulated (four axle) bus based on the same modular components but for track use only. These vehicles can be coupled to form trains. The overall family covers a wide range of capacity requirements. This approach provides a family of vehicles based on standard production line bus technology at a relatively low cost compared to light rail vehicles and at low vehicle weight per seat, offering energy savings.

O-Bahn buses can be equipped with any type of power source. Those currently in use are diesel powered, electric powered, or a combination of both (the duo-bus) which permits diesel operation beyond the overhead power lines. The easy change from manual to automatic steering permits a simple design without switches. However, switches have been developed for use with the system if required, such as for bus-train formation.

For tunnel operation with diesel buses, an extraction system for exhaust gases has been developed which provides for discharge directly from a roof-mounted exhaust outlet into a longitudinal channel.

Systems have been developed to provide a fully automatic bus operation on track if required.

Operating Experience

Patronage has exceeded the capacity initially provided by the operator and has necessitated additions to the schedules. Current daily two-way patronage is about 16,000, of which just over 50% occurs in the two-hour morning and evening peak periods. Surveys show that the effect of the busway has been to generate a 24% increase in patronage from new riders. The overall increase on busway routes is about 30%, indicating that there has also been a transfer from other bus routes. This has occurred during a period when overall metropolitan public transport usage has declined by about 10%.

Access to buses can be made either at ordinary street bus stops or at stations. The majority of passengers board prior to the stations, as was expected, with approximately 81% of passengers boarding at street stops, 19% at stations.

Surveys undertaken after the first six months of operation indicated some marked differences between the characteristics of previous users of the old street bus system and new users attracted to the busway. The overall impression is that new users tend to be older, working males with greater transportation mode choice. This, and other evidence, suggests that new users have been attracted to the busway because of a perceived difference and improvements in the bus system in general. As travel time savings are small until the second stage is opened, this perception must be based on such factors as comfort, frequency, convenience of access and overall image of an off-street operation. The ability to use a single vehicle from home to central business district, with part of the trip on a fast, exclusive facility with rail characteristics, also seems to offer significant attraction to passengers.

Changed Characteristics of Users

Characteristic	Previous %	Current %
Trip purpose		
work	52	56
shopping	11	13
school	28	21
Purchased cash ticket	65	76
Car available for trip	44	67
Gender		
Female	60	54
Male	40	46
Car parked at station	17	38
Frequency of use		
Every weekday	46	36
3-4 weekly	11	13
1-2 weekly	7	9
1-2 monthly	3	4
Age		
Under 24 yrs	55	36
25-54 yrs	38	51
Over 55 yrs	7	6

A feature of the demand has been the higher than anticipated use of car park facilities at the interchanges. It was expected that the availability of bus routes close to trip origins would result in a lower demand than for comparable rail system because no vehicle transfer is required when the bus reaches the busway station. In practice, it appears that the high frequency of buses at the stations (significantly higher than for a rail system) generates the opposite effect, putting pressure on car parking facilities. Total commuter parking space provision (following an expansion program after opening) is now about 600 at two stations. The number of parked cars varies but is normally around 750. The overflow is accommodated by parking in the kiss-and-ride areas, on access roads, landscaped areas and local streets.

Expansion of parking after operations commenced did not result in less illegal parking, but more cars, suggesting that actual demand is still not being satisfied. Carparks are normally full by the end of the morning peak, leaving little room for daytime passengers. Measures to alleviate the problem are being reviewed, ranging from further car-park extension to the introduction of parking charges at peak hours. The situation will be improved when the second stage is completed and more carparks are opened at the terminal station.

Land Use Impacts

The principal effect on land use has been to accelerate and modify the development of the Regional Center and to encourage expansion of residential areas, all generally in accordance with the Metropolitan Development Plan. The site for a major new tertiary education institution for the northeast was changed to a location adjacent to the terminal interchange. The developers of the retail center have modified their plans to ensure direct links into the interchange and propose to develop the air rights over the interchange for commercial office use. This interchange is proposed to be built partly on land owned by the developer, and the cost of its construction is to be met by the developer. A major new residential area for 30,000 people is now under development with direct bus access to the busway terminal.

Future Busway Development in Adelaide

The technical success and passenger appeal of the busway has encouraged preliminary studies for application in other parts of the metropolitan area. A new route to the south is being evaluated against alternative options and consideration has been given to the replacement of parts of the existing rail system by guided busway.

The planning and decision process that resulted in the construction of this first application of the O-Bahn dual mode bus guidance technology to a high speed operation was a long and controversial exercise. By contrast, the construction and operation have been remarkably problem free for a new technology, and the system has proved to be highly attractive to its potential users.

Bus Technology as Rapid Transit in Ottawa-Carleton

Ian Stacey
Director, Transportation Planning and Transitway Programme, Regional Municipality of Ottawa-Carleton

The Regional Municipality of Ottawa-Carleton is one of eleven regional governments in the Province of Ontario. It comprises 11 individual municipalities with a combined population of approximately 600,000 people. In this two-tiered system, the regional government is responsible for common services such as public transport, major roads, water purification and distribution, waste disposal, social services, and overall land use planning.

The public transit system in Ottawa-Carleton radiates from the downtown but also includes approximately six transfer stations where connections are made to and from suburban areas. Transit service by the Ottawa-Carleton Regional Transit Commission, called OC Transpo, is provided within the urban area comprising approximately 550,000 people. Transit services to rural areas are not provided except by special contract.

The area that is covered by public transit service is approximately 380 square kilometers. OC Transpo's fleet consists of approximately 760 vehicles that traverse 48 million service kilometers annually. Ridership is approximately 87 million persons per year; on any given day, up to 400,000 trips are made on the public transit system, about 160 rides per capita, which is one of the highest rates for any city of comparable size in North America.

Public transit improvements in Ottawa-Carleton began in the early 1970s with the formation of the Regional Municipality of Ottawa-Carleton. Prior to 1969 only the city of Ottawa had public transit service, and the townships surrounding Ottawa in Carleton County were growing with little mutual coordination. The provincial government stepped in to form a regional government to coordinate services that were common to all. One of the first tasks of the new Regional Council was the development of an official master plan for the area. The planning process took four to five years, and in October 1974 the Council adopted an overall regional plan, the Official Plan for the Ottawa-Carleton Planning Area.

Transit Policy, Funding and Improvements

The formation of and responsibility for transit policy in Ottawa-Carleton is different from that typically found in the United States. The Regional Council is made up to 32 representatives who are first elected to one of the 11 municipal local governments. The Council is responsible for all regional transportation and land use planning. It also funds both capital and operating programs with respect to transit services run by OC Transpo. Simultaneous consideration of land development and the provision of public transit service is therefore logical.

In the Province of Ontario, transit funding for major capital programs is split: 75% is provided by the government of Ontario and 25% by the municipal government. Major capital expenditures include the construction of the transitway, bus purchases and construction of garages. The 25% local share is raised through property taxes from the urban transit area (raising funds from sales taxes is prohibited). Regarding operations, together with the Province of Ontario, a target revenue cost ratio is set based on a population formula, in this case 65%. The shortfall of 35% comes half from the Province of Ontario and half from the property tax base within the urban transit area of Ottawa-Carleton.

With respect to the 25% contribution for capital funding, Regional Council has set up a transit reserve fund for future expenditures. Each year an amount ($10 million in 1987) is added to the fund and, as construction of the transitway program proceeds, funds are withdrawn, enabling us to "pay as we go" and keep borrowing to a minimum. Regional Council has taken a very progressive approach to the financial planning side of the transitway program.

One of the major elements of the Regional Official Plan, apart from designating the location and size of major residential, employment and activity areas, is transportation policy, which gives precedence to public transit over all forms of road construction or road widening. In the 1960s a transportation study was undertaken which was not unlike studies done in other cities at that time. Since no one felt that public transit could satisfy transportation needs, these studies led to transportation plans that were heavily roadway dependent, recommending construction of an increasing number of freeways. Yet, during the five years (1969-74) leading to the development of the Regional Official Plan, public sentiment indicated that residents did not want more roads but instead preferred improvements in public transit in the area. Consequently, with the adoption of the Official Plan, a two-pronged approach to public transit servicing was established, calling for an operational improvement program and the development of rapid transit.

The ridership gains achieved since the formation of the Regional Government are a result of many factors, not just the construction of the rapid transit system. In fact, most of the ridership growth that has occurred in Ottawa-Carleton took place before the first sections of transitway were opened. With regard to operational improvements, a major change was the dramatic service increase following the establishment of the Regional Transit Commission in 1972. Not only was provision of regular service increased, a significant amount of express service was added. A number of bus priority measures were implemented, including a transit mall on a major retail street downtown, reserved bus lanes on two other streets, and an exclusive transit terminal in almost every regional suburban shopping center. Flexible hours were established in downtown Ottawa-Carleton, helping to reduce peak loadings on express bus routes. Consequently, more people were able to be seated and the fleet was more efficiently used. Automated passenger information systems were installed, offering ready access to information for patrons, and all service is controlled and monitored through a main computerized control center.

Suburban Development Planning

Since 1974 with the adoption of the Official Plan, all new residential and industrial subdivisions must satisfy certain conditions in order to be approved. Density and zoning matters are determined by individual municipal governments, but the Regional Council grants final approval. Conditions relating particularly to transit services are always included.

An attempt is made to establish the collector road system so that transit can efficiently serve the most people, and higher density land uses are encouraged to locate close to planned bus routes and bus stops. Our goal is to provide public transit service within 400 meters of every resident. Walkways are provided, and, in some cases, transit-only roadways are provided in lieu of collector roads, thus discouraging automobile traffic through neighborhoods.

One of the most important conditions of approval, and the one most difficult to achieve, is development staging. We attempt to incorporate staging conditions so that, as a subdivision grows, it does so from the transit service outwards. We try to avoid having small, isolated pockets of development far from the transit service, even though it may be the most desirable first-stage location for the developer due to, perhaps, its scenic nature or its close proximity to piped sewer and water services.

Rapid Transit Development Program

With the establishment of the Official Plan of land use and transportation in 1974, a concept for rapid transit was approved that was based on five corridors radiating to and from downtown, with a projected area population of one million. (At the time of the establishment of the Plan, the population was about 500,000.) During 1975 and 1976, the Regional Council considered what it could afford, which technologies might be employed, and which implementation strategies for providing rapid transit services might be appropriate.

One was the "inside-out" strategy, which would build the downtown portion of the rapid transit system first and then head out the first-priority corridor. The other was an "outside-in" strategy through which the rapid transit lines would be built outside the downtown first, thereby delaying construction of the expensive downtown portion as long as possible. A comparison of the two strategies showed that the entire system outside downtown could be built for about the same cost as a downtown tunnel. Since studies indicated that the volume of riders in the downtown could be handled on a surface street system for at least a 10- to 15-year period, it was decided to adopt the "outside-in" strategy and defer the high cost of construction in the central area. This finding, which meant operating the system on surface streets in the downtown, suggested two possible technologies—busways and light rail. Each could provide the necessasry capacity of 15,000 passengers per hour per direction.

Following approval of the implementation strategy in 1976, all possible route options were identified. Over 100 public meetings were held at which citizens suggested routes. A variety of routes was considered, including abandoned railways, hydro corridors, and open space corridors, and a rigorous evaluation of all route options was undertaken, taking into consideration environmental impact, level of service, and cost. It took almost five years to establish preferred routes in four of the five different corridors identified in the Official Plan.

Following route selection, the Official Plan was amended to contain a schedule of land use, major roads and rapid transit. Having originated as a lumber town, with railways running through the central area, Ottawa had a legacy of 100-foot-wide railway rights-of-way that radiated out of the city center. During the 1950s a railway relocation study was undertaken that resulted in the federal government moving the main railway terminal, both freight and passenger, out of downtown, thus making the abandoned rights-of-way available. At the same time, the federal government, through its planning agency, the National Capital Commission, undertook a study that resulted in the formation of the Greenbelt (a 2-1/2-mile-wide band of public land around the urban area) and a proposal to build a system of scenic parkways. Some of the parkways were built, but the program was discontinued before completion, so the corridors remain as open space which available, where required, for the transitway system. The Regional Council paid for the land and, through use of the corridors, major social disruption caused by acquisition of homes and businesses has been avoided.

Technology Selection

Busways and light rail transit were compared in terms of three basic factors: total system cost, level of service, and staging flexibility. (Environmental impact was not considered to be a factor in choosing technologies.) In the area of capital costs, it was determined that a bus system would be about 70% of the cost of a light rail system, based on data from Calgary's light rail construction experience. Regarding operational cost, the City of Edmonton's experience was considered, which showed that an entirely bus-based system would operate at about 80% of the cost of a rail and bus system. With respect to level of service, a busway was considered superior. In a city like ours with low densities and where very few people can walk to rapid transit stations, buses could operate through residential areas and enter the transitway directly, thus avoiding transfers which are inherent in a rail-based system. From a staging flexibility standpoint, congestion problems were expected outside of downtown in the suburban areas. A busway could avoid these heavily congested areas. A rail system would have required much more extensive construction before any relief could be provided.

In summary, a busway system was selected based on its superior level of service, its lower construction and operating costs, and its staging flexibility. This decision was controversial, however, so to keep options open, all system geometry is compatible with light rail if in the future there is a requirement to change to that technology. However, if rail is introduced in Ottawa-Carleton, it will be heavy rail due to its greater capacity.

The Busway System

The busway consists of a two-lane roadway 8 meters wide with a 2-1/2-meter shoulder on each side that serves as a refuge area for disabled buses and maintenance vehicles as well as a snow storage area. Through the stations the transitway widens to four lanes, or 15 meters of pavement, to allow for express services to bypass stoppped vehicles. Buses operate on the transitway at 80 kilometers (50 miles) per hour; speed through the stations is 50 kilometers (30 miles) per hour.

The transitway operates in a number of different ways, depending on the time of day. Some services operate on the transitway in a line-haul fashion all day, every day. The headways vary by the time of day, typically four minutes in peak periods and five minutes in off-peak. In addition, express services operate in the peak periods, originating in suburban residential areas, entering the transitway, and becoming express service to downtown and all other destinations on the transitway. In the off-peak periods, feeder buses operate to and from the transitway stations where people then transfer, a service typical of a rail based system.

Following completion of the planning studies (1974-80), the Regional Council approved the initiation of construction. A ten-year construction program of approximately 31 kilometers (20 miles) was established. The first sections were opened in December 1983 to the east and west of downtown. In 1985 another section of transitway was opened along Scott Street in the west area, and a further extension of the southeast transitway was completed towards downtown. In 1987 another three kilometers was added, bringing the total system in operation by the end of 1987 to over 17 kilometers.

Downtown, the transitway runs on surface streets using with-flow bus lanes. When and how to grade-separate in the downtown area is under consideration. Extensions of the transitway to the east, west and south are planned. Running east-west through the city is a freeway known as the Ottawa Queensway. Presently, express buses from the outlying areas operate in mixed traffic on the freeway at speed; at the point where the freeway becomes congested the buses exit via special ramps onto the transitway and continue their journey. The justification for extending the transitway parallel to the freeway will be the level of congestion.

Today there are 17 kilometers of transitway in use. Peak hour, peak direction ridership is approximately 9,000 passengers per hour, and the daily ridership is approximately 200,000 people. The total system cost for the 31-kilometer, 26-station facility will be approximately $400 million (Canadian) or about C$13 million per kilometer. The actual cost of the first sections in 1983-84 dollars was about C$10 million per kilometer. Included in that figure are stations costing an average of C$4 million each. The maintenance cost for right-of-way is primarily for snow removal, about C$60,000 per kilometer. Annual station maintenance is about C$40,000.

Rapid Transit and Land Use Development

Since the opening of sections of the busway, land development activity around the stations has occurred. More than $800 million in new development is either being considered or under construction near stations in areas which in some cases have been designated on the Official Plan as activity nodes. These areas have been dormant for many years but, with construction of the transitway, development is occurring. Near Baseline Station a new office tower has been built (the first of five office towers), and the new City Hall for the City of Nepean, the second largest municipality in the area, is under construction. Located across from the station is Algonquin College, a community college in Ottawa. The Tunney's Pasture Station was built to serve a 10-12,000 person employment area of the government of Canada. Partially as a result of the public transit level of service provided by the station, the transformation of an adjacent former foundry site to a $100-million mixed use development comprising four towers of residential, office and retail uses is occurring. The first tower was sold out before ground breaking. Around the Hurdman Station, high rise residential development is emerging. Two luxury condominium buildings of an eventual three-tower complex have been built.

One of the more interesting transit/land use developments is at the St. Laurent Station located at a suburban shopping center. The transitway runs parallel to the crosstown freeway adjacent to the center. The station was built by the system, and the developer provided the land. In addition, the developer built an expansion to the shopping center that links directly to the station. The developer also built a new office tower on one corner of the property within walking distance of the station.

At the present time the East Transitway Extension is under construction. In anticipation of the benefits of public transit, new development is being built and planned. At the future Blair Station a new City Hall for the city of Gloucester, a new shopping center and numerous office complexes are being developed, and all will tie into the station.

Benefits

The transitway system provides Ottawa-Carleton with major financial benefits. In any bus system, about 80% of the costs of providing service are time related. Faster service reduces the number of buses, maintenance, equipment, maintenance areas, and so forth. Operating cost savings due to the transitway will, within about 10-15 years, pay for the systems portion of the capital cost of building the transitway. The annual operating cost of OC Transpo is around $100 million, and even a 10% saving goes a long way towards paying the 25% share of the capital cost of the transitway.

In 1982 the total fleet numbered 780 buses. Since that time ridership has increased 7-8%. Yet because of the transitway and higher speeds, over 820 buses would have been needed to move the same number of people. By 1995, the fleet will be reduced even more. Without the transitway, an additional 200 buses would have been needed, which would have required capital outlays not only for buses and maintenance but for additional garages to store and house them.

In addition to the operating cost savings, the development occurring around the stations is bringing added tax revenues to the municipal governments in the area.

The lesson from Ottawa-Carleton is that bus transit can compete successfully with the automobile. Its success, however, is a result of a whole series of actions, not just the result of a single, major program.

Denver's 16th Street Mall

Gary Zehnpfennig
Project Manager,
Civic Design Team,
The Denver Partnership

The 16th Street Mall is situated in downtown Denver's retail district. It runs parallel with and is immediately adjacent to the main financial and employment center of the city. As a transit facility, the Mall is a 13-block-long, two-way surface shuttle system that operates between two regional bus intercept stations. It was designed as a high quality downtown transit circulator/distributor to relieve regional buses from having to operate inefficiently on congested downtown streets, especially at peak hours.

The 16th Street Mall distributor system has permitted increased efficiency for the regional/express bus system and has allowed many buses to make the desired "second and third trips" during the peak hour. It has also resulted in the removal of 501-624 bus trips from downtown streets. The two transfer terminals at the ends of the Mall are well used and presently accommodate 640 regional and express buses per day. Over 15,300 passengers transfer to the 16th Street Mall from these buses via the Civic Center and Market Street stations.

Shuttle ridership is greatest during morning and evening peak regional transit commute hours, but at noontime and throughout much of the day ridership is unexpectedly high. When designed, the Mall was expected to carry 8,000 to 10,000 shuttle passengers per day. The figure is now averaging 45,000 per day, and growing. We are rapidly approaching

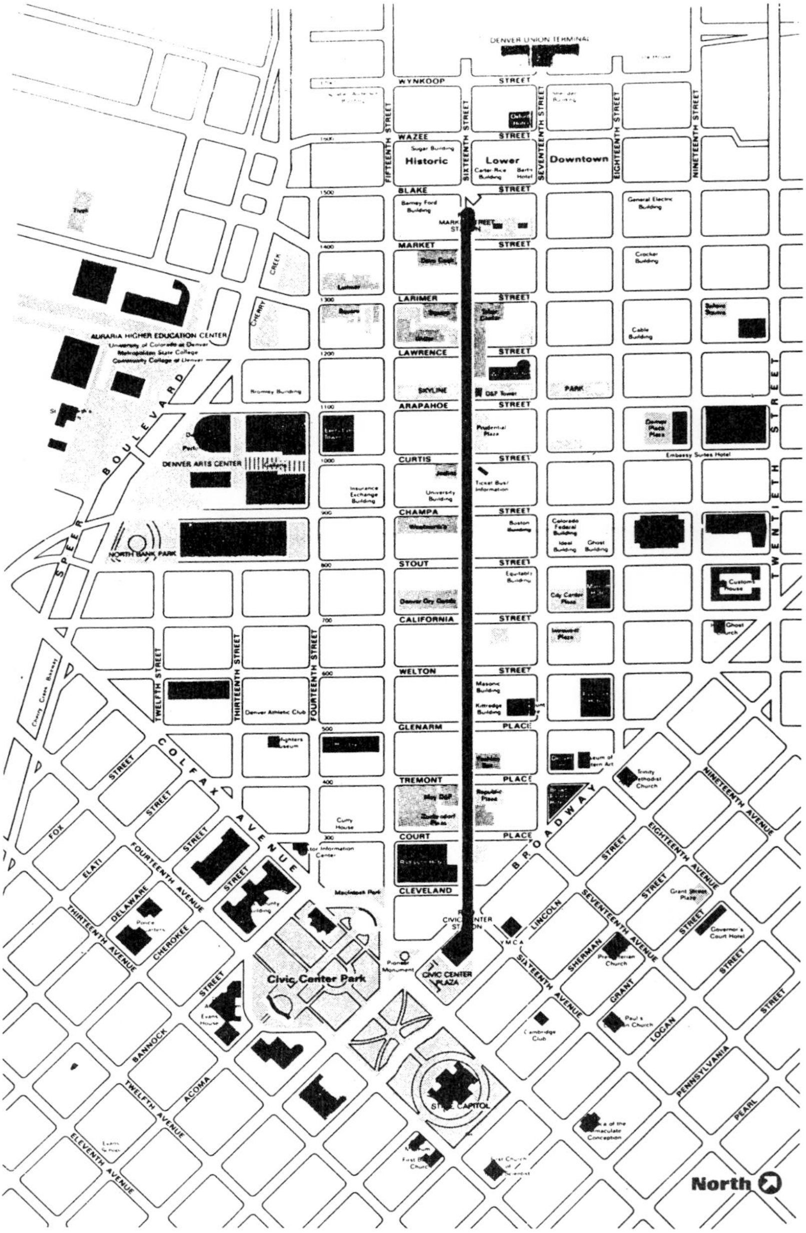

In Denver the 16th Street transit mall offers shuttle bus connections to two regional bus stations.

the point where the shuttles and the Mall itself are being filled to capacity at peak periods during the day. The Regional Transportation District feels it is a success because it has allowed them to operate a conventional bus fleet more efficiently and effectively and has mitigated some of the severe impacts of through-bus trips that crowded many of the streets around the core area.

In addition to this role as the hub of Denver's regional transit system, the 16th Street Mall is the city's key urban design amenity and the primary pedestrian and transit connection between major downtown activity centers. It is uniquely designed to accommodate both transit and

16th Street Mall, Denver. Photo by Tessa Dalton, The Denver Partnership.

pedestrian functions within the same overall space. The Mall and adjacent privately-owned spaces have also become successful locations for special events, festivals, vending operations and outdoor cafes. Denver's 16th Street Mall has become an international symbol of a successful transit mall.

The 16th Street Mall is an appropriate way, at least in Denver, to accommodate conventional regional bus transit when it gets into the downtown corridor area. This urban design/transit development currently seems to work very well for Denver, but it is not sufficient for future needs. Thus, we are beginning to work on a mall to cross 16th Street as well as improvements to the overall regional transit system. The Mall has been a mechanism for increasing the attractiveness of transit to the consumer, and it also helps us gain public acceptance for a new transitway outside downtown.

While the Mall was paid for with 80% federal funds, future projects of this magnitude will need major private sector involvement. Given the success of the 16th Street Mall, I expect the business community to make that kind of a commitment.

History

Since the 1890s, 16th Street has been the primary retail shopping corridor, but beginning in the 1960s, 16th Street and downtown businesses began feeling the increased competition from the rapidly expanding suburban shopping parks and malls. In 1971, the Board of Directors of Downtown Denver, Inc. (DDI), the predecessor to The Denver Partnership, announced it favored the concept of a mall. Several proposals were studied, debated and eventually scrapped for lack of support or sufficient funding.

Meanwhile, the Regional Transportation District (RTD) was investigating ways to relieve the massive congestion of buses downtown and more efficiently provide transportation services. RTD and DDI began to explore ways to meet the objectives of both groups. In August 1977, I. M. Pei and Associates, consultants to RTD, unveiled a model of a 13-block 16th Street Mall with bus transfer centers at each end. The entire 80-foot right-of-way would be dedicated to pedestrian and transit usage, flanked by retail shops. The plan received the enthusiastic support of most businesses. An economic benefit study estimated the new mall would increase sales by 7.5-10% and that gains would be higher if it were properly maintained and managed. The proposal also received the blessing of the Urban Mass Transportation Administration (UMTA), which approved funding of 80% of the construction cost.

Construction of the Mall began in February 1980. For 2-1/2 years, work progressed on a block-by-block basis with considerable disruption to the normal flow of commerce in downtown Denver. The 13-block transitway opened in October 1982. All express, intercity and regional bus routes were revised to end at the two transfer facilities from which a free-fare, custom-designed shuttle bus system distributes passengers along the Mall. Capital costs for the entire project totaled $76.1 million.

One of the reasons that the 16th Street Mall took ten years to materialize was that people felt it would hurt retailing by taking traffic out of the downtown core. In fact, it has increased person trips in the downtown about tenfold because of its much greater efficiency over the private automobile. The Mall has done a lot of things, but it has not significantly decreased commuter trips by autos; it does have a carryover influence regionally because people view it as a very successful transit facility. It is a great marketing tool.

Management

Denver's leadership learned early that brick and mortar were not enough to ensure success. They concluded that issues like transportation, environmental impact, downtown management, and revitalization must be addressed in a coordinated and comprehensive manner over the entire core of downtown. In order to ensure the necessary coordination and fund the supplemental services, downtown business leaders initiated a 1978 amendment to Denver's City Charter creating a special Mall Benefit District to pay for the care, management and operation of the Mall. The district currently encompasses approximately 865 property owners within a 70-block area of downtown. Its five-member Board is appointed by the mayor and headed by the City's manager of public works. The district's approximately $1.8 million annual budget is raised through an assessment of the property owners. The current assessment ranges from 10 cents to 58 cents per square foot of land, depending on proximity to the Mall. There is an eight-point system that determines what the assessment should be.

The day-to-day management, maintenance and promotion of the Mall is directed by the staff of The Denver Partnership. The shuttle-bus service and transfer stations are operated and funded by RTD. The Mall's public/private structure has provided an evolving and ambitious program of festivals, events, promotions, sidewalk cafes, festive vendors and special Mall activities, adopted to entice people to the Mall and maintain an atmosphere of color and excitement downtown. This is in most cases a natural adjunct to the transit function, which itself animates the area. Pushcart vendors, entertainers, sidewalk cafes and other special users of the Mall are regulated by The Denver Partnership on behalf of the City and the District. In addition to events specifically sponsored by The Denver Partnership, individual businesses and community organizations are actively encouraged and assisted in staging their own activities and special events. A full schedule of downtown festivals has attracted thousands of visitors. A restored 1946 London Double Decker bus has been transformed into an eye-catching Ticket Bus which dispenses information and bus passes and sells both advance and half-price, day-of-performance tickets for over 30 of the city's performing arts groups.

An essential element of the Mall's success has been the quality of its day-to-day maintenance and security programs. A full-time unit of ten Denver police officers patrols the Mall on motorcycle, horseback and foot and participates in a variety of special crime prevention and pedestrian safety programs. Crime impact studies have revealed a significant drop in virtually every category of crime in downtown Denver each year since the Mall opened in 1982. The Mall district supplements normal city maintenance services by providing daily sweeping and scrubbing of sidewalks, snow removal, landscaping and repairs. The Mall and its companion management district have convinced private business that it is worthwhile to pay into a management and maintenance district for the upkeep of the Mall and provision of special security and special needs, promotions, and so forth that keeps the Mall well used and a popular place as well as a very effective transit facility. The Mall's success has led to additional agreements to help manage adjacent public spaces, such as the Denver Arts Center complex, two mall transfer stations, and a three-block-long city park that intersects the Mall.

The Mall District has taken the lead in seeking creative solutions to downtown parking management. A strategy is currently being implemented to improve parking, particularly as it relates to retail needs.

The Denver Partnership's Civic Design and Development Team reports to the Mall District Board. The team's persistent encouragement of the use of sound design principles and mixed use development has been crucial to preserving the retail vitality and pedestrian environment of the Mall. An ongoing streetscape and signage program directs the development of civic directories, banners and temporary sculptures and the drafting of standards for parking and directional signs. Active involvement with city officials and business leaders in identifying potential tenants and working closely with them in obtaining necessary financing has helped pave the way for numerous new residential, commercial and retail investments along the Mall.

Catalyst

There is little question that the 16th Street Mall has contributed greatly to the revitalization and improvement of downtown Denver. Sales figures exceed $400 per square foot for some stores in the Mall corridor. More than 90,000 persons per day enjoy the Mall, its shops and restaurants, and traffic officials come from around the world to study the innovative shuttle bus and transfer system.

The construction of the 16th Street Mall has stimulated significant new commercial development in the downtown core. Since the decision was made to proceed with the Mall, there have been four major commercial complexes built adjacent to it and a fifth within walking distance. These developments include approximately 300,000 square feet of retail and nearly 2.5 million square feet of leasable office space. The major retail developments are Tabor Center (120,000 square feet of retail space), Writer Square (76,000 square feet), Republic Plaza (50,000 square feet), and the Masonic Building Renovation (50,000 square feet). The Mall has also been the basis for active development plans to build a major (500,000 square feet), "high-end" retail center slated for construction within the next two years or after the end of our major economic slump, whichever comes first.

It was estimated by the Gladstone Associates Economic Benefits study that 1987 retail sales would be $52 million within the downtown area as a result of the Mall project. This translates into $3.7 million in sales taxes per year. The projected increase in sales due to the Mall in 1992 is $68 million. According to Gladstone Associates, "Energy savings, reduced traffic congestion, reduced air polution, a more effective transportation system, and a strengthened and revitalized downtown area affect all Denver metropolitan residents, businesses, employees, shoppers and visitors."

Other benefits attributed to the construction and operation of the Mall include improved access to employment and retail areas of the center city; higher net income for nearby properties; lower employee costs; increased security via the Mall Management District; improved maintenance of the area via the Mall Management District; increased demand for downtown housing and cultural activities; the probability of value increase in property; and recreational betterments.

It took two false starts and nearly a decade of controversy and discussion before work began, but even the critics consider the 16th Street

Mall a success. The 13-block-long pedestrian mall and transitway has become the downtown gathering place for tourists, residents, suburban shoppers and the city's 114,000-member workforce, with a consequent dramatic revitalization of retail activity. The Mall has spurred economic development and restoration throughout downtown and is the catalyst for an ambitious effort to link all downtown amenities and open spaces into a cohesive and distinctive urban environment. The public/private partnership established to maintain and manage the Mall has become a national model for effective downtown management.

The Vancouver Skytrain

L. E. Miller
Executive Director, Regional Public Transportation Authority, Phoenix, Arizona

Rarely where full disclosure is required does a new project, especially one that involves new technology, come in on time and on budget and fulfill the original expectations of its proponents. The Vancouver Regional Rapid Transit (ALRT) project is, therefore, atypical. The advent of the current rapid transit system has roots that go back to the mid-1960s when governments, academics and interested observers began the study of rail transit alternatives. Political decisions to halt freeway construction before it had begun added to a general and popular consensus that "rapid transit" was to be an important part of the region's future transportation network.

The process of developing the necessary commitments for decisions, funding and contract authority to begin the rapid transit project in Greater Vancouver was carried out in an atmosphere of concern regarding mounting transit expenses, new technology and the economy in general. As we see evidenced today by the highly successful operation of the Skytrain system, the result is due in large part to the groundwork that was documented in regional planning studies and the approach taken in developing the project proposal.

It is important to understand the regional context within which the deliberations took place. The time frame was, by all standards, relatively short, approximately 10 years from the official origins to beginning of revenue service. From the standpoint of community awareness of transit and its potential, however, the history of public transport in British Columbia spans nearly 95 years.

Background Perspective

The Greater Vancouver metropolitan region is situated on Canada's west coast, between the Pacific Ocean and the coastal mountain range, and extends south to the international border. The region has 15 municipal jurisdictions within its 1,150 square kilometer area, each with a locally elected mayor and council. As the major urban center in the province of British Columbia, Greater Vancouver is home to approximately 1.2 million people (two-thirds of the province's population) and is the business, cultural and social capital of the province.

The history of public transport in Greaver Vancouver dates back to 1891 when a private company, the Westminster and Vancouver Tramway Company, began operating. The interurban tramway was expanded over the years and further streetcar lines were added to serve the growing regional population. However, by the mid-1950s the operator, BC Electric, had phased out all passenger rail services and replaced them with electric trolley and diesel bus routes.

Little has changed in the system since the 1950s. Expansion has taken place and the fleet has grown; service has become more frequent in certain corridors, and a cross-harbor passenger ferry system (SeaBus) was introduced in 1977. The statistics in Table I provide an overview of the current status of the transit system in Greater Vancouver.

Rapid Transit Revival

The citizens of the Vancouver region successfully opposed the construction of urban freeways during the 1960s and attention turned to rapid transit for solutions to accommodate travel demands from the suburbs to the downtown business core. Studies during the late 1960s and early 1970s drew several conclusions:

- Potential patronage would not warrant construction of a heavy rail/subway system.
- Two corridors from downtown Vancouver to the suburbs were candidates for a light rail system.
- Without transit improvements, major roadway construction would be required.

Table I: Transit Facts: Greater Vancouver 1985-86

	Conventional Transit	Custom Transit*
Area Served	1150 sq. km.	1150 sq. km.
Population	1.2 million	
Operators	BC Transit (formerly MTOC) and West Vancouver Municipal Transit	12 separate operators
No. of Passengers (annually)	93 million	370,000
Km Operated (annually)	60 million	n/a
Hrs. of Service (annually)	3 million	171,000
Employment		
Drivers	1802	
Mechanical	383	
Administrative	475	
Fleet Description	245 electric trolleys; 640 diesel buses; 2 SeaBuses	80 vans and autos
Total Cost	$154.5 million	$4.1 million
Revenue	$70.9 million	$400,000
Fare Structure	3-zone fare structure with peak/off-peak fares and concession fares; monthly passes and single-ride tickets available; single-zone adult fare of $1	4 zones

*Door-to-door parallel services for disabled persons who cannot use conventional systems.

In 1975 the light rail concept received Provincial support in the form of a policy statement from the Minister of Municipal Affairs. A long range transit plan was unveiled that included a light rail system as part of the improvement program. Immediate steps to implement the plan included only bus fleet expansion, bus service improvements, and the cross-harbor ferry (SeaBus). Although the acquisition of a single prototype LRT (light rail) vehicle from Germany was authorized for the purposes of building support for the LRT component of the plan, implementation was stalled with the defeat of the presiding government in late 1975. A Seimans-Duway light rail vehicle was delivered to Vancouver in January 1976 and still sits in storage unused. Succeeding Ministers have initiated attempts to sell the vehicle, although none has been successful.

In the late 1970s the Province's transit agency, the Urban Transit Authority (now BC Transit), authorized funding and participated in a detailed LRT planning study with the Greater Vancouver Regional District, which represented the municipal jurisdictions on the basis of a federation approach to regional government. This agency's regional planning studies, which culminated in 1975-76 with the publication of the "Liveable Region Program," identified transit as a cornerstone of the plan. At least 50% of the plan's objectives depended on transit improvements for full or partial achievement. Public transportation, including the introduction of advanced systems (such as LRT) was identified and supported as the most efficient means of alleviating the region's transportation problems. The plan documented benefits of the "transit solution," namely, greater level of mobility for all segments of the population; more efficient use of existing resources such as roads, parking facilities, bridges, etc.; use of transit to help shape communities and support land use/development strategies; and energy conservation.

The joint study undertaken after the local endorsement of the "Livable Region Program" culminated in final reports being published in 1979. Identification of rail corridors and priorities, station location and some preliminary engineering analyses were part of the final report. Public meetings throughout the region helped solidify support for the rail proposals and bus system improvements.

At the conclusion of the local political approval process, the report and its capital program for implementation were forwarded to the Provincial agency (BC Transit) for funding authority. By this time, early 1980, several significant developments had occurred that ultimately affected timely action on the recommended plan. The transit plan did not and could not have forecast the changes that were taking place. They were to play a major role in determining the provincial government's position regarding rapid transit.

The first, probably most significant and least understood change was the worldwide economic downturn. British Columbia was somewhat sheltered during the early years of the international cycle, but there were indications that the province's economy was not lagging far behind that of the rest of the world. Although predictions were vague, the impact of the economic slide was going to be felt in declining government revenue and increasing unemployment (leading to higher social welfare costs).

The second factor was also not perceived by many at the time as important. During the late 1970s and early 1980s the national and provincial governments were wrestling with the problems of independence, unity and repatriation of the Constitution. In the country as a whole there was a growing concern for national unity culminating in the Constitutional debates. The provincial government leaders and their governments endeavored to build linkages that would support the unity concepts and constitutional reforms.

Finally, more visible and closer to home, three specific projects altered the decision-making environment. The initial decisions concerned the location of a new 60,000-seat, domed stadium for Vancouver. After a detailed study of alternatives, a site adjacent to the downtown core was recommended which was a considerable distance from the proposed rail alignment. Following quickly on the heels of that recommendation, the provincial government announced its intention to host a World's Fair with a transportation theme in Vancouver in 1986. The EXPO '86 site was adjacent to the downtown Vancouver business district on abandoned railway yards. It encompassed the stadium location and was planned for urban redevelopment (the largest such project in North America) following completion of the World's Fair. This announcement added three new factors to the process:

- The site was to be a major origin/destination (both during and after 1986) and yet would not be well served by the LRT proposal.
- By abandoning the rail yards, a rail tunnel beneath the business district became available for use.
- The date of the opening of the World's Fair, May 2, 1986, became a new deadline for in-service status of any rail system.

Ultimately, as part of the federal/provincial agreement with respect to the World's Fair, but initially as a stand-alone project, a trade and convention center was planned for the Vancouver waterfront. It, too, would not be served by rail in the initial plan, and ultimately it was necessary to provide a high capacity link between the EXPO site and the trade and convention facility. During the World's Fair, the waterfront site housed the Canadian Pavilion and business center.

The Famous "Second Look"

Traffic congestion was not decreasing and bottlenecks at bridges were becoming increasingly long, so pressure mounted for a solution. The Province's transit agency, now BC Transit, had not taken a formal position on the original proposal, although participation in the study had been instrumental in its execution. However, with the changes noted previously, increasing editorial comment about the absence of any action, and a proposal from the City of Vancouver to build (with provincial funding) a second system to serve the stadium, EXPO sites, convention center and other downtown activities, the Board initiated its own review. The two factors that influenced the "second look" decision were related to the interest by the Canadian Government in having a Canadian Transit Showcase at EXPO '86 and the availability of the existing, now abandoned, railway tunnel.

Conducted by BC Transit staff in light of new information, the study of the rapid transit proposals tended to confirm the major corridors and order of priority for construction. Questions were raised, however, with respect to the feasibility of the southern portion of the first priority line and with the current vs. future ridership levels with regard to an at-grade LRT system.

The ridership analysis tended to be on the conservative side, with a prediction of 21-27 million rides per year in the first stable year of operation. (The year 1986, with EXPO and its expected 15 million visitors, was not considered a typical or stable time on which to base capacity requirements.) The estimates of 8,000 people per hour in rush hours for the early years were confirmed, based simply on the transfer of riders from bus to train. However, the studies questioned the ability of the LRT proposal to meet capacity requirements in the future due to conflicts with traffic. It was noted that price, ride quality and travel time would attract new riders and that a capacity of more than 20,000 people per hour could be required with the building of park-and-ride facilities and the maturity of the service.

The effects of the at-grade system on the 36 street crossings was of concern to highway engineers as well. These streets, many of them arterial roads, were at or approaching their capacity, and rush hour train services would cause massive disruption. However, the availability of the tunnel below downtown Vancouver would, subject to detailed engineering studies, provide an alternative that was less expensive and provide better service through a single system than previous routings.

Finally, the analysis identified the potential for job creation and economic stimulation in a project of such magnitude. The opportunity to use a capital project that would build urban infrastructure to help shape development and create new economic initiatives at a time when the economy was most in need was identified and quantified.

Weighing the alternatives and the analysis of the initial plan and priorities, transit matters were discussed. It must be stressed that these transport-related issues were only one part of the deliberations and ultimately formed a minor part of the decision. In summary, the significant issues were as follows:

Capacity: The fact that the system would have to accommodate at least double the initial traffic volume without incurring major capital cost or disrupting the service was recognized.

Operating Budget: Every effort to reduce the impact of inflation on operating costs both in labor and fuel cost should be pursued.

Service Levels: Acknowledgment was given that the more frequent the service and the shorter the travel times, the more attractive the system would be.

Traffic Disruptions: Level crossings would create traffic tie-ups and/or slower train speeds, and ultimately both would be unacceptable.

Construction Disruption: Construction plans that would force closure of streets or massive relocation would also be unacceptable.

Economic Spin-Offs: Jobs and industrial diversification opportunities for local firms must be guaranteed as part of the project.

Time: The opening of EXPO '86, May 2, was the critical date.

Project Control: Sufficient attention had to be given to the issue of project schedule and budget control to avoid cost overruns and public skepticism.

Public Information: The impact that the project would have on the lives and travel habits of local residents along the route and on those in the remainder of the region who would pay a portion of the cost warranted an aggressive information campaign.

System Choice

The decision to negotiate contracts and build the ALRT system was reached in two stages. The first dealt with the description of the acceptable transit system, almost a performance specification, and the second involved the review of technologies and products that met the specific requirements. With the issues, opportunities and constraints noted above, the definition of the acceptable system took the following path. In order to achieve the necessary capacity without creating chaos in the traffic system, it was obvious that the transit line would have to be grade-separated, either elevated or underground. It also meant that the system could be automated, thus achieving the objective of reducing some of the inflationary impacts on operating costs.

There was very little available information on soils, etc., along the alignment that could provide a base for costing the underground option. However, based on data obtained from other transit construction projects, the cost of an underground system was judged to exceed reasonable funding limits. In addition, the time to construct an underground system and the disruption to neighborhoods and businesses during construction led to the decision to plan an elevated system.

Two new issues then had to be examined if the system was to be elevated. The first dealt with the visual intrusion of structures and stations along the alignment. Conventional light rail or heavy rail systems both required considerable civil works in terms of columns and guideways, and stations would have to be very long to accommodate full trains. However, in both cases the capacity requirements could be met. Monorail posed safety questions (evacuation procedures, for example) that eliminated this option. Therefore, the decision was made to pursue a system with smaller and lighter vehicles than the typical +16 meter light rail vehicle. The smaller vehicle (at 12 meters in order to achieve capacity requirements) would permit the design of a smaller guideway beam and more compact stations.

The next question, however, concerned noise levels that an elevated system would generate. Three options were examined. The first was to consider noise baffles along the guideway to reduce the penetration of track noise into surrounding neighborhoods. This scheme was discounted since it ran counter to the visual intrusion specification. The two remaining options were then examined. Rubber-tired vehicles were studied first since the noise levels would be low. The complexity of switching systems and speed limitations of the rubber-tired vehicles were factors that constrained the ultimate system capacity. Therefore, a steel wheel and steel rail option was sought. The steel-wheel, steerable truck was judged to provide system speed and capacity and acceptable levels of sound emission.

Other considerations, such as the smaller vehicles being more compatible with the existing tunnel beneath downtown Vancouver, reinforced the major element of the specification for the new system. On this basis a detailed review of technical options was initiated. At least 11 automated systems were examined and several new technologies studied in some detail. In the end, the ALRT system proposed by the Urban Transportation Development Corporation (UTDC) of Toronto was favored. Since this was a new system, albeit tested for some ten years in Kingston at the UTDC's test facility, it presented new risks as well as economic opportunities. Sourcing of components and internal systems for the ALRT vehicles could involve BC firms and address the issue of industrial diversification. However, the new technology brought with it factors of risk that required analysis.

Addressing the Risks

Having defined the appropriate configuration of the regional transit system and a potential supplier, BC Transit began an analysis of the risks associated with introducing different technology from earlier studies and the magnitude of the task. These risk factors were divided into four areas—technology, design and construction, project management (including operations), and community reaction. The major outcome of the analysis was the provision that required the construction of a short section of the guideway soon after the contract was signed.

The "Prebuild" section, as this became known, involved the construction of 1.2 kilometers of elevated guideway, one station and delivery of two prototype vehicles a mere 14 months after the notice to proceed. For five months the two vehicles shuttled forward and back along the guideway, making in excess of 13,000 trips and carrying more than 300,000 people. Although "Prebuild" carried with it a premium price tag, the concept was aimed at dealing with the four areas of defined risk. The successful completion of the major portion of the project—the remaining 20 kilometers of guideway, 14 stations and the maintenance facility—both on time and under budget, resulted from the lessons learned and support generated by the "Prebuild."

A "fast-track" project, the "Prebuild" was the first test of the civil design and construction methods planned for the remainder of the project. The production of the precast beams, their transport and creation had a number of potential difficulties which were easily ironed out before a significant problem arose. The design of the final cross heads, at alternate columns, was ultimately changed in response to the "Prebuild" experience, and the utility and communications companies satisfied themselves that the system would not interface with their adjacent facilities.

During the demonstration period, the train control system and the vehicles performed well, with reliability of the system in excess of 99%. However, the tracking of component failures and maintenance procedures provided refinements that were incorporated into the final production vehicle. The operation of "Prebuild" by engineering and technical staff from the manufacturer/designer permitted the fast and efficient flow of technical feedback to the head office design group in Kingston, Ontario.

From the institutional standpoint, "Prebuild" allowed the participating agencies, as well as the principals in the System Contract, to develop working relationships and practices that would carry on throughout the project. From the BC Transit point of view it provided the opportunity to show the utility companies, etc., that ALRT was a serious project and that business arrangements would be made and invoices paid on time. With the tight time schedule that "Prebuild" was on, much of the "red tape" often associated with mobilizing agencies, such as those necessary for utility relocation, building approvals, etc., was eliminated. Therefore, in addition to the political support and public endorsement of the project, the participants at the technical level gained a renewed enthusiasm for the whole project when the "Prebuild" was successful.

The most significant impact of "Prebuild" was felt in exactly the constituency for which it was initiallly targeted, the public in Vancouver and British Columbia. The experience of riding the two-vehicle shuttle was enjoyed by over a quarter of a million people during the five months. Comments from these passengers indicated support for the choice of technology, system and design standards that were exhibited, and impatience for the time it would take to complete the whole first phase line. The information presented at the "Prebuild" center conveyed two messages, that the new system (ALRT) was an improvement over previously recommended technologies and that this improvement meant travel times would be reduced by almost 30% and service levels would be higher. And the risks associated with new systems (BART in San Francisco being the scapegoat) were being dealt with. Technical explanations of linear motors and automated train control systems, for example, were translated into descriptions that could be understood by everyone. The purpose was to defuse critics who used the apparent mystery of the technology to foresee future problems.

In the end, however, it was the actual experience that convinced people. "Prebuild" was tangible evidence that the rapid transit system that had been talked about for almost two decades had actually taken shape in only 14 months.

The Urban Form Consequence

During the "Prebuild" project an opportunity to build confidence in another major constituency was presented. Throughout the debate on the timing, choice and cost of rapid transit, the development community had been silent. Like many others, it was skeptical of the project ever being launched. However, a development company purchased a piece of property adjacent to the proposed "Prebuild" alignment, in close proximity to the Main Street Station site. Originally, the station site was to have straddled the Main Street/Terminal Avenue intersection in line with the guideway that would utilize a median on Terminal Avenue. However, negotiations with the developer resulted in an attractive alternative that relocated the station to a portion of the developer's site. The new site provided better access to and from the ALRT system and buses and the EXPO site, direct connection to a proposed commercial/office/hotel development planned for the property, a new transit market on which to draw, and a contribution in cash from the developer for the station.

In total, the agreement became a landmark in the ongoing project's right-of-way acquisition program. BC Transit demonstrated a willingness to come to a business agreement with the development industry, recognizing that the cash contribution was payable on occupancy of the building. Furthermore, the contribution formula was based on the size of the ultimate project, making BC Transit a partial recipient of any beneficial zoning and floor space ratio increases granted by the City of Vancouver. The publicity given this agreement in the business press heralded a new relationship with the development community. Subsequent agreements

Six-car "Skytrain" moving past the east gate of EXPO '86. Transportation was the theme of the Vancouver World's Fair. Photo by Perspective 5 Photography.

have been negotiated at major stations along the alignment, and a joint venture project with a developer is being considered.

As the project was being formulated and contracts negotiated, it was apparent that the most significant risk factor to be overcome was that of the possible loss of public support. With big lead times for completion and potentially disruptive activities affecting people's lives, it was critical that there be a way to solidify public support. "Prebuild" was judged to be the most significant statement that could be made.

Project Execution

In December 1980 the Government of British Columbia acted on the recommendation of the Board of the Urban Transit Authority (now BC Transit) and announced funding authorization for the construction of the Vancouver Regional Rapid Transit System using the ALRT technology. The Provincial Government instructed the Authority to do the following:

(1) negotiate a contract with UTDC for the supply of ALRT vehicles and services for the first phase of a rapid transit system for downtown Vancouver to New Westminster;
(2) investigate further the costs associated with crossing the Fraser River to Surrey (i.e., Phase 2);
(3) develop a management system to oversee the ALRT contract;
(4) pursue opportunities for federal government financial support of ALRT based on present Regional Economic Investment Strategies; and
(5) encourage UTDC to enter into economic/manufacturing agreements with existing British Columbian firms relative to the ALRT vehicle product program.

As a result of negotiations with representatives of UTDC, the two companies entered into a systems contract in late May 1981 that covered the following major items:

- the division of responsibility for design/engineering between the Urban Transit Authority and MCL (a UTDC subsidiary);
- the supply of and fixed price for vehicles, train control, other related hardware, and testing/commissioning;
- the completion of the entire first phase system (to Westminster) by January 1, 1986;
- completion and operation of a demonstration section (one kilometer) by May 1, 1983;
- system performance standards, bonds and guarantees; and
- provision for supplementary agreements to deal with the development of an operating company and the initial two years of revenue service.

The system that was to be built was broadly defined in planning documents by the Greater Vancouver Regional District and the supplier's (UTDC's) evaluation of the application of ALRT to Vancouver's requirements. Table II outlines the principal features of the ultimate system.

Project Organization

BC Transit was authorized and empowered to establish the ALRT system. The Board was responsible for the design, construction, commissioning and initial (two year) operation of the project, retaining sole authority for planning, alignment approval, land acquisition, communications and finance. The system contract called for Metro Canada to manage design,

Table II: Principal Features: Vancouver Regional Rapid Transit

Phase I

Inauguration	January 1986	
Length	22 km	12 km elevated, 7 km at-grade, 2 km underground; double track
No. of Stations	15	9 elevated, 2 at-grade, 4 underground
Vehicles	114 cars	supplied by Venturtrans, Kingston, Ontario; capacity 90 persons (40 seated); operated in married pairs in 2-, 3-, or 6- car trains
Operations	Seltrac Train Control; planned minimum headway 1.75 minutes; schedule speed 72 km/hr; passenger capacity initially 10,000 persons per peak hour direction; passenger capacity design 21,600 persons per peak hour direction	
Maintenance Service Facility	system control center; site size 20 acres; track for storage 1.5 km; storage capacity ultimately 250 vehicles; maintenance shop 4 tracks; vehicle inspection and cleaning 2 tracks	
Guideway	elevated; trapezoidal beam, precast and prestressed concrete; double-track width 6.02 meters; nominal span 30 meters; depth 2.97 meters	

Table III: Organization: Vancouver Regional Rapid Transit Project Joint Project Office (1983)

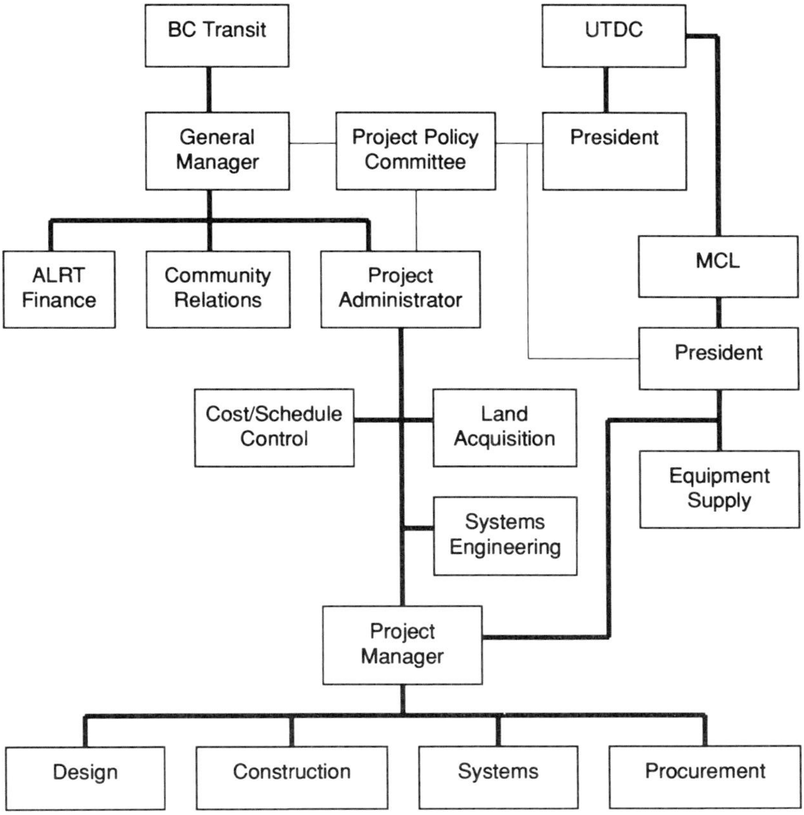

construction, procurement, etc. A small organization representing the owner (BC Transit) was established to administer the system contract.

By mid-1983, after the "Prebuild" construction was completed and the final alignment well defined, the management structure was re-evaluated and a Joint Project Office (JPO) established. The melding of the professional staff that represented the owner (BC Transit) and the prime contractor (MCL) and its major subcontractors was designed to ensure that the group functioned efficiently. A new atmosphere developed within the working levels based on a single vision of the definition of project success. Contractual matters (which had previously interfered in the quick resolution of outstanding issues) were referred to senior management of both BC Transit and MCL, while the work progressed. Table III shows the ultimate project organization.

BC Transit retained the sole corporate responsibility for finance and community relations. However, the merging of the two, previously separate, project teams created significant efficiencies in the design specification, cost control and project management. The unity of purpose (building the project on time and on budget) became the most outstanding aspect of the Joint Project Office.

Operations and Beyond

On January 3, 1986, the ALRT System connecting downtown Vancouver with New Westminster began revenue service. From the outset, the ALRT line had been planned to be an integral part of the regional transit system. This integration took two forms: a single fare system covering all public transit modes (SeaBus, Bus and ALRT); and bus schedules and routes redesigned to feed ALRT stations and allow convenient transfer between modes. The restructuring of the bus system was based on the concept that the ALRT line would provide the line haul function, while buses would collect and distribute passengers in local areas. Operational cost savings would accrue as a result of the shortened bus routes and the efficient use of the rail system to carry greater volumes of passengers at lower cost. In addition, to achieve the maximum operational savings, duplicate or overlapping bus/train services were eliminated. Finally, the bus system restructuring provided for the greatest potential of local transit trips. Not only were the local bus routes scheduled to coincide with train schedules, but the integration of the bus network with local centers and the timed-transfer schedules for station bus loops were designed to support increased transit patronage in local communities as well as on the line haul service.

Ridership during the initial year was skewed quite substantially due to the influence of the visitors to EXPO. Early estimates of ridership without fair-goers (that is, before May 2) indicated the following:

- 80% of ALRT users will be bus patrons;
- 67% of ALRT users will come from Burnaby, New Westminster and Vancouver; and

- peak hour ridership is projected to be on the order of 6,000-8,000 persons per hour peak direction.

These projections were based solely on demand projections from within the corridor and did not take into account any significant shift from auto to transit or a reverse commute. The latter is a particularly interesting variable if you combine North Shore residents using SeaBus and ALRT to commute to jobs in central Burnaby.

Revenue service, which began on January 3, 1986, has been a great success. The overwhelming response by the local residents caused long queues, especially during weekends during the early service weeks. Daily ridership averaged 70,000 people through April. The opening of EXPO '86 brought new ridership levels well beyond early projections. The World Exposition Skytrain carried 150,000 people daily, with some days approaching 200,000. Statistically, the system operated at 150% of its capacity. Revenue during the Fair exceeded expenses and the transit system actually recorded profits for the term of the Fair.

The ongoing public response has been high and the government has authorized the extension of the system to Surrey, across the Fraser River. A new cable-stayed bridge is being built to accommodate this initial 2.5 kilometer extension. It is intended that this new addition be available for revenue service in February 1990.

Conclusion

The view that transit will play a major part in the future of urban centers is commonly held throughout North America, notwithstanding articles in contemporary magazines that attempt to alter this perception. The transit "solutions" are not all as radical as the Vancouver rapid transit project, nor will they be as expensive. However, with rising expectations on the part of the constituencies and the complex set of needs that each has for project success, even the simplest of improvement efforts requires consensus prior to initiation. At the heart of the "solution" must be the political will necessary to begin the project. A comprehensive transit program needs to identify innovative approaches to technical administration and financial issues that today inhibit the delivery of better, faster and more comfortable service at the lowest possible price. Solutions need not be "big ticket" projects.

Support for any solution has to include the benefits that will accrue to the various constituencies that transit serves. These constituencies include the community in its broadest sense, transit employees, business, and politicians at all levels. It is critical that the needs/goals of each of these broadly defined constituencies be identified and common routes found to satisfy them. The credibility of the agency, and the potential for future support, depends on an ongoing dialogue and timely delivery of project successes.

The impact of the positive relationship developed in Vancouver between the project team and the various constituencies is a good measure of the consensus-building process. There are several broad and more specific lessons that were learned and can be summarized as follows:

(1) The program must deal with the needs of all of the constituent groups.
(2) Before a project is launched, consensus must be well established on a broad base.
(3) Any program aimed at consensus-building will cost money to administer, and it should be administered by the most senior levels of organization.
(4) The process of obtaining consensus is an ongoing one and is not simply related to single projects.

Materials generated to answer questions and educate and dispel myths combined with those received from other sources will provide the best source of visionary direction for decision makers. Linkages forged to support transit initiatives within any community are likely to influence the future, even for issues that are not transit-related. This is an unexpected side effect of transit development.

In Vancouver, through the project's development and execution, transit has been viewed as a priority. The need has been identified, accepted and internalized as a belief. BC Transit was able to position the "transit solution"—Skytrain—as serving that need. It is important to note once more, however, that by itself the Skytrain system (as simply a transit solution) would not have generated and sustained the level of support that was experienced. Understanding the other issues and targeting information campaigns, etc., at broader matters such as unemployment contributed to its success.

Part III

Issues in Implementation

The sudden blossoming of new transit systems in American cities has been accompanied by an explosion of techniques and strategies for funding transit development and for marshalling economic, legal and political tools for shaping the urban impacts such development can have. Sharing the responsibilities, risks and benefits is a theme common to these essays; their central lesson is that imagination, coordination and enlightened leadership are needed to turn opportunities associated with transit development into workable, livable and affordable urban design.

Community Involvement and Planning for Transit

Thomas C. Parker, Jr.
Economic Development,
Division Chief,
Arlington County, Virginia

Arlington, Virginia, is one of the smallest counties in the United States, with an area of 25 square miles and a population of 159,000. As a close-in suburb to the District of Columbia, Arlington was the fastest growing jurisdiction in the United States between 1930 and 1950 when its population increased by 400%, from 26,600 to 135,400. Commercial growth in the past 25 years has been explosive, with private office space increasing from one million square feet to 22 million square feet and hotel rooms increasing from 1,000 to 8,000.

With this tremendous growth, one would have expected significant citizen/developer conflicts. The opposite is true, however, because of Arlington's rich tradition of citizen involvement. In a region where controlling growth has become the catchword, Arlington is a "sea of tranquility" where citizens and developers not only sit down and reason together, but jointly market the county's development areas through public/private partnerships. The lesson is that dedicated political leadership, sound planning and citizen involvement can bring this diverse community together. In Arlington's experience, the most important of these elements is citizen involvement.

We have all seen cases in which strong political leadership has made a dramatic difference in a community. In most cases, however, this leadership occurs sporadically as individuals come and go in the political process. What is more important with respect to political leadership is a long term commitment to solid community goals and objectives. Sound planning is also important. Unfortunately, we have all lost count of the imaginative and thoughtful plans that have ended up on coffee tables or, worse, buried in a bookcase. On-going citizen involvement is the solution, but it must be more than a band-aid or pro forma planning commission/developer involvement. What is required is the long term commitment of an involved, informed citizenry.

A Cautionary Tale of Two Counties

The importance of citizen involvement is well illustrated with a tale of two counties. In the early 1960s, local governments in the Washington region began planning the regional rapid rail system that is now Metro. All jurisdictions participated equally, but in the final analysis, some were more committed to the task than others. Arlington took the opportunity to lobby for the most extensive and best system that could be afforded, a system that would not only serve the county's existing commercial corridors but in the future provide a catalyst for the selective redevelopment of what were, even then, declining commercial areas.

It was a simple plan, based on the premise that if growth were to occur in the county, the community would target that growth, and the growth would be concentrated in limited areas where development would serve the community's objectives. Through targeting and concentrating development, the balance of the established community would be preserved and protected. This concept was popular with the Transit Authority planners because it concentrated employees and residents close to transit, using the subway to maximum advantage.

This was an expensive plan for Arlington. With the exception of stations on federal property, all the Metro stations in Arlington are heavy-rail subway construction through established commercial corridors. In fact, with 11 Metro stations, Arlington has more stations per capita than any jurisdiction in the region, and within the United States we are second only to Decatur, Georgia (two stations for 25,000 residents).

In contrast, the other county chose to use rapid transit as a commuter rail system. It was the least expensive alternative, surface rail through lightly-developed residential corridors. This jurisdiction has been extremely successful in attracting corporate and high-tech development over the past several years and has experienced an explosive growth of commercial office space from two million square feet in 1960 to 44 million square feet today. Unfortunately, none of that development is served by rapid transit. As a direct consequence, that jurisdiction is now involved in the hottest political growth debate in the Washington area. The focus of that debate is the impact of traffic from development and the issue of how to, or whether to, develop around the Metro stations that are now in place and operational.

The Tradition of Involvement

Arlington's history of community participation in transit planning dates from 1961, when the community was involved in the early planning of the Metro rail alignment and station locations in Arlington. When the regional system was approved in 1968, this involvement continued over a six-year period with issue papers and background planning studies on the county's two transit corridors. In 1974 over 200 community groups participated in an intensive community goals process that focused on defining the kind and character of development that would be appropriate for each transit station. In 1975 and 1976, those goals were translated into general plans for each Metro station. Following the adoption of general plans, the Planning Commission and staff and community groups began working out the definitive sector plans for implementing development.

During this 25-year period, when virtually every other jurisdiction in the region experienced heated debates over the location of transit and transit-related development, never once was the original alignment and location of stations in Arlington questioned. Although there was significant community discussion over the character and type of development, the basic premise that each Metro station would serve as a focus of intensive residential and commercial development was never challenged. This can be fully appreciated only when one understands that Arlington's most intensely developed transit corridor, with five stations in three miles, is surrounded by established residential neighborhoods, all within a one- to three-block walk of the stations.

Specific examples of citizen involvement are informative. In 1979, Arlington was faced with the major problem of how to achieve true mixed-use development, residential and office, in Ballston, an older, marginal commercial area in the center of the county. A citizen/developer/government committee was formed. Following an extensive study, the committee proposed a new zoning district that nearly doubled the effective density permitted elsewhere in the county and, in addition, proposed building heights that would be the highest in the county. In most communities this would have resulted in a major citizen/developer confrontation. When the public hearing was held on the adoption of the zoning district, the first three speakers were representatives of neighboring civic associations speaking in favor of the proposed district. Their message was that if the community was to achieve a true mix of residential and office development, density and height were absolutely necessary. The district proposal was adopted with no opposition.

For a number of reasons, however, development in Ballston did not proceed as planned. In 1985, the Economic Development Commission brought together developers and citizens to discuss how the government, developers and citizens might work together to bring about the planned development. What followed was one of the most remarkable success stories in the Washington region. Before the end of that meeting, the Ballston Partnership had been formed with two active committees of

Ballston Partnership Briefing Center. L to R: Krista Amason, John Shooshan Company; Gisela Miller, Executive Director, The Ballston Partnership; Tom Carr, The Oliver T. Carr Company.

developers and citizens, one to address marketing and the other to address urban design issues. Both committees began meeting on a weekly basis. Within six months, $25,000 had been raised, and the Partnership had successfully petitioned the Arlington County Board to match on a dollar-for-dollar basis the Partnership's first year work program of $200,000. Within 18 months, a staffed, fully-equipped marketing center was in operation.

The spontaneous enthusiasm continued with committees and subcommittees meeting on an almost daily basis. In the spirit of true partnership, every issue has been resolved through consensus. Citizens joining the Partnership at $10 per year carry equal weight with developers joining at $5,000 per year. Citizen members are active on all committees, drafting the by-laws and editing the newsletter. The results have been dramatic. In the Partnership's first two years, two office buildings and an 800,000-square-foot, four-level shopping center were constructed and heavily pre-leased; three additional office buildings are under construc-

Ballston Partnership Briefing Center Model.

tion in addition to new apartments and condominiums and two hotels. The Ballston Partnership's success has encouraged others. Two additional partnerships in Arlington are now actively organizing.

When we are asked by those outside of Arlington, "How do you do it?", the response has been, "Involve citizens." We often hear the answer, "We can't do that." Unfortunately, that is probably true. If there is no history of citizen involvement, if the citizen involvement is a charade or a Trojan horse, and if all parties do not accept citizen involvement as a true partnership of the community, development and government interests, success will be difficult to achieve.

Infrastructure Financing and Joint Development

Jon W. Martz
Director of Research,
Rice Center for Urban Mobility
Research, Houston, Texas

The philosophy which holds that public transportation is a social service to be provided by government is being superceded by a philosophy of benefit-based financing of transit: those who benefit pay. The list of beneficiaries of public transit service is being expanded to reflect this new philosophy, which identifies not only riders as beneficiaries but also the public at large, employers, retail businesses and private developers. While this re-examination of public transit finance has focused on techniques for assessing the private sector for its share of the benefits of transit services and facilities, another approach to private sector involvement is gaining momentum. Public transit agencies are increasingly examining opportunities to involve the private sector in all facets of construction, maintenance and operation of transit services and facilities. These can provide significant cost savings and reduced risk to the public agency.

The value of joint development opportunities is different for each participant. From the transit agency's perspective, the benefits are expanded ridership, increased revenues, and/or savings in construction or the costs of transit. While the transit agency's primary objective is to develop and operate a transit system, its actions affect property values and area development. To capitalize on this potential, it may have to assume a more aggressive role. From the developer's perspective, maximizing a return on investment or improving accessibility to a specific site or area may be the primary objective. Working cooperatively with the transit agency can facilitate this. From a community standpoint, improved transit service and area revitalization may be the objectives. Nonusers benefit through increased local sales taxes. All participants in the process seek to "capture value" that appears with the advent of transit in a locale.

The following are mechanisms for public/private partnerships discussed in terms of the following evaluation criteria: revenue impacts, political feasibility, equity issues and questions, administrative costs and feasiblity, simplicity, ridership enhancements, timing, transit precedents, opportunity costs, and legal barriers. There are many advantages to engaging the private sector in transit development, but to succeed, transit agencies may need to be entrepreneurial. In some joint development efforts, the public transit agency might even be the developer. That is quite different from the mission that most transit agencies assume, which is merely to provide transportation service.

Acquiring Land

Land acquisition involves the assembly of land by the transit agency at a station site or along a corridor and the subsequent leasing of portions of this acquired property right to a private developer. Land acquisition can be achieved on a substantial cost sharing basis or on a sale-leaseback basis to an owner on a long term ground lease. Moreover, land acquired may be used for immediate transit needs or leased for private use and held for future transit needs. The political feasibility of this mechanism depends on the reactions of landowners and competing developers in the project area. If a transit agency seeks to assemble land over the objections of area landowners, serious problems can result. If land assembly is carried out in an area where developers see little prospect for development, or if the stimulant effect of such a project is actively sought by other landowners, such objections might be muted.

Serious objections to a transit agency's power to acquire land might result in a legal challenge to its authority, a power that has not been thoroughly tested. Some transit agencies are legally prohibited from acquiring land too far in advance of actual use. Equity questions can arise in this process when competing landowners object to the involvement of a public agency in private development. Landowners who gave up land to the transit agency to see it developed by another private developer might argue that the private developer is being given unfair advantage.

Public agencies have not typically undertaken extensive land acquisition without immediate justification for such actions. The timing of these actions could make sense to the general public when property values are considered depressed, as mentioned, and the infusion of revenue to the development community could be beneficial to the timing of private sector projects.

Land banking in transit most often takes the form of right-of-way donations for future corridors, less often for station sites. The MBTA in Boston borrowed funds to make large purchases in 1973 and 1976; and in Philadelphia the transit property purchased suburban right-of-way from the Pennsylvania Central Railroad. In acquiring land early, there is an opportunity cost, but since advance property acquisition is an early action needed in implementing a major transit project, the cost will likely be deemed acceptable to the degree that the property is ultimately used for the project. Land and property donation may decrease up-front land cost for transit stations while giving locational advantages to other property owners by the donor. The sharing of parking with the transit station would be included under this mechanism.

Donation of land is most likely to occur when such an action by a private landowner can alter a station location to the landowner's advantage. There have been many examples of land donations that have altered the location of freeway routes, suggesting that the legal feasibility of this mechanism would be high. Revenue impacts of land donation are direct in that they offset one major cost of station and route development at an early stage. However, the full cost of right-of-way includes relocation costs and possible payment of damages that may occur if the site is donated. Political feasibility is often a problem. The suspicion that a landowner making a donation is benefiting disproportionately or disadvantaging competition by changing a station location is an issue likely to be faced.

Land donations are an easily understood concept and an accepted way of implementing transportation improvememts. Opportunity costs for land donations occur only if the donation is tied to specific requirements of the transit agency. It is likely that such donations will have strings attached, such as location of stations or architectural or urban design requirements. There are some legal and procedural conditions to accepting donated land, the most prominent being the use of the Federal Uniform Property Acquisition and Relocation standards, so real estate and legal staff need to be consulted early in employing this mechanism.

Development Rights

Leasing and selling development rights by a transit agency is a way of turning surplus or under-utilized property to the agency's advantage; for example, leasing air rights over a station or a transit center, or the development of land held for a park-and-ride facility at a suburban location would fall under this category. The revenue aspects of air or surface development can be substantial for an individual station.

The feasiblity of this mechanism depends in part on the method by which it is accomplished. If development rights are put up for bid, feasiblity is likely to increase due to the perception that this method eliminates unfair advantages while realizing the greatest revenue for the transit agency. As a site owner, the transit agency may be placing itself in a position of competing with surrounding property owners. Again, this involves the entrepreneurial aspects of joint development for a public agency. To ease concerns, this issue must be dealt with in a general policy context, not on a site-specific basis. Large cities are involved in a limited way through economic and development agencies and authorities.

Air rights leasing is used for both transit stations and highway development. Boston's Copley Place uses air rights over a major highway and rail right-of-way. WMATA in Washington, D.C., leases 1-1/2 acres to Prudential and receives a percentage of net profit from the development. The State of California leases right-of-way sites to private concerns for parking and storage as well as for active development. In Sparks, Nevada, a casino leases air space and space underneath the highway. In Santa Cruz, the transit district is leasing office and retail space in a downtown transfer facility. In Tacoma, six sites have been leased for 20 to 30 years at a low annual rate. In Phoenix, a similar arrangement exists at a shopping center where a transfer center will be built. There are many legal considerations including questions about the involvement of a public agency in private development. For example, can a transit agency use its power of eminent domain to acquire property to be sold and leased for air rights development when the property has been acquired for the specific purpose of providing transit service?

Station Cost Sharing

The cost of constructing a station can be shared with a developer or owner of property benefiting from station development. It can change station locations or can upgrade the quality of the station. It seems politically feasibile to upgrade the quality of a station; cost sharing to relocate a station, however, carries the possibility of controversy regarding equity questions, including the disadvantage to other landowners and possible impacts on ridership. Cost sharing could also be approached in a straightforward partnership with area businesses sharing directly in the cost of building the transit system. One advantage of cost sharing is that it is easily understood. This makes it easier to describe and discuss publicly.

The perceived motivation for cost sharing becomes a primary issue. Is the donor buying a decision from the transit agency or is he paying for a higher quality or better location than the transit agency could achieve by itself? Are they helping or are they paying off? This will depend on the donor, the reaction of other people in the immediate area, and the procedures used by the transit agency. Cost sharing seems to benefit the public good, in that the private sector often contributes directly to support other public benefit types of projects, even where direct benefit is difficult to gauge. If community leadership adopts this type of approach, a significant percentage of station costs might be paid by this form of cost sharing.

The method of payment takes several forms, including assessments, ownership arrangements, direct donations, etc. Examples of cost sharing include St. Louis, where a local advertising company paid for bus shelters. In Secaucus, New Jersey, a local industry constructed a rail station for New Jersey Transit as part of a multi-use development. Elsewhere, private developers have contributed portions of the cost of interchanges in order to expedite highway improvement. In Houston, a bridge spanning a major highway was built with private funds at Greenway Plaza. In Dallas, Southland Corporation is negotiating with the city to assist in cost sharing of the light rail program as part of Southland's major development project along the North Central Expressway. In New Orleans, developers and property owners have made commitments to build the ten stations for a proposed streetcar system.

As in any situation in which a private sector entity is paying for part of a public project, the administrative and legal requirements can be complex, but they are not overwhelming. Private sector donors may attach conditions to their donations, and they may want provisions for extended payment of these donations. As in any partnership agreement, these conditions would have to be set forth as part of a final agreement, a contract.

Lease of Concessions

Through lease agreements, transit stations can provide income to transit agencies and retail opportunities to the private sector. Such concessions are a convenience to transit patrons and are appropriate especially in regional transit centers, where the size of such facilities is large enough to accommodate retail activity. The range of commercial concessions is partially a policy choice. Do current station guidelines preclude commercial activity such as convenience stores, cleaning or auto services, and major vending operations? Are auto-teller banking machines, hot dog stands and similar activities permissible? It is common for local governments to recoup that part of the cost of facilities associated with the provision of these concessions. But competing businesses often complain about the disadvantages they suffer when concessions on public property are granted to others. Such equity issues usually require that transit agencies request bids for concessions.

Within activity centers it is conceivable that transit center concessions be managed by area businesses as part of their efforts to maintain a high standard of operation. Concessionaires and related businesses might object to this approach, but such objections can be minimized when it is part of a larger strategy for transit center operations. Concession needs vary with the level of activity found at each station. These can range from simple vending operations for basic transit user needs to more complex food stands, commercial stalls and open air markets. The amount of revenue will depend on ridership and the retail traffic actually generated.

A transit agency may participate with a private developer on public land that the agency has acquired, share in the equity and receive a percentage of the income. A variation on this approach is participation of the transit agency in the redevelopment of privately- or publicly-owned property. The revenue potential of co-development is similar to that of leasing and selling development rights but carries with it possible risks if such risks are shared equally by the public and private partners. Much of the public/private partnership rhetoric that we hear is based on trying to spread risk, taking more of the risk away from the public and putting it on the private investor. In fact, the public may take a greater risk than it would typically be expected to handle.

Shared right-of-way has been used to establish transitways within existing rights-of-way of state-owned freeways, for example, in Houston. It has allowed transit agencies to reduce development cost by incorporating transit facilities into existing roadways.

System Interface

Property can provide a direct physical connection between adjoining private or public development and a transit station, thereby improving pedestrian access and improving the safety, aesthetics and convenience of the development. Revenue potential could vary substantially, depending on the type of interface, including pedestrian, bus or auto connections such as parking. The private sector could contribute to these costs and benefits with more direct connections or with strategic placement. System interface improvements can involve complex urban design and transit engineering to achieve solutions, for system interface is critical to achieving good ridership on a transit facility. Real and even perceived delays and inconvenience, such as lack of sidewalks and parking, inadequate signage, dangerous walkways, a sense of insecurity, and poor appearance, can cause potential riders to choose their personal autos. Many of these problems lie beyond the immediate environment of the transit station, so the transit agency has little power to resolve them.

Pedestrian and other system linkages have been built by the private sector. Skywalks, tunnels, bus stops and shelters, park-and-ride and kiss-and-ride locations have been provided through private agreements to use excess parking at shopping centers. Provisions for private sector contributions to publicly-implemented projects are the primary legal and administrative concerns for this mechanism. Where service contributions are made, there may be no need for a formal agreement unless contributions are contingent on some requirement or action on the part of the public agency.

Connector fees may be paid by private concerns to the transit agency to connect their development to a station. These fees are usually above and beyond the cost incurred by the developer in building the connecting facility. Connector fees should be sufficient to cover the construction costs of the physical connection plus an additional amount reflecting some part of the benefit to the property owner or business. It is likely this amount would not be a major revenue source but supplementary to the basic cost of a station.

Benefit Assessment Fees

Local governments with the active support of area property owners can establish an assessment district to fund a portion of the cost of transit or other improvement or services. These districts would assess a fee on properties within a district to pay for all or part of the expense for specific improvement in the district. The boundaries of the district would be defined to include all properties benefiting from the transit improvement. The intent of this technique is to equably assess those who gain special benefit from transit improvements in the district at a level commensurate with the value of the benefits being realized. Assessment districts have one of the greatest revenue potentials of all the private sector mechanisms. They provide significant annual revenues which can be pledged to finance debt for capital projects or for operations. The use of assessment districts is also an equable way of financing transit improvements.

Areas that are unable to make such payments are generally not included in these districts. Questions may arise regarding use of a district for one activity but not for another. The basis for establishing or not establishing a district should be set forth early in the process, if this mechanism is to be selected. Benefit assessment districts have been used in Miami, Denver, Madison, Seattle, and Los Angeles, to name a few.

Tax Increment Financing

Another method of funding public project construction is based on increases in property tax revenues. Districts usually are created in declining areas where public projects are used to encourage redevelopment. In tax increment financing districts, a base-year property value is determined. Taxes collected on any increases in property value above that base are dedicated to needed improvements. This mechanism provides a dedicated source of tax revenues without levying additional taxes on property owners. Funds raised from these districts are spent on improvements within the district.

Competitive Service Contracting

Revenue from competitive service contracting can occur as cost savings and by cost stabilization through multi-year, fixed-fee contracts. The amount of money saved depends on the amount of service being contracted and the relative cost of privately-provided vs. publicly-provided service. Research on cost savings shows varying results but indicates that private sector involvement fairly consistently results in noticeable savings. The savings range from 10-60%, although in some cases, like San Antonio, there will not be any. It is not unusual to find realistic savings in the 20-30% range at most transit properties.

Equity issues are generally not a problem with service contracting if handled in a competitive environment. Some cities have created competitive situations where the transit agency itself competed to provide service under a formal contract arrangement against private providers. There are administrative costs for contracting that require suitable knowledge and skills to negotiate, administer, monitor and evaluate services. Service contracting is a straightforward way of providing service that most people readily understand, although how a private contractor can provide service at a lower cost than a public provider is sometimes a matter of curiosity. While private companies cannot always achieve such savings, the strong incentives working in the private marketplace help to assure those savings and probably account for the greatest percentage of cases where private providers can provide service more cheaply. Ridership will not be affected to any extent through the use of service contracting; in fact, studies of service quality have shown that service under private contracting is equal to or even better than publicly provided services. In New York City, about one-third of the buses that come into the city are privately owned and operated.

Turnkey-Plus

This mechanism involves the private sector in design, construction, service start-up, maintenance and service operation. It can include transit center operations, security communications, financing and even marketing. This is more commonly known as privatization. Construction is a familiar use of turnkey programs now, but the actual ownership of the system, the contractual relationship with a private entity to design, build and construct a system, is something new. Revenues are likely to be substantially affected by this type of mechanism, but there are many unknowns that make consequences difficult to estimate. The extent to which increased revenue and cost savings are experienced would depend partially on the degree to which the public sector is able to tap the characteristics of private sector involvement, which means competition, management and labor efficiency, profit motivation and productivity. Private firms can build heavy rail systems. Ridership incentives can also be placed. Whether the mechanism is a cost based proposal from a private firm or performance based contract may affect the technologies involved and could even affect ridership.

The use of development controls can have a substantial impact on private development revenue. In many cases of joint development, the zoning allowances near the rail stations influence the success of joint development opportunities as much as anything else. For example, where development bonuses are given in terms of extra density provisions, developers are able to more intensely develop properties than would otherwise be possible. The tradeoff may be participation in transit and transportation management programs or the provision for certain physical improvements which complement transportation. By increasing the intensity of development at identified locations, the transit system is able to provide better services and increased ridership. Both parties gain. The private developer is allowed to do more than existing zoning and deed restrictions allow; the transit agency benefits from access to a higher use facility.

These examples show that it takes a very active role in property and land development by the transit agency to realize these benefits. To succeed, an agency must participate actively and early, openly and equally with private interests. It must make sure that the facilities meet certain criteria of quality and still benefit the developer. It must specify performance incentives or disincentives to ensure that the system operates correctly.

Value Capture and Benefit Sharing for Public Transit Systems

Jane Howard
Howard/Stein-Hudson Assoc.
Boston, Massachusetts

Benefit sharing is the distribution of public and private costs and benefits associated with transit facility construction, rehabilitation, or operation. The types of benefits extend beyond financial advantages to the realms of urban design, urban planning, ridership, perceptions of the system, etc. The objective is to achieve the broadest benefits for all of the participants, public and private at a reasonable cost to each. The possiblities cover all types of transit facilities from the smallest para-transit, dial-a-ride systems, to big fixed rail systems; and from small bus shelters and ad benches to the multi-million dollar office development like that which has funded improvements to Times Square Station in Manhattan.

A study I completed for the Transportation Research Board in 1984 tried to determine what transit agencies in various cities in the United States were doing to generate private sector funding for transit projects by reviewing case studies. New York City (the MTA and the Transit Authority in New York) used incentive zoning, joint development, system interface connections, and voluntary contributions—a variety of techniques—to fund a complete rebuilding of the Times Square Station at 42nd Street with private sector funds and major contributions to new platforms for their transit stations in midtown. In Los Angeles, the Southern California Rapid Transit District is experimenting with a sophisticated process of joint land use and transit planning for their proposed metrorail project. Boston's MBTA is an older transit system which has a history of joint development successes in various projects but which is also considering reuse of some of its under-utilized property like outmoded power stations and large parking lots in prospering suburban development areas, and is revamping their lease, concession and advertising structure to gain more revenue.

In Washington, D.C., the transit agency has pioneered some of the concepts of joint development and system interface. We examined the New Carrollton station where Metro took an active role in planning for joint development on a parcel it owned and the Bethesda Metro Center which worked the other way to achieve joint development with city involvement on property that the city owned. For Portland's Banfield Light Rail Line we focused on the station area and master planning process and some of the benefit assessment districts which were implemented in the downtown to fund light rail special amenities. It is reported that these assessment districts have proven very successful, partly because the system is so appealing to the public that the development community is jumping on the bandwagon to expand the scope of that program. Toledo has an innovative downtown bus loop, showing that you can attract developer support through good design and careful attention to detail, even for buses. Finally, our study considered a range of intercity-intermodal terminal projects in Michigan, ranging from large cities like Flint to little Dowagiac. These projects tried to fund the operations of their transit centers through lease of ancillary space, office space and concession space, and to a large degree they have been very successful.

Other Factors Affecting Benefit Sharing

These cases show that the type and size of the transit system can determine the scale of benefit sharing and the strategies available. But other factors also are important. The type of development or investment is a factor in terms of scale and expected investment. You cannot expect Times Square-scale private investment at small city neighborhood bus stops. Also, the goals of the participants in the process—public agencies, private agencies, developers—will vary with location of the station, guideway or transit facility, and with the character of the neighborhood—residential suburban, neighborhood commercial, urban residential, or downtown.

Market conditions are very important in determining how you negotiate with the private sector, as are the timing of achievable development and the distance of the development from the station. Distance affects benefits assessment and the extent of fixed financial contributions. Finally, the stage of construction or operation of the system is crucial in negotiations.

Perhaps the most surprising lesson from this study is that benefit sharing strategies can be successful at every stage of the transit planning and operation process. These include initial system planning, location of transit corridors, station location, site selection and facility design. During construction, cost savings can be achieved by coordinating construction processes. Of course, it is in the operation of facilities that benefits are usually expected. Assistance can be negotiated in rehabilitating older facilties once they are outmoded. Finally, disposing of disused facilities can produce benefit to the transit agency, the public sector and the development community as well.

We found that agencies that succeeded in coordinating land use and transit planning and in working with the private sector effectively had incorporated a "benefit sharing" philosophy into their ongoing planning and implementation process. Transit agencies that succeeded were those willing to take a point of view that went beyond the traditional transit provision function of the agency by looking at the broader land use implications and the development implications and by taking the extra step or risk necessary for capturing opportunities while still carrying out the service mandate. This requires a continuing cooperative process between the transit property, local planning and development agencies, elected officials and key actors in the private sector.

Reflections on the Benefit Sharing Process

Benefit sharing can be defined differently at various stages of the planning and design process; and it must be kept in mind continually and flexibly as the planning process proceeds. But first, the transit agency must decide whether to include development-related considerations as criteria in their planning and design, or whether to take the course of many transit agencies and seek the lowest cost and most convenient right-of-way, letting development proceed later.

Once the decision is made to employ an interrelated process, the transit agency has to establish a professional capability for dealing with the land-use issues. Some of the larger agencies (SCRTD, WMATA) have established huge land use and development planning departments. In smaller agencies—in Toledo, for example—a planner was assigned the role. Other agencies have used consultants very effectively. It does not have to be a major organizational decision or terribly costly, but whatever capability is established, it has to be funded. In Los Angeles, this meant local UMTA funding.

As planning progresses, another consideration is obtaining reliable "before and after" data upon which to base some of the benefit calculations. This is an area that sometimes is overlooked. Finally, transit planners must assess the long term tradeoffs between density bonuses or other considerations that are granted to developers in return for transit contributions. This also is true where incentive zoning and similar strategies are used. New York is trying to deal with this right now. Some of its incentive zoning grants a number of extra square feet to developers in return for, say, an escalator. The New York City Planning Commission is now starting to question whether an escalator or other contributions are an equal tradeoff against the impacts caused by the extra density.

How does this all fit into the transit planning process? The traditional UMTA transit system planning process includes the phases of system planning, alternative analysis, draft environmental impact survey preparation, preliminary engineering, final design of the project, construction and operation. Traditionally, what might happen is that consideration of joint development opportunities or land use interactions occur incidentally. At some point, the planning agency review may decide that a building should have been located on a particular parcel, or well into the preliminary engineering stage a developer may express interest in building a connection to a station. The transit agency typically must decline because the process is too far along. So opportunities are lost. But in the ideal process, steps are coordinated, and general system planning steps for transit are paralleled by logical steps in land use planning and development.

At the system planning stage, land use agencies look at regional plan development, decisions about growth corridors and subcenter business districts, etc. Developers conduct their own regional market analyses of growing communities, where office space could be absorbed.

During the alternatives analysis/draft environmental impact survey stage, corridor level master plans are worked out. This was done successfully in Portland and in Los Angeles where corridors were assigned on the various land use roles: a growth corridor, a medium growth corridor, etc. The land use characteristics were considered on a corridor basis along with the transit planning. It is at this point that basic land use regulatory strategies and financial strategies can most effectively be considered comprehensively.

Rice Center did some good work in Denver at this phase for a proposed light rail line, looking at various corridors and the financial strategies appropriate at the various types of stations. They concluded that at some inner city, higher density stations a more aggressive financial strategy might be applied. At some lower density, residential area stations the major financial strategies were limited to concessions within the station or to advertising only. At this stage, developers look at alternative sites and concept plans for land use mixes. Basic land use regulation decisions must be made—whether to use incentive zoning, development rights transfer, etc.—and financial strategies must be targeted to specific corridor locations.

During preliminary engineering, when the transit agency is finalizing details like stair and entrance locations, land use agencies are working simultaneously on station area and master plans. Again, in Portland and Los Angeles, this process was carried out effectively. Plans for each station in terms of land use, density and design guidelines for development were detailed so that land use regulations could actually be implemented. Similarly, specific information about projected benefits of stations at varied distances could be developed so the formulas for assessment districts that were to be implemented later could be fixed. On the private side, a developer in a transit impact area is typically finalizing his development program at this time, locating tenants, trying to make investment and financing decisions, and preparing initial site planning.

During final system design, city agencies conduct design reviews and reach "nitty gritty" agreements with developers and the transit agency. The developer is actively involved in the process, negotiating taxes, leases, etc. The Urban Land Institute has stressed the importance of the transit agency having a lead person to deal in a professional manner with the private sector to establish credibility.

In coordinating public and private elements during construction, it is possible to achieve time and cost savings. And it is during construction that implementation of financial techniques occurs and benefits assessment districts proceed. There is an inevitable credibility gap, especially with new systems, until construction proceeds to a point where the private sector is certain the system will be built. In Boston's southwest corridor, which has been in the planning stage for 15 years, there were a number of well conceived development plans, but it was only in the last year or so, when the system was three-quarters built and stations were rising from the ground, that city agencies started issuing kits to developers for their parcels.

Finally, once the system is built and operating, transit and development project impacts should be monitored to determine positive impacts on property values, etc. Benefit district boundaries or fees can be modified, if necesssary. This happened in the case of the the Denver Transit Mall; assessment district boundaries were changed after the initial operation when the actual benefit levels became clear. During operation, the developer is also conducting the same sort of "fine tuning" in terms of property management, finding tenants, altering lease rates, etc.

Benefit sharing is a continuing process. There is a perception that benefit sharing techniques are fixed and that choices are limited. But this is not the case. Some techniques are appropriate at one point of the planning and design process, but they may not be later. For this reason, the transit agency needs the staff capability for dealing with benefit sharing on a continuing basis.

Toledo

Toledo shows how a small transit agency can implement a very high quality project with significant private sector participation in a style suited to the community. This project was part of a downtown revitalization plan initiated in 1976. TARTA, the regional transit agency for Toledo, rerouted all of its bus service into a 1.1-mile downtown bus loop which has five stations integrated into adjacent structures. When TARTA started out in 1976, however, a downtown transit mall was planned. When a new general manager examined the transit mall and plans for Toledo, he saw that new development was not to be located anywhere near the transit mall. In fact, the development was focused on the fringes of downtown; the mall had very little relationship to the actual development plans. The general manager initiated discussions with the City and major employers like Owens-Illinois, Toledo Trust and Toledo Edison, to rethink the plan and determine the possibility of the transit agency serving the new development more logically. He put together the loop plan with a 1979 Urban Initiatives grant which, as it turns out, was the impetus for much of the benefit sharing that occurred in those years.

Instead of running buses on the mall, the transit agency looped routes around the downtown area with five stations that each serve all bus routes. Fares are set up on a "pay-as-you-enter-inbound, pay-as-you-leave-outbound" system such that anybody who gets on the bus downtown and gets off downtown does not have to pay at all. Effectively, downtown buses provide free downtown circulation.

The land for Seagate Station was purchased by TARTA from the City for about $25,000. In anticipation of future development on the site, the City reserved an aerial easement above the station, so the station was built first and the building later. The station is connected directly to the building through an enclosed pedestrian concourse, an element of the plan that helped to sell it. As with Portland, local architects familiar with the city's scale and style were retained for the station designs. The concourse was paid for by TARTA but was built by the City at its cost; Owens-Illinois funded the 20% local share and now maintains the concourse and escalators. The City is responsible for security.

The Promenade Station is linked directly to the Toledo Trust Building and was built in the right-of-way made available by the City through closing the street. Pedestrian concourses were added to the existing Toledo Edison Building and again cost sharing for capital and operating costs was used.

The Perry Station is incorporated into the street level of the City parking garage. It will eventually connect with the City's new convention center. In this case, construction of the station was delayed until the parking garage was under way so that construction contracts could be coordinated.

Park Street Station is leased from the City for $1 per year. Since it is not linked to any development project and therefore does not benefit from surveillance by private maintenance forces, it is the only station with the typical problems of loiterers, etc.

Finally, Government Station is also connected to a parking garage. Again, the transit authority leases land for about $1 per year. Unique to this case is the fact that the transit agency took a lead role in tailoring its plans to development plans. The agency's general manager was the key actor in selling the plan to the private sector. He met with the Chamber of Commerce and other groups at breakfasts and other events and did the kind of marketing and public relations necessary for gaining widespread support for a project.

Summary

In identifying factors leading to successful implementation of benefit sharing, the willingness of the transit agency to go beyond its traditional function is most important. The first step should be a systemwide review of the opportunities for benefit sharing within the agency. Some type of continuing structure should be established that is appropriate to the agency's size. The benefit sharing philosophy should be inculcated in all of the operations of transit. The private sector should be approached in a businesslike manner. Their timing schedules and requirements are valid, and the developer should not be treated in a punitive way or with an exploitative attitude. Similarly, public agencies must maintain credibility in terms of timing, funding and meeting deadlines, because time is the one element that costs the private sector the most. Careful attention must be paid to market analysis in terms of what is appropriate and to every detail of the design. Phasing, construction coordination, and maintenance are extremely important. Legal agreements in successful cases were used to expedite, not delay, implementation. In the Toledo case, because of pressing deadlines, many agreements were reached through letters of cooperation in order to expedite progress. Detailed legal contracts were worked out later.

Finally, it is necessary to be both realistic and flexible in determining transit agency costs on projects and benefits to be realized. Actual financial returns to the transit agency vary. They are not always immediate and the benefits are not always financial. In particular, private developers seem more likely to fund enhancements to the system, which they see as a direct fringe benefit to them, rather than systemwide operating improvements which are better done through sales taxes or other mechanisms. Our basic conclusion in studying these cases was that benefit sharing, while certainly worthy of pursuit, should not be pursued as a substitute for public funding. In most cases, it is impossible to gain private sector commitment up front, although the commitment may occur at some point. Our findings were that UMTA and federal policy should be more appropriately directed toward rewarding the agencies that experiment with benefit sharing approaches rather than penalizing people who cannot package things into fixed, arbitrary formulas.

Joint Development at Transit Stations

King Cushman
Director of Development
and Community Affairs,
Pierce Transit,
Tacoma, Washington

This discussion focuses on a range of contemporary activities involving joint public-private financial participation in development on or around transit centers and transit stations in a few U.S. transit systems. The examples are certainly not all-inclusive, and the greatest activity of this sort is typically related to major heavy or metrorail transit systems. But to show that there are other, more modest opportunities sometimes overlooked, examples from a medium-sized bus system like Pierce Transit and a couple of west coast light rail systems are also noted.

The interest from the private sector in some degree of financial participation through direct or indirect development activity near transit stations is because successful transit stations handle thousands of people every day and begin to perform like freeway interchanges. Lots of people and traffic means economic opportunity, and we all see endless examples of the desire for development to be well situated with good access in the vicinity of busy freeway interchanges. However, there is a caution about having great expectations for joint development at station sites. It is neither certain nor automatic. One must understand and respect the marketplace and conditions for economic development and carefully coordinate transit station development with economic and land use realities. Transit system line ridership for a given station or transit center has to be impressive to attract the interest of the private sector. What is considered "impressive"? No standards have been suggested or developed in the literature, but it seems the transit systems that receive substantial financial interest from the private sector have demonstrated daily transit ridership that approaches or exceeds equivalent traffic volumes on a well utilized interstate freeway. This might suggest that a transit system needs to achieve something greater than 100,000 daily riders on its rail system as a minimum threshold to attract private commercial interests who will find it beneficial to invest in areas in and around transit stations.

This discussion begins with examples of modest financial activity related to bus transit centers and light rail stations and includes some impressive investment/financial examples relating to four heavy or metrorail system activities. The discussion ends with a summary of advice and suggestions for approaches to joint development. This advice is gleaned from interviews with experienced transit and community development agency staff who have demonstrated that positive things can happen with creative joint development (and who learned a few "negatives" along the way) and who have graciously shared their experiences with me for the preparation of this paper.

Pierce County, Washington—Transit Centers

Pierce Transit is a medium-sized bus system operating in a classic post-1950s suburban sprawl setting. Pierce County has only one high density activity center to support transit ridership, the Tacoma central business district, about 35 miles south of Seattle. Although Pierce County is 1,675 square miles in area, Pierce Transit's smaller service area of about 275 square miles encompasses a population of approximately 435,000 people, taking in nearly 90% of the county's population and most of the developed portions. To provide effective transit service for this dispersed and unfocused suburban sprawl form of development, Pierce Transit adopted a plan in 1980 committed to development of an integrated network of six transit centers throughout the urban area. Five of these six transit centers are now in operation in conjunction with regional shopping centers, community college/commercial centers, etc. Three transit centers are fully constructed and have been in operation for several years. Two others remain in an interim operating status at temporary sites without complete passenger amenities while final sites are being planned. The sixth transit center is due to come on line for operation in 1989.

Modeled after similar facilities in many Canadian cities, transit centers perform a valuable function for the dispersed development in Pierce County's suburban areas. This operational concept involves a "timed transfer" mode of operation, bringing buses from many routes together at the transit centers once or twice an hour and holding them for a few minutes around one or more off-street passenger loading islands to allow for convenient transfers between buses. This type of service configuration provides reliable and hassle-free transferring for the passengers, interconnecting many origins to many destinations with a relatively low level of transit service (30- and 60-minute frequencies), but a level commensurate with less dense forms of suburban development.

Through careful planning and negotiations, Pierce Transit has achieved some modest joint development successes. Each of Pierce Transit's three completed transit centers has required from one to three acres of land for its development, obtained with 20-30 year leases for only $1 per year. The construction costs for the transit centers have been funded with capital assistance grants from the UMTA which has determined that long term leases (at least 10-20+ years, depending on the nature and cost of the project) are able to satisfy the need for "continuing control." A new transit center will go into construction in 1988 on about one acre of land, also with a 30-year lease for $1 per year.

Two of these four centers are on private property in regional shopping centers, each requiring that the shopping center owner give up from 110-130 parking spaces to allow for development of the transit centers. The other two transit centers were developed on publicly owned land, one on State property in conjunction with a community college, the other on a local school district's surplus property that is adjacent to a private university and a neighborhood commercial center. Obtaining essentially "contributed" land for these transit centers under long term leases was a major accomplishment; the alternative arrangement of fee simple acquisition would have required from $250,000 to over $500,000 per site.

The key to obtaining these joint development land leases for transit centers was great patience in the identification of elements of mutual benefit and interest to the transit system and the property owner. The first regional shopping center project required almost three years of interim operation of the transit center along sidewalks in the shopping center's driveways. Once it could be demonstrated that, of the 8,000 people per day riding the six routes that came into the shopping center, nearly 1,500 people stopped in to shop, the value of a permanent transit center as a means to increase shopping center patronage while minimizing parking space development became clear to the land owner. After the need and benefit could be established, the negotiations for the actual leases went fairly quickly, at least with the private property owners.

The longest and most arduous lease negotiation was the public-to-public lease between Pierce Transit and the State of Washington on community college property. This involved the State Attorney General's Office where creative joint development concepts, especially the $1 per year aspect, were particularly foreign and suspect; but it finally worked out after more than six months of drafts and discussions.

Portland and San Diego—Light Rail Stations

Ridership has been good on the Portland and San Diego light rail transit systems, even better than originally expected, with just over 20,000 and 23,000 daily passengers, respectively, in 1987. Neither of these two light rail systems has seen the degree of economic or joint development activity at stations that is occurring with the higher volume operations of the metrorail type systems, but there has been some interest. Tri-Met in Portland, the newer of the two systems, reviewed a proposal for air rights development to place a YMCA facility over one of its major stations. Unfortunately, the proposal was shelved due to other financial problems of the developer and not from lack of interest for the station development concept. Given Portland's strong commitment to the concept of growth management and its formal State-mandated metropolitan control of land use, it is likely to generate greater attention and interest in the long run for joint development at stations than other light rail systems in metropolitan areas that typically lack metropolitan land use planning controls.

To San Diego's credit, the Metropolitan Transit Development Board (MTDB) developed its light rail system in a quick, efficient and spartan manner, obtaining little land for the stations or along the right-of-way that was not absolutely necessary for immediate system access and operations. While this served the MTDB well in minimizing costs for rapid system development, they now find they have little to bargain with in terms of useful real estate of interest to the private sector at the outlying stations. They want to market joint development at stations and are beginning to see greater interest in the downtown area. The MTDB recently proposed a refined alignment for a modest trolley system expansion in the northern part of the downtown San Diego area. During this process, they are working on a package with developers for a significant mixed use building project (around 800,000 square feet) that may include office, retail, hotel and government uses and will enable the MTDB to obtain an important station as part of the project. With growing public use and acceptance of San Diego's trolley system, additional opportunities are presenting themselves for joint development at new stations as plans are being developed for a more extensive north line extension. San Diego's light rail system initially developed minimal park-and-ride lots adjacent to stations and thus will not have access to potentially significant future lease/participation revenues.

Los Angeles—Metrorail Station Development

The Southern California Rapid Transit District (SCRTD) has been working on development of its future metrorail system with the focus of initial activity in downtown Los Angeles. The Community Redevelopment Agency (CRA) of the City of Los Angeles has the expertise to work out land development packages and has the lead responsibility for working out details with private land owners for station development agreements for the SCRTD. Significant progress was made in this area through CRA's negotiation of an agreement with Home Savings Bank for the development of the Home Savings Tower on a key block in the Los Angeles central business district along the metrorail alignment. The agreement provides for Home Savings to pay for development of the metrorail station entrance utilizing three basement levels under their property and to make the basement area available for the staging of construction of the metrorail system, a valuable factor in the crowded underground space of Los Angeles. This package was worth approximately $25 million in private sector participation towards metrorail development for the Los Angeles system.

San Francisco—BART Station Developments

The San Francisco Bay Area Rapid Transit District (BART) had the vision to see financial opportunities related to economic activity in and around its stations and so created a Joint Development Division in 1983-84 to be responsible for pursuing joint development and similar creative financial options. It has found private market interest that parallels the pace of economic development in general in its heavy rail corridors. With a consistent pattern of ridership on the BART system running close to 200,000 daily passengers, it clearly has the volume to attract the private sector. BART was paid a one-time fee of about $300,000 from a developer in downtown San Francisco to allow direct station access for the "388 Market" high rise development project on Market Street. In Pleasant Hill with several acres in use for a park-and-ride lot, it is exploring proposals for joint use of this land in conjunction with a major commercial development (1-2 million square feet) and may even consider an equity position in such a development. At the Concord Station, BART is considering proposals to allow development of a 250-300 room hotel which would involve a long term lease plus participation rights. This proposal would generate basic ground rent involving the park-and-ride lot property and, consistent with other such BART leases, it would provide BART with a percentage of gross revenue from the income of the commercial project.

Atlanta—MARTA Station Developments

Like San Francisco's BART, the Metropolitan Atlanta Rapid Transit Authority (MARTA), Atlanta's heavy rail rapid transit system, has had a successful record of public patronage, about 180,000 daily rail passengers. MARTA has been experiencing quite a bit of joint development activity at its stations over the past several years, although not always of the character or balance considered most desirable. Prior to the 1979 opening of the MARTA rail system in Atlanta, the Atlanta Regional Commission (ARC) conducted a series of transit station area development studies that resulted in many changes in zoning and even modified some system alignments and station locations. The ARC is a seven-county, State-created metropolitan planning organization with a history of regional land use planning in the Atlanta region dating back to the 1940s. In the early 1980s, after the start-up of MARTA's rail system (MARTA includes only two metropolitan counties), the ARC initiated a major development review procedure to try to help balance and coordinate the growing pressures placed on elements of the local infrastructure (water, roads, sewers and schools) that were resulting from major regional developments. Twenty of MARTA's 30 stations are in the city of Atlanta and rezoning incentives for higher density development around Atlanta's rail stations was undertaken with special public interest zones.

Companies now want to be near direct transit system access, so MARTA has been able to obtain a great deal of commercial office space activity around a number of its stations, including a couple of 50-story buildings on its air rights at two stations. For example, Georgia Pacific built its corporate headquarters over a MARTA station, IBM put its regional high rise headquarters (one million square feet) on MARTA air rights, and some major utility companies also have been interested in joint development with MARTA station property. Much of this development did not occur with the original opening of the rail lines, as development commitments seem to typically follow about 3-5 years behind the completion of construction at the stations. Not surprisingly, private investment often needs the demonstration of success before making major financial moves.

Some joint development activity is beginning to generate respectable income for MARTA. Lease income for 1987 was estimated at around $700,000, with a potential increase to about $900,000 in 1988. Long range lease income projections from joint development are optimistically expected to be greater than $10 million annually. The only down side of MARTA's otherwise positive success story is that it has not been able to realize development of some of the broader mixed use and residential projects that were originally planned around several stations. Some station area plans assumed that other public improvements like libraries and community centers were being developed that would support more mixed use and privately developed residential activities. Most did not happen, since the general economic recession of the mid-1970s (during MARTA construction) depressed associated private development, and the changing and more conservative federal funding philosophy dried up other potential financial sources.

MARTA and Atlanta are aiming for development of more residential projects around stations, since they have been losing some of their older apartment housing supply as new higher density commercial developments have been constructed. Overall, the existence of a well established regional planning framework in Atlanta along with a very visible and strong transit plan have given local governments a better opportunity for achieving desired land use control while also providing MARTA with a positive source of revenues related to managed growth and development.

Washington, D.C.—WMATA/Metro Stations

Of the newer U.S. heavy or metrorail transit systems, the Washington Metropolitan Area Transit Authority (WMATA) clearly leads in joint development efforts and ridership. WMATA, which is called Metro, has daily rail system ridership of well over 460,000 passengers, with a July 1987 peak ridership of about 490,000 passengers. Consistent growth in transit ridership has created the beneficial economic climate for WMATA where developers see Metro transit stations as important "selling" components of their commercial projects. WMATA estimates receiving about $14 million to date in capital contributions towards construction of joint development facilities such as bus-rail transfer areas, park-and-ride lots, elevators, escalator ways, chiller plants for station air conditioning, etc. Metro, like BART, has also been paid for connection agreements that allow direct underground access from buildings to Metro stations. By 1986, these earned about $1.2 million, and another $775,000 is estimated for current connection activity in fiscal year 1988. These are a category of one-time capital contributions related to development/connection agreements. Greater long term financial gains for Metro are coming from annual rent and participation agreements. Although WMATA policies do not allow commercial activity within Metro operating facilities, they do allow commercial enterprises within the broader definition of Metro properties that are outside of their shell operating areas. Metro's annual rental income for fiscal year 1988 is estimated at $3.6 million.

Additionally, Metro's "participation" agreements with developers using and/or connecting to their property provide for a percentage of profits related to the commercial income. The problem has been that no revenues were coming in with these earlier agreements, as they were pegged to a percentage of net income, and the "net" never seemed to materialize after costs. Metro has now changed to a formula of seeking about 5-8.5% of gross revenue on large scale projects (generally buildings with greater than 100,000 square feet), but even then this income distribution clause takes effect only after the commercial enterprise or development achieves an agreed-upon cash flow level in order to recognize the time needed to achieve successful development. For smaller projects (those that may yield rental income in the $15-20,000 range), the joint development participation agreements are now simply requiring automatic income escalators of about 4.5% per year. Although the minimum rent aspects of joint development agreements have gone well, the "participation elements" have been disappointing to Metro. It is felt that the recent changes in the newer agreements will reverse the trend over the next few years. Over the longer term, WMATA expects to be receiving $8-10 million per year through its joint development agreements. Ironically, the greatest challenge to Metro in achieving some of these creative joint transit and land development agreements has been obtaining the cooperation from local governments for necessary approvals.

Advice and Suggestions

From the wide range of approaches and activities that can be seen in joint development efforts around the country, some common threads of advice emerge:

(1) Do not underestimate land acquisition requirements. To have a future position for bargaining in joint development, one must first have the property base. It is far easier to sell surplus land later (if necessary) than to find you have no room for system/station expansion and no bargaining position for air rights/connection access development. Given the dispersed patterns of contemporary urban development and the high costs of labor for feeder services, obtain adequate park-and-ride lot property in suburban areas and develop these lots in phases as demand warrants.

(2) Given a choice, try for co-development with original transit line or station construction. Developing access for buildings or adjacent developments is far cheaper and less disruptive to system operations if done at the time of initial system/station development.

(3) Design/build with flexibility/adaptability to accommodate future modifications and access connections. Stretch the concepts of the transit plan and anticipate adapting to future development access requests so as to minimize disruption to system operations when new stations or access connections are made.

(4) Get the collective "public act" together for joint development negotiations. Usually there are diverse transit and other local or regional jurisdictional concerns at stake when looking at joint development interests for a regional transit system. It is essential to structure the respective relationships of all parties before entering into any negotiations with private sector interests. Each party needs to decide what it truly wants; that is, the transit agency should set objectives and priorities for the transit system *and* the general purpose government or land use control agency should define its objectives for the nature of desired development—local circulation/traffic access at each individual station, etc. Then there needs to be "trust" between the public agencies and an agreement to let *one* party negotiate the deal. Public "committee" negotiations with private sector inter-

ests can do more damage than good and drag things out to the point of "no deal." The actual negotiations must focus on mutual enlightened self interest; both public and private interests have to "win" for success.

(5) Consider each station separately for joint development. Although there need to be standard and common design components for the transit system itself, this is not so with development. Some station areas may need to take a neighborhood "preservation" approach while others may warrant any of the many varieties of development approaches (for example, pure commercial office, retail, high density residential, mixed use, civic, educational, health care institutional, suburban parking, etc.). To sell the station and the joint development approach, one must recognize the needs and objectives of each of the local areas within a sphere of influence/impact around a given proposed station.

(6) Seek a percentage of gross, not net, revenue for participation rights in addition to establishing straight rent/lease revenues with developers. Unless one wants to get lost and frustrated trying to duplicate Internal Revenue Service audits, seek a percentage of basic gross revenues on larger scale developments and avoid the morass of never seeing net profits because of the creative ways that costs can be calculated to avoid profits.

(7) Expect and accept nothing less than a quality productive public enterprise. Success in joint development activity means greater revenues for the transit system and an offset to the tax subsidies that are inevitable in operating and maintaining an urban transportation system. Maximize utilization of the transit system to achieve transit and land use objectives. Seek coordinated local and regional development strategies to manage growth and development to create a "transit-friendly" and supportive environment. Then make sure the system design and operation delivers attractive, high quality productive service.

Transit, Urban Life and Development in the Sunbelt

Anthony James Catanese

Dean, College of Architecture,
University of Florida,
Gainesville, Florida

Three issues are crucial in considering transit impacts on urban form: the effects of transit on development; how transit affects the quality of life; and what mechanisms are available for implementation. In a city like Austin, Texas, the quality of life impact may be the most important politically; but we also need to understand the impact on urban form and development, as well as how to marshall political support.

Urban Form and Development

Sunbelt cities are exhibiting a new urban form, no longer just a central city, but a central city with subcenters. In Atlanta, the central city comprises 40% of development, primarily work destinations, and that is rapidly decreasing. There are four or five major subcenters which soon will be as large as the Atlanta central city. Its circumferential road has drastically changed the urban form of that region. New jobs are located in subcenters, office parks, and retail and mixed use centers outside of downtown. What this means is that the initial concept of rapid transit from the British, the idea of conflux and dispersion around a single urban center, is not sufficient as a concept to accommodate the multi-directional trips created by multi-nucleated urban form.

Atlanta's Lenox Station is near one of the largest shopping malls in the southeast. Courtesy MARTA.

Circulation does not create urban form; the two work together. Perhaps, through planning, circulation can affect urban form, but what tends to happen in many Sunbelt cities is that urban form occurs and transportation follows. What we must do in planning, given this new kind of urban form, is use rapid transit to guide it or at least to help shape it. That is not easy because it requires transit and highways to be built in advance. Canada has been doing this for a long time, but it is not the custom of Sunbelt cities to use transportation to shape urban form. It is time a Sunbelt city tried to apply this basic planning principle.

Atlanta tried to use its transit system to shape a linear city, a strong downtown with major subcenters at midtown, Lenox Square, Perimeter Center, and the airport area, thus reinforcing the central business district and recreating that spine of development. But there were several suburban counties in the region that were not part of the Metropolitan Atlanta Rapid Transit Authority (MARTA) and continued to sprawl in typical Sunbelt fashion. Rapid transit had no effect whatsoever on development in those suburbs. When developers could not get what they wanted in the city, they went to Cobb County or Gwinnett County. The only things that have affected that competition are water and sewerage facilities, which in the Atlanta case turned out to be a lot more important than rapid transit.

Miami made a big mistake. It was far more concerned with using existing rail rights-of-way than with trying to shape urban form. The priority there was to reduce costs by using rights-of-way already assembled by the railroad. Consequently, the rapid transit system is in the wrong place and, in my opinion, that is the major reason for the failure of rapid transit in Miami. The system does not serve the needs of the elderly, the poor, and children—the major markets. This failure is important because it has affected the entire nation's policy at the federal level. The Reagan Administration points to Miami as a reason why the federal government should *not* be involved in rapid transit.

It takes a sizeable ridership to affect real estate development, at least 100,000 trips per day. Atlanta has about 180,000 riders per day. In Miami, on a good day, ridership is about 45,000-50,000. Consequently, in Atlanta, transit can affect real estate development; in Miami it cannot. Thus, we must be realistic about numbers when we try to use rapid transit to create urban form and development in the Sunbelt.

Rapid transit is not a quick fix nor an alternative to automobile-dominant transportation in the United States. Nor is it a panacea for urban growth patterns. In Atlanta it was argued that a major program of transit construction had to be undertaken if traffic was to be no *worse* in the future than it is now. But in Atlanta, a very successful system, ridership is only 10% of trips taken. By comparison, in Toronto one of the goals in the new suburban centers was to achieve a 50% modal split. In the Sunbelt, successes will be relatively modest in comparison to foreign cities; modal splits are more likely to be 1 in 10 trips.

Quality of Life

I strongly suspect that in Sunbelt cities transit is really a quality of life issue requiring long term politial decisions based upon social and economic conditions. The concern is much bigger than getting large numbers of people from place to place. So at very early stages, the rapid transit development process must be incorporated into the political process. Transit development must be seen as strategically important. One argument is that transit development is one of the greatest opportunities a growing city has if it is serious about affecting its future quality of life. Politicians and those in the private sector must be convinced that the quality of life issue is not for today, but for 20 years from now. We are planning for the quality of life of our children and grandchildren. Even if you started tomorrow in most Sunbelt cities, it would take a decade for significant rapid transit to develop. So it is difficult to get political support because it is such a long term commitment. People cannot relate to the long term, especially where there is no tradition of long term planning and development. Politicians predictably look to re-election in a few years, not for votes a decade away.

Transit development is not without social costs. There were some negative impacts on neighborhoods as a result of rapid transit in Atlanta, Miami and elsewhere. Obviously, construction of a station and associated real estate development are going to have an impact on neighborhoods. Toronto demonstrates a long term, sensitive way of doing that. In some cities in the Sunbelt we have not had that luxury of time or experience. In Atlanta, the phenomenon of residential buyouts has been occurring around the new transit stations. The transit authority or developers go into surrounding neighborhoods—sound, good condition residential areas—and buy out 50 to 100 homes, tear them down and redevelop the area for commercial, retail or mixed uses. The process creates difficult political problems in the Atlanta area. Good planning, logical land economics it might be, but with difficult impacts on neighborhoods.

There is a further caveat, based primarily on the Atlanta experience. While sound transit planning aims at long term quality of life, short term effects are not positive. Atlanta was torn up for about five years in the 1970s and was a terrible place to be. But Atlanta did not take the quick fix by using existing railway rights-of-way; it put transit where Atlantans thought that the development should occur. The downtown was torn up, and major streets were closed by construction, and it was very unpleasant. But Atlanta got through it, sacrifices were made, and today the system is very successful. But I suspect if you were to ask people in the mid-1970s what they thought about rapid transit in Atlanta, you would not have gotten quite the glowing reports that you hear today. We must understand that there is short term pain in order to secure long term gains. In the Sunbelt, that is a major consideration.

One other point about quality of life is crime. In the Sunbelt we do not have much experience with rapid transit, but we have clear perceptions of crime. We read about what happens in New York and Chicago and we think these are dangerous systems, not the kind of places people will go to. The concerns are real. The first time there was a murder on the transit system in Atlanta—which happened to be a domestic dispute that could have just as easily occurred in a home or anywhere else—it had a major impact on ridership in Atlanta. It was as though people knew it would happen, and as soon as it did ridership decreased drastically. In Atlanta one of the mistakes was choosing a high tech, computerized video surveillance system. Although cameras and monitors are found everywhere in the Atlanta system, in the control room there is only one person watching nearly fifty screens. Can he have the slightest idea what is going on? It is almost a random event if that surveillance system really catches any perpetrators in action. Thus it is important to design for concepts like "eyes on the street," defensible space, and crime-free passenger zones to mitigate crime. Still, one of the things we learned in Atlanta was a very old idea: It takes a police presence to reduce and prevent crime. A combination of crime prevention design and police presence is absolutely essential in the Sunbelt to eliminate perceptions of crime in transit facilities.

Cooperation

Rapid transit can have a major impact on shaping real estate development in the future, so development interests must be involved in strategic planning from the start. But rapid transit is not the only thing that will affect development. Transit must relate to highways since it is part of a transportation package—multi-modal, intermodal. Rapid transit must also relate to open space and parks, service infrastructure (especially water and sewer) and land use controls. It must not be presented as an isolated issue, but as part of a comprehensive approach to improving future development.

Can a transit authority, a planning department, a parks department, and water and sewerage authorities work hand-in-hand? How can we relate all the key elements of land use development, especially in the Sunbelt context, which has no tradition of control upon land development and will not in the foreseeable future (outside of Florida). We have to do it through the political process and we have to bring in the private sector. We cannot do it as technicians. It has to be directive, and that requires political leadership. We need political leadership, but often that means bringing in private leadership as well. Public/private partnerships are the future in the Sunbelt; in fact, they are the future in the entire country. None of this is going to happen in isolation on its own.

The free market economy is part of our heritage, part of our tradition. But we must realize that there is a strong need for government to set up systems of control over where development occurs, with business cooperation and involvement. This requires a new way of doing business with government support. There have been some stunning successes in Canada which show that new public/private partnerships can have highly visible results around transit stations.

This is related to paying for the transit system. Atlanta used excess condemnation, acquiring most of the land around the stations and selling it in a bidding process, through negotiations. In Washington, D.C., seven of the stations are already in joint developments of public/private partnerships. When I say public/private partnerships, I mean true partnerships, sharing the risks and sharing the rewards. That is what Atlanta, Miami and Washington, D.C., are doing. In Atlanta it was a very simple process, using excess condemnation for future development. Areas around stations in Atlanta were changed to "public interest zones," a category lending itself to interpretation and negotiation, and attractive when you own the land. There was no question that the goal was to create high density development around stations which would both create new ridership and generate income.

Some observers assume floor area ratios (FARs) of 2 in suburban centers, yet there are some FARs of 10 in Atlanta. The prospect of 40-story buildings on some suburban sites offers developers an incentive to plan as partners with government. This also affects parking policy. Miami did not coordinate its parking policy with its transit station development, while Atlanta did. As a matter of fact, some of the high rise buildings negotiated around the transit stations in Atlanta had no parking requirements. Imagine buildings of 500,000 square feet with little required parking! There were incredible concessions given to developers in Atlanta, which worked successfully, and helped transit ridership to grow. It took a long time in Toronto for these clusters, these subnodal developments, to occur. In Atlanta it has been occurring within the last ten years.

Steps in the Transit and Development Planning Process

In summary, the following are key steps in the transit/land use planning process. First, develop a comprehensive plan that integrates transit goals with highway, land development, infrastructure and open space agendas. It must all relate comprehensively. Somehow the leadership has to emerge that says, "We shall do it this way." Then transit can seriously affect comprehensive planning.

Second, create a 20-year political commitment. It is not easy. Do not expect it to be unanimous. Atlanta passed rapid transit by about two percentage points; three of the five counties that were involved did not pass MARTA and still have not. Rapid transit also squeaked by in Miami. Overwhelming mandates cannot be expected, so do not look to the political process in terms of elections to get this commitment and will. Instead, form a political coalition that has transit as one of its major goals, but not the only one. A political coalition will get you only part of the way, but it is needed to press for the quality of life issue.

Unfortunately, in most Sunbelt cities, such coalitions are limited to "yuppies," but the base must be broader. Rapid transit would never have emerged in Atlanta without the support of the Black community. It never would have been realized in Miami without the support of the Black, Hispanic and elderly communities. Unfortunately, those groups tend to be left out when the system is finally built. The elderly, who were so supportive of rapid transit in Miami, were left out because the line was never built to Miami Beach, rather, to a white, middle class residential neighborhood. In Atlanta, one of the promises made to the Black community was that major Black residential areas and public housing projects would be served by the transit line. Twenty years later those areas are still not served. But the promises were made, so now the mayor of Atlanta has to deliver on those promises. It would have been better to have all of this done as the result of a multi-cultural, diversified political constituency's effort. Another aspect of that long range commitment is a financial plan cognizant of the capability and willingness of a community to pay. That often means that upper income people will have to subsidize lower income people.

Third, the private sector must be involved. Real estate developers understand the impact of rapid transit, but the broader private sector must be involved as well. To a certain extent, especially at this time in history, rapid transit is an economic development issue. It may be as related to jobs as it is to transportation.

Part of the economic development issue is downtown redevelopment. Much has been made of the downtown renaissance, and we should recognize that redevelopment of American cities, especially in the Sunbelt, is an economical way to use infrastructure and invested capital. But a great deal of the downtown renaissance was a result of federal tax policy. It was a result of tax write-offs and, to a major extent, historic rehabilitation tax credits. No question about it, in the last 10 years there has been a tremendous renaissance in downtowns. There is also a lot of unused space in downtowns. The 1986 Tax Reform Act takes away all those incentives, and we will not see that kind of development in the near future. Regarding the renaissance of downtown, caution is advised, especially in commercial uses. What we should do, however, is to view downtown redevelopment in its proper sense—redevelopment of the primary urban center. That is a major economic development undertaking that can affect new job creation. It can be a positive result of a rapid transit system.

Transit can have a major impact upon urban life and development in the Sunbelt. We must be forewarned that there may be short term negative impacts that muddle the goals of long term positive impacts. Yet we know that Sunbelt residents, both native and transplants, are acutely aware of the growing problems of transportation. Transit can and should play a role in resolving some of the problems, but let us not overpromise or exaggerate the effects.

Reflections

Simon Atkinson
Mike Hogg Professor,
School of Architecture,
The University of Texas at
Austin; Principal, Black
Atkinson Vernooy

One of the main issues raised by the preceding essays is about the kind of world we wish to live in. What values should guide the city, and what are their social and economic ramifications? The role of the car is an unavoidable consideration: Is the automobile a liberator or an inhibitor of life in cities?

A number of studies have indicated that the widespread and low net density of a city is the basis for offering greater opportunity and choice, as well as freedom through personal auto-mobility. In this "urban non-place realm," it is argued that people can select their lifestyles with ample space and a rich marketplace. Land is opened up and served by federal or state subsidized roads and adequate infrastructure. It is this footloose marketplace that home builders, new industries, offices and retail outlets are said to be looking for. As an amenity, patches of open space are given back for public use as part of the development process. While this is not the design dream of the compact city we were taught to love, I am constantly assured that this *is* the newly emerging American landscape, and, given time, we will learn to appreciate its new aesthetic as much as the new opportunities it offers.

But the very nature of low density creates sprawling development and, ultimately, an unworkable traffic equation. Total automobile usage is practical only at relatively low densities, somewhat below the densities of current suburban development. Congestion in American cities is moving away from downtown, peak hour bottlenecks to suburban areas and extends over wider time spans.

Traffic analysts are finding remedies to these problems somewhat difficult to achieve. Regional networks are being used heavily for local traffic, and trip distributions are far more complex in suburban areas, with outer loop roads of a number of American cities congested through large parts of the day. This, however, may be a small price to pay for the opportunities that auto-mobility presents. By learning the alternatives in this flexible urban system, perhaps one can get around adequately, despite congestion.

In this laissez faire view of the city, it is argued that transit planners pay far too much attention to those few who may lose out. After all, minorities and the urban poor have cars and drive everywhere, however hard up they may be. Even the very young are encouraged to give priority to automobile literacy over real literacy, and now that we have a wide range of medical programs for the elderly, then surely they will go on driving forever. If segments of groups such as these have mobility problems, then low key programs, often privatized, can respond in an effective and efficient manner. America has dealt adequately with the busing of school kids; similarly, any special group with a low mobility level could be singled out for special attention and service.

This view of the emerging American city is based on optimism about the marketplace. The land and development market, if left alone, will effectively work it all out on our behalf. What is not needed is the interference of government, in whatever form, except perhaps in the provision of a back-up of some urban services. Development enterprise will build the range of houses people wish to and can afford to live in, large and small corporations will seek out their own patches of opportunity, and the spread of the road network will link them together. In the next phase, strip centers and small commercial centers will appear adjacent to roads and provide further services and amenities.

To prove that the equation is working, higher density urban concentrations will then occur at key locations, normally adjacent to the intersection of main highways. The basis of public/private cooperation in this venture is relatively simple. In a number of cases, the city is not even called upon to provide road infrastructure, as this can be implemented through a road district. Most other development can be facilitated through municipal utility districts and planned unit developments. It is unlikely that the huge dollar investment in highways will ever be carried by the private sector, but in a number of cities this is the only infrastructure that municipalities are directly called upon to supply, the remainder being indirectly supplied through electricity and water districts.

The apparent advantages, in terms of dollars to the city, appear seductive, and even though the spread of services is covering a vastly wider area per head of population than in the older parts of the city, it is still seen to be a bargain. Cities scrape for concessions in the planning of such areas and there is evidence that many cities have taken care in their ordinances to control rain water runoff from buildings, the scope and quality of car parking provision, a high quality of access roads, and at times have gained concessions in terms of open space and land to be given to school districts and for other urban services.

If this is a model for city development in the future, why is Houston moving away from its loose-fit, "string bag" road system in a bid to implement an effective transit system, and why is city policy now firmly focused upon attracting life and enterprise into the downtown? Why also is the other archetypal new American city, Los Angeles, also seriously reassessing its entire mode of movement in the city, placing substantial emphasis on public transit, and devising a series of incentives for inner city development? These cases indicate the need to modify our conception of the American city and its form. It is not enough to say that cities are formed in a compact manner around the central place, then transformed through a loose fitting suburban extension. A third phase, restructuring, is needed to accommodate new levels of traffic. If, however, the major components of the second phase of growth are dotted across a vast spatial area, and if some of the suburban growth centers become as dominant as the original city center, then it is difficult to imagine how these cities can be effectively restructured. What they can do to combat the dying economy of the city core, and to make a token gesture towards relieving suburban road congestion, is to restructure a preferred transit corridor focused upon the downtown. It would appear to be an uncomfortable and expensive way of restructuring the city, but it may be the only opportunity available.

I suggest that most of the case study cities have searched further than this last model and are informed by a broader set of beliefs. They have not been directed by transit considerations alone. Quality of life considerations appear to have been paramount in opting for public transit development.

From speaking informally with the authors of the case studies, and attempting to ascertain the sentiments in their cities that indicated support for transit development, a rich mixture of both emotional incentives and hard-headed political and econonic arguments are evident. What was difficult to establish in many cases was one dominant variable that triggered other actions. Undoubtedly, traffic congestion was a key concern, but this appeared to have stimulated a dialogue that went beyond to issues such as social justice, economic revitalization, quality of life, and principles of good urban form. It might be useful to list the main points used to justify a more public transit-based city.

The Arguments for Transit Development

Highway Extension/Expansion Doesn't Work. Cities have no way of mitigating the worst effects of traffic congestion. Mitigation through further enlargement and extension of the city's road network has not brought the desired results and has had many unfortunate side effects. As a general principle, traffic projections rapidly become a self-fulfilling prophecy; the more you open up the opportunities for the free flow of traffic, the earlier those new links become congested as they are sought out by persons who would not normally have used them.

Nearly all road networks, particularly at higher capacities, have had damaging side effects on adjacent neighborhoods. Often located through a "weak seam" of the city, they, in fact, separate the city from its shoreline or a significant inner urban community from its downtown. In nearly all cases, the economy adjacent to these major arteries was weakened, if not sterilzed, and living became impossible. Recent studies have determined that it is not simply the physical barrier and intrusion of roadways that is so objectionable, but that they also foster a perceived severance of the city's structure.

In addition to being an unreliable solution to traffic congestion and harmful to neighboring districts, highway extension is not a force that can revitalize the decayed parts of a city. Many city plans sought a newly restructured downtown with upgraded radials and inner loop roads, only to find after a vastly expensive implementation program that this did not precipitate related urban investment. Furthermore, there is still the issue of the substantial amount of space to be given over to car parking. In a number of cases, improved automobile access to the downtown has meant the removal of the city's smaller scale and historic fabric simply to provide parking.

It Makes Social Justice and Economic Sense. An effective public transit system is more egalitarian than individual automobiles because it does not discriminate against any single group in the city. It is a safe mode of transportation for the young, old and disadvantaged, as well as any other citizen. Proponents stress that with a generally aging population, we need to enable the elderly to be mobile. There are also strong sentiments that younger people should not be coerced into having to be automobile drivers, which can place an undue pressure on their need to earn money while at school or college.

The idea of social justice in the city can be expanded to include economic factors. Road accidents and drunken driving have become major causes of death in the United States, particularly among the young, which have serious impacts on the national economy. The declining efficiency of the automobile in certain city locations is bound to have an indirect impact on its economy. What, for example, is the cost of frustration and lost time due to traffic jams and clogged freeways?

Possibly the most important point in these arguments is the principle of a built-in economic insurance policy. Transit systems have a greater chance of working for a greater segment of the population over a longer period of time with less cost than our current automobile dependency does. More than redressing the problems of automobile congestion, we are anticipating energy shortages and the greater need for energy efficiency in the future.

The Classic City is Still a Valid Ideal. The car city takes up too much room, and the laissez faire city lacks the critical mass that makes the city a city. Instead, downtown again can be a focus of city life. Streets can again be places of activity, with cars as our servants, not our masters. Transit can help replace parking lots with useful buildings and allow us to replace intrusive highway structures. Transit can, over time, redress an uncomfortable imbalance and encourage movement back to a more integrated city.

Transit Supports Neighborhood Consolidation. Whereas most road improvements are detrimental to inner urban neighborhoods, transit systems can reinforce neighborhood identity and desirability. They support local commerce and, by offering accessible transportation, encourage both home-owning and stable rentals. People are prepared to live with less space in a higher density location in return for increased accessibility and a high level of urban facility and amenity. Inner urban communities can be reestablished and reinforced.

It Fosters Identity. The development of transit focused upon the downtown indicates a strong commitment to the city's most significant and unique area; it is accompanied by commitments to develop peripheral nodes with their own identities. These commitments are, broadly, to a "sense of place." In nearly all case studies cited, there is a strong motivation to create significant and identifiable landmarks within the city. This was the case in both the revitalization of downtown and the restructuring of a new subcenter around a station.

One strong argument for making places identifiable is to help people develop "mental maps" of the city. People begin to comprehend the city in terms of its districts, parts, and local characteristics. Furthermore, the transit system itself becomes an orientor within the city and the stations become points of orientation, for meaningful cities are ones that are also navigable. Citizens begin to read the city through a series of identifiable features that not only make a city familiar, but act as a series of signposts, creating a greater sense of accessibility. If access to a city is viewed as one of the cornerstones of democracy, the city itself opens up a greater range of opportunity for all sections of its community.

The city center is reestablished as a major focus of this concentration. With less space being taken up by cars, there is more scope to increase its density, and thus its opportunity for interaction. The key argument here is that a sense of real city activity and interaction cannot, and will not, take place without both increased density and variety.

Transit development helps reestablish a sense of streetscape, a rich mix of urban uses, and the importance of sidewalk activity. In many cases, it has been a major support to the night life of the city with its ability to concentrate eating and entertainment within walking distance of transit stops. It is also seen as a foundation upon which a number of the future key land use decisions can be based. For example, major performing arts centers, arts districts, convention centers, city parks, hotels, centers of high tech industry, airports, and retail malls can be located near transit stops.

There are Economic Payoffs. Possibly one of the most exciting insights from the case studies is the ability of public transit to work in unison with progressive land use planning and, in turn, precipitate substantial development benefits. This is perhaps an obvious point, bearing in mind the substantial costs of these enterprises, but it is gratifying to know that in many cases they were planned as a means to channel direct development opportunities that would not normally have been realized, and certainly not in those locations. In a number of cases, direct development bonuses were worked out in connection with the location of transit stops. In a number of other cases, developers were prepared to contribute public benefits as a spinoff from their development. It is unfortunate that there is not strong evidence from cost/benefit analysis to indicate the range and types of return from these substantial infrastructure investments. This should be the next stage of study, but it would be reasonable to assume that with the careful planning of transit and land use, the transit system can act as a significant development catalyst, and as such may well be a good investment towards the future prosperity of the city.

It is difficult to measure the degree of development and economic benefit that is generated by transit, as some of it may have taken place anyway, and it is even more difficult to measure the wider economic benefits coming from this commitment. If, for example, experiencing of the transit system becomes memorable as well as convenient for visitors to the city, then it may well enhance tourism or a city's importance as a convention center. The New Orleans trolley system is viewed in this way, and Vancouver saw tourism as an important consideration in the recent implementation of its transit system. Another aspect of this wider economic benefit could be those economic enterprises and people who elect to stay within the locality because of the benefits of such a system. Proximity to transit can increase housing values, and it may also be a stabilizing factor.

A less tangible benefit relates to that of the system itself. A number of the case studies indicated that the quality of the transit experience attracted people away from the automobile. This measure of seduction undoubtedly brings substantial benefits over time, whether it is reducing the dependence on gasoline or reducing levels of road maintenance.

Transit Systems Can Be Customized for Each Region. Some transit systems offer rapid, high-capacity service but are inflexible and expensive. Others are more modest in scale and service, but better suited to their contexts. The point is that transit is not an alien system but one that can be customized to the city and neighborhoods it serves. Part of my own experience has been with very low key and low budget public transit systems. Two projects featured bus-separated lanes "stolen" from cars, bus-only streets in parts of the downtown, and intercept parking. In both cases, we found the image of the bus to be negative, so we attempted to present the bus in a more congenial manner to a wider section of the citizenry. In one scheme, we started all routes within neighborhoods, using smaller buses with regular, friendly drivers. These small buses then became expresses, using a bus lane to the downtown.

In another project, we offered free ridership within the downtown. This proved to be a considerable convenience to shoppers and at the same time made them more familiar with bus usage. In recent years, the privatized operation of microbuses has been introduced, many of which are the size of large taxis, offering space for shopping bags and operating at all hours. The drivers work on a franchise basis and can often provide greater service and efficiency than cars. One way mid-sized American cities could make a more serious commitment to public transit is to separate bus lanes in key parts of the city and use small, attractive buses. My point is that the stigma of any system—bus, rail or whatever—can be overcome through careful and thoughtful design and planning.

Visionary Leadership

I found a particular sense of leadership in most of the cities that I have visited and worked with, which is reinforced by the case studies. This leadership could be described as visionary and supported by a deeply rooted consensus moving through many parts of the leadership structure of each city. In many cases, one group had taken upon itself the task of communicating a brave notion of how the city could see itself in the future. In all cases, this was not a question of just selling public transit as a way to cope with traffic congestion, but more the communication of a broader set of ideals. Most of the reports embodied values related to land use, quality of life, and environmental quality. Most also predicted that they were not adopting the easiest passage towards the city's future, and that there would be a number of difficult, and often expensive, decisions, particularly in the short term. The most amazing aspect of the visions is that the cities were not only able to foresee but also to convince an electorate that there was a more widely beneficial city development model than that of simple laissez faire. Possibly the strength of this vision, as evidenced by all the case studies, is that it did not attempt to be comprehensive. In other words, as opposed to some of the earlier European ideal models of city transformation, the vision left much of the city to evolve like any other city, but selected key locations for investment and special land use policies. People could see an enrichment of the city that did not prescribe a lifestyle or directly force them into an unpalatable equation.

Another important and related aspect of this vision was an understanding of the need for cooperation between different sectors involved in decision making. In many urban situations, private and public sectors see themselves as in conflict, and bureaucracies are structured on functional principles of departmental decision making. Nearly all the examples we have witnessed have been based upon a close working relationship between the public and private sectors, with the private sector occasionally taking over much of the leadership, and a necessary restructuring of key actors in the local government machinery.

It was obvious in those cities from the very beginning that such a major commitment from the public sector, with extensive capital commitments up front, would not work without an equal balance of commitment and repositioning of investment capital within the private sector. In many cases, joint teams were established that worked from the basic premise that change could be positive and that development could be good, compatible and well integrated with the new transit system. A number of the teams establishing the first parts of the planning and urban design framework were comprised of representatives from both the public and private sectors. From the point of view of the public sector, a dynamism appears to have been evident from the beginning. In a number of cases, special agencies were formed with committed capital; these developed a necessary sense of mission. In no case was this merely the haven of transit planners, but more the meeting of many minds concerned with the rich complexity of the city development process. This might indicate a message that planning endeavors are more aggressive and successful when task oriented, committed to the private sector, and judged by the results. Put another way, one could see that all sides of the equation constituted a development agency destined to bring direct economic returns from the major committed investment capital, and in so doing significantly enhanced the quality of life in that city.

The second aspect of particular note is clearly one of leadership. In most of the case studies, it was clear that people had adopted very high levels of leadership commitment from the onset. Key figures were often drawn from the private sector where their known leadership skills could be committed to this special project as well as their ability to communicate its ideals to a wider section of the citizenry. It is also evident that leadership is not on and off, single-minded effort, but more the enterprise of several leaders respected for their roles and expertise, working in unison as a team.

Possibly the most important aspect of the decision-making character of these cities is that they have moved largely towards a consensus form of politics. This does not mean that many of the arguments were not agonized over through public debate, but that the larger ideal of moving forward and accomplishing the overall task should always predominate. I also witnessed leadership that was positively self-critical and prepared to correct mistakes made along the way. At the same time, each had a future-oriented perspective triggered by the ambitions of the next phases of their projects and the greater benefits and accomplishments that would lie ahead. Perhaps it is too glib to suggest that a key feature in the successful evolution of significant American cities is a model of consensus leadership, but that is what the evidence indicates.

Transit and Urban Design

Two studies for Texas cities demonstrate some of the potential urban design impacts of transit. Each also indicates that neither the scale of operation nor the population base needs to be large to make things happen. One study was of the small, historic downtown area of Galveston, while the other addressed a development opportunity adjacent to a possible transit corridor in Austin.

Galveston is a small, nationally-known island city off the coast of Texas with a residential population of only about 40,000, but with a large tourist population. It has an historic, although neglected, downtown, which has become the focus of a number of urban design studies and considerable enterprise aimed at turning around both its economy and urban character. One study focused on integrating a new public transit system into the fabric of the downtown. A particular concern was not only the fact that downtown had a stagnant economy, but that tourists were going to the beach, not experiencing the city and spending money there. A low key light rail system was proposed to link the beach with downtown and its many attractions. In doing this the transit system would strengthen one's sense of the city as a whole and, more practically, it would tie together a number of other developments as well. Gateways were designed to define entrances to the downtown, streets were given more character and related to the trolley, and each of the new places around stations was given a distinct character. These places and their predominant activities are to become the new place markers of the downtown and a

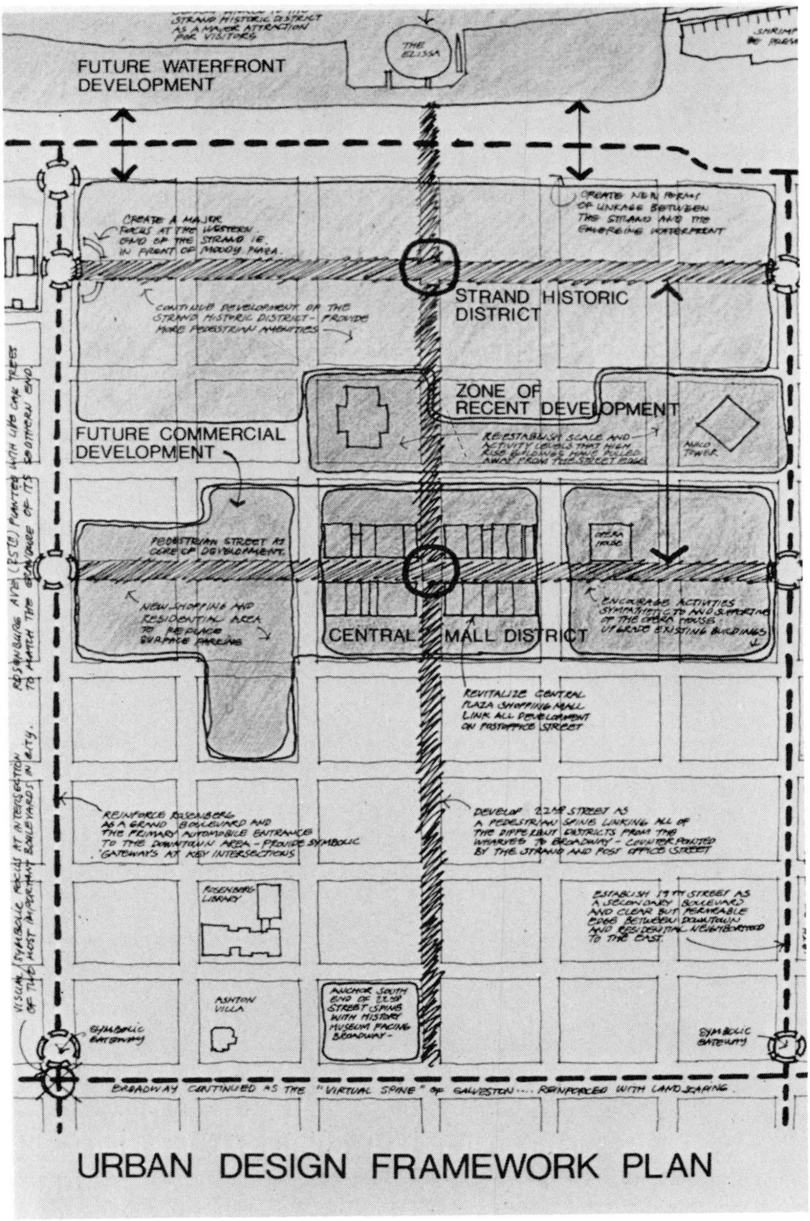

The transit system is integral to the urban design plan for Galveston.

catalyst for its future economic revival. Coupled with these actions, many of the historic buildings are being both renovated and converted to new uses. An attractive bus system has been operating for some time now, and the light rail system, financed from the city's tax base, is close to implementation. This initial design plan gives strong emphasis to a number of streets in the downtown as a means of creating a visual focus and as the base for new urban attractions and land uses. It is proposed that the trolley will link a new downtown shopping mall with an arts district, harbor area and the historic Strand. A variety of street identities developed in relation to the transit system is intended to reinforce, visually, one's understanding of place and harmonize the trolley line and adjacent development. In a small way, this strategy aims to shape visitors' "mental maps" of Galveston.

(Regarding visionary leadership, the Galveston Historical Foundation has shown remarkable leadership throughout the entire development and preservation process, often supported by forward-looking charitable foundations and developers. Two of the most important figures in this

Transit development will reinforce the overall plan for making downtown an attraction.

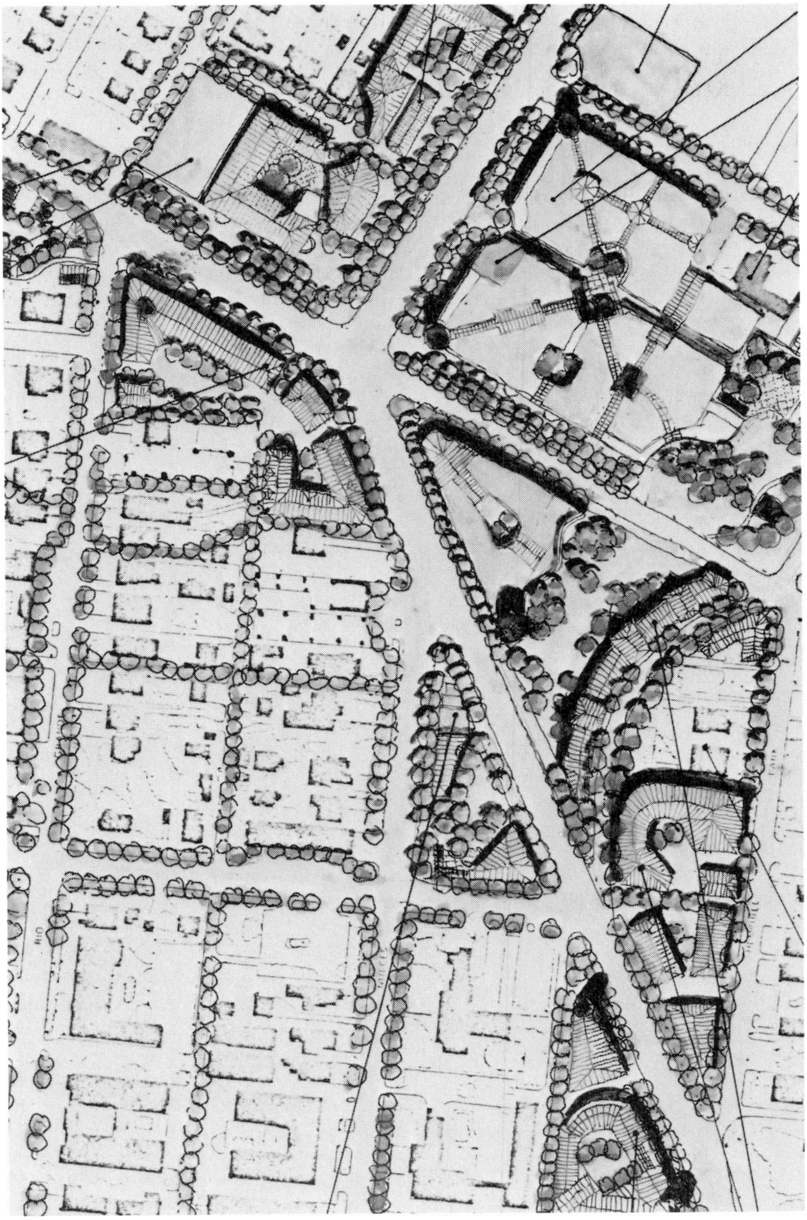

In one instance, investigated by Santiago Abasolo, the character of nearby neighborhoods can be drawn upon to give identity to transit-related development.

overall enterprise are Peter Brink, the Director of the Historical Foundation, and George Mitchell, who has been a key investor and developer throughout much of the revitalization.)

The second study relates to the proposed first phase of transit implementation for Austin. Capital Metro has designated a first priority corridor that links the downtown to the immediate northern neighborhoods and then to the new residential and industrial suburbs of the northwest. An important feature of this first link is that it combines into one corridor the medium density inner urban neighborhoods, the University of Texas campus, the State Capitol complex and downtown. System technology, whether an exclusive bus lane or light rail, has not been selected, and for the purposes of the study was not centrally significant; emphasis was on the route's land use consequences.

Surveys undertaken found that, whereas residents of inner urban neighborhoods were generally pleased with the opportunities offered by a transit system close to their homes, they were at the same time wary of its impact. Having been used to the threat of both road expansion and rezoning, residents anticipated that the transit system could bring about an incompatible and dramatic change of land uses. Therefore, design scenarios that would realize benefits for the transit system, while remaining neighborhood compatible, were worked out with Capital Metro, the City of Austin, and neighborhood groups. Two projects that represent urban designs arising from this type of consensus decision making are presented here. Each has in common the principle that a station can become a significant place marker in the city, with its own particular identity and act as a focus for adjacent neighborhoods. Each is not only a meeting point, but an opportunity to knit together the fabric of the city in an attractive mixed use development.

The first project has a garden city character, suggesting a return to the theme of the original close-knit suburbs which were based on transit. Lines of trees are used to bind the area together and pick up the theme of the adjacent neighborhoods, pulling that existing quality into the new transit center. Similarly, vegetation links a series of well-articulated street edge buildings and small pavilion structures. The central point of the project is to build considerable variety into a small area and create an identity distinct from other areas of the city.

The second design has a much more urban character. Buildings are used not only to define and harness the transit and road corridors, but also to form edges and links with the adjacent neighborhoods and define a

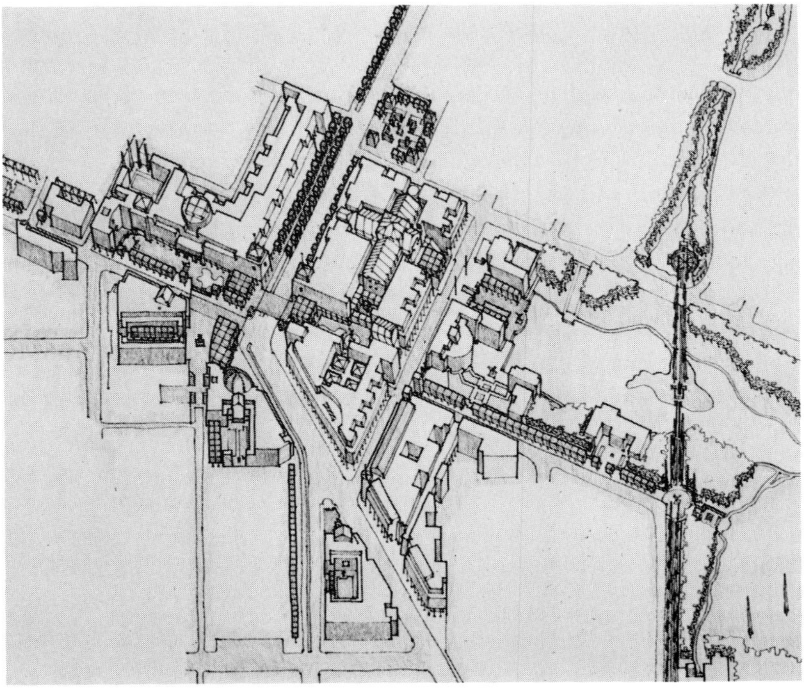

By contrast, Dean Almy shows how a design that is more classically urban could make a transit-related development memorable.

series of small, urban "people places" that become well populated because new buildings offer connections between adjacent areas and the transit stop. This design is also rich in the variety it offers, in its ability to attract a local night life, and the relationship it forms between the station, its adjacent development, and the neighborhood park.

Perhaps the more important aspect of these two projects is that they demonstrate an attitude towards quality urban form that will not come from zoning practices alone. They indicate an opportunity so seldom realized, whereby public and private agencies can work closely together towards common benefits and at the same time be sensitized towards neighborhood compatibility and support. Even if these designs hold no more than a clue, they point to a new kind of professional expertise based on enterprise, design sensitivity and the ability to work with a variety of related professional groups.

I suggest that the American city cannot be planned, and that it is almost an anacronism to have attempted to do so in the first place. What has been shown is that the development of the American city must be guided by a clear and robust set of incentives and priorities if good, rational results are to be achieved. Just as a city can plan a new airport to serve the needs of certain members of its population and sectors of its economy, it can plan to implement a public transit and land use partnership to identify and enhance the key components of the city. This, in turn, can act as the basis for control of new, key components in the city's future growth. In other words, this view of the city is not totally planned, nor does it resist enterprise. Instead, it provides a series of incentives and opportunities and in doing so steers us towards an alternative direction for the city.